Lecture Notes in Mathematics

Editors:
J.-M. Morel, Cachan
F. Takens, Groningen
B. Teissier, Paris

Subseries:
Institut de Mathématiques, Université de Strasbourg
Adviser: J.-L. Loday

M. Émery   M. Ledoux   M. Yor   (Eds.)

Séminaire
de Probabilités XXXVIII

 Springer

Editors

Michel Émery
Institut de Recherche
  Mathématique Avancée
Université Louis Pasteur
7, rue René Descartes
67084 Strasbourg Cedex, France
e-mail: emery@math.u-strasbg.fr

Marc Yor
Laboratoire de Probabilités
  et Modèles Aléatoires
Université Pierre et Marie Curie
Boîte courrier 188
4 place Jussieu
75252 Paris Cedex 05, France

Michel Ledoux
Laboratoire de Statistiques
  et Probabilités
Université Paul Sabatier
118, route de Narbonne
31601 Toulouse Cedex, France
e-mail: ledoux@cict.fr

Library of Congress Control Number: 2004115715
Mathematics Subject Classification (2000): 60Gxx, 60Hxx, 60Jxx

ISSN 0075-8434 Lecture Notes in Mathematics
ISSN 0720-8766 Séminaire de Probabilités
ISBN 3-540-23973-1 Springer Berlin Heidelberg New York
DOI: 10.1007/b104072

Springer is a part of Springer Science + Business Media
http://www.springeronline.com
© Springer-Verlag Berlin Heidelberg 2005
Printed in Germany

Typesetting: Camera-ready TeX output by the authors

41/3142/ du - 543210 - Printed on acid-free paper

C'est avec gratitude et admiration que nous dédions ce volume à

Jacques Azéma,

à l'occasion de son 65e anniversaire. Ses travaux, parmi lesquels ceux sur le retournement du temps, le balayage, les fermés aléatoires et bien sûr la martingale d'Azéma, ont prolongé, toujours avec originalité et élégance, la théorie générale des processus.

Son apparente décontraction, sa réelle rigueur et ses incessantes questions ("his healthy skepticism", comme l'écrivait J. Walsh dans *Temps Locaux*), ont été indissociables du Séminaire de Probabilités pendant de nombreuses années.

———————

We are also indebted and grateful to Anthony Phan, whose patient and time-consuming work behind the scene, up to minute details, on typography, formatting and TEXnicalities, was a key ingredient in the production of the present volume.

———————

Volume XXXIX, which consists of contributions dedicated to the memory of P. A. Meyer, is being prepared at the same time as this one and should appear soon, also in the Springer LNM series. It may be considered as a companion to the special issue, also in memory of Meyer, of the *Annales de l'Institut Henri Poincaré*.

———————

Finally, the Rédaction of the Séminaire is thoroughly modified: J. Azéma retired from our team after Séminaire XXXVII was completed; now, following his steps, two of us—M. Ledoux and M. Yor—are also leaving the board.

From volume XL onwards, the new Rédaction will consist of Catherine Donati-Martin (Paris), Michel Émery (Strasbourg), Alain Rouault (Versailles) and Christophe Stricker (Besançon). The combined expertise of the new members of the board will be an important asset to blend the themes which are traditionally studied in the Séminaire together with the newer developments in Probability Theory in general and Stochastic Processes in particular.

M. Émery, M. Ledoux, M. Yor

Table des Matières

Tanaka's Construction for Random Walks and Lévy Processes

Ronald A. Doney

Department of Mathematics, University of Manchester
Oxford Road, Manchester, UK M13 9PL
e-mail: rad@maths.man.ac.uk

Summary. Tanaka's construction gives a pathwise construction of "random walk conditioned to stay positive", and has recently been used in [3] and [8] to establish other results about this process. In this note we give a simpler proof of Tanaka's construction using a method which also extends to the case of Lévy processes.

1 The random walk case

If S is any rw starting at zero which does not drift to $-\infty$, we write S^* for S killed at time $\sigma := \min(n \geqslant 1 : S_n \leqslant 0)$, and $S^\uparrow$ for the for the harmonic transform of S^* which corresponds to "conditioning S to stay positive". Thus for $x > 0, y > 0$, and $x = 0$ when $n = 0$

$$P(S_{n+1}^\uparrow \in dy \mid S_n^\uparrow = x) = \frac{V(y)}{V(x)} P(S_{n+1} \in dy \mid S_n = x) = \frac{V(y)}{V(x)} P(S_1 \in dy-x),$$
(1)

where V is the renewal function in the weak increasing ladder process of $-S$. In [10], Tanaka showed that a process R got by time-reversing one by one the excursions below the maximum of S has the same distribution as $S^\uparrow$; specifically if $\{(T_k, H_k), k \geqslant 0\}$ denotes the strict increasing ladder process of S (with $T_0 = H_0 \equiv 0$) then R is defined by

$$R_0 = 0, \ R_n = H_k + \sum_{i=T_{k+1}+T_k+1-n}^{T_{k+1}} Y_i, \qquad T_k < n \leqslant T_{k+1}, \ k \geqslant 0. \quad (2)$$

If S drifts to $+\infty$, then it is well known (see [9]) that the post-minimum process

$$\overrightarrow{S} := (S_{J+n} - S_J, \ n \geqslant 0), \quad \text{where} \quad J = \max\left\{n : S_n = \min_{r \leqslant n} S_r\right\} \quad (3)$$

also has the distribution of $S^\uparrow$. In this case a very simple argument was given in [7] to show that the distributions of R and $\overrightarrow{S}$ agree, thus yielding a proof of

Tanaka's result in this case. The first point of this note is to show that a slight modification of this argument also yields Tanaka's result in the oscillatory case, without the somewhat tedious calculations in [10].

To see this, let S be any random walk with $S_0 \equiv 0$, $S_n = \sum_1^n Y_r$ for $n \geqslant 1$, introduce an independent Geometrically distributed random time G with parameter ρ and put $J_\rho = \max\{n \leqslant G : S_n = \min_{r \leqslant n} S_r\}$. In [7] a time-reversal argument was used to show that

$$(S_{J_\rho+n} - S_{J_\rho}, 0 \leqslant n \leqslant G - J_\rho) \overset{D}{=} [\hat{\delta}_K(\rho), \ldots, \hat{\delta}_1(\rho)] \qquad (4)$$
$$\overset{D}{=} [\hat{\delta}_1(\rho), \ldots, \hat{\delta}_K(\rho)],$$

where $\hat{\delta}_1(\rho), \ldots, \hat{\delta}_K(\rho)$ are the time reversals of the completed excursions below below the maximum of $\hat{S}(\rho) := (S_G - S_{G-n}, 0 \leqslant n \leqslant G)$, and $[\ldots]$ denotes concatenation. Note that the post-minimum process on the left in (4) has the same distribution as $(S_n, 0 \leqslant n \leqslant G \,|\, \sigma > G)$. Now in Theorem 1 of [4] it was shown that $S^\uparrow$ is the limit, in the sense of convergence of finite-dimensional distributions, of $(S_n, 0 \leqslant n \leqslant k \,|\, \sigma > k)$ as $k \to \infty$. (Actually [4] treated the case of conditioning to stay **non-negative**, and minor changes are required for our case). However it is easy to amend the argument there to see that as $\rho \downarrow 0$ this post-minimum process also converges in the same sense to $S^\uparrow$. Specifically a minor modification of Lemma 2 therein shows that

$$\liminf_{\rho \downarrow 0} \frac{P\{S_n \geqslant -x, n \leqslant G\}}{P\{S_n \geqslant 0, n \leqslant G\}} \geqslant V(x), \qquad x \geqslant 0,$$

and the rest of the proof is the same. Noting that $\hat{\delta}_1(\rho), \ldots, \hat{\delta}_K(\rho)$ are independent and identically distributed and independent of K, and that $\hat{\delta}_1(\rho) \overset{D}{\to} \hat{\delta}_1$ and $K \overset{P}{\to} \infty$ as $\rho \downarrow 0$, we conclude that $S^\uparrow \overset{D}{=} [\hat{\delta}_1, \hat{\delta}_2, \ldots] \overset{D}{=} R$, which is the required result.

2 The Lévy process case

The main point of this note is that, although the situation is technically more complicated, exactly similar arguments can be used to get a version of Tanaka's construction for Lévy processes.

We will use the canonical notation, and throughout this section $\mathbb{P}$ will be a measure under which the coordinate process $X = (X_t, t \geqslant 0)$ is a Lévy process which does not drift to $-\infty$ and is regular for $(-\infty, 0]$. For $x > 0$ we can use a definition similar to (1), with V replaced by the potential function for the decreasing ladder height subordinator to define a measure $\mathbb{P}_x^\uparrow$ corresponding to conditioning X starting from x to stay positive. But for $x = 0$ we need to employ a limiting argument. The following result is an immediate consequence of results in Bertoin [2]; see also Chaumont [5].

Theorem 1 (Bertoin). *Let τ be an $\mathrm{Exp}(\rho)$ random variable independent of X, and put*

$$J_\rho = \sup\{s < \tau : X_s = \underline{X}_s\} \quad where \quad \underline{X}_s = \inf\{X_u : u < s\}.$$

Write $\underset{\rightarrow}{\mathbb{P}}^{(\rho)}$ for the law of the post-minimum process $\{X_{J_\rho} + s - X_{J_\rho}, 0 \leqslant s < \tau - J_\rho\}$ under $\mathbb{P}_0$; then for each fixed t and $A \in \mathcal{F}_t$

$$\lim_{\rho \downarrow 0} \underset{\rightarrow}{\mathbb{P}}^{(\rho)}\{A\} = P^\uparrow\{A\},$$

where $P^\uparrow$ is a Markovian probability measure under which X starts at 0 and is such that the conditional law of $X_{t+\cdot}$, given $X_t = x > 0$, agrees with $P_x^\uparrow$.

Remark 1. It has recently been shown that, under very weak assumptions, $P_x^\uparrow$ converges to $P^\uparrow$ as $x \downarrow 0$ in the sense of convergence of finite-dimensional distributions. See [6].

Next, we recall another result due to Bertoin which is the continuous time analogue of the result from [7] which we have stated as (4). Noting that (2) can be written in the alternative form

$$R_n = \bar{S}_{T_{k+1}} + \left(\bar{S} - S\right)_{T_k + T_{k+1} - n}, \qquad T_k < n \leqslant T_{k+1},$$

we introduce $\bar{X}_s = \sup_{u < s} X_u$ and

$$g(t) = \sup\left(s < t : X_s = \bar{X}_s\right), \qquad d(t) = \inf\left(s > t : X_s = \bar{X}_s\right),$$

the left and right endpoints of the excursion of $\bar{X} - X$ away from 0 which contains t, and define $R_t = \bar{X}_{d(t)} + \tilde{R}_t$, where

$$\tilde{R}_t = \begin{cases} \left(\bar{X} - X\right)_{(d(t)+g(t)-t)-} & \text{if } d(t) > g(t), \\ 0 & \text{if } d(t) = g(t). \end{cases}$$

We also introduce the future infimum process for X killed at time τ by

$$\underset{=}{X}_t = \inf\{X_s : t \leqslant s \leqslant \tau\},$$

and note that $\underset{=}{X}_0 = X_{J_\rho}$. The following result is established in the proof of Lemme 4 in [1]; note that, despite the title of the paper, this Lemme 4 is valid for any Lévy process which drifts to $+\infty$, and the result for the killed process, which is what we need is clearly valid for any Lévy process.

Theorem 2 (Bertoin). *Under $\mathbb{P}_0$ the law of $\{(\tilde{R}_t, \bar{X}_{d(t)}), 0 \leqslant t < g(\tau)\}$ coincides with that of*

$$\left\{\left((X - \underset{=}{X})_{J_\rho + t}, \underset{=}{X}_{J_\rho + t} - \underset{=}{X}_0\right), 0 \leqslant t < \tau - J_\rho\right\}.$$

Of course, an immediate consequence of this is the equality in law of

$$\{R_t, 0 \leqslant t < g(\tau)\} \quad and \quad \{X_{J_\rho + t} - \underset{=}{X}_0, 0 \leqslant t < \tau - J_\rho\}.$$

Letting $\rho \downarrow 0$ and appealing to Theorem 1 above we deduce

Theorem 3. *Under $\mathbb{P}_0$ the law of $\{R_t, t \geqslant 0\}$ is $\mathbb{P}^\uparrow$.*

References

1. Bertoin, J. Sur la décomposition de la trajectoire d'un processus de Lévy spectralement positif en son minimum. Ann. Inst. H. Poincaré, **27**, 537-547, (1991).
2. Bertoin, J. Splitting at the infimum and excursions in half-lines for random walks and Lévy processes. Stoch. Proc. Appl., **47**, 17-35, (1993).
3. Biggins, J. D. Random walk conditioned to stay positive. J. Lond. Math. Soc., **67**, 259-272, (2003).
4. Bertoin, J. and Doney, R. A. Random walks conditioned to stay positive. Ann. Probab., **22**, 2152-2167, (1994).
5. Chaumont, L. Conditionings and path decompositions for Lévy processes. Stoch. Proc. Appl., **64**, 39-54, (1996).
6. Chaumont, L. and Doney, R. A. On Lévy processes conditioned to stay positive. Preprint, (2003).
7. Doney, R. A. Last exit times for random walks. Stoch. Proc. Appl., **31**, 321-331, (1989).
8. Hambly, B. M., Kersting, G., and Kyprianou, A. E. Law of the iterated logarithm for oscillating random walks conditioned to stay positive. Preprint. (2001).
9. Millar, P. W. Zero-one laws and the minimum of a Markov process. Trans. Amer. Math. Soc., **226**, 365-391, (1975).
10. Tanaka, H. Time reversal of random walks in one dimension. Tokyo J. Math., **12**, 159-174, (1989).

Some Excursion Calculations
for Spectrally One-sided Lévy Processes

Ronald A. Doney

Department of Mathematics, University of Manchester
Oxford Road, Manchester, UK M13 9PL
e-mail: rad@maths.man.ac.uk

1 Introduction.

Let $X = (X_t, t \geqslant 0)$ be a spectrally negative Lévy process and write Y and $\widehat{Y}$ for the reflected processes defined by

$$Y_t = X_t - I_t, \qquad \widehat{Y}_t = S_t - X_t, \qquad t \geqslant 0,$$

where

$$S_t = \sup_{0 \leqslant s \leqslant t} (0 \vee X_s), \qquad I_t = \inf_{0 \leqslant s \leqslant t} (0 \wedge X_s).$$

In recent works by Avram, Kyprianou and Pistorius [1] and Pistorius [8] some new results about the times at which Y and $\widehat{Y}$ exit from finite intervals have been established. The proofs of these results in the cited papers involve a combination of excursion theory, Itô calculus, and martingale techniques, and the point of this note is to show that these results can be established by direct excursion theory calculations. These calculations are based on the known results for the two-sided exit problem for X in Bertoin [3], together with representations for the characteristic measures n and $\hat{n}$ of the excursions of Y and $\widehat{Y}$ away from zero. The representation for n has been established by Bertoin in [2] and that for $\hat{n}$ follows from results in Chaumont [4], (for a similar result for general Lévy processes see [5]), and are described in the next section.

2 Preliminaries

Throughout we assume that $X = (X_t, t \geqslant 0)$ is a Lévy process without positive jumps which is neither a pure drift nor the negative of a subordinator, and we adopt without further comment the notation of Chapter VII of [2]. In particular ψ and Φ denote the Laplace exponent of X and its inverse, and W denotes the scale function, the unique absolutely continuous increasing function with Laplace transform

$$\int_0^\infty e^{-\lambda x} W(x)\, dx = \frac{1}{\psi(\lambda)}, \qquad \lambda > \Phi(0).$$

The scale function determines the probability of X exiting at the top or bottom of a 2-sided interval, and the q-scale function $W^{(q)}$, which informally is the scale function of the process got by killing X at an independent $\text{Exp}(q)$ time, determines also the distribution of the exit time. Specifically $W^{(q)}$ denotes the unique absolutely continuous increasing function with Laplace transform

$$\int_0^\infty e^{-\lambda x} W^{(q)}(x)\, dx = \frac{1}{\psi(\lambda) - q}, \qquad \lambda > \Phi(q),\ q \geqslant 0, \tag{1}$$

and for convenience we set $W^{(q)}(x) = 0$ for $x \in (-\infty, 0)$. We also need the "adjoint scale function" defined by $Z^{(q)}(x) = 1$ for $x \leqslant 0$ and

$$Z^{(q)}(x) = 1 + q \int_0^x W^{(q)}(y)\, dy \qquad \text{for } x > 0. \tag{2}$$

Extending previous results due to Emery [6], Takacs [11], Rogers [9], and Suprun [10], in [3] Bertoin gave the full solution to the 2-sided exit problem in the following form:

Proposition 1. *Define for $a \geqslant 0$ the passage times*

$$T_a = \inf(t \geqslant 0 : X_t > a), \qquad \widehat{T}_a = \inf(t \geqslant 0 : -X_t > a).$$

Then for $0 \leqslant x \leqslant a$ we have

$$\mathbb{E}_x\big(e^{-qT_a}; T_a < \widehat{T}_0\big) = \frac{W^{(q)}(x)}{W^{(q)}(a)}, \tag{3}$$

and

$$\mathbb{E}_x\big(e^{-q\widehat{T}_0}; \widehat{T}_0 < T_a\big) = Z^{(q)}(x) - \frac{W^{(q)}(x) Z^{(q)}(a)}{W^{(q)}(a)}. \tag{4}$$

Furthermore let $U^{(q)}$ denote the resolvent measure of X killed at the exit time $\sigma_a := T_a \wedge \widehat{T}_0$; then $U^{(q)}$ has a density which is given by

$$u^{(q)}(x, y) = \frac{W^{(q)}(x)}{W^{(q)}(a)} W^{(q)}(a - y) - W^{(q)}(x - y), \qquad x, y \in [0, a]. \tag{5}$$

Remark 1. Suppose that $\mathbb{P}^a$ is a measure under which X is a Lévy process having the same characteristics as under $\mathbb{P}$ except that Π is replaced by

$$\Pi^a(dx) = \Pi(dx)\, \mathbf{1}_{\{x \geqslant -a\}} + \Pi\big((-\infty, -a)\big)\, \delta_{-a}(dx),$$

where $\delta_{-a}(dx)$ denotes a unit mass at $-a$. Then it is clear that up to time σ_a, X behaves the same under $\mathbb{P}^a$ as it does under $\mathbb{P}$. Thus the righthand sides of (3) and (4) are unchanged if $W^{(q)}$ is replaced by $W^{(q)a}$, the scale function for X under $\mathbb{P}^a$. It then follows that we must have the identity

$$W^{(q)}(x) \equiv W^{(q)a}(x) \qquad \text{for } 0 \leqslant x \leqslant a.$$

Note that the behaviour of Y and $\widehat{Y}$ up to the time that they exit the interval $[0, a]$ is also the same under $\mathbb{P}^a$ as it is under $\mathbb{P}$.

Remark 2. The probability measures $\mathbb{P}$ and $\mathbb{P}^\#$ are said to be *associates* if X is also a spectrally negative Lévy process under $\mathbb{P}^\#$ and there is a constant $\delta \neq 0$ such that

$$\mathbb{P}^\#(X_t \in dx) = e^{\delta x}\, \mathbb{P}(X_t \in dx), \qquad -\infty < x < \infty.$$

It is known that if X drifts to $-\infty$ under $\mathbb{P}$ then $\mathbb{P}^\#$ exists, is unique, and $\delta = \Phi(0) > 0$ is a zero of ψ. (See [2], p. 193.) On the other hand, if X drifts to ∞ under $\mathbb{P}$ then $\mathbb{P}^\#$ may or may not exist; if it does it is unique, and δ is a negative zero of ψ. In both cases the corresponding scale functions are related by $W^\#(x) = e^{\delta x}W(x)$. Note that if the Lévy measure is confined to a finite interval, as Π^a is in Remark 1, then $\psi(\lambda)$ exists for all real λ, and if $\mathbb{E}\, X_1 \neq 0$ then it has 2 real zeros, so the associate measure exists.

We also need some information about the excursion measures n and $\hat{n}$ of Y and $\widehat{Y}$ away from zero. (n.b. this notation is the opposite of that in [2]). In what follows it should be noted that whereas 0 is always regular for $(0, \infty)$ for Y, it is possible for 0 to be irregular for $(0, \infty)$ for $\widehat{Y}$. (This situation was excluded in [8].) In this case we adopt the convention outlined on p. 122 of [2], which allows us to assume that $\widehat{Y}$ has a continuous local time at 0.

In the following result ζ denotes the lifetime of an excursion and $\mathbb{Q}_x$ and $\mathbb{Q}_x^*$ denote the laws of X and $-X$ killed on entering $(-\infty, 0)$ respectively.

Proposition 2. *Let* $A \in \mathcal{F}_t, t > 0$, *be such that* $n(A^o) = 0$ *(respectively* $\hat{n}(A^o) = 0$*), where* A^o *is the boundary of* A *with respect to the J-topology on* D. *Then there are constants* k *and* $\hat{k}$ *(which depend only on the normalizations of the local time at zero of* Y *and* $\widehat{Y}$*) such that*

$$n(A, t < \varsigma) = k \lim_{x \downarrow 0} \frac{\mathbb{Q}_x(A)}{W(x)}, \tag{6}$$

and, assuming further that if X *drifts to* $+\infty$ *under* $\mathbb{P}$ *then the associate measure* $\mathbb{P}^\#$ *exists,*

$$\hat{n}(A, t < \varsigma) = \hat{k} \lim_{x \downarrow 0} \frac{\mathbb{Q}_x^*(A)}{x}. \tag{7}$$

Proof. According to Propositions 14 and 15, p. 201–202 of [2] for any $A \in \mathcal{F}_t$ we have

$$n(A, \, t < \varsigma) = k \, \mathbb{E}^{\uparrow}\big(W(X_t)^{-1}; \, A\big), \tag{8}$$

where $\mathbb{P}^{\uparrow}$ is the weak limit in the Skorohod topology as $x \downarrow 0$ of the measures $\mathbb{P}_x^{\uparrow}$ which correspond to "conditioning X to stay positive", and are defined by

$$\mathbb{P}_x^{\uparrow}(X_t \in dy) = \frac{W(y)}{W(x)} \, \mathbb{Q}_x(X_t \in dy), \qquad x > 0, \, y > 0.$$

Combining these results and using the assumption on A gives (6). The proof of (7) is similar. If X does not drift to $+\infty$ under $\mathbb{P}$ the potential function of the increasing ladder height process is given by

$$V(x) = \begin{cases} x & \text{if } \mathbb{E} \, X_1 = 0, \\ 1 - e^{-x\Phi(0)} & \text{if } \mathbb{E} \, X_1 < 0, \end{cases}$$

so that $V(x) \frown cx$ as $x \downarrow 0$ in both cases. The analogue of (8) is

$$\hat{n}(A, \, t < \varsigma) = k \, \mathbb{E}^{*\uparrow}\big(V(X_t)^{-1}; \, A\big)$$

where, by Theorem 6 of [4], $\mathbb{P}^{*\uparrow}$ is the weak limit of the measures

$$\mathbb{P}_x^{*\uparrow}(X_t \in dy) = \frac{V(y)}{V(x)} \, \mathbb{Q}_x^*(X_t \in dy), \qquad x > 0, \, y > 0.$$

If X does drift to $+\infty$ under $\mathbb{P}$ then it is easy to check that, with $\varepsilon = (\varepsilon(t), t \geqslant 0)$ denoting a generic excursion and $\hat{n}^{\#}$ denoting the excursion measure of $\hat{Y}$ under the associate measure $\mathbb{P}^{\#}$,

$$\hat{n}\big(A, \varepsilon(t) \in dy, \, t < \varsigma\big) = e^{-\delta y} \, \hat{n}^{\#}\big(A, \, \varepsilon(t) \in dy, \, t < \varsigma\big). \tag{9}$$

Since X drifts to $-\infty$ under $\mathbb{P}^{\#}$ we can apply the previous result and the fact that

$$\mathbb{Q}_x^*(X_t \in dy) = e^{-\delta(y-x)} \, \mathbb{Q}_x^{\#*}(X_t \in dy)$$

to complete the proof. □

Remark 3. One way to check (9) is to use our knowledge of the Wiener–Hopf factors and equation (7), p. 120 of [2] to compute the double Laplace transforms of $\hat{n}(\varepsilon(t) \in dy, \, t < \varsigma)$ and $e^{-\delta y}\hat{n}^{\#}(\varepsilon(t) \in dy, \, t < \varsigma)$.

We also need some facts about $W^{(q)}$:

Lemma 1. (i) $\lim_{x \downarrow 0} \frac{W^{(q)}(x)}{W(x)} = 1$;

(ii) *If X has unbounded variation then $W^{(q)'}(x)$, the derivative with respect to x of $W^{(q)}(x)$ exists and is continuous for all $x > 0$.*

(iii) *If X has bounded variation let $\mathcal{D}$ denote $\{x : \Pi$ has positive mass at $-x\}$. Then $W_+^{(q)'}(x)$ and $W_-^{(q)'}(x)$, the right and lefthand derivatives of $W^{(q)}(x)$ exist at all $x > 0$, agree off $\mathcal{D}$, and*

$$\lim_{y \downarrow x} W_+^{(q)'}(y) = W_+^{(q)'}(x) \qquad \text{for all } x \in \mathcal{D}.$$

Proof. (i) This follows from the expansion

$$W^{(q)}(x) = \sum_{k=1}^{\infty} q^{k-1} W^{(k*)}(x), \tag{10}$$

where $W^{(k*)}$ denotes the k-fold convolution of W, together with the bound

$$W^{(k*)}(x) \leqslant \frac{x^{k-1} W(x)^k}{(k-1)!}, \qquad k \geqslant 1, \, x \geqslant 0.$$

(ii) Provided X does not drift to $-\infty$ under $\mathbb{P}$, we have the representation

$$W(x) = c \exp\left(-\int_x^\infty \hat{n}(h(\varepsilon) > t) \, dt\right),$$

(see [2], p. 195). As pointed out in [8], this implies that

$$W_+'(x) = W(x) \, \hat{n}(h(\varepsilon) > x), \qquad W_-'(x) = W(x) \, \hat{n}(h(\varepsilon) \geqslant x),$$

and the result follows when $q = 0$ since $\hat{n}$ has no atoms in the case of unbounded variation, (see [7]). If X does drift to $-\infty$ under $\mathbb{P}$ we use the device of the associate measure $\mathbb{P}^\#$ introduced in Remark 2. Since X drifts to ∞ and has unbounded variation under $\mathbb{P}^\#$, it is easy to check that the result also holds in this situation. The case when $q > 0$ again follows easily, using (10).

(iii) In this case excursions of $\hat{Y}$ away from 0 start with a jump, and then evolve according to the law of $-X$. Since 0 is irregular for $(-\infty, 0)$, this shows that $\hat{n}(h(\varepsilon) = x) > 0$ for all $x \in \mathcal{D}$, but the fact that X has an absolutely continuous resolvent means that $\hat{n}(h(\varepsilon) = x) = 0$ for all $x \notin \mathcal{D}$, and this implies the stated results for $q = 0$. Again the results for $q > 0$ follow easily, using (10). □

To demonstrate the use of the above result, we calculate below the n and $\hat{n}$ measures of a relevant subset of excursion space. Put $h(\varepsilon) := \sup_{t < \varsigma} \varepsilon(t)$ and $T_a(\varepsilon) = \inf\{t : \varepsilon(t) > a\}$ for the height and the first passage time of a generic excursion ε whose lifetime is denoted by $\varsigma(\varepsilon)$, and with Λ_q denoting an independent Exp(q) random variable set $A = B \cup C$, where

$$B = \{\varepsilon : h(\varepsilon) > a, T_a(\varepsilon) \leqslant \varsigma(\varepsilon) \wedge \Lambda_q\} \quad \text{and} \quad C = \{\varepsilon : h(\varepsilon) \leqslant a, \Lambda_q < \varsigma(\varepsilon)\}.$$

Since we will only be concerned with ratios of n and $\hat{n}$ measures in the following we will assume that $k = \hat{k} = 1$.

Lemma 2. *In all cases*

$$\alpha := n(A) = \frac{Z^{(q)}(a)}{W^{(q)}(a)}, \tag{11}$$

and, provided that if X drifts to $+\infty$ under $\mathbb{P}$ then the associate measure $\mathbb{P}^\#$ exists,

$$\hat{\alpha} := \hat{n}(A) = \frac{W_+^{(q)'}(a)}{W^{(q)}(a)}. \tag{12}$$

Proof. Since ([2], p. 202)

$$n\big(h(\varepsilon) > x\big) = c/W(x)$$

is continuous, we see from (6) that

$$n\big(h(\varepsilon) > a, T_a(\varepsilon) \in dt\big) = \lim_{x\downarrow 0} \frac{\mathbb{Q}_x\{T_a \in dt\}}{W(x)} = \lim_{x\downarrow 0} \frac{\mathbb{P}_x\{T_a < \widehat{T}_0, T_a \in dt\}}{W(x)},$$

and

$$n(\varepsilon_t \in dy, T_a(\varepsilon) > t) = \lim_{x\downarrow 0} \frac{\mathbb{Q}_x\{X_t \in dy, T_a > t\}}{W(x)} = \lim_{x\downarrow 0} \frac{\mathbb{P}_x\{X_t \in dy, \sigma_a > t\}}{W(x)}.$$

Thus

$$\alpha = n(B) + n(C) = \lim_{x\downarrow 0} \frac{1}{W(x)} \left(\mathbb{E}_x\{e^{-qT_a}; T_a < \widehat{T}_0\} + \mathbb{P}_x\{\Lambda_q < \sigma_a\} \right)$$

$$= \lim_{x\downarrow 0} \frac{1}{W(x)} \left(1 - \mathbb{E}_x\{e^{-q\widehat{T}_0}; \widehat{T}_0 < T_a\} \right).$$

Combining this with (4) gives

$$\alpha = \lim_{x\downarrow 0} \frac{1 - Z^{(q)}(x)}{W(x)} + \frac{Z^{(q)}(a)}{W^{(q)}(a)} \lim_{x\downarrow 0} \frac{W^{(q)}(x)}{W(x)} = \frac{Z^{(q)}(a)}{W^{(q)}(a)}.$$

Note next that the results of Lemma 2.3 show that the right-hand side of (12) is a cadlag function of a; since it is easy to see that the same is true of the left-hand side, it suffices to establish these results, for $a \notin \mathcal{D}$. In this case we have $\hat{n}(h = a) = 0$, so the required J-continuity holds and by a similar argument we can use (7) and (3) to get

$$\hat{a} = \lim_{x\downarrow 0} \frac{1}{x} \left(\mathbb{E}_{a-x}\{e^{-q\widehat{T}_0}; \widehat{T}_0 < T_a\} + \mathbb{P}_{a-x}\{\Lambda_q < \sigma_a\} \right)$$

$$= \lim_{x\downarrow 0} \frac{1}{x} \left(1 - \mathbb{E}_{a-x}\{e^{-qT_a}; T_a < \widehat{T}_0\} \right)$$

$$= \lim_{x\downarrow 0} \frac{1}{x} \left(\frac{W^{(q)}(a) - W^{(q)}(a - x)}{W^{(q)}(a)} \right) = \frac{W_+^{(q)'}(a)}{W^{(q)}(a)}. \qquad \square$$

3 Exit results for the reflected processes

We use the following notation for passage times of Y and $\widehat{Y}$: for $a > 0$

$$\tau_a = \inf\{t \geqslant 0 : Y_t > a\}, \qquad \hat{\tau}_a = \inf\{t \geqslant 0 : \widehat{Y}_t > a\}.$$

Note that when $x \leqslant 0$ the initial value of $\widehat{Y}$ under $\mathbb{P}_x$ is $-x$.

Our main result gives the q-resolvent measures $R^{(q)}(x, A)$ and $\widehat{R}^{(q)}(x, A)$ of Y and $\widehat{Y}$ killed on exiting the interval $[0, a]$.

Theorem 1 (Pistorius [8]). (i) *The measure $R^{(q)}(x, A)$ is absolutely continuous with respect to Lebesgue measure and a version of its density is*

$$r^{(q)}(x, y) = \frac{Z^{(q)}(x)}{Z^{(q)}(a)} W^{(q)}(a - y) - W^{(q)}(x - y), \qquad x, y \in [0, a). \quad (13)$$

(ii) *For $0 \leqslant x \leqslant a$ we have $\widehat{R}^{(q)}(x, dy) = \hat{r}^{(q)}(x, 0)\, \delta_0(dy) + \hat{r}^{(q)}(x, y)\, dy$ where*

$$\hat{r}^{(q)}(x, 0) = \frac{W^{(q)}(a - x) W^{(q)}(0)}{W_+^{(q)\prime}(a)}, \quad (14)$$

$$\hat{r}^{(q)}(x, y) = \frac{W^{(q)}(a - x) W_+^{(q)\prime}(y)}{W_+^{(q)\prime}(a)} - W^{(q)}(y - x), \qquad 0 < y \leqslant a, \quad (15)$$

and δ_0 denotes a unit mass at 0.

Proof. (i) We start with the obvious decomposition

$$r^{(q)}(x, y) = u^{(q)}(x, y) + \mathbb{E}_x\{e^{-q\widehat{T}_0}; \widehat{T}_0 < T_a\} r^{(q)}(0, y),$$

where $u^{(q)}(x, y)$, the resolvent density of X killed at time σ_a, is given by (5). Together with (4) this means that we need only show that

$$r^{(q)}(0, y) = \frac{W^{(q)}(a - y)}{Z^{(q)}(a)}.$$

To do this we first establish that

$$r^{(q)}(0, y)\, dy = \frac{n\{\Lambda_q < \zeta, \varepsilon(\Lambda_q) \in dy, \bar{\varepsilon}(\Lambda_q) \leqslant a\}}{\alpha}, \quad (16)$$

where $\bar{\varepsilon}(t) = \sup_{s \leqslant t} \varepsilon(s)$. For this, note first that the lefthand side of (16) is

$$\mathbb{P}\{Y(\Lambda_q) \in dy, \overline{Y}(\Lambda_q) \leqslant a\}.$$

Next, with L denoting the local time at zero of Y and ε_s the excursion of Y at local time s, we can write

$$\{Y(\Lambda_q) \in dy, \overline{Y}(\Lambda_q) \leqslant a\}$$
$$= \bigcup_{t>0}\left(\bigcap_{s<t}\{\varepsilon_s \in D(L^{-1}(s-))\} \cap \{\varepsilon_t \in E(L^{-1}(t-))\}\right),$$

where

$$D(x) = \{\varepsilon : h(\varepsilon) \leqslant a,\, x + \varsigma(\varepsilon) \leqslant \Lambda_q\},$$

and

$$E(x) = \{\varepsilon : \Lambda_q \in (x, x + \varsigma(\varepsilon)),\, \varepsilon(\Lambda_q - x) \in dy,\, \overline{\varepsilon}(\Lambda_q - x) \leqslant a\}.$$

Using the lack of memory property of Λ_q we see that $L\{\inf(s : \varepsilon_s \notin D(L^{-1}(s-))\}$ is exponentially distributed with parameter $n(D(0)^c) = n(A) = \alpha$, and noting that $E(x) \subset D(x)^c$ we get

$$\mathbb{P}\{Y(\Lambda_q) \in dy, \overline{Y}(\Lambda_q) \leqslant a\} = \int_0^\infty \alpha\, \mathbb{E}^{-\alpha t} n\big(E(0)\mid D(0)^c\big)\, dt = \frac{n\big(E(0)\big)}{\alpha},$$

and this is (16). The result now follows, since $\alpha = Z^{(q)}(a)/W^{(q)}(a)$ by (11) and

$$n\{\Lambda_q < \varsigma,\, \varepsilon(\Lambda_q) \in dy,\, \overline{\varepsilon}(\Lambda_q) \leqslant a\} = \lim_{x\downarrow 0}\frac{\mathbb{P}_x\{\Lambda_q < \sigma_a,\, X(\Lambda_q) \in dy\}}{W(x)}$$
$$= \lim_{x\downarrow 0}\frac{u^{(q)}(x, y)\, dy}{W(x)} = \lim_{x\downarrow 0}\frac{W^{(q)}(a-y)W^{(q)}(x)\, dy}{W(x)W^{(q)}(a)} = \frac{W^{(q)}(a-y)\, dy}{W^{(q)}(a)}.$$

(ii) First note that if X drifts to ∞ under $\mathbb{P}$ then by Remark 1 we can assume that $\Pi((-\infty, -a)) = 0$, and then by Remark 2 we know that the associate measure $\mathbb{P}^{\#}$ exists, and we are free to use the results of Proposition 2.2 and Lemma 2.4.

For $y > 0$ a calculation similar to that in (i) gives

$$\hat{r}^{(q)}(0, y)\, dy = \frac{\hat{n}\{\Lambda_q < \varsigma,\, \varepsilon(\Lambda_q) \in dy,\, \overline{\varepsilon}(\Lambda_q) \leqslant a\}}{\hat{\alpha}},$$

where $\hat{\alpha} = W_+^{(q)'}(a)/W^{(q)}(a)$ by (12). But

$$\hat{n}\{\Lambda_q < \varsigma,\, \varepsilon(\Lambda_q) \in dy,\, \overline{\varepsilon}(\Lambda_q) \leqslant a\} = \lim_{x\downarrow 0}\frac{\mathbb{P}_{a-x}\{\Lambda_q < \sigma_a,\, X(\Lambda_q) \in a - dy\}}{x}$$
$$= \lim_{x\downarrow 0}\frac{u^{(q)}(a - x, a - y)\, dy}{x}$$
$$= \lim_{x\downarrow 0}\frac{\{W^{(q)}(y)W^{(q)}(a-x) - W^{(q)}(a)W^{(q)}(y-x)\}\, dy}{xW^{(q)}(a)}$$
$$= \frac{\{W^{(q)}(a)W_+^{(q)'}(y) - W^{(q)}(y)W_+^{(q)'}(a)\}\, dy}{W^{(q)}(a)}.$$

Thus

$$\hat{r}^{(q)}(0,y) = \frac{W^{(q)}(a)W_+^{(q)'}(y)}{W_+^{(q)'}(a)} - W^{(q)}(y).$$

We conclude by substituting this into the decomposition

$$\hat{r}^{(q)}(x,y) = u^{(q)}(a-x,a-y) + \mathbb{E}_{a-x}\{e^{-qT_a}; T_a < \widehat{T}_0\}\hat{r}^{(q)}(0,y),$$

to get (15).

For the case $y = 0$ the fact that

$$\widehat{R}^{(q)}(x,\{0\}) = \mathbb{E}_{a-x}\{e^{-qT_a}; T_a < \widehat{T}_0\}\widehat{R}^{(q)}(0,\{0\})$$

means we need only consider the case $x = 0$. Recall that $B = \{\varepsilon : h(\varepsilon) > a, T_a(\varepsilon) \leqslant \varsigma(\varepsilon) \wedge \Lambda_q\}$; then by arguments similar to those we have already used we see that

$$
\begin{aligned}
\hat{n}(B) &= \lim_{x \downarrow 0} \frac{1}{x} \mathbb{E}_{a-x}\{e^{-q\widehat{T}_0}; T_a > \widehat{T}_0\} \\
&= \lim_{x \downarrow 0} \frac{1}{x}\left\{ Z^{(q)}(a-x) - \frac{W^{(q)}(a-x)Z^{(q)}(a)}{W^{(q)}(a)} \right\} \\
&= \lim_{x \downarrow 0} \frac{1}{x}\left(\frac{\{W^{(q)}(a) - W^{(q)}(a-x)\}Z^{(q)}(a)}{W^{(q)}(a)} + Z^{(q)}(a-x) - Z^{(q)}(a) \right) \\
&= \frac{W^{(q)'}(a)Z^{(q)}(a)}{W^{(q)}(a)} - qW^{(q)}(a) = \frac{W_+^{(q)'}(a)Z^{(q)}(a)}{W^{(q)}(a)} - qW^{(q)}(a),
\end{aligned}
$$

so that

$$\mathbb{P}\{\hat{\tau}_a \leqslant \Lambda_q\} = \frac{\hat{n}(B)}{\hat{\alpha}} = \frac{\frac{W_+^{(q)'}(a)Z^{(q)}(a)}{W^{(q)}(a)} - qW^{(q)}(a)}{W_+^{(q)'}(a)/W^{(q)}(a)} = Z^{(q)}(a) - \frac{q(W^{(q)}(a))^2}{W_+^{(q)'}(a)}. \tag{17}$$

Using these facts and a straight forward integration gives

$$
\begin{aligned}
q\widehat{R}^{(q)}(0,\{0\}) &= \mathbb{P}(\Lambda_q < \hat{\tau}_a) - q\int_0^a \hat{r}^{(q)}(0,y)\,dy \\
&= 1 - Z^{(q)}(a) + \frac{q(W^{(q)}(a))^2}{W_+^{(q)'}(a)} \\
&\quad - \left(\frac{qW^{(q)}(a)\{W^{(q)}(a) - W^{(q)}(0)\}}{W_+^{(q)'}(a)} - \{Z^{(q)}(a) - 1\} \right) \\
&= \frac{qW^{(q)}(a)W^{(q)}(0)}{W_+^{(q)'}(a)},
\end{aligned}
$$

and this is (14) with $x = 0$. (Note that $W^{(q)}(0) = W(0) = 0$ whenever X has unbounded variation.) $\qquad\square$

14   Ronald A. Doney

An immediate consequence of the Theorem is the following, which is a result in [1]:

Corollary 1. *For $0 \leqslant x \leqslant a$ and $q \geqslant 0$ we have*

$$\mathbb{E}_x\{e^{-q\tau_a}\} = \frac{Z^{(q)}(x)}{Z^{(q)}(a)} \tag{18}$$

and

$$\mathbb{E}_{-x}\{e^{-q\hat{\tau}_a}\} = Z^{(q)}(a - x) - \frac{qW^{(q)}(a - x)W^{(q)}(a)}{W_+^{(q)'}(a)}. \tag{19}$$

Proof. For (18) just use (13) to compute $\mathbb{P}_x\{\tau_a > \Lambda_q\}$. For (19) use (17), the fact that

$$\mathbb{E}_{-x}\{e^{-\hat{\tau}_a}\} = \mathbb{E}_{-x}\{e^{-q\widehat{T}_{-a}}; \widehat{T}_{-a} < T_0\} + \mathbb{E}_{-x}\{e^{-qT_0}; T_0 < \widehat{T}_{-a}\}\,\mathbb{E}\{e^{-q\hat{\tau}_a}\}$$

$$= \mathbb{E}_{a-x}\{e^{-q\widehat{T}_0}; \widehat{T}_0 < T_a\} + \mathbb{E}_{a-x}\{e^{-qT_a}; T_a < \widehat{T}\}\,\mathbb{P}\{\hat{\tau}_a \leqslant \Lambda_q\},$$

and (3) and (4).   □

Remark 4. Using (15) and the standard formula

$$\mathbb{E}_{-x}\{e^{-q\hat{\tau}_a}; \widehat{Y}(\hat{\tau}_a-) \in dy, \widehat{Y}(\hat{\tau}_a) \in dz\} = \Pi\{-dz + y\}\,\hat{r}^{(q)}(x,y)\,dy$$

yields a refinement of (19).

Acknowledgement. The author is grateful to Andreas Kyprianou for several interesting discussions, and to the referee for some valuable comments.

References

1. Avram, F., Kyprianou, A. E., Pistorius, M. R. Exit problems for spectrally positive Lévy processes and applications to (Canadized) Russian options. Ann. Appl. Probab., 14, 215–238, (2003).
2. Bertoin, J. *Lévy processes*, Cambridge University Press, Cambridge, (1996).
3. Bertoin, J. Exponential decay and ergodicity of completely assymmetric Lévy processes in a finite interval. Ann. Appl. Probab., 7, 156–169, (1997).
4. Chaumont, L. Conditionings and path decompositions for Lévy processes. Stoch. Proc. Appl. 64, 39–54, (1996).
5. Chaumont, L. and Doney, R. A. On Lévy processes conditioned to stay positive. Preprint, (2003).
6. Emery, D. J. Exit problem for a spectrally positive process. Adv. Appl. Probab., 5, 498–520, (1973).
7. Lambert, A. Completely assymmetric Lévy processes confined in a finite interval. Ann. Inst. H. Poincaré Prob. Stat. 36, 251–274, (2000).
8. Pistorius, M.R. On exit and ergodicity of the spectrally negative Lévy process reflected at its infimum. J. Theoret. Probab. 17 (2004), no. 1, 183–220.

9. Rogers, L. C. G. The two-sided exit problem for spectrally positive Lévy processes. Adv. Appl. Probab., 22, 486–487, (1990).
10. Suprun, V.N. The ruin problem and the resolvent of a killed independent increment process. Ukrainian Math. J. 28, 39–45, (1976).
11. Takács, L. *Combinatorial methods in the theory of stochastic processes.* Wiley, New York, (1966).

A Martingale Review
of some Fluctuation Theory
for Spectrally Negative Lévy Processes

Andreas E. Kyprianou[1] and Zbigniew Palmowski[1,2]

[1] Mathematical Institute, Utrecht University
   P.O. Box 80.010, 3508 TA Utrecht, The Netherlands
[2] Mathematical Institute, University of Wrocław
   pl. Grunwaldzki 2/4, 50-384 Wrocław, Poland
e-mail: kyprianou@math.uu.nl, zpalma@math.uni.wroc.pl

Summary. We give a review of some fluctuation theory for spectrally negative Lévy processes using for the most part martingale theory. The methodology is based on the techniques found in Kyprianou and Palmowski (2003) which deal with similar issues for a general class of Markov additive processes.

1 Introduction

Two and one sided exit problems for spectrally negative Lévy processes have been the object of several studies over the last 40 years. Significant contributions have come from Zolotarev (1964), Takács (1967), Emery (1973), Bingham (1975) Rogers (1990) and Bertoin (1996a, 1996b, 1997). The principal tools of analysis of these authors are the Wiener–Hopf factorization and Itô's excursion theory.

In recent years, the study of Lévy processes has enjoyed rejuvenation. This has resulted in many applied fields such as the theory of mathematical finance, risk and queues adopting more complicated models which involve an underlying Lévy process. The aim of this text is to give a reasonably self contained approach to some elementary fluctuation theory which avoids the use of the Wiener–Hopf factorization and Itô's excursion theory and relies mainly on martingale arguments together with the Strong Markov property. None of the results we present are new but for the most part, the proofs approach the results from a new angle following Kyprianou and Palmowski (2003) who also used them to handle a class of Markov additive processes.

2 Spectrally negative Lévy processes

We start by briefly reviewing what is meant by a spectrally negative Lévy process. The reader is referred to Bertoin (1996a) and Sato (1999) for a complete discussion.

Suppose that $(\Omega, \mathcal{F}, \mathbb{F}, P)$ is a filtered probability space with filtration $\mathbb{F} = \{\mathcal{F}_t : t \geqslant 0\}$ satisfying the usual conditions of right continuity and completion. In this text, we take as our definition of a Lévy process for $(\Omega, \mathcal{F}, \mathbb{F}, P)$, the strong Markov, $\mathbb{F}$-adapted process $X = \{X_t : t \geqslant 0\}$ with right continuous paths having the properties that $P(X_0 = 0) = 1$ and for each $0 \leqslant s \leqslant t$, the increment $X_t - X_s$ is independent of $\mathcal{F}_s$ and has the same distribution as X_{t-s}. In this sense, it is said that a Lévy process has stationary independent increments.

On account of the fact that the process has stationary independent increments, it is not too difficult to show that

$$E\left(e^{i\theta X_t}\right) = e^{t\Psi(\theta)},$$

where $\Psi(\theta) = \log E(\exp\{i\theta X_1\})$. The Lévy–Khinchine formula gives the general form of the function $\Psi(\theta)$. That is,

$$\Psi(\theta) = i\mu\theta - \frac{\sigma^2}{2}\theta^2 + \int_{(-\infty,\infty)} \left(e^{i\theta x} - 1 - i\theta x\, \mathbf{1}_{|x|<1}\right) \Pi(dx) \qquad (1)$$

for every $\theta \in \mathbb{R}$ where $\mu \in \mathbb{R}$, $\sigma > 0$ and Π is a measure on $\mathbb{R}\backslash\{0\}$ such that $\int(1 \wedge x^2)\, \Pi(dx) < \infty$.

Finally, we say that X is spectrally negative if the measure Π is supported only on $(-\infty, 0)$. We exclude from the discussion however the case of a descending subordinator, that is a spectrally negative Lévy process with monotone decreasing paths. Included in the discussion however are descending subordinators plus an upward drift (such as one might use when modelling an insurance risk process, dam and storage models or a virtual waiting time process in an $M/G/1$ queue) and a Brownian motion with drift. Also included are processes such as asymmetric α-stable processes for $\alpha \in (1, 2)$ which have unbounded variation and zero quadratic variation. By adding independent copies of any of the above (spectrally negative) processes together one still has a spectrally negative Lévy process.

For spectrally negative Lévy processes it is possible to talk of the Laplace exponent $\psi(\lambda)$ defined by

$$E\left(e^{\lambda X_t}\right) = e^{\psi(\lambda)t}, \qquad (2)$$

in other words, $\psi(\lambda) = \Psi(-i\lambda)$. Since Π has negative support, we can safely say that $\psi(\lambda)$ exists at least for all $\lambda \geqslant 0$. Further, it is easy to check that ψ is strictly convex and tends to infinity as λ tends to infinity. This allows us to define for $q \in \mathbb{R}$,

$$\Phi(q) = \sup\{\lambda \geqslant 0 : \psi(\lambda) = q\},$$

the largest root of the equation $\psi(\lambda) = q$ when it exits. Note that there exist at most two roots for a given q and precisely one root when $q > 0$. Further we can identify $\psi'(0^+) = E(X_1) \in [-\infty, \infty)$ which, as we shall see in the next section, determines the long term behaviour of the process.

Suppose now the probabilities $\{P_x : x \in \mathbb{R}\}$ correspond to the conditional version of P where $X_0 = x$ is given. We simply write $P_0 = P$. The equality (2) allows for a Girsanov-type change of measure to be defined, namely via

$$\left.\frac{\mathrm{d}P_x^c}{\mathrm{d}P_x}\right|_{\mathcal{F}_t} = \frac{\mathcal{E}_t(c)}{\mathcal{E}_0(c)}$$

for any $c \geqslant 0$ where $\mathcal{E}_t(c) = \exp\{cX_t - \psi(c)t\}$ is the exponential martingale under P_x. Note that the fact that $\mathcal{E}_t(c)$ is a martingale follows from the fact that X has stationary independent increments together with (2). It is easy to check that under this change of measure, X remains within the class of spectrally negative processes and the Laplace exponent of X under P_x^c is given by

$$\psi_c(\theta) = \psi(\theta + c) - \psi(c)$$

for $\theta \geqslant -c$.

3 Exit problems

Let us now turn to the one and two sided exit problems for spectrally negative Lévy processes. The exit problems essentially consist of characterizing the Laplace transforms of τ_a^+, τ_0^- and $\tau_a^+ \wedge \tau_0^-$ where

$$\tau_0^- = \inf\{t \geqslant 0 : X_t \leqslant 0\} \qquad \text{and} \qquad \tau_a^+ = \inf\{t \geqslant 0 : X_t \geqslant a\}$$

for any $a > 0$. Note that X will hit the point a when crossing upwards as it can only move continuously upwards. On the other hand, it may either hit 0 or jump over zero when crossing 0 from above depending on the components of the process.

It has turned out (cf. Zolotarev (1964), Takács (1967), Emery (1973), Bingham (1975), Rogers (1990) and Bertoin (1996a, 1996b, 1997)) that one and two sided exit problems of spectrally negative Lévy processes can be characterized by the exponential function together with two families, $\{W^{(q)}(x) : q \geqslant 0, x \in \mathbb{R}\}$ and $\{Z^{(q)}(x) : q \geqslant 0, x \in \mathbb{R}\}$ known as the scale functions which we defined in the following main theorem of this text.

Theorem 1. *There exist a family of functions* $W^{(q)} : \mathbb{R} \to [0, \infty)$ *and*

$$Z^{(q)}(x) = 1 + q \int_0^x W^{(q)}(y)\,\mathrm{d}y, \qquad for\ x \in \mathbb{R}$$

defined for each $q \geqslant 0$ *such that the following hold (for short we shall write* $W^{(0)} = W$*).*

One sided exit above. *For any $x \leqslant a$ and $q \geqslant 0$,*

$$E_x\big(e^{-q\tau_a^+}\mathbf{1}_{(\tau_a^+<\infty)}\big) = e^{-\Phi(q)(a-x)}. \tag{3}$$

One sided exit below. *For any $x \in \mathbb{R}$ and $q \geqslant 0$,*

$$E_x\big(e^{-q\tau_0^-}\mathbf{1}_{(\tau_0^-<\infty)}\big) = Z^{(q)}(x) - \frac{q}{\Phi(q)}\,W^{(q)}(x), \tag{4}$$

where we understand $q/\Phi(q)$ in the limiting sense for $q = 0$, so that

$$P_x(\tau_0^- < \infty) = \begin{cases} 1 - \psi'(0)W(x) & \text{if } \psi'(0) > 0 \\ 1 & \text{if } \psi'(0) \leqslant 0. \end{cases}$$

Two sided exit. *For any $x \leqslant a$ and $q \geqslant 0$,*

$$E_x\big(e^{-q\tau_a^+}\mathbf{1}_{(\tau_0^->\tau_a^+)}\big) = \frac{W^{(q)}(x)}{W^{(q)}(a)}, \tag{5}$$

and

$$E_x\big(e^{-q\tau_0^-}\mathbf{1}_{(\tau_0^-<\tau_a^+)}\big) = Z^{(q)}(x) - Z^{(q)}(a)\frac{W^{(q)}(x)}{W^{(q)}(a)}. \tag{6}$$

Further, for any $q \geqslant 0$, we have $W^{(q)}(x) = 0$ for $x \leqslant 0$ and $W^{(q)}$ is characterized on $(0,\infty)$ by the unique left continuous function whose Laplace transform satisfies

$$\int_0^\infty e^{-\beta x}W^{(q)}(x)\,\mathrm{d}x = \frac{1}{\psi(\beta) - q} \qquad \text{for } \beta > \Phi(q). \tag{7}$$

Remark 2. Let us make a historical note on the appearance of these formulae. Identity (3) can be found in Emery (1973) and Bertoin (1996a). Identity (4) appears in the form of a Fourier transform again in Emery (1973). Identity (5) first appeared for the case $q = 0$ in Zolotarev (1964) followed by Takács (1967) and then with a short proof in Rogers (1990). The case $q > 0$ was first given in Bertoin (1996b) for the case of a purely asymmetric stable process and then again for a general spectrally negative Lévy process in Bertoin (1997) (who refered to a method used for the case $q = 0$ in Bertoin (1996a)). Finally (6) belongs originally to Suprun (1976) with a more modern proof given in Bertoin (1997).

Remark 3. By changing measure using the exponential martingale, one may extract identities from the above expressions giving the joint Laplace transform of the time to overshoot and overshoot itself. For example we have for any v with $\psi(v) < \infty$, $u \geqslant \psi(v) \vee 0$ and $x \in \mathbb{R}$,

$$E_x\left(e^{-u\tau_0^- + vX_{\tau_0^-}}\mathbf{1}_{(\tau_0^-<\infty)}\right) = e^{vx}\left(Z_v^{(p)}(x) - \frac{p}{\Phi(p)}\,W_v^{(p)}(x)\right)$$

where $W_v^{(p)}$ and $Z_v^{(p)}$ are scale functions with respect to the measure P^v, $p = u - \psi(v)$ and $p/\Phi(p)$ is understood in the limiting sense if $p = 0$, as in (3). In fact, it was shown in Bertoin (1997) that for each $x \in \mathbb{R}$, $W^{(q)}(x)$ is analytically extendable, as a function in q, to the whole complex plane; and hence the same is true of $Z^{(q)}(x)$. In which case arguing again by analytic extention one may weaken the requirement that $u \geqslant \psi(v) \vee 0$ to simply $u \geqslant 0$.

The proof we give of (3) is not new and follows as an easy consequence of Doob's optional stopping theorem applied to the exponential martingale; a technique traditionally attributed to *Wald*. The proof of the remaining results in Theorem 1 are a direct consequence of a special martingale which we shall discuss in Section 5. The proofs of (5), (4) and (6) are given in Sections 6, 7 and 8 respectively. The structure of this text is based on new results and methodology for a general class of Markov additive processes given in Kyprianou and Palmowski (2003).

4 Proof: one sided exit above

Assume that $x \leqslant a$ and $q > 0$. Since $t \wedge \tau_a^+ \leqslant t$ is a bounded stopping time and $X_{t \wedge \tau_a^+} \leqslant a$, it follows from Doob's Optional Stopping Theorem that

$$E_x \left(\frac{\mathcal{E}_{t \wedge \tau_a^+}(\Phi(q))}{\mathcal{E}_0(\Phi(q))} \right) = E_x \left(e^{\Phi(q)(X_{t \wedge \tau_a^+} - x) - q(t \wedge \tau_a^+)} \right) = 1.$$

By dominated convergence and the fact that $X_{\tau_a^+} = a$ on $\tau_a^+ < \infty$ we have,

$$E_x \left(e^{-q\tau_a^+} \mathbf{1}_{(\tau_a^+ < \infty)} \right) = e^{-\Phi(q)(a-x)}. \tag{8}$$

The case for $q = 0$ is dealt with by taking the limit as $q \downarrow 0$ in the above identity.

5 The Kella–Whitt martingale

As already mentioned in the introduction, we shall base our proofs for the most part on martingale arguments. A martingale which plays a fundamental role in our calculations is the Kella–Whitt martingale, introduced in Kella and Whitt (1992). This martingale has close links to so called Kennedy martingales (cf. Kennedy (1976)). For completeness we shall introduce the Kella–Whitt martingale in the following theorem.

Theorem 4. *Let* $\overline{X}_t = \sup_{0 \leqslant u \leqslant t} X_u$, *and* $Z_t = \overline{X}_t - X_t$, *then for* $\alpha \geqslant 0$

$$M_t := \psi(\alpha) \int_0^t e^{-\alpha Z_s} \, ds + 1 - e^{-\alpha Z_t} - \alpha \overline{X}_t, \qquad t \geqslant 0 \tag{9}$$

is a martingale.

Proof. Let $\mathcal{E}_t(\alpha) = \exp\{\alpha X_t - \psi(\alpha)t\}$ and note that

$$d\mathcal{E}_t(\alpha) = \mathcal{E}_{t-}(\alpha)\big(\alpha\,dX_t - \psi(\alpha)\,dt\big) + \frac{1}{2}\alpha^2\mathcal{E}_{t-}(\alpha)\,d[X,X]_t^c$$
$$+ \{\Delta\mathcal{E}_t(\alpha) - \alpha\mathcal{E}_{t-}(\alpha)\Delta X_t\}.$$

Note also that

$$dM_t = \psi(\alpha)e^{-\alpha Z_{t-}}\,dt + \alpha e^{-\alpha Z_{t-}}\,dZ_t - \frac{1}{2}\alpha^2 e^{-\alpha Z_{t-}}\,d[X,X]_t^c$$
$$- \{\Delta e^{-\alpha Z_t} + \alpha\Delta Z_t\} - \alpha\,d\overline{X}_t$$
$$= e^{-\alpha\overline{X}_t + \psi(\alpha)t}\Big[\psi(\alpha)\mathcal{E}_{t-}(\alpha)\,dt + \alpha\mathcal{E}_{t-}(\alpha)\big(d\overline{X}_t - dX_t\big)$$
$$- \frac{1}{2}\alpha^2\mathcal{E}_{t-}(\alpha)\,d[X,X]_t^c$$
$$- \mathcal{E}_{t-}(\alpha)\{e^{\alpha\Delta X_t} - 1 - \alpha\Delta X_t\}$$
$$- \alpha e^{\alpha\overline{X}_t - \psi(\alpha)t}\,d\overline{X}_t\Big]$$
$$= e^{-\alpha\overline{X}_t + \psi(\alpha)t}\Big\{-d\mathcal{E}_t(\alpha) + \alpha\big(\mathcal{E}_t(\alpha) - e^{\alpha\overline{X}_t - \psi(\alpha)t}\big)\,d\overline{X}_t\Big\},$$

where we have used that $\overline{X}_{t-} = \overline{X}_t$. Since $\overline{X}_t = X_t$ if and only if $\overline{X}_t$ increases, we may write

$$dM_t = e^{-\alpha\overline{X}_t + \psi(\alpha)t}\Big\{-d\mathcal{E}_t(\alpha) + \alpha\big(\mathcal{E}_t(\alpha) - e^{\alpha\overline{X}_t - \psi(\alpha)t}\big)\mathbf{1}_{(\overline{X}_t = X_t)}\,d\overline{X}_t\Big\}$$
$$= -e^{-\alpha\overline{X}_t + \psi(\alpha)t}\,d\mathcal{E}_t(\alpha)$$

showing that M_t is a local martingale since $\mathcal{E}_t(\alpha)$ is a martingale. To prove that M is a martingale, it suffices to show that for each $t > 0$,

$$E\left(\sup_{s\leqslant t}|M_s|\right) < \infty.$$

To this end note that since the events $\{\overline{X}_{e_q} > x\}$ and $\{\tau_x^+ < e_q\}$ are almost surely equivalent where e_q is an exponential distribution with intensity $q > 0$ independent of X, it follows from (8)

$$P\big(\overline{X}_{e_q} > x\big) = E\big(e^{-q\tau_x^+}\mathbf{1}_{(\tau_x^+ < \infty)}\big) = e^{-\Phi(q)x}$$

showing that $\overline{X}_{e_q}$ is exponentially distributed with parameter $\Phi(q)$. It follows that

$$E\big(\overline{X}_{e_q}\big) = \int_0^\infty q e^{-qt}\,E\big(\overline{X}_t\big)\,dt = \frac{1}{\Phi(q)} < \infty$$

and hence, since $\overline{X}_t$ is an increasing process, we have $E\big(\overline{X}_t\big) < \infty$ for all t. Now note by the positivity of the process Z and again since $\overline{X}$ increases,

$$E\left(\sup_{s\leqslant t}|M_s|\right) \leqslant \psi(\alpha)t + 2 + \alpha\,E(\overline{X}_t) < \infty$$

for each finite $t > 0$.   □

An application involving this martingale, brings us to an identity which is effectively the Wiener–Hopf factorization in disguise. Alternatively one may say that the Wiener–Hopf factorization for spectrally negative Lévy processes brings one to the same conclusion.

Theorem 5. *Let $\underline{X}_t = \inf_{0\leqslant u\leqslant t} X_u$ and suppose that $\mathbf{e}_q$ is an exponentially distributed random variable with parameter $q > 0$ independent of the process X. Then for $\alpha > 0$,*

$$E\left(e^{\alpha\underline{X}_{\mathbf{e}_q}}\right) = \frac{q(\alpha - \Phi(q))}{\Phi(q)(\psi(\alpha) - q)}. \tag{10}$$

Proof. We begin by noting some facts which will be used in conjunction with the martingale (9). Recall that $\mathbf{e}_q$ is an exponentially distributed random variable with parameter $q > 0$ independent of the process X.

First note that by an application of Fubini's theorem,

$$E\int_0^{\mathbf{e}_q} e^{-\alpha Z_s}\,\mathrm{d}s = \int_0^\infty e^{-qs}\,E\left(e^{-\alpha Z_s}\right)\mathrm{d}s = \frac{1}{q}E\left(e^{-\alpha Z_{\mathbf{e}_q}}\right).$$

Next we recall a well known result, known as the Duality Lemma, which can best be verified with a diagram. That is by defining the process $\{\widetilde{X}_s = X_{(t-s)^-} - X_t : 0 \leqslant s \leqslant t\}$ as the time reversed Lévy process from the fixed moment, t, the law of $\widetilde{X}$ and $\{-X_s : 0 \leqslant s \leqslant t\}$ are the same. In particular, this means that

$$-\inf_{0\leqslant s\leqslant t} X_s \overset{d}{=} \sup_{0\leqslant s\leqslant t} \widetilde{X}_s = \overline{X}_t - X_t.$$

From Theorem 4 we have that $E(M_{\mathbf{e}_q}) = E\,M_0 = 0$ and hence using the last two observations we obtain

$$\frac{\psi(\alpha) - q}{q}\,E\left(e^{\alpha\underline{X}_{\mathbf{e}_q}}\right) = \alpha\,E(\overline{X}_{\mathbf{e}_q}) - 1.$$

Recall from the proof of Theorem 4 that $\overline{X}_{\mathbf{e}_q}$ is exponentially distributed with parameter $\Phi(q)$. It now follows that

$$\frac{\psi(\alpha) - q}{q}\,E\left(e^{\alpha\underline{X}_{\mathbf{e}_q}}\right) = \frac{\alpha - \Phi(q)}{\Phi(q)} \tag{11}$$

and the theorem is proved.   □

Remark 6. Recall that $\overline{X}_{e_q}$ is exponentially distributed with parameter $\Phi(q)$. It thus follows that for $\alpha < \Phi(q)$

$$E\left(e^{\alpha \overline{X}_{e_q}}\right) = \frac{\Phi(q)}{\Phi(q) - \alpha} \tag{12}$$

and hence (11) reads

$$E\left(e^{\alpha \overline{X}_{e_q}}\right) E\left(e^{\alpha \underline{X}_{e_q}}\right) = \frac{q}{q - \psi(\alpha)} = E\left(e^{\alpha X_{e_q}}\right).$$

which is a conclusion that also follows from the Wiener–Hopf factorization.

In the previous section it was remarked that $\psi'(0^+)$ characterizes the asymptotic behaviour of X. We may now use the results of the previous Remark and Theorem to elaborate on this point. We do so in the form of a Lemma.

Lemma 7. *We have that*

(i) $\overline{X}_\infty$ and $-\underline{X}_\infty$ *are either infinite almost surely or finite almost surely,*
(ii) $\overline{X}_\infty = \infty$ *if and only if* $\psi'(0^+) \geq 0$,
(iii) $\underline{X}_\infty = -\infty$ *if and only if* $\psi'(0^+) \leq 0$.

Proof. On account of the strict convexity ψ it follows that $\Phi(0) > 0$ if and only if $\psi'(0^+) < 0$ and hence

$$\lim_{q \downarrow 0} \frac{q}{\Phi(q)} = \begin{cases} 0 & \text{if } \psi'(0^+) \leq 0 \\ \psi'(0^+) & \text{if } \psi'(0^+) > 0. \end{cases}$$

By taking q to zero in the identity (10) we now have that

$$E\left(e^{\alpha \underline{X}_\infty}\right) = \begin{cases} 0 & \text{if } \psi'(0^+) \leq 0 \\ \psi'(0^+)\alpha/\psi(\alpha) & \text{if } \psi'(0^+) > 0. \end{cases}$$

Next, recall from (12) that for $\alpha > 0$

$$E\left(e^{-\alpha \overline{X}_{e_q}}\right) = \frac{\Phi(q)}{\Phi(q) + \alpha}$$

and hence by taking the limit of both sides as q tends to zero,

$$E\left(e^{-\alpha \overline{X}_\infty}\right) = \begin{cases} (\alpha/\Phi(0) + 1)^{-1} & \text{if } \psi'(0^+) < 0 \\ 0 & \text{if } \psi'(0^+) \geq 0. \end{cases}$$

Parts (i)–(iii) follow immediately from the previous two identities by considering their limits as $\alpha \downarrow 0$. □

6 Proof: two sided exit above

Our proof first deals with the case that $\psi'(0^+) > 0$ and $q = 0$, then the case that $q > 0$ (no restriction on $\psi'(0^+)$) or $q = 0$ and $\psi'(0) < 0$. Finally the case that $\psi'(0^+) = 0$ and $q = 0$ is achieved by passing to the limit as q tends to zero.

Assume then that $\psi'(0^+) > 0$ so that $-\underline{X}_\infty$ is almost surely finite. As earlier seen in the proof of Lemma 7, by taking q to zero in (10) it follows that

$$E\big(e^{\alpha \underline{X}_\infty}\big) = \psi'(0) \frac{\alpha}{\psi(\alpha)}.$$

Integration by parts shows that

$$
\begin{aligned}
E\big(e^{\alpha \underline{X}_\infty}\big) &= \int_{[0,\infty)} e^{-\alpha x} P(-\underline{X}_\infty \in \mathrm{d}x) \\
&= \alpha \int_0^\infty e^{-\alpha x} P(-\underline{X}_\infty < x)\, \mathrm{d}x \\
&= \alpha \int_0^\infty e^{-\alpha x} P_x(\underline{X}_\infty > 0)\, \mathrm{d}x.
\end{aligned}
$$

Now define the function

$$W(x) = \frac{1}{\psi'(0^+)} P_x(\underline{X}_\infty > 0). \tag{13}$$

Clearly $W(x) = 0$ for $x \leqslant 0$, is left continuous since it is also equal to the left continuous distribution function $P(-\underline{X}_\infty < x)$ and therefore is uniquely determined by its Laplace transform, $1/\psi(\alpha)$ for all $\alpha > 0$. [Note: this shows the existence of the scale function when $\psi'(0^+) > 0$ and $q = 0$]. A simple argument using the law of total probability and the Strong Markov Property now yields for $x \in (0, a)$

$$
\begin{aligned}
&P_x(\underline{X}_\infty > 0) \\
&\quad = E_x P_x(\underline{X}_\infty > 0 \,|\, \mathcal{F}_{\tau_a^+}) \\
&\quad = E_x \Big(1_{(\tau_a^+ < \tau_0^-)} P_a(\underline{X}_\infty > 0)\Big) + E_x \Big(1_{(\tau_a^+ > \tau_0^-)} P_{X_{\tau_0^-}}(\underline{X}_\infty > 0)\Big) \tag{14} \\
&\quad = P_a(\underline{X}_\infty > 0) P_x(\tau_a^+ < \tau_0^-),
\end{aligned}
$$

where the second term in the second equality disappears as $X_{\tau_0^-} \leqslant 0$ and $P_x(\underline{X}_\infty > 0) = 0$ for $x \leqslant 0$. That is to say

$$P_x(\tau_a^+ < \tau_0^-) = \frac{W(x)}{W(a)} \tag{15}$$

and clearly the same equality holds even when $x \leqslant 0$.

Now assume that $q > 0$ or $\psi'(0) < 0$ and $q = 0$. In this case, by convexity of ψ, we know that $\Phi(q) > 0$ and hence $\psi'_{\Phi(q)}(0) = \psi'(\Phi(q)) > 0$ (again by convexity). Changing measure using the Girsanov density, we have for $x \in (0, a)$

$$E_x\left(e^{-q\tau_a^+}1_{(\tau_a^+ < \tau_0^-)}\right) = E_x\left(\frac{\mathcal{E}_{\tau_a^+}(\Phi(q))}{\mathcal{E}_0(\Phi(q))}1_{(\tau_a^+ < \tau_0^-)}\right)e^{-\Phi(q)(a-x)}$$

$$= e^{-\Phi(q)(a-x)}P_x^{\Phi(q)}\left(\tau_a^+ < \tau_0^-\right).$$

According to our previous calculations for the case that $q = 0$ and $\psi'(0^+) > 0$, we can now identify

$$E_x\left(e^{-q\tau_a^+}1_{(\tau_a^+ < \tau_0^-)}\right) = \frac{W^{(q)}(x)}{W^{(q)}(a)} \tag{16}$$

such that $W^{(q)}(x) = e^{\Phi(q)x}W_{\Phi(q)}(x)$ where $W_{\Phi(q)}(x)$ is identically zero on $(-\infty, 0]$, is left continuous and has Laplace transform $1/\psi_{\Phi(q)}(\alpha)$ for all $\alpha > 0$. Taking Laplace transforms of $W^{(q)}(x)$ it appears now that for $\alpha > \Phi(q)$,

$$\int_0^\infty e^{-\alpha x}W^{(q)}(x)\,dx = \int_0^\infty e^{-(\alpha-\Phi(q))x}W_{\Phi(q)}(x)\,dx$$

$$= \frac{1}{\psi_{\Phi(q)}(\alpha - \Phi(q))}$$

$$= \frac{1}{\psi(\alpha) - q}, \tag{17}$$

where in the last equality we have used the fact that for $c > 0$, $\psi_c(\theta) = \psi(\theta+c) - \psi(c)$. [Note again that this last calculation again justifies that $W^{(q)}$ exists for the regime that we are considering.]

As mentioned at the beginning of the proof, the final missing case of X not drifting to infinity (ie $\psi'(0^+) = 0$) and $q = 0$ is achieved by passing to the limit as $q \downarrow 0$. Since $W_{\Phi(q)}$ has Laplace transform $1/\psi_{\Phi(q)}$ for $q > 0$, integration by parts reveals that

$$\int_{(0,\infty)} e^{-\beta x}W_{\Phi(q)}(dx) = \frac{\beta}{\psi_{\Phi(q)}(\beta)}. \tag{18}$$

One may appeal to the Extended Continuity Theorem for Laplace Transforms, see Feller (1971) Theorem XIII.1.2a, and (18) to deduce that since

$$\lim_{q\downarrow 0}\int_{(0,\infty)} e^{-\beta x}W_{\Phi(q)}(dx) = \frac{\beta}{\psi(\beta)}$$

then there exists a measure W^* such that in the weak sense $W^* = \lim_{q\downarrow 0} W_{\Phi(q)}$ and

$$\int_{(0,\infty)} e^{-\beta x}W^*(dx) = \frac{\beta}{\psi(\beta)}.$$

Integration by parts shows that its left continuous distribution,

$$W(x) := W^*(-\infty, x) = \lim_{q \downarrow 0} W^{(q)}(x)$$

satisfies

$$\int_0^\infty e^{-\beta x} W(x)\, dx = \frac{1}{\psi(\beta)} \qquad (19)$$

for $\beta > 0$. Considering the limit as $q \downarrow 0$ in (16) and remembering that $W^{(q)}(x) = e^{\Phi(q)x} W_{\Phi(q)}(x)$ we recover the required identity (15).

7 Proof: one sided exit below

Taking (17) and (18) into account, we can interpret (10) as saying that

$$P\big(-\underline{X}_{e_q} \in dx\big) = \frac{q}{\Phi(q)} W^{(q)}(dx) - q W^{(q)}(x)\, dx$$

and hence with an easy manipulation, for $x > 0$

$$
\begin{aligned}
E_x\big(e^{-q\tau_0^-} \mathbf{1}_{(\tau_0^- < \infty)}\big) &= P_x\big(e_q > \tau_0^-\big) \\
&= P_x\big(\underline{X}_{e_q} < 0\big) \\
&= 1 + q \int_0^x W^{(q)}(y)\, dy - \frac{q}{\Phi(q)} W^{(q)}(x) \\
&= Z^{(q)}(x) - \frac{q}{\Phi(q)} W^{(q)}(x). \qquad (20)
\end{aligned}
$$

Note that since $Z^{(q)}(x) = 1$ and $W^{(q)}(x) = 0$ for all $x \in (-\infty, 0]$, the statement is valid for all $x \in \mathbb{R}$. The proof is now complete for the case that $q > 0$.

Recalling that $\lim_{q \downarrow 0} q/\Phi(q)$ is either $\psi'(0^+)$ or zero, the proof is completed by taking the limit in q.

8 Proof: two sided exit below

Fix $q > 0$. The Strong Markov Property together with the identity (20) give us that

$$
\begin{aligned}
&P_x\Big(\underline{X}_{e_q} < 0 \;\Big|\; \mathcal{F}_{t \wedge \tau_a^+ \wedge \tau_0^-}\Big) \\
&= e^{-q(t \wedge \tau_a^+ \wedge \tau_0^-)} P_{X_{t \wedge \tau_a^+ \wedge \tau_0^-}}\big(\underline{X}_{e_q} < 0\big) \\
&= e^{-q(t \wedge \tau_a^+ \wedge \tau_0^-)} \Big(Z^{(q)}\big(X_{t \wedge \tau_a^+ \wedge \tau_0^-}\big) - \frac{q}{\Phi(q)} W^{(q)}\big(X_{t \wedge \tau_a^+ \wedge \tau_0^-}\big)\Big)
\end{aligned}
$$

showing that the right hand side is a martingale for $t \geqslant 0$. Note also that with a similar methodology we have (using that $W^{(q)}\big(X_{\tau_0^- \wedge \tau_a^+}\big) = \mathbf{1}_{(\tau_a^+ < \tau_0^-)} W^{(q)}(a)$)

$$E_x\left(e^{-q\tau_a^+}1_{(\tau_a^+<\tau_0^-)}\,\Big|\,\mathcal{F}_{t\wedge\tau_a^+\wedge\tau_0^-}\right)$$

$$= 1_{(t<\tau_0^-\wedge\tau_a^+)}e^{-qt}E_{X_t}\left(e^{-q\tau_a^+}1_{(\tau_a^+<\tau_0^-)}\right) + 1_{(t>\tau_0^-\wedge\tau_a^+)}e^{-q\tau_a^+}1_{(\tau_a^+<\tau_0^-)}$$

$$= 1_{(t<\tau_0^-\wedge\tau_a^+)}e^{-qt}\frac{W^{(q)}(X_t)}{W^{(q)}(a)} + 1_{(t>\tau_0^-\wedge\tau_a^+)}e^{-q(\tau_0^-\wedge\tau_a^+)}\frac{W^{(q)}(X_{\tau_0^-\wedge\tau_a^+})}{W^{(q)}(a)}$$

$$= e^{-q(t\wedge\tau_a^+\wedge\tau_0^-)}\frac{W^{(q)}(X_{t\wedge\tau_a^-\wedge\tau_0^+})}{W^{(q)}(a)}$$

showing again that the right hand side is a martingale for $t \geq 0$.

Now it follows by linearity that

$$e^{-q(t\wedge\tau_a^+\wedge\tau_0^-)}\left(Z^{(q)}\left(X_{t\wedge\tau_a^+\wedge\tau_0^-}\right) - \frac{Z^{(q)}(a)}{W^{(q)}(a)}W^{(q)}\left(X_{t\wedge\tau_a^+\wedge\tau_0^-}\right)\right)$$

is also a martingale for $t \geq 0$. In fact it is a uniformly integrable martingale and hence its terminal expectation is equal to its initial expectation. That is to say

$$E_x\left(e^{-q(\tau_a^+\wedge\tau_0^-)}\left(Z^{(q)}\left(X_{\tau_a^+\wedge\tau_0^-}\right) - \frac{Z^{(q)}(a)}{W^{(q)}(a)}W^{(q)}\left(X_{\tau_a^+\wedge\tau_0^-}\right)\right)\right)$$

$$= E_x\left(e^{-q\tau_0^-}1_{(\tau_a^+>\tau_0^-)}\right)$$

$$= Z^{(q)}(x) - \frac{Z^{(q)}(a)}{W^{(q)}(a)}W^{(q)}(x),$$

where as usual we have used the fact that

$$Z^{(q)}\left(X_{\tau_a^+\wedge\tau_0^-}\right) = 1 \quad \text{and} \quad W^{(q)}\left(X_{\tau_a^+\wedge\tau_0^-}\right) = 0 \qquad \text{if } \tau_0^- < \tau_a^+,$$

and

$$Z^{(q)}\left(X_{\tau_a^+\wedge\tau_0^-}\right) = Z^{(q)}(a) \quad \text{and} \quad W^{(q)}\left(X_{\tau_a^+\wedge\tau_0^-}\right) = W^{(q)}(a) \qquad \text{if } \tau_0^- > \tau_a^+.$$

For the case that $q = 0$, we again take limits as q tends to zero.

9 Final Remarks

We conclude with some final remarks concerning some more subtle points of the calculations we have made which are not necessarily immediately obvious. The definition of $\tau_x^- = \inf\{t \geq 0 : X_t \leq x\}$ requiring *weak* first passage below the level x forces the definition of W proportional to $P_x(\underline{X}_\infty > 0)$ in the case that $q = 0$ and $\psi'(0^+) > 0$ in (13). This in turn determines the left continuity of $W^{(q)}$ for all $q \geq 0$, a fact which is seen to be of importance in the calculation (14) as well as later, for example in Section 8, where it is stated that

$W^{(q)}(X_{\tau_0^- \wedge \tau_a^+}) = \mathbf{1}_{(\tau_a^+ < \tau_0^-)}$. However, Bertoin (1997) works with a definition of *strong* first downward passage equivalent to $\tau_x^- = \inf\{t \geqslant 0 : X_t < x\}$. Following the analysis here one sees in (13) that W should then be taken as

$$W(x) = \frac{1}{\psi'(0^+)} P_x(\underline{X}_\infty \geqslant 0) = \frac{1}{\psi'(0^+)} P(-\underline{X}_\infty \leqslant x).$$

But then, if 0 is irregular for $(-\infty, 0)$ for X we have $P(\tau_0^- > \tau_a^+) > 0$, which itself is a result of the definition of τ_0^- in the strong sense. The effect of this definition is that $W^{(q)}$ is now right continuous. None the less, with very subtle adjustments, all the arguments go through as presented. An example of a calculation which needs a little extra care is (14).

In this case, it is possible that $X_{\tau_0^-} = 0$ with positive probability, that is to say X may creep downwards over zero, and hence in principle the second term in (14) may not be zero. However, it is known that spectrally negative Lévy processes may only creep downwards if and only if a Gaussian component is present (cf. Bertoin (1996a) p. 175). In this case $P(\underline{X}_\infty \geqslant 0) = 0$ anyway and the calculation goes through.

To some extent, it is more natural to want work with the right continuous version of $W^{(q)}$ because one captures the probability of starting at the origin and escaping at a before entering $(-\infty, 0)$ in the expression $W(0)/W(a)$ as opposed to $W(0^+)/W(a)$ for the left continuous case. However we promised in the introduction a self contained approach to our results which avoids the use of the Wiener–Hopf factorization. Hence we have opted to present the case of left continuity in $W^{(q)}$ thus avoiding the deeper issue of creeping, which is intimately connected to the Wiener–Hopf factorization.

For other recent perspectives and new proofs of existing results concerning fluctuation theory of spectrally negative Lévy processes see Doney (2004), Pistorius (2004) and Nguyen-Ngoc and Yor (2004).

Acknowledgements

We are very grateful to the referee for insightful comments on the original version of this document which lead to the final remarks in this paper. The first author would like to thank Risklab, ETH Zürich where lectures based on this text were given in April 2003. The second author would like to thank EU-RANDOM for their support. In addition, both authors gratefully acknowledge grant nr. 613.000.310 from Nederlandse Organisatie voor Wetenschappelijk Onderzoek.

References

1. Avram, F., Kyprianou, A.E. and Pistorius, M. (2004) Exit problems for spectrally negative Lévy processes and applications to (Canadized) Russian options, *Ann. Appl. Probab.* **14**, 215-238.

2. Bertoin, J. (1996a) *Lévy processes*, Cambridge University Press.

3. Bertoin, J. (1996b) On the first exit time of a completely asymmetric stable process from a finite interval, *Bull. London Math. Soc.* **28**, 514–520.

4. Bertoin, J. (1997) Exponential decay and ergodicity of completely asymmetric Lévy processes in a finite interval, *Ann. Appl. Probab.* **7**, 156–169.

5. Bingham, N. H. (1975) Fluctuation theory in continuous time, *Adv. Appl. Probab.* **7**, 705–766.

6. Doney, R. A. (2004) Some excursion calculations for spectrally one-sided Lévy processes, *in this volume.*

7. Emery, D.J. (1973) Exit problems for a spectrally positive process, *Adv. Appl. Probab.* **5**, 498–520.

8. Kella, O. and Whitt, W. (1992) Useful martingales for stochastic storage processes with Lévy input, *J. Appl. Probab.* **29**, 396–403.

9. Kennedy, D. (1976) Some martingales related to cumulative sum tests and single server queues, *Stoch. Proc. Appl.* **29**, 261–269.

10. Kyprianou, A.E. and Palmowski, Z. (2003) Fluctuations of spectrally negative Markov additive processes, *submitted.*

11. Nguyen-Ngoc, L. and Yor. M. (2004) Some martingales associated to reflected Lévy processes, *in this volume.*

12. Pistorius, M. R. (2004) A potential-theoretic review of some exit problems of spectrally negative Lévy processes, *in this volume.*

13. Rogers, L. C. G. (1990) The two-sided exit problem for spectrally positive Lévy processes, *Adv. Appl. Probab.* **22**, 486–487.

14. Sato, K. (1999) *Lévy processes and infinitely divisible distributions*, Cambridge University Press.

15. Suprun, V.N. (1976) Problem of destruction and resolvent of terminating processes with independent increments, *Ukranian Math. J.* **28**, 39–45.

16. Takács, L. (1967) *Combinatorial methods in the theory of stochastic processes*, John Wiley & Sons, Inc.

17. Zolotarev, V.M. (1964) The first passage time of a level and the behaviour at infinity for a class of processes with independent increments, *Theory. Prob. Appl.* **9**, 653–661.

A Potential-theoretical Review of some Exit Problems of Spectrally Negative Lévy Processes

Martijn R. Pistorius

King's College London,
Department of Mathematics,
Strand,
London WC2R 2LS, UK
e-mail: pistoriu@mth.kcl.ac.uk

Summary. In this note we consider first exit problems of completely asymmetric (reflected) Lévy processes and present an alternative derivation of their Laplace transforms essentially based on potential theory of Markov processes.

Key words: Potential theory, first passage, Wiener–Hopf factorisation, Lévy processes

1 Introduction and main results

Let X be a spectrally negative Lévy process, i.e. a stochastic process with càdlàg paths without positive jumps that has stationary independent increments defined on some probability space $(\Omega, \mathcal{F}, P)$ that satisfies the usual conditions. By $(P_x, x \in \mathbb{R})$ we denote the family of measures under which the Lévy process X is translated over a constant, that is P_x denotes the measure P conditioned on $\{X_0 = x\}$. We exclude the case that X has monotone paths. By the absence of positive jumps, the moment generating function of X_t exists for all $\theta \geqslant 0$ and is given by

$$E\big[\mathrm{e}^{\theta X_t}\big] = \exp\big(t\,\psi(\theta)\big), \qquad \theta \geqslant 0,$$

for some function $\psi(\theta)$ which is well defined at least on the positive half axis, where it is convex with the property $\lim_{\theta \to \infty} \psi(\theta) = +\infty$. Let $\Phi(0)$ denote its largest root. On $[\Phi(0), \infty)$ the function ψ is strictly increasing and we denote its right-inverse function by $\Phi : [0, \infty) \to [\Phi(0), \infty)$. Denote by I and S the past infimum and supremum of X respectively, that is,

$$I_t = \inf_{0 \leqslant s \leqslant t}(X_s \wedge 0), \qquad S_t = \sup_{0 \leqslant s \leqslant t}(X_s \vee 0)$$

and write $Y = X - I$ for X reflected at its past infimum I. By T_a^-, T_a^+ we denote

$$T_a^- = \inf\{t \geqslant 0 : X_t < a\}, \qquad T_a^+ = \inf\{t \geqslant 0 : X_t > a\},$$

the first passage times of X into the sets $(-\infty, a)$ and (a, ∞), respectively. Similarly, we write

$$\tau_a^+ = \inf\{t \geqslant 0 : Y_t > a\}$$

for the first passage time of Y into the set (a, ∞). The following theorem gives the form of the Laplace transforms of these passage times:

Theorem 1. (i) *For $q > 0$, the q-potential measure of X*

$$U^q(\mathrm{d}x) = \int_0^\infty \mathrm{e}^{-qt} P(X_t \in \mathrm{d}x)\, \mathrm{d}t \tag{1}$$

is absolutely continuous with respect to the Lebesgue measure and a version of its density on $[0, \infty)$ is given by

$$u^q(x) = \Phi'(q) \exp\left(-\Phi(q)x\right). \tag{2}$$

(ii) *For $q \geqslant 0$, there exists a continuous increasing function $W^{(q)} : [0, \infty) \to [0, \infty)$ with Laplace transform*

$$\int_0^\infty \mathrm{e}^{-\lambda x} W^{(q)}(x)\, \mathrm{d}x = \left(\psi(\lambda) - q\right)^{-1}, \qquad \lambda > \Phi(q)$$

and, denoting by u^q a version of $U^q(\mathrm{d}x)/\mathrm{d}x$, it holds that for $q > 0$

$$W^{(q)}(x) = \Phi'(q) \exp(\Phi(q)x) - u^q(-x) \qquad \text{for a.e. } x \geqslant 0. \tag{3}$$

(iii) *(Exit from a half-line) For $q \geqslant 0$, $x \leqslant a$ and $y \geqslant 0$ we have*

$$E_x\left[\mathrm{e}^{-qT_a^+} I_{(T_a^+ < \infty)}\right] = \mathrm{e}^{\Phi(q)(x-a)}; \tag{4}$$

$$E_y\left[\mathrm{e}^{-qT_0^-} I_{(T_0^- < \infty)}\right] = Z^{(q)}(y) - q\,\Phi(q)^{-1} W^{(q)}(y), \tag{5}$$

where

$$Z^{(q)}(x) = 1 + q \int_0^x W^{(q)}(y)\, \mathrm{d}y$$

and for $q = 0$, $q\Phi(q)^{-1}$ is understood in the limiting sense, $\lim_{q \downarrow 0} q\Phi(q)^{-1}$.

(iv) *(Exit from a finite interval) For $x \in [0, a]$ and $q \geqslant 0$ we have*

$$E_x\left[\mathrm{e}^{-qT_a^+} I_{(T_a^+ < T_0^-)}\right] = \frac{W^{(q)}(x)}{W^{(q)}(a)}; \tag{6}$$

$$E_x\left[\mathrm{e}^{-qT_0^-} I_{(T_a^+ > T_0^-)}\right] = Z^{(q)}(x) - Z^{(q)}(a) \frac{W^{(q)}(x)}{W^{(q)}(a)}; \tag{7}$$

$$E_x\left[\mathrm{e}^{-q\tau_a^+}\right] = \frac{Z^{(q)}(x)}{Z^{(q)}(a)}. \tag{8}$$

Remark. The function $W^{(q)}$ is called the q-scale function of X in the literature. In particular, one calls $W = W^{(0)}$ the scale function of the Lévy process, in analogy with the theory of diffusions.

Remark (probabilistic derivation of formula (3)). Noting that, for $q > 0$, $\{\exp(\Phi(q)X_t - qt), \, t \geqslant 0\}$ is a martingale, we define the tilted measure $P^{\Phi(q)}$ by

$$P^{\Phi(q)}(A) = E\big[\exp\big(\Phi(q)X_t - qt\big)I_A\big], \qquad A \in \mathcal{F}_t.$$

Under the measure $P^{\Phi(q)}$ the process X is still a Lévy process and its characteristic exponent $\psi_{\Phi(q)}$ can be checked to be given by

$$\psi_{\Phi(q)}(\lambda) = \psi\big(\Phi(q) + \lambda\big) - \psi\big(\Phi(q)\big) = \psi\big(\Phi(q) + \lambda\big) - q.$$

We write $W_{\Phi(q)}$ for the scale function of X under $P^{\Phi(q)}$. Comparing Laplace transforms yields the identity

$$W^{(q)}(x) = \mathrm{e}^{\Phi(q)x} W_{\Phi(q)}(x).$$

Since $q u^q(x)\,\mathrm{d}x = P(X_{\eta(q)} \in \mathrm{d}x)$, where $\eta(q)$ denotes an independent exponential time, the strong Markov property yields for $x < 0$

$$
\begin{aligned}
&q^{-1} P(X_{\eta(q)} \in \mathrm{d}x) \\
&= \int E\Big[\mathrm{e}^{-qT_x^-} \,;\, X_{T_x^-} \in \mathrm{d}y\Big] u^q(x-y) = \int E\Big[\mathrm{e}^{-qT_x^-} \,;\, X_{T_x^-} \in \mathrm{d}y\Big] \Phi'(q)\,\mathrm{e}^{\Phi(q)(y-x)} \\
&= \Phi'(q)\,\mathrm{e}^{-\Phi(q)x} E\Big[\mathrm{e}^{-qT_x^- + \Phi(q)X_{T_x^-}}\Big] = \Phi'(q)\,\mathrm{e}^{-\Phi(q)x} P^{\Phi(q)}(T_x^- < \infty) \\
&= \Phi'(q)\,\mathrm{e}^{-\Phi(q)x}\left(1 - \frac{W_{\Phi(q)}(-x)}{W_{\Phi(q)}(\infty)}\right) = \Phi'(q)\,\mathrm{e}^{-\Phi(q)x} - W^{(q)}(-x),
\end{aligned}
$$

where in the first line we used the explicit form (2) and in the second line a change of measure. The third line follows by letting $a \to \infty$ and taking $q = 0$ in (6) and noting next that $W_{\Phi(q)}(\infty) = \lim_{x \to \infty} W_{\Phi(q)}(x)$ is equal to $1/\psi'(\Phi(q)) = \Phi'(q)$ by a Tauberian theorem applied to the Laplace–Stieltjes transform $\lambda/\psi_{\Phi(q)}(\lambda)$ of $W_{\Phi(q)}$.

The explicit form of the potential density given, in Theorem 1 allows one to determine whether X is transient of recurrent. Let U^0 denote the potential measure of X, given by (1) with $q = 0$.

Definition. The process X is called *transient* if $U^0(K) < \infty$ for every compact set $K \subset \mathbb{R}$ and it is called *recurrent* if $U^0(B) = \infty$ for every open interval of the form $B = (-r, r)$, $r > 0$.

Corollary 1. *The process X is recurrent if*

$$\Phi'(0^+) := \lim_{q\downarrow 0} \frac{\Phi(q) - \Phi(0)}{q} = \infty.$$

Otherwise, $\Phi'(0^+) < \infty$ and X is transient, and the potential measure $U^0(\mathrm{d}x)$ of X is given by

$$U^0(\mathrm{d}x) = \left(\Phi'(0^+)\exp(-\Phi(0)x) - W(-x)I_{(x<0)}\right)\mathrm{d}x. \qquad (9)$$

Another consequence from Theorem 1 is the following result on the downward 'creeping' of X. The process X is said to creep across the level $x < 0$ if X first enters $(-\infty, x)$ continuously, that is if $X_{T_x^-} = x$. Recall that we excluded the case where X is a negative deterministic drift and denote by $\sigma^2 = 2\lim_{\lambda\to\infty}\lambda^{-2}\psi(\lambda)$ the Gaussian coefficient of X.

Corollary 2. *The process X creeps across $x < 0$ if and only if X has a nonzero Gaussian coefficient σ^2 and then*

$$P\left(X_{T_x^-} = x\right) = \frac{\sigma^2}{2}\left[W'(-x) - \Phi(0)W(-x)\right], \qquad x < 0. \qquad (10)$$

In the literature there exist already several proofs for the statements in Theorem 1 and Corollaries 1 and 2. The one-sided exit identities (4) – (5) were first studied by Zolotarev [17], although formulated in a different form. The existence and properties of the scale function were proved by Bingham [4] and Emery [8]. The well established identities (4) – (5) and (6) – (7) are related to the two-sided exit problem to which among others Takács [16], Rogers [14], Emery [8] and Bertoin [2] made significant contributions. In its current form it was first formulated by Bertoin [2, 3]. The given proofs rely on (complex-) analytic and combinatorial methods or invoke Itô-excursion theory applied to the excursions of X away from its supremum S. The identity (8) was first proved in [13] using a martingale argument. Recently, these identities received more attention in the literature and several short proofs were given. Kyprianou and Palmovski [12] gave proofs for the identities invoking the Kella–Whitt martingale [11] and Doney [7] used excursion theory to prove the identity (8).

Here we follow yet another approach which exploits the connection between potential analysis and Markov processes: We show that, essentially, potential theory allows us to give simple proofs of the above results. For a deeper analysis of the relationship between Markov processes on the one hand and potential analysis on the other hand, we refer the reader to the classical works by Blumenthal and Getoor [5] and Dellacherie and Meyer [6].

The rest of this note is organised as follows. In the next section, we state and prove a first hitting time identity for a certain class of continuous time Markov processes in terms of their potential density. In the third section, we then derive explicit expressions for the potential densities of X killed upon entering a negative half-line and of X reflected at its infimum and give the proofs of Theorem 1 and Corollaries 1 and 2.

2 Potential theory and first hitting

Denote by $(\mathcal{S}, \mathcal{B}(\mathcal{S}))$ a measurable space consisting of some interval $\mathcal{S}$ of the real line and its Borel sigma-algebra $\mathcal{B}(\mathcal{S})$ and fix $a \in \mathcal{S}$. Let Z be a continuous time strong Markov process with state space $(\mathcal{S}, \mathcal{B}(\mathcal{S}))$ defined on a probability space $(\Omega, \mathcal{F}, P)$ that satisfies the usual conditions. In the sequel we restrict ourselves to processes Z which are *quasi-left continuous*, that is, if $(T_n, \ n \in \mathbb{N})$ is an increasing sequence of stopping times with $T = \lim_{n \to \infty} T_n$ almost surely, then $Z_T = \lim_{n \to \infty} Z_{T_n}$ almost surely on the event $\{T < \infty\}$. For $x \in \mathcal{S}$ we denote by $U^q(x, \cdot)$ the potential measure of Z

$$U^q(x, A) = \int_0^\infty e^{-qt} P_x(Z_t \in A) \, dt, \qquad A \in \mathcal{B}(\mathcal{S}),$$

where P_x denotes the measure P conditioned on $\{Z_0 = x\}$. Assume that for any $x \in \mathcal{S}$, $U^q(x, \cdot)$ restricted to some open interval containing a is absolutely continuous with respect to the Lebesgue measure with density $u^q(x, \cdot)$, say. Denote the first passage time of Z into the set A by

$$T_A' = \inf\{t \geqslant 0 : Z_t \in A\}, \qquad A \in \mathcal{B}(\mathcal{S}). \tag{11}$$

Below the Laplace transform of $T'_{\{a\}}$ is expressed in terms of known quantities. To formulate the result we define for any $\epsilon > 0$ the open sets $B_a(\epsilon) = (a - \epsilon, a + \epsilon)^2$ and $D_a(\epsilon) = \{(x, y) \in \mathbb{R}^2 : x < y, \ a - \epsilon < y < a + \epsilon\}$.

Proposition 1. *Let $x \in \mathcal{S}$ and $q \geqslant 0$.*

(i) *If, for some $\epsilon_0 > 0$, u^q restricted to $B_a(\epsilon_0)$ is continuous, we have*

$$E_x\left[e^{-qT'_{\{a\}}} I_{(T'_{\{a\}} < \infty)}\right] = \frac{u^q(x, a)}{u^q(a, a)}, \tag{12}$$

provided $u^q(a, a) > 0$.

(ii) *If Z has no positive jumps, for some $\epsilon_0 > 0$, u^q restricted to $D_a(\epsilon_0)$ is continuous and $u^q(a^-, a) = \lim_{\epsilon \downarrow 0} u^q(a - \epsilon, a) > 0$, the identity (12) holds for $x < a$ with $u^q(a, a)$ replaced by $u^q(a^-, a)$.*

Proof. Write T'_ϵ as shorthand for $T'_{(a-\epsilon, a+\epsilon)}$. The strong Markov property of Z yields that, for $\epsilon > 0$, $\frac{1}{\epsilon} U^q(x, (a - \epsilon, a + \epsilon))$ is equal to

$$\frac{1}{\epsilon} \int_{a-\epsilon}^{a+\epsilon} u^q(x, y) \, dy = \int E_x\left[e^{-qT'_\epsilon}; Z_{T'_\epsilon} \in dz\right] \frac{1}{\epsilon} \int_{a-\epsilon}^{a+\epsilon} u^q(z, y) \, dy. \tag{13}$$

If ϵ tends to zero, T'_ϵ increases to a stopping time, $\tilde{T}$ say, with $\tilde{T} \leqslant T'_{\{a\}}$. By quasi-left continuity of Z we find that $Z_{T'_\epsilon}$ tends to $Z_{\tilde{T}} = a$ on $\{\tilde{T} < \infty\}$ almost surely and thus $\tilde{T} = T'_{\{a\}}$ on $\{\tilde{T} < \infty\}$. If we let ϵ tend to zero the measures $E_x[e^{-qT'_\epsilon}; Z_{T'_\epsilon} \in dz]$ vaguely converge to $E_x[e^{-qT_{\{a\}}}] \delta_0$, where δ_0 denotes the unit mass in zero. Combined with the continuity of u^q in an open neighbourhood of (a, a) we end up with (12) if we let $\epsilon \downarrow 0$ in (13). In the second case, by the fact that Z has no positive jumps, (13) reduces to

$$\frac{1}{\epsilon} \int_{a-\epsilon}^{a+\epsilon} u^q(x, y) \, dy = E_x\left[e^{-qT'_\epsilon}\right] \frac{1}{\epsilon} \int_{a-\epsilon}^{a+\epsilon} u^q(a - \epsilon, y) \, dy.$$

Letting again ϵ tend to zero, the assumed continuity of u^q leads to the required identity. $\square$

Consider now the set $\{a, b\}$ for some $a, b \in S$. A related question that arises then is: what is the probability that Z hits $\{a\}$ before $\{b\}$? To be more precise, can we find an expression for

$$t^q_{a,b}(x) = E_x\left[e^{-qT_{\{a,b\}}} I_{(T_{\{a\}} < T_{\{b\}},\, T_{\{a,b\}} < \infty)}\right], \qquad x \in [a, b],$$

in terms of known quantities? The answer in terms of the potential density u^q is given in the following result.

Corollary 3. *Let, for some $\epsilon_0 > 0$, u^q restricted to $B_a(\epsilon_0) \cup B_b(\epsilon_0)$ be continuous. Then we have for $q > 0$*

$$t^q_{a,b}(x) = \frac{u^q(x, a)u^q(b, b) - u^q(x, b)u^q(b, a)}{u^q(a, a)u^q(b, b) - u^q(a, b)u^q(b, a)}, \qquad x \in [a, b], \qquad (14)$$

provided $u^q(a, a)u^q(b, b) > 0$. If $q = 0$, the identity (14) remains valid, where the right-hand side of (14) is to be understood in the limiting sense of $q \downarrow 0$ if $u^0(a, a)u^0(b, b) = u^0(a, b)u^0(b, a)$.

Proof. If $u^q(a, a) > 0$, the strong Markov property combined with Proposition 1 yields that for $q \geqslant 0$

$$u^q(x, a)/u^q(a, a) = t^q_{a,b}(x) + t^q_{b,a}(x)u^q(b, a)/u^q(a, a), \qquad x \in [a, b].$$

By interchanging the role of a and b, we can derive a similar second identity. For $q > 0$, this system of two equations is non-singular. Indeed, since $T_{\{a\}} > 0$ P_b-a.s., $E_b\left[e^{-qT_{\{a\}}}\right] < 1$ and it follows from (12) that $u^q(b, a) < u^q(a, a)$. Interchanging a and b, we find that $u^q(a, b)u^q(b, a) < u^q(a, a)u^q(b, b)$. Solving this system finishes the proof for $q > 0$. Note that $t^q_{a,b}(x)$ increases to $t^0_{a,b}(x)$ if $q \downarrow 0$. Hence $t^0_{a,b}(x)$ is equal to the limit of $q \downarrow 0$ of the right-hand side of (14). If $u^0(a, a)u^0(b, b) \neq u^0(a, b)u^0(b, a)$ then this limit is given by (14) for $q = 0$ as the previously derived system is non-singular. $\square$

Example. For a Brownian motion Z the potential density u^q is given by $u^q(x, a) = (2q)^{-1/2}e^{-\sqrt{2q}|x-a|}$. By Corollary 3 and de l'Hôpital's rule we find back the well known identity

$$P_x\left(T_{\{a\}} < T_{\{b\}}\right) = (b - x)/(b - a), \qquad x \in [a, b].$$

3 Proofs of Theorem 1 and Corollaries 1 and 2

Let $\eta(q)$ denote an independent exponential random variable with parameter $q > 0$. We start recalling the following results which we will frequently use in the proof of Theorem 1:

Lemma 1. *Let X be a Lévy process.*

(i) *For each fixed $t > 0$, $(S_t - X_t, S_t)$ has the same law as $(-I_t, X_t - I_t)$.*

(ii) *The processes $X - I$ and $S - X$ are strong Markov process.*

(iii) *For $q > 0$, $S_{\eta(q)} - X_{\eta(q)}$ is independent of $S_{\eta(q)}$.*

Proof. (i) This result follows as consequence of the duality lemma (see e.g. Lemma II.2 and Proposition VI.3 in [2]).

(ii) This follows straightforwardly from the independence and stationarity of the increments of X. See e.g. [2, Prop. VI.1] for a proof.

(iii) The independence can for example be proved using Itô-excursion theory applied to the excursions of the Markov process $S - X$ away from zero, see Greenwood and Pitman [10].   □

Proof of Theorem 1(i)–(iii). We divide the proof in several steps.

Step 1: Absolute continuity of the potential measure U^q. By the strong Markov property of X and the spatial homogeneity we note that

$$P\big(S_{\eta(q)} > t + s\big) = P\big(S_{\eta(q)} > s\big)P\big(S_{\eta(q)} > t\big) \qquad \text{for all } t, s \geqslant 0, \, q > 0.$$

Hence we deduce that $S_{\eta(q)}$ is exponentially distributed with parameter $\lambda(q)$, say. Using then Lemma 1(iii), we get that

$$P\big(X_{\eta(q)} \in \mathrm{d}x\big) = \int_{0 \vee x}^{\infty} P\big(S_{\eta(q)} \in \mathrm{d}(z - x)\big)P\big((S - X)_{\eta(q)} \in \mathrm{d}z\big) \leqslant \lambda(q)\,\mathrm{d}x.$$

Step 2: Existence of the scale function $W^{(q)}$ for $q \geqslant 0$. Next we show, following Bingham [4], that, for $\theta > \Phi(q)$, the function $\theta/(\psi(\theta) - q)$ can be represented as a Laplace–Stieltjes transform. To be more precise, we prove that there exists a measure $\mathrm{d}W^{(q)}$ on $[0, \infty)$ such that

$$\int_0^{\infty} \theta\,\mathrm{e}^{-\theta x} W^{(q)}(x)\,\mathrm{d}x = \int_0^{\infty} \mathrm{e}^{-\theta x} W^{(q)}(\mathrm{d}x) = \theta/\big(\psi(\theta) - q\big), \quad \theta > \Phi(q). \tag{15}$$

By the Lévy–Khintchine formula and by partial integration we have the following representation for ψ and $\theta \in [0, \infty)$

$$\begin{aligned}
\psi(\theta) &= a\theta + \frac{\sigma^2}{2}\theta^2 + \int_0^{\infty} \big(\mathrm{e}^{-\theta x} - 1 + \theta x I_{(x<1)}\big)\,\Lambda(\mathrm{d}x) \\
&= \theta\left(a' + \frac{\sigma^2}{2}\theta - \int_0^{\infty} \big(\mathrm{e}^{-\theta x} - I_{(x<1)}\big)\Lambda\big((x, \infty)\big)\,\mathrm{d}x\right),
\end{aligned} \tag{16}$$

where a, σ are constants and $a' = a - \Lambda((1, \infty))$ with Λ a measure satisfying $\int_0^{\infty}(1 \wedge x^2)\,\Lambda(\mathrm{d}x) < \infty$. The measure Λ is related to the Lévy measure ν of X by $\nu(\mathrm{d}x) = \Lambda(-\mathrm{d}x)$. From the previous display, we see that

$$\frac{\mathrm{d}}{\mathrm{d}\theta}\frac{\psi(\theta) - q}{\theta} = \frac{\sigma^2}{2} + \int_0^{\infty} x\,\mathrm{e}^{-\theta x}\Lambda\big((x, \infty)\big)\,\mathrm{d}x + \frac{q}{\theta^2}\,.$$

Hence $(\psi(\theta) - q)/\theta$ has derivatives that oscillate in sign and has thus a completely monotone derivative. Since also $1/\theta$ is completely monotone it follows (e.g. Feller [9, XIII.4, Criterion 2]) that $\theta/(\psi(\theta)-q)$ itself is completely monotone. By Feller [9, XII.4, Thm. 1a] a function on $[0,\infty)$ is completely monotone if and only if it can be represented as Laplace–Stieltjes transform of a measure. Partial integration yields then also the first identity of (15) and the claim is proved.

Step 3: Form of the potential density u^q. Denote by u^q a version of the density of U^q with respect to the Lebesgue measure. For $q > 0$ the Fourier transform of u^q, $\mathcal{F}u^q$, is given by

$$\mathcal{F}u^q(\xi) = \int e^{ix\xi} u^q(x)\, \mathrm{d}x = q^{-1} E\big[e^{i\xi X_{\eta(q)}}\big] = \big(q - \psi(i\xi)\big)^{-1}.$$

Note that $\xi \mapsto \psi(i\xi)$ is an analytic function in $\Im(\xi) < 0$. By the independence from Lemma 1(iii) and the fact from part 1 above that $S_{\eta(q)}$ has an exponential distribution with mean $\lambda(q)^{-1}$, we see that for $\xi \in \mathbb{R}$

$$q\big(q - \psi(i\xi)\big)^{-1} = E\big[e^{i\xi X_{\eta(q)}}\big] = E\big[e^{i\xi S_{\eta(q)}}\big] E\big[e^{-i\xi(S-X)_{\eta(q)}}\big]$$
$$= \lambda(q)\big(\lambda(q) - i\xi\big)^{-1} E\big[e^{-i\xi(S-X)_{\eta(q)}}\big].$$

Since $\xi \mapsto E\big[e^{-i\xi(S-X)_{\eta(q)}}\big]$ can be analytically extended to $\Im(\xi) < 0$, this identity remains valid for ξ in $\Im(\xi) < 0$. In particular, we see that $(q-\psi(i\xi))^{-1}$ is meromorphic in $\Im(\xi) < 0$ with one pole in $\xi = -i\lambda(q)$. Since $\psi(\lambda) = q$ has only one positive real root in $\lambda = \Phi(q)$, we deduce that $\lambda(q) = \Phi(q)$. The inversion formula for characteristic functions yields now that for $a > 0$

$$U^q([0,a]) = \lim_{T\to\infty} \frac{1}{2\pi} \int_{-T}^{T} \frac{1 - e^{-i\xi a}}{i\xi} \frac{1}{q - \psi(i\xi)}\, \mathrm{d}\xi = \frac{1 - e^{-\Phi(q)a}}{\Phi(q)\psi'(\Phi(q))}, \quad (17)$$

where the second equality can be seen as follows. Let C_T be the (clockwise) contour in the complex plane that consists of the interval $[-T, T]$ on the real axis joined to the semi-circle R_T of radius T in the lower half of the complex plane and set $f(\xi) = \frac{1-e^{-i\xi a}}{2\pi i\xi}(q - \psi(i\xi))^{-1}$. Then by Cauchy's theorem,

$$\int_{C_T} f(t)\, \mathrm{d}t = 2\pi i \cdot \mathrm{Res}_{t=-i\Phi(q)} f(t) = -2\pi \left(\frac{1 - e^{-\Phi(q)a}}{\Phi(q)}\right)\big(-\psi'(\Phi(q))^{-1}\big).$$

On the other hand, since we also have that

$$\int_{C_T} f(t)\, \mathrm{d}t = \int_{-T}^{T} f(t)\, \mathrm{d}t + \int_{R_T} f(t)\, \mathrm{d}t,$$

where, by Jordan's lemma, $\int_{R_T} f(t)\, \mathrm{d}t$ converges to zero as T tends to infinity, the result (17) follows.

Noting that $\psi'(\Phi(q)) = \Phi'(q)^{-1}$ and differentiating (17) with respect to a we find that $u^q(a) = \Phi'(q)\exp(-\Phi(q)a)$ for $a > 0$.

Step 4: An identity between Laplace transforms. Note that for $q, \lambda > 0$ with $q > \psi(\lambda)$ (or equivalently $\Phi(q) > \lambda$) one has that

$$
(q - \psi(\lambda))^{-1} = q^{-1} E\big[e^{\lambda X_{\eta(q)}}\big] = \int_0^\infty e^{\lambda x} u^q(x)\, dx + \int_0^\infty e^{-\lambda x} u^q(-x)\, dx
$$

$$
= \Phi'(q)\big/\big(\Phi(q) - \lambda\big) + \int_0^\infty e^{-\lambda x} u^q(-x)\, dx. \tag{18}
$$

By analytic continuation in λ, the identity (18) remains valid for $\Re(\lambda) > 0$ except for $\lambda = \Phi(q)$ and then by continuity for all λ with $\Re(\lambda) \geqslant 0$. Inverting the Laplace transforms in λ leads then to equation (3).

Step 5: Wiener–Hopf factorisation. Since a Lévy process is quasi left continuous (e.g. [2, Proposition I.7]) and satisfies the strong Markov property (e.g. [2, Proposition I.6]), we deduce from Proposition 1(ii) that

$$
P\big(S_{\eta(q)} > x\big) = E\Big[e^{-qT_x^+} I_{(T_x^+ < \infty)}\Big] = u^q(x)/u^q(0^+) = e^{-x\Phi(q)}. \tag{19}
$$

Lemma 1(i), (iii) imply then that

$$
E\big[\exp(\lambda I_{\eta(q)})\big] = E\big[e^{\lambda X_{\eta(q)}}\big] E\big[e^{\lambda S_{\eta(q)}}\big]^{-1} = \frac{q}{q - \psi(\lambda)} \times \frac{\Phi(q) - \lambda}{\Phi(q)}.
$$

Using (15) to invert this transform we find that

$$
P\big(-I_{\eta(q)} \in dx\big) = \frac{q}{\Phi(q)} W^{(q)}(dx) - qW^{(q)}(x)\, dx, \qquad x \geqslant 0. \tag{20}
$$

Step 6: The function $q \mapsto W^{(q)}(x)$ is analytic for $x \geqslant 0$. Following Bertoin [3], we invert the Laplace transform (term wise)

$$
(\psi(\lambda) - q)^{-1} = \sum_{k \geqslant 0} q^k \psi(\lambda)^{-k-1}
$$

to find the series expansion

$$
W^{(q)}(x) = \sum_{k \geqslant 0} q^k W^{\star k+1}(x) \qquad x, q \geqslant 0, \tag{21}
$$

where $W^{\star k}$ denotes the kth convolution power of W, $W^{\star k} = W \star \cdots \star W$. This series converges since

$$
W^{\star k}(x) \leqslant W(x)^k x^k/k!, \qquad k \geqslant 1, x \geqslant 0
$$

as W is increasing (recalling that $dW = dW^{(0)}$ is a nonnegative measure).

Step 7: Continuity of the function $x \mapsto W^{(q)}(x)$. The next step is to prove that $P(-I_{\eta(q)} \in dx)$ has no atoms. Applying the strong Markov property we get that for $x, q > 0$

$$
P\big(-I_{\eta(q)} = x\big) = E\Big[e^{-qT_{-x}^-} I_{(X_{T_{-x}^-} = -x, T_{-x}^- < \infty)}\Big] P\big(I_{\eta(q)} = 0\big). \tag{22}
$$

If 0 is regular for $(-\infty, 0)$, the second factor of the right-hand side of (22) is zero, whereas if 0 is irregular for $(-\infty, 0)$, the paths of the infimum form step functions almost surely and the first factor of the right-hand side of (22) is zero. Combining with (20) we see that the measure $dW^{(q)}$ has no atoms and thus $x \mapsto W^{(q)}(x)$ is continuous for $q > 0$. Since $W = W^{(0)}$ is increasing, a discontinuity of W at $a > 0$ would imply $\lim_{x \downarrow a} W(x) > \lim_{x \uparrow a} W(x)$. In view of the continuity of $x \mapsto W^{(q)}(x)$ for $q > 0$ on the one hand and the expansion (21) on the other hand, this would yield a contradiction. $\qquad\square$

To prove the identities in Theorem 1(iv), we express the resolvents of the strong Markov processes $X^\dagger$ (X killed upon entering the negative half line $(-\infty, 0)$) and $Y = X - I$ (Lemma 1(ii)) in terms of the scale functions $W^{(q)}, Z^{(q)}$ and then invoke Proposition 1(ii).

Lemma 2. *For $x, y > 0$, we have*

$$\frac{P_x\big(X_{\eta(q)} \in dy, \eta(q) < T_0^-\big)}{dy} = q\, e^{-\Phi(q)y} W^{(q)}(x) - 1_{\{x>y\}} q W^{(q)}(x-y). \quad (23)$$

$$\frac{P_x\big(Y_{\eta(q)} \in dy\big)}{dy} = \Phi(q)\, e^{-\Phi(q)y} Z^{(q)}(x) - 1_{\{x>y\}} q W^{(q)}(x-y). \quad (24)$$

Proof. A proof of the first identity can be found e.g. in Bertoin [3] or Suprun [15] and of the second identity in [13]. In order to be self-contained we provide the proofs here.

(i) Invoking the identities (19) and (20) and the independence and duality (Lemma 1(iii,i)) and noting that $\eta(q) < T_{-x}^-$ iff $I_{\eta(q)} > -x$ we find that $q^{-1} P_x(X_{\eta(q)} \in dy, \eta(q) < T_0^-)$ is equal to

$$q^{-1} \int_0^x P\big(-I_{\eta(q)} \in dz\big) P\big((X-I)_{\eta(q)} \in d(y-x+z)\big)$$

$$= \int_{(x-y)\vee 0}^x e^{-(y-x+z)\Phi(q)} W^{(q)}(dx) - \int_{(x-y)\vee 0}^x \Phi(q) e^{-(y-x+z)\Phi(q)} W^{(q)}(x)\, dx.$$

The identity (23) follows now by performing a partial integration on the first integral in the second line of above display.

(ii) The strong Markov property of Y at the stopping time $\tau_0 = \inf\{t \geqslant 0 : Y_t = 0\}$ implies that

$$P_x\big(Y_{\eta(q)} \in dy\big) = P_x\big(Y_{\eta(q)} \in dy, \eta(q) < \tau_0\big) + E_x\big[e^{-q\tau_0}\big] P_0\big(Y_{\eta(q)} \in dy\big) \quad (25)$$

$$= P_x\big(X_{\eta(q)} \in dy, \eta(q) < T_0^-\big) + E_x\big[e^{-qT_0^-}\big] P_0\big(Y_{\eta(q)} \in dy\big),$$

where in the second line we used that $(Y_t, t \leqslant \tau_0)$ has the same law as $(X_t, t \leqslant T_0^-)$. By integrating (20) we find the Laplace transform of T_0^- to be equal to

$$P\big(I_{\eta(q)} < -x\big) = E_x\big[e^{-qT_0^-}\big] = Z^{(q)}(x) - q\Phi(q)^{-1} W^{(q)}(x), \qquad x > 0. \quad (26)$$

Substituting (26) and (23) into (25) and recalling that $Y_{\eta(q)}$ has an exponential distribution with mean $\Phi(q)^{-1}$, we end up with the required identity (24). $\quad\square$

Now we can finish the proof of Theorem 1.

Proof of Theorem 1(iv). Since a Lévy process is a quasi left continuous strong Markov process and $W^{(q)}$ is continuous (proved above in part 6), it follows by combining with Lemma 2 that the conditions of Proposition 1 are met for the Markov processes $X^{\dagger}$ and Y (Lemma 1(ii)). Taking for u^q in (12) the resolvents (23), (24) we find (6) and (8) respectively. Finally, the strong Markov property yields that

$$E_x\left[e^{-qT_0^-}\right] = E_x\left[e^{-qT_0^-} I_{(T_0^- < T_a^+)}\right] + E_x\left[e^{-qT_a^-} I_{(T_0^- > T_a^+)}\right] E_a\left[e^{-qT_0^-}\right]. \quad (27)$$

Inserting (26) and (6) in (27) completes the proof. □

Proof of Corollary 1. Since ψ is differentiable, convex and increasing on $(0, \infty)$, it follows that the right-derivative of ψ in $\Phi(0)$ is finite and non-negative and equal to $(\Phi'(0^+))^{-1}$. Thus $\lim_{q\downarrow 0}(\Phi(q) - \Phi(0))/q$ is positive and finite or equal to $+\infty$. Suppose first that the latter is the case. By the identity (2) it then follows, taking $q \downarrow 0$, that $U^0((-r, r))$ is infinite for $r > 0$.

In the case $\Phi'(0^+) < \infty$, we show that the identity (9) holds true. The fact that $\psi(\cdot)$ is C^1 on $(0, \infty)$ in conjunction with the implicit function theorem applied to $\psi(\lambda) = q$ implies that Φ and Φ' are continuous. Combining with the continuity of $q \mapsto W^{(q)}$ we find that, for any compact set K, $U^0(K) = \lim_{q\downarrow 0} \int_K u^q(x)dx$ by dominated convergence. On the other hand, monotone convergence implies that $U^0(K) = \lim_{q\downarrow 0} U^q(K)$. Using then the explicit formulas (2) and (3), equation (9) follows and the proof is finished. □

Proof of Corollary 2. Let $q > 0$ and let $T'_{\{x\}}$ be as in (11). If X creeps across $x < 0$, this implies that $T'_{\{x\}}$ is smaller than T_b^- for all $b < x$ by right-continuity of its sample paths. On the other hand, if $X_{T_x^-} < x$, X enters $(-\infty, x)$ by a jump and it follows that there exists an $\epsilon_0 > 0$ such that $T'_{\{x\}} > T_b^-$ for all $b \in (x - \epsilon_0, x)$. Thus $\{T'_{\{x\}} < T_b^-\}$ increases to $\{X_{T_x^-} = x\}$ as $b \uparrow x$ and we have

$$E\left[e^{-qT'_{\{x\}}} I_{(X_{T_x^-}=x)}\right] = \lim_{b\uparrow x} E\left[e^{-qT'_{\{x\}}} I_{(T'_{\{x\}} < T_b^-)}\right]. \quad (28)$$

Invoking Proposition 1 applied to the Markov process $\tilde{X}^{\dagger}$ (X killed upon entering $(-\infty, b)$) in conjunction with Lemma 2 we see that the right-hand side of (28) is equal to

$$\lim_{b\uparrow x} \frac{W^{(q)}(-b) - e^{-\Phi(q)(x-b)} W^{(q)}(-x)}{W^{(q)}(x-b)} = \frac{\sigma^2}{2}\left(W_+^{(q)'}(-x) - \Phi(q)W^{(q)}(-x)\right), (29)$$

where $W_+^{(q)'}$ is the right-derivative of $W^{(q)}$ (see e.g. Lemma 1 in [13] for a proof of the right-differentiability of $W^{(q)}(\cdot)$) and we used that $z/W^{(q)}(z)$ converges to $\sigma^2/2$ for $z \downarrow 0$ (e.g. Lemma 4 in [13]). Letting now $q \downarrow 0$ in (29) and using the differentiability of $W^{(q)}$ if $\sigma > 0$ (e.g. Lemma 1 in [13]) and continuity of the maps $q \mapsto W^{(q)}(x)$, $W_+^{(q)'}(x)$, we end up with (10) and the proof is complete. □

Acknowledgement. The author would like to thank an anonymous referee for his careful reading and useful remarks.

References

1. Avram, F., Kyprianou, A.E. and Pistorius, M.R. (2004). Exit problems for spectrally negative Lévy processes and applications to (Canadized) Russian options. *Ann. Appl. Probab.* **14**, 215–238.
2. Bertoin, J. (1996). *Lévy processes*, Cambridge University Press.
3. Bertoin, J. (1997). Exponential decay and ergodicity of completely asymmetric Lévy processes in a finite interval. *Ann. Appl. Probab.* **7**, 156–169.
4. Bingham, N.H. (1975). Fluctuation theory in continuous time. *Adv. in Appl. Probab.* **7**, 705–766.
5. Blumenthal, R. M., Getoor, R. K. (1968). *Markov processes and potential theory.* Pure and Applied Mathematics, Vol. 29, Academic Press.
6. Dellacherie, C.; Meyer, P. (1988). *Probabilities and potential. C. Potential theory for discrete and continuous semigroups.* North-Holland Mathematics Studies, 151, North-Holland Publishing Co.
7. Doney, R.A. (2004). Some excursion calculations for spectrally one-sided Lévy processes. *In this volume.*
8. Emery, D.J. (1973). Exit problem for a spectrally positive process. *Adv. in Appl. Probab.* **5**, 498–520.
9. Feller, W. (1971). *Introduction to probability theory and its applications.* Vol 2, 2nd. ed.
10. Greenwood, P. and Pitman, J. (1980). Fluctuation identities for Lévy processes and splitting at the maximum. *Adv. in Appl. Probab.* **12**, 893–902.
11. Kella, O.; Whitt, W. (1992). Useful martingales for stochastic storage processes with Lévy input. *J. Appl. Probab.* **29**, 396–403.
12. Kyprianou, A.E. and Palmovski, Z. (2003). A martingale review of some fluctuation theory for spectrally negative Lévy processes. *In this volume.*
13. Pistorius, M.R. (2004). On exit and ergodicity of the spectrally negative Lévy process reflected at its infimum. *J. Theor. Probab.* **17**, 183–220.
14. Rogers, L. C. G. (1990). The two-sided exit problem for spectrally positive Lévy processes. *Adv. in Appl. Probab.* **22**, 486–487.
15. Suprun, V.N. (1976). The ruin problem and the resolvent of a killed independent increment process. *Ukrainian Math. J.* **28**, 39–45.
16. Takács, L. (1966). *Combinatorial Methods in the Theory of Stochastic Processes*, Wiley, New York.
17. Zolotarev, V.M. (1964). The first passage time of a level and the behaviour at infinity for a class of processes with independent increments. *Theory Prob. Appl.* **9**, 653–661.

Some Martingales Associated to Reflected Lévy Processes

Laurent Nguyen-Ngoc[†] and Marc Yor

Laboratoire de Probabilités et Modèles Aléatoires, Universités Paris 6 et 7,
4 place Jussieu, F-75252 Paris Cedex 5
[†]e-mail: laurent.nguyenngoc@free.fr

Summary. We introduce and describe several classes of martingales based on re-flected Lévy processes. We show how these martingales apply to various problems, in particular in fluctuation theory, as an alternative to the use of excursion methods. Emphasis is given to the case of spectrally negative processes.

Key words: Lévy processes, Kella–Whitt and Kennedy martingales, fluctuation theory, Wiener–Hopf factorization

1 Introduction and notations

Let us fix a probability space $(\Omega, \mathcal{A}, \mathbb{P})$. Let X denote a real Lévy process, started at 0 under $\mathbb{P}$. We call $\mathbb{P}_x$ the law of $x + X$ under $\mathbb{P}$. Also, we let $\mathcal{F}_t$ denote the usual right-continuous, universal augmentation of the filtration generated by X. We shall be concerned with the process X reflected at its supremum: if $S_t = \sup_{s \leqslant t} X_s$, the reflected process is defined by $R_t = S_t - X_t$. We recall that R is a Feller process with respect to $(\mathcal{F}_t, \mathbb{P})$, and we denote by $\mathbb{P}_{R_0 = r}$ the probability under which R starts at r.

Our goal in this paper is to present various martingales related to the process R; these martingales generalize the corresponding ones in the case of Brownian motion, where they have proved quite useful in the study of a number of questions. Also for Lévy processes, some of the martingale properties we present are already known in special cases, and have been used in various contexts; some other seem to be new, though quite simple. For example, we recover the following special case of Kella–Whitt [13]: for every $u \in \mathbb{R}$, the process $(M_t^{(u)}, t \geqslant 0)$ is a local martingale, where

$$
M_t^{(u)} = \mathrm{e}^{\mathrm{i}u(X_t - S_t)} - 1 + \Psi(u) \int_0^t \mathrm{e}^{\mathrm{i}u(X_r - S_r)} \, \mathrm{d}r \\
+ \mathrm{i}u S_t^c - \sum_{r \leqslant t} \mathrm{e}^{\mathrm{i}u(X_r - S_r)} \left(1 - \mathrm{e}^{\mathrm{i}u \Delta S_r} \right),
$$

Ψ being the Lévy exponent of X as defined below; in fact, we shall prove under mild assumptions, that $M^{(u)}$ is a class (D) martingale on every interval $[0, t]$. We also show that if X is spectrally negative, and a martingale, then for any locally bounded Borel function g, the process

$$\left(G(S_t) - (S_t - X_t)g(S_t), \, t \geqslant 0\right)$$

with $G(x) = \int_0^x g(y)\, \mathrm{d}y$, is a local martingale, which extends a previous result of Azéma and Yor [5] in the Brownian case.

We shall give various applications of these and other martingales; in particular, they can be used as an alternative to excursion methods to derive some results in fluctuation theory.

In our presentation, we have tried to keep the results as general as possible, but the case of spectrally negative processes is still particularly interesting, due mostly to the existence of their scale function, which naturally arise in fluctuation theory.

Let us introduce some further notations that will be needed throughout. We write the Lévy–Itô decomposition of X in the form

$$X_t = at + \xi_t + \sum_{s \leqslant t} \Delta X_s 1_{|\Delta X_s| > 1}; \tag{1}$$

here, $a \in \mathbb{R}$ and ξ is the square integrable martingale $\sigma B_t + \int_0^t \int_{|x| \leqslant 1} x\,(\mu - n)(\mathrm{d}s, \mathrm{d}x)$ with $\sigma \geqslant 0$ and μ the random measure associated to the jumps of X. Lastly n is the compensator of μ, and is given by $n(\mathrm{d}s, \mathrm{d}x) = \mathrm{d}s\, \nu(\mathrm{d}x)$ where ν is the Lévy measure of X.

We denote by Ψ the Lévy exponent of X, defined by $\mathbb{E}\left[\mathrm{e}^{\mathrm{i}\lambda X_t}\right] = \mathrm{e}^{-t\Psi(\lambda)}$ ($\lambda \in \mathbb{R}$), and recall that Ψ is given by the Lévy–Khintchine formula

$$\Psi(\lambda) = -\mathrm{i}a\lambda + \frac{\sigma^2}{2}\lambda^2 - \int \left(\mathrm{e}^{\mathrm{i}\lambda x} - 1 - \mathrm{i}\lambda x 1_{|x| \leqslant 1}\right) \nu(\mathrm{d}x). \tag{2}$$

Given $x \in \mathbb{R}$, we denote by T_x^+ and T_x^- the first time X goes above or below x:

$$T_x^{\pm} = \inf\{t : X_t \gtrless x\}.$$

We now turn to the introduction of some objects which are specific to the situation when X is spectrally negative. Hence, until the end of this introduction, we assume that the Lévy measure ν is carried by $(-\infty, 0)$.

In this case, we denote by ψ the Laplace exponent of X, which is defined by $\mathbb{E}\left[\mathrm{e}^{\lambda X_t}\right] = \mathrm{e}^{t\psi(\lambda)}$ ($\Re(\lambda) \geqslant 0$), and is related to the Lévy exponent by $\psi(\lambda) = -\Psi(-\mathrm{i}\lambda)$. Moreover, the function ψ is strictly convex and strictly increasing on the interval $(\Phi(0), \infty)$ where $\Phi(0)$ is the largest root of ψ, and we denote by Φ the inverse function of ψ, defined on $(0, \infty)$.

Our martingale methods allow us, in this case, to recover the existence of the q-scale function $W^{(q)}$, which is characterized by: $W^{(q)}(x) = 0$ if $x \leqslant 0$, and

$$\int_0^\infty e^{-\lambda x} W^{(q)}(x) \, dx = \frac{1}{\psi(\lambda) - q}, \qquad \lambda > \Phi(q),$$

and which satisfies the following property: for all $a < x < b$, the process $e^{-q(t \wedge T)} W^{(q)}(X_{t \wedge T} - a)$ is a $\mathbb{P}_x$-martingale, where $T = \inf\{t : X_t \notin (a, b)\}$ (see Example 3). Lastly, the following primitive of $W^{(q)}$ will play an important role:

$$Z^{(q)}(x) = \begin{cases} 1, & x \leqslant 0 \\ 1 + q \displaystyle\int_0^x W^{(q)}(y) \, dy, & x > 0. \end{cases}$$

Note that there exists no scale function unless X is spectrally negative—or spectrally positive—see e.g. [17].

The rest of the paper is organized as follows. In Section 2, we show how a martingale property related to X translates into a martingale property related to the reflected process R. In Section 3, we give some martingale properties involving both X and R, which generalize those of Kennedy [14] for Brownian motion. Examples are given at the end of the section in the spectrally negative case.

Some applications are presented in Section 4. Precisely, under mild assumptions on X, we give a proof of the Wiener–Hopf factorization, each step of which is elementary in that it uses only the property that martingales have constant expectation. Moreover, only simple processes are involved, in particular we never refer to any path-valued process as in excursion theory. At this stage, a warning is in order: that the proof is elementary does by no means imply that the result is easy; actually avoiding the use of excursion theory leads us to quite long computations. While this is done in a general setting, the next application concerns only spectrally negative processes. For such processes, we specify the behavior of X at the first passage times of R and we also study some related first passage problems involving R and S.

These problems have already been studied via excursion theory (see e.g. [6] for the Wiener–Hopf factorization, [4] for the first exit of R), which is the natural tool to use, since it is the continuous time counterpart of the renewal property at ladder times for random walks. However, it is interesting to examine these problems via martingale theory. First, this leads to more elementary proofs of the results, if perhaps less intuitive. Second, it may shed light on the relationship between martingale theory and excursion theory in this context.

2 From X-martingales to R-martingales

In this section, we show how a martingale property for X "translates" into a martingale property for R. Before turning to this topic, we recall the following well-known martingale property related to the jumps of a Lévy process:

Lemma 1 (Basic compensation). *Let Z be a d-dimensional Lévy process with Lévy measure ν_Z. Let $f : \mathbb{R}_+ \times \mathbb{R}^d \to \mathbb{R}$ be a Borel measurable function such that, for every $t > 0$, $\int_0^t \mathrm{d}s \int_{\mathbb{R}^d \setminus \{0\}} |f(s, z)| \, \nu_Z(\mathrm{d}z) < \infty$. Then the process*

$$\left(\sum_{s \leqslant t} f(s, \Delta Z_s) 1_{\Delta Z_s \neq 0} - \int_0^t \mathrm{d}s \int_{\mathbb{R}^d \setminus \{0\}} f(s, z) \nu_Z(\mathrm{d}z), \quad t \geqslant 0 \right)$$

is a martingale.

The above lemma will be used a lot in this paper, and we shall refer to it as "basic compensation".

We now turn to the main result of this section, which is the following

Proposition 1. *Let $f : \mathbb{R}_+ \times \mathbb{R} \to \mathbb{R}$ be a $C^{1,1}$, function and for $0 < a < b$, set $T = \inf\{t : X_t \notin (a, b)\}$. Assume that $(f(t \wedge T, X_{t \wedge T}), t \geqslant 0)$ is a $\mathbb{P}_x$-martingale for all $x \in (a, b)$. Set*

$$M_t^f = f(t, X_t - S_t) + \int_0^t f_x(s, 0) \, \mathrm{d}S_s^c + \sum_{s \leqslant t} (f(s, \Delta S_s) - f(s, 0)) 1_{\Delta S_s > 0} \quad (3)$$

Then for all $x \in (a, b)$, $(M_{t \wedge \tau}^f, t \geqslant 0)$ is a martingale under $\mathbb{P}[. \,|\, R_0 = x]$, where $\tau = \inf\{t : R_t \notin (a, b)\}$.

Remarks 1.

1. Derivatives with respect to t, x, etc., will be denoted whenever convenient either by f_t or by $\partial_t f$ and so on...

2. If $f \in C^{1,2}(\mathbb{R}_+ \times \mathbb{R}, \mathbb{R})$, Prop. 1 is an immediate consequence of Itô's formula. The proof only consists in showing that one can actually weaken the regularity requirement on f to: $f \in C^{1,1}(\mathbb{R}_+ \times \mathbb{R}, \mathbb{R})$. This is needed later in applications (see 4.2).

3. Of course, there is a similar result with $X - I$ instead of $S - X$. Specifically, assume that $f(t, X_t)$ is a martingale; then

$$f(t, X_t - I_t) + \int_0^t f_x(s, 0) \, \mathrm{d}I_s^c + \sum_{s \leqslant t} (f(s, \Delta I_s) - f(s, 0)) 1_{\Delta I_s < 0} \quad (4)$$

is a martingale. The fact that f_x is evaluated at 0 in the integrals with respect to $\mathrm{d}S^c$ and $\mathrm{d}I^c$ is intuitively clear from the fact that S^c and I^c are local times at 0 for the processes $S - X$ and $X - I$ respectively (see Section 4 for an explanation of this fact). The measures $\mathrm{d}S^c$ and $\mathrm{d}I^c$ are then respectively carried by $\{S = X\}$ and $\{I = X\}$.

4. Here we have treated the special case with S or I as the bounded variation process which perturbs X, but it is clear from the proof that one could work as well with a general bounded variation process Y (in the spirit of [13]). In this case, the only difference would be that one cannot say anything about the value of $X + Y$ on the support of $\mathrm{d}Y$.

Proof. Let $(\phi_n, n \geqslant 1)$ be an approximation of unity (e.g. $\phi_n(x) = n\phi(nx)$, where $\phi \in C^\infty$ has compact support containing 0 and satisfies $\int \phi = 1$), and set

$$f_n(t, x) = \int \phi_n(y) f(t, y + x) \, dy.$$

Then f_n is smooth in x, and the functions f_n, $\partial_t f_n$ and $\partial_x f_n$ converge uniformly on compacts to f, $\partial_t f$ and $\partial_x f$ respectively.

Denote by $[\alpha_n, \beta_n]$ the support of ϕ_n ($\alpha_n < 0 < \beta_n$) and introduce $T_n = \inf\{t : X_t \notin (a_n, b_n)\}$ with $a_n = a - \alpha_n$ and $b_n = b - \beta_n$. If n is large enough, we have $a \leqslant a - \alpha_n \leqslant b - \beta_n \leqslant b$. Since the support of ϕ_n shrinks to $\{0\}$ as $n \uparrow \infty$, we have $T_n \uparrow T$ a.s.

We note that $f_n(t \wedge T_n, X_{t \wedge T_n})$ is a $\mathbb{P}_x$-martingale, for large enough n. In fact, we have

$$f_n(t, X_t) = \int \phi_n(y) f(t, y + X_t) \, dy$$

and if n is large enough, the support of ϕ_n is so small that $y + x \in (a, b)$ for all $y \in \mathrm{Supp}(\phi_n)$. It follows from our assumption that $f(t \wedge T_n, y + X_{t \wedge T_n})$ is a martingale for all $y \in [\alpha_n, \beta_n]$, hence so is $f_n(t \wedge T_n, X_{t \wedge T_n})$.

Since f_n is smooth with compact support, it is in the domain of the generator $\mathcal{A}^X$ of X, and we have

$$\frac{\partial f_n}{\partial t}(t, x) + \mathcal{A}^X f_n(t, x) = 0, \qquad t \geqslant 0, \, x \in (a_n, b_n). \tag{5}$$

Now set

$$N_t^n = f_n(t, X_t - S_t) + \int_0^t \frac{\partial f_n}{\partial x}(s, 0) \, dS_s^c$$
$$+ \sum_{s \leqslant t} (f_n(s, \Delta S_s) - f_n(s, 0)) 1_{\Delta S_s > 0}. \tag{6}$$

By Itô's formula and (5), we obtain that $N_{t \wedge \tau_n}^n$ is a martingale under $\mathbb{P}_{R_0 = x}$, where $\tau_n = \inf\{t : R_t \notin (a_n, b_n)\}$. In the sequel, we denote $P \equiv \mathbb{P}_{R_0 = x}$ and $E \equiv \mathbb{E}_{R_0 = x}$.

We now show that $(N_{t \wedge \tau_n}^n, t \geqslant 0)_{n \geqslant 1}$ converges in $\mathbb{H}^2$ (the space of square-integrable martingales up to any fixed time), as $n \to \infty$, to a martingale N. To see this, let n, $p \geqslant q$ and note that by Doob's L^2 inequality

$$E\left[\sup_{s \leqslant t \wedge \tau_q} (N_s^n - N_s^p)^2 \right] \leqslant 4E\left[[N^n - N^p]_{t \wedge \tau_q} \right].$$

To simplify the notation, we set $f_{n,p} = f_n - f_p$, and, if $f : \mathbb{R}_+ \times \mathbb{R} \to \mathbb{R}$ is some regular function, we denote $\Delta f(s) = f(s, X_s - S_s) - f(s, X_{s-} - S_{s-})$. We have:

$$[N^n - N^p]_t = [f_{n,p}]_t + A_1(t) + A_2(t),$$

where $[f_{n,p}]_t$ means $[f_{n,p}(\,.\,, X_. - S_.)]_t$, and

$$A_1(t) = \sum_{s \leqslant t} (\Delta f_{n,p}(s))(f_{n,p}(s, \Delta S_s) - f_{n,p}(s, 0))1_{\Delta S_s > 0}$$

$$A_2(t) = \sum_{s \leqslant t} (f_{n,p}(s, \Delta S_s) - f_{n,p}(s, 0))^2 1_{\Delta S_s > 0}.$$

Note that $X_s = S_s$ if $\Delta S_s > 0$ and apply the basic compensation formula to get

$$A_1(t) \sim \int_0^t \mathrm{d}s \int_{S_s - X_s}^\infty \nu(\mathrm{d}z)[f_{n,p}(s, 0) - f_{n,p}(s, X_s - S_s)]$$
$$\times [f_{n,p}(s, X_s - S_s + z) - f_{n,p}(s, 0)],$$

and

$$A_2(t) \sim \int_0^t \mathrm{d}s \int_{S_s - X_s}^\infty \nu(\mathrm{d}z)(f_{n,p}(s, X_s - S_s + z) - f_{n,p}(s, 0))^2,$$

where we write $A \sim B$ to mean that $A - B$ is a local martingale. Now, using Itô's formula, we see that

$$\mathrm{d}[f_{n,p}]_t = \sigma^2 \left(\frac{\partial f_{n,p}}{\partial x}(t, (X - S)_{t-})\right)^2 \mathrm{d}t + \mathrm{d}A_3(t) + \mathrm{d}A_4(t) + \mathrm{d}A_5(t) + 2\,\mathrm{d}A_6(t)$$

where

$$A_3(t) = \sum_{s \leqslant t} \left(\frac{\partial f_{n,p}}{\partial x}(s, (X - S)_{s-})\right)^2 (\Delta X_s)^2 1_{|\Delta X_s| \leqslant 1}$$

$$A_4(t) = \sum_{s \leqslant t} 1_{|\Delta X_s| \leqslant 1} \left(\Delta f_{n,p}(s) - \left(\frac{\partial f_{n,p}}{\partial x}(s, (X - S)_{s-})\right)\Delta X_s\right)^2$$

$$A_5(t) = \sum_{s \leqslant t} (\Delta f_{n,p}(s))^2 1_{|\Delta X_s| > 1}$$

$$A_6(t) = \sum_{s \leqslant t} 1_{|\Delta X_s| \leqslant 1} \left[\frac{\partial f_{n,p}}{\partial x}(s, (X - S)_{s-})\right]$$
$$\times \left[\Delta f_{n,p}(s) - \left(\frac{\partial f_{n,p}}{\partial x}(s, (X - S)_{s-})\right)\Delta X_s\right]$$

Applying Lemma 1 to each term, we replace A_i by its compensator and obtain that $E[A_i(t \wedge \tau_q)] \to 0$ as n and p go to ∞ $(i = 1, \ldots, 6)$. So $(N_{t \wedge \tau_q}^n)_{t \geqslant 0}$ is a Cauchy sequence in $\mathbb{H}^2$ and therefore converges as $n \to \infty$ to a process $^{(q)}N$ which is a martingale. Let us set

$$N_t = f(t, X_t - S_t) + \int_0^t \frac{\partial f}{\partial x}(s, 0)\,\mathrm{d}S_s^c + \sum_{s \leqslant t} 1_{\Delta S_s > 0}(f(s, \Delta S_s) - f(s, 0)). \quad (7)$$

From (6) and since $f_n \to f$ and $\partial_x f_n \to \partial_x f$ uniformly on compacts, we have clearly ${}^{(q)}N_t = N_{t \wedge \tau_q}$.

We now show that $N_{t \wedge \tau} - N_{t \wedge \tau_q} \to 0$ in L^2 as $q \uparrow \infty$, which finishes the proof of the proposition as it implies that $(N_{t \wedge \tau})_{t \geqslant 0}$ is a $\mathbb{H}^2$ martingale. Indeed, it follows from inspection of (7) that there are constants C_1, C_2 and C_3 such that:

$$E\big[(N_{t \wedge \tau} - N_{t \wedge \tau_q})^2\big] \leqslant 3\bigg\{ C_1 \, P[\tau_q < \tau] + C_2 \, E\Big[(S_{t \wedge \tau}^c - S_{t \wedge \tau_q}^c)^2 1_{\tau_q < \tau}\Big]$$
$$+ C_3 \, E\bigg[\bigg(\sum_{t \wedge \tau_q < s \leqslant t \wedge \tau} 1_{\Delta S_s > 0}\bigg)^{\!2}\,\bigg]\bigg\}$$

Each of these three expectations goes to 0 as $q \to \infty$: it is clear for the two first ones; here are some details for the latter. Set

$$\Gamma_t = \sum_{s \leqslant t} 1_{\Delta S_s > 0}.$$

We have, since Γ is an increasing process, $(\Gamma_{t \wedge \tau} - \Gamma_{t \wedge \tau_q})^2 \leqslant \Gamma_{t \wedge \tau}^2 - \Gamma_{t \wedge \tau_q}^2$, and

$$\Gamma_{t \wedge \tau}^2 - \Gamma_{t \wedge \tau_q}^2 = 2 \int_{t \wedge \tau_q}^{t \wedge \tau} \Gamma_{s-} \, d\Gamma_s + \sum_{t \wedge \tau_q < s \leqslant t \wedge \tau} 1_{\Delta S_s > 0}$$

A basic compensation gives $\Gamma_t = M_t + \int_0^t \bar{\nu}(R_s) \, ds$, where M is a local martingale and $\bar{\nu}(x) = \nu(x, \infty)$. In fact, $(M_{t \wedge \tau})_{t \geqslant 0}$ is a square integrable martingale, since $E\big[[M, M]_{t \wedge \tau}\big] = E\big[\Gamma_{t \wedge \tau}\big]$ and

$$E[\Gamma_{t \wedge \tau}] = E\bigg[\int_0^{t \wedge \tau} \bar{\nu}(R_s) \, ds\bigg] \leqslant \bar{\nu}(a)t.$$

As a consequence, we obtain:

On the one hand,

$$E\bigg[\sum_{t \wedge \tau_q < s \leqslant t \wedge \tau} 1_{\Delta S_s > 0}\bigg] = E\bigg[\int_{t \wedge \tau_q}^{t \wedge \tau} \bar{\nu}(R_s) \, ds\bigg] \leqslant \bar{\nu}(a) \, E[(t \wedge \tau) - (t \wedge \tau_q)]$$

and the foregoing goes to 0 as $q \uparrow \infty$.

On the other hand,

$$E\bigg[\int_{t \wedge \tau_q}^{t \wedge \tau} \Gamma_{s-} \, d\Gamma_s\bigg] = E\bigg[\int_{t \wedge \tau_q}^{t \wedge \tau} \Gamma_s \bar{\nu}(R_s) \, ds\bigg]$$
$$\leqslant \bar{\nu}(a) \, E\big[\Gamma_{t \wedge \tau-} \big((t \wedge \tau) - (t \wedge \tau_q)\big)\big]$$
$$\leqslant \bar{\nu}(a) \, E\big[\Gamma_{t \wedge \tau-}^2\big]^{1/2} E\big[((t \wedge \tau) - (t \wedge \tau_q))^2\big]^{1/2}$$

by Cauchy–Schwarz inequality. The last term in the right-hand side goes to 0 as $q \uparrow \infty$, and all we need to show is that $E\big[\Gamma_{t \wedge \tau}^2\big] < \infty$. However,

$$E[\Gamma_{t\wedge\tau}^2] \leqslant 2\left\{E[M_{t\wedge\tau}^2] + E\left[\left(\int_0^{t\wedge\tau} \bar{\nu}(R_s)\,\mathrm{d}s\right)^2\right]\right\}$$

$$\leqslant 2\left\{E[\Gamma_{t\wedge\tau}] + E\left[\int_0^{t\wedge\tau} \bar{\nu}(R_s)\,\mathrm{d}s\right]\right\}$$

$$\leqslant 2\{\bar{\nu}(a)t + \bar{\nu}(a)^2 t^2\}.$$

Our claim follows readily.   □

Remark 2. For fixed a and b, the martingales we have just constructed are uniformly integrable, since f is uniformly bounded on each $[0, t] \times [a, b]$. If we let $a \to -\infty$ and/or $b \to +\infty$, we get local martingales, and we should be careful if we wish to apply the optional stopping theorem.

Example 1. In [13] (see [1] or [3] for a generalization), the authors introduced a local martingale which has proved quite useful in a number of applications. Specifically, if Y has bounded variation and is adapted, and if $Z_t = X_t + Y_t$, the process $(K_t, t \geqslant 0)$ defined by

$$K_t = \mathrm{e}^{\mathrm{i}\lambda Y_0} - \mathrm{e}^{\mathrm{i}\lambda Z_t} - \Psi(\lambda)\int_0^t \mathrm{e}^{\mathrm{i}\alpha Z_s}\,\mathrm{d}s + \mathrm{i}\lambda\int_0^t \mathrm{e}^{\mathrm{i}\lambda Z_s}\,\mathrm{d}Y_s^c + \sum_{s\leqslant t}\mathrm{e}^{\mathrm{i}\lambda Z_s}\left(1 - \mathrm{e}^{-\mathrm{i}\lambda\Delta Y_s}\right)$$

is a local martingale. If $Y = -S$, this follows from Prop. 1. Indeed, consider the function $f(t, x) = \mathrm{e}^{\mathrm{i}\lambda x + t\Psi(\lambda)}$, which generates the exponential martingale $f(t, X_t)$. By Prop. 1, the process

$$M_t = \mathrm{e}^{t\Psi(\lambda)+\mathrm{i}\lambda(X_t - S_t)} - \mathrm{i}\lambda\int_0^t \mathrm{e}^{s\Psi(\lambda)}\,\mathrm{d}S_s^c - B_t$$

is a local martingale, where

$$B_t = \sum_{s\leqslant t}\left(\mathrm{e}^{s\Psi(\lambda)} - \mathrm{e}^{s\Psi(\lambda)+\mathrm{i}\lambda\Delta S_s}\right)1_{\Delta S_s > 0}.$$

K is then recovered as $K_t = \int_0^t \mathrm{e}^{-s\Psi(\lambda)}\,\mathrm{d}M_s$.

If X is spectrally negative, the result of Prop. 1 simplifies since S is continuous.

Corollary 1. *Suppose that X is spectrally negative. Let $0 < a < b$ and $f : \mathbb{R}_+ \times \mathbb{R} \to \mathbb{R}$ be in $C^{1,1}$. Set $Z_t = f(t, X_t)$, and assume that $Z_{t\wedge T_a^- \wedge T_b^+}$ is a $\mathbb{P}_x$-martingale for all $x \in (a, b)$. Then, if*

$$\bar{Z}_t = f(t, X_t - S_t) + \int_0^t f_x(s, 0)\,\mathrm{d}S_s,$$

$\bar{Z}_{t\wedge\tau}$ is a $\mathbb{P}_{R_0 = x}$-martingale, where $\tau = \inf\{t : X_t - S_t \notin (a, b)\}$.

3 Kennedy martingales

In this section, we describe an analogue, for Lévy processes, of the two-parameter family of martingales introduced by Kennedy [14] for Brownian motion. Here, both the reflected process R and the supremum process S come into play independently.

3.1 Brownian motion with drift

Let us first recall the particular case of Brownian motion (with drift). The martingale property of the next proposition, in the Brownian motion case, was used in [5] to obtain a solution to Skorokhod's embedding problem. As we shall see in Section 4, this particular application cannot be extended to Lévy processes.

Let hence B be a standard Brownian motion and $\mu \in \mathbb{R}$; set $X_t = B_t + \mu t$ and $S_t = \sup_{s \leqslant t} X_s$. We then have the following result, which follows immediately from Itô formula, and the fact that dS_t is carried by $\{t : X_t = S_t\}$.

Proposition 2 ([5]). Let $f : \mathbb{R}_+ \times \mathbb{R}_+ \times \mathbb{R}_+ \to \mathbb{R}$ be a smooth function. If

$$
\begin{aligned}
&\frac{1}{2} f_{xx} - \mu f_x + f_t = 0 \\
&f_x(t, 0, y) + f_y(t, 0, y) = 0
\end{aligned}
\tag{8}
$$

then $(f(t, S_t - X_t, S_t), t \geqslant 0)$ is a local martingale.

In relation to the exponential martingales, it is natural to look for functions f that satisfy (8) in the form $f(t, x, y) = h(x)e^{-\alpha y - \beta t}$. If $\mu = 0$, the associated martingales as stated in Prop. 2 are due to Kennedy [14].

The function h must then satisfy $\frac{1}{2} h'' - \mu h' - \beta h = 0$, together with the boundary condition $h'(0) = \alpha h(0)$. Hence, we get

$$
h(x) = C e^{\mu x} \big((\alpha - \mu) \sinh(\delta x) + \delta \cosh(\delta x) \big)
\tag{9}
$$

for some constant C, where $\delta = \sqrt{\mu^2 + 2\beta}$. As an application, these martingales allow to recover immediately a formula of H. M. Taylor [20]. Indeed let $T = \inf\{t : S_t - X_t = a\}$, then the local martingale $h(S_{t\wedge T} - X_{t\wedge T})e^{-\alpha S_{t\wedge T} - \beta(t\wedge T)}$ is bounded, so that

$$
\mathbb{E}\big[e^{\alpha X_T - \beta T}\big] = e^{-\alpha a} \frac{h(0)}{h(a)} = \frac{e^{-(\alpha + \mu)a}\delta}{\delta \cosh(a\delta) - (\alpha + \mu)\sinh(a\delta)}
\tag{10}
$$

(see also [21]).

Later on, we shall see to what extent this situation can be generalized to Lévy processes, and an example where no interesting such function h can be found.

Remark 3. If $\mu = 0$, by the well-known equivalence theorem of Paul Lévy, $S - B$ and S in the proposition above can be replaced respectively with $|B|$ and L, where L is the local time of B at 0.

3.2 Kennedy martingales for Lévy processes

In this paragraph we assume that our Lévy process X is integrable, and we denote $\mu = \mathbb{E}[X_1]$. Under this condition, we have the following analogue to Prop. 2:

Proposition 3. Let $f(t, x, y) : \mathbb{R}_+ \times \mathbb{R}_+ \times \mathbb{R}_+ \to \mathbb{R}$ be a smooth function, such that, for every t, x, $y \geqslant 0$:

$$f_t(t, x, y) - \mu f_x(t, x, y) + \frac{\sigma^2}{2} f_{xx}(t, x, y)$$
$$+ \int_{-\infty}^{x} \left(f(t, x - z, y) - f(t, x, y) + z f_x(t, x, y) \right) \nu(dz) \tag{11}$$
$$+ \int_{x}^{\infty} \left(f(t, 0, y + z - x) - f(t, x, y) + z f_x(t, x, y) \right) \nu(dz) = 0$$
$$f_x(t, 0, y) + f_y(t, 0, y) = 0.$$

Then $(f(t, S_t - X_t, S_t), t \geqslant 0)$ is a local martingale.

Proof. As in Prop. 2, this follows immediately from Itô's formula, and the "support property" that $X_{t-} = X_t = S_t = S_{t-}$ on $\mathrm{Supp}(dS_t^c)$. $\qquad \square$

Remark 4. If f does not depend on y, this can also be directly deduced from Prop. 1.

Just as in the Brownian case, we can look for f in the form $f(t, x, y) = h(x) e^{-\alpha y - \beta t}$. The conditions of Prop. 3 then imply that h is a solution to the integro-differential equation ($x \geqslant 0$):

$$-\beta h(x) - \mu h'(x) + \frac{\sigma^2}{2} h''(x) + \int_{-\infty}^{x} \left(h(x - z) - h(x) + z h'(x) \right) \nu(dz)$$
$$+ \int_{x}^{\infty} \left(h(0) e^{-\alpha(z-x)} - h(x) + z h'(x) \right) \nu(dz) = 0$$

with the boundary condition $h'(0) = \alpha h(0)$.

Example 2. Let us look at a simple example. Assume that X is the compensated Poisson process $X_t = N_t - t$. In this case, the function h must satisfy

$$h'(x) - \beta h(x) + 1_{0 \leqslant x \leqslant 1} \left(h(0) e^{-\alpha(1-x)} - h(x) \right) + 1_{x > 1} \left(h(x - 1) - h(x) \right) = 0. \tag{12}$$

On $[0, 1]$, we obtain

$$h(x) = C_0 \exp\left((1 + \beta)x \right) + \frac{h(0)}{1 + \beta - \alpha} \exp\left(-\alpha(1 - x) \right), \qquad 0 \leqslant x \leqslant 1.$$

By putting $x = 0$ in the formula above, we get

$$h(0) = C_0 + \frac{h(0)}{1 + \beta - \alpha}\, e^{-\alpha}. \tag{13}$$

On the other hand, the boundary condition $h'(0) = \alpha h(0)$ gives

$$C_0(1 + \beta) + \frac{\alpha h(0)}{1 + \beta - \alpha}\, e^{-\alpha} = \alpha h(0).$$

From the last two formulas, we deduce that

$$\frac{h(0)(1 + \beta) e^{-\alpha}}{1 + \beta - \alpha} - \frac{h(0)\alpha e^{-\alpha}}{1 + \beta - \alpha} = (1 + \beta)h(0) - \alpha h(0).$$

If we rule out the trivial case $h(0) = 0$, we then obtain $(1 + \beta - \alpha) e^{-\alpha} = (1 + \beta - \alpha)^2$. Note also that if $1 + \beta - \alpha = 0$, then solving (12) on $[0, 1]$ together with $h'(0) = \alpha h(0)$ yields $\alpha = e^{-\alpha}$. Hence we also rule out this case. We thus obtain $\beta = e^{-\alpha} + \alpha - 1 = \psi(-\alpha)$, where ψ is the Laplace exponent of X, $\mathbb{E}[e^{\lambda X_t}] = e^{t\psi(\lambda)}$.

Plugging this into (13), we obtain $h(x) = h(0)e^{\alpha x}$, $x \leqslant 1$.

Let us now turn to the case $x > 1$. Then h must satisfy $h' - \beta h + h(x - 1) - h(x) = 0$. Set $\hat{h}(\lambda) = \int_1^\infty e^{-\lambda x} h(x)\, dx$. The difference-differential equation (12) yields

$$-e^{-\lambda} h(1) + \lambda \hat{h}(\lambda) - \beta \hat{h}(\lambda) + e^{-\lambda} \int_0^1 e^{-\lambda x} h(0)e^{\alpha x}\, dx + e^{-\lambda}\hat{h}(\lambda) - \hat{h}(\lambda) = 0$$

which can be rewritten, taking into account the previous results,

$$(\lambda - \beta + e^{-\lambda} - 1)\hat{h}(\lambda) - e^{-\lambda} h(0)\left(e^\alpha - \frac{e^{\alpha - \lambda} - 1}{\alpha - \lambda}\right) = 0$$

or, since $\beta = \psi(-\alpha)$,

$$\big(\psi(-\lambda) - \psi(-\alpha)\big)\hat{h}(\lambda) = e^{\alpha - \lambda}\frac{h(0)}{\alpha - \lambda}\big(\psi(-\alpha) - \psi(\lambda)\big)$$

This gives:

$$\hat{h}(\lambda) = h(0)\frac{e^{\alpha - \lambda}}{\lambda - \alpha} = h(0)\int_1^\infty e^{-\lambda x} e^{\alpha x}\, dx.$$

Hence, $h(x) = h(0)e^{\alpha x}$ for all $x \geqslant 0$, and we have found again the classical Wald martingale.

We now turn, until the end of this section, to the special case when X is spectrally negative. From Prop. 3, we obtain:

Proposition 4. Let $f(t, x, y) : \mathbb{R}_+ \times \mathbb{R}_+ \times \mathbb{R}_+ \to \mathbb{R}$ be a smooth function, such that

$$f_t - \mu f_x + \frac{\sigma^2}{2} f_{xx} + \int_{-\infty}^{0} \big(f(t, x - z, y) - f(t, x, y) + z f_x(t, x, y)\big) \nu(\mathrm{d}z) = 0$$
$$f_x(t, 0, y) + f_y(t, 0, y) = 0.$$

(14)

Then $(f(t, S_t - X_t, S_t), t \geqslant 0)$ *is a local martingale.*

As a corollary, we obtain the analogue of Cor. 2.2.2') in [5]:

Corollary 2. *Suppose that* $\mu = 0$, *so that* X *is a martingale. Let* g *be a locally bounded Borel function, and set* $G(x) = \int_0^x g(y) \, \mathrm{d}y$. *Then,*

$$\big(G(S_t) - (S_t - X_t)g(S_t), \quad t \geqslant 0\big) \tag{15}$$

is a local martingale.

Proof. If g is C^1, one checks that $f(t, x, y) := G(y) - xg(y)$ satisfies conditions (14). The result for a general function g then follows from the monotone class theorem. $\quad\square$

The above corollary has interesting consequences, which will be explored in Section 4.

Let us look again for functions f in Prop. 4 of the form $f(t, x, y) = h(x)\mathrm{e}^{-\alpha y - \beta t}$. In addition to the boundary condition $h'(0) = \alpha h(0)$, h must satisfy: for all $x \geqslant 0$,

$$-\beta h(x) - \mu h'(x) + \frac{\sigma^2}{2} h''(x) + \int_{-\infty}^{0} \big(h(x - z) - h(x) + z h'(x)\big) \, \mathrm{d}\nu(z) = 0. \tag{16}$$

Setting $\tilde{h}(x) = -h(-x)$, this can be rewritten

$$\beta \tilde{h}(x) - \mathcal{A}^X \tilde{h}(x) = 0, \qquad x \leqslant 0.$$

Hence $\mathrm{e}^{-\beta(t \wedge T_0^+)} \tilde{h}(X_{t \wedge T_0^+})$ must be a $\mathbb{P}_x$-martingale for all $x < 0$, where $T_0^+ = \inf\{t : X_t > 0\}$. Equivalently, $\mathrm{e}^{\beta(t \wedge T_0^-)} h(X_{t \wedge T_0^-})$ must be a $\mathbb{P}_x$-martingale for all $x > 0$, where $T_0^- = \inf\{t : X_t < 0\}$. In other words, the function $\mathrm{e}^{\beta t} h(x)$ is time-space harmonic for X in $\mathbb{R}_+ \times \mathbb{R}_+$. If we ask moreover that h be positive, then by a result of Küchler and Lauritzen [15]:

$$\mathrm{e}^{\beta t} h(x) = \int \pi(\mathrm{d}\alpha) \, \mathrm{e}^{\alpha x - t \psi(\alpha)}$$

where π is a finite positive measure (conversely, any function of the form above is a time-space harmonic function for X).

Example 3 (Scale functions). In the case of a spectrally negative process X, the scale functions $W^{(q)}$ and $Z^{(q)}$ give rise to other natural examples of martingales. Indeed, as is well-known (see e.g. [4]), for any $q \geqslant 0$ and $a < x < b$, the processes

$$e^{-q(t\wedge T)}W^{(q)}(X_{t\wedge T} - a) \quad \text{and} \quad e^{-q(t\wedge T)}Z^{(q)}(X_{t\wedge T} - a) \tag{17}$$

are $\mathbb{P}_x$-martingales, where $T = \inf\{t : X_t \notin [a,b]\}$. Therefore, any linear combination is also a martingale; this will be used in Paragraph 4.2.

For the sake of completeness, we provide a simple derivation of the existence and martingale property of $W^{(q)}$. Set $I_t = \inf_{s\leqslant t} X_s$ and assume first that X drifts to $+\infty$, so that its overall infimum I_∞ is finite. Define the function W by

$$W(x) = c\,\mathbb{P}[I_\infty \geqslant -x], \qquad x \geqslant 0$$

where c is an arbitrary positive constant and $W(x) = 0$ for $x \leqslant 0$. An application of the strong Markov property at T_y^+ ($y > 0$) yields

$$\mathbb{P}[I_{T_y^+} \geqslant -x] = \frac{W(x)}{W(x+y)}. \tag{18}$$

On the other hand, it is proved independently in Section 4 (see (27)) that, if θ is an independent exponential variable with parameter $r > 0$, then

$$\mathbb{E}[e^{\lambda I_\theta}] = \frac{r}{\Phi(r)}\frac{\Phi(r) + \lambda}{r - \psi(\lambda)}.$$

Letting $r \downarrow 0$, we get

$$\mathbb{E}[e^{\lambda I_\infty}] = \frac{1}{\Phi'(0+)}\frac{\lambda}{\psi(\lambda)},$$

and choosing $c = \Phi'(0+)$, we obtain finally

$$\int_0^\infty e^{-\lambda x}W(x)\,\mathrm{d}x = \frac{1}{\psi(\lambda)}.$$

Moreover, it follows immediately from the definition of W and (18) that the process $(W(X_{t\wedge\tau}), t \geqslant 0)$ is a $\mathbb{P}_x$-martingale for all $x > 0$, where $\tau = \inf\{t : X_t < 0\}$ (see also [6, Lemma VII.11 p. 198]). The above discussion is easily extended:

- to oscillating processes, by adding a small positive drift which we let tend to 0;
- to processes which drift to $-\infty$, by "conditioning‚them to drift to $+\infty$" via an Esscher transform.

For the details, we refer to [6].

Lastly, consider the probability $\mathbb{P}^{\Phi(q)}$ defined by

$$\mathbb{P}^{\Phi(q)}|_{\mathcal{F}_t} = e^{\Phi(q)X_t - qt}\,\mathbb{P}|_{\mathcal{F}_t},$$

and let $W_{\Phi(q)}$ denote the scale function relative to $\mathbb{P}^{\Phi(q)}$. Setting

$$W^{(q)}(x) = e^{\Phi(q)x}W_{\Phi(q)}(x),$$

we have

$$\int_0^\infty e^{-\lambda x} W^{(q)}(x)\, dx = \frac{1}{\psi(\lambda) - q}, \qquad \lambda > \Phi(q)$$

and for $x > 0$, since $W_{\Phi(q)}(X_{t \wedge \tau})$ is a $\mathbb{P}_x^{\Phi(q)}$-martingale, it follows that $e^{-q(t \wedge \tau)} W^{(q)}(X_{t \wedge \tau})$ is a $\mathbb{P}_x$-martingale. The case of a general interval follows by translation.

Now if $Z^{(q)}$ is defined by

$$Z^{(q)}(x) = 1 + q \int_0^x W^{(q)}(y)\, dy = 1 + q \int W^{(q)}(x - y) 1_{0 < y < x}\, dy$$

the martingale property for $Z^{(q)}$ is simply a consequence of the martingale property we have just seen for $W^{(q)}$.

To end this discussion, we recall a few examples where the function $W^{(q)}$ is known explicitly:

Lévy process X	$W^{(q)}$	$Z^{(q)}$
Brownian with drift μ $\delta = \sqrt{\mu^2 + 2q}$	$\dfrac{2}{\delta} e^{-\mu x} \sinh(\delta x)$	$e^{-\mu x}\left(\cosh(\delta x) + \dfrac{\mu}{\delta} \sinh(\delta x)\right)$
Compound Poisson, unit drift, negative exponential jumps	$k(q) e^{h(q)x}$ $+ \left(1 - k(q)\right) e^{\frac{h(q)k(q)-1}{k(q)} x}$	$1 + \dfrac{qk(q)}{h(q)}\left(e^{h(q)x} - 1\right)$ $+ \dfrac{qk(q)(1 - k(q))}{h(q)k(q) - 1}\left(e^{\frac{h(q)k(q)-1}{k(q)} x} - 1\right)$
Stable, index $\alpha > 1$	$\dfrac{1}{q} \dfrac{d}{dx}\{E_\alpha(qx^\alpha)\}$	$E_\alpha(qx^\alpha)$

On the second line, $X_t = t - \sum_{k=1}^{N_t} U_k$ where N is a Poisson process with intensity λ and U_k are independent, also independent of N and have an exponential distribution with parameter c; the functions h and k are defined by

$$h(q) = \frac{1}{2}\left(\lambda + q - c + \sqrt{(\lambda + q - c)^2 + 4qc}\right),$$

$$k(q) = \frac{h(q) + c}{(h(q) + c)^2 - \lambda c}.$$

On the third line, due to [7], E_α is the Mittag-Leffler function of index α (see [22] or [10]).

The functions $e^{-qt} W^{(q)}(x)$ and $e^{-qt} Z^{(q)}(x)$ provide us with examples of time-space harmonic functions in $\mathbb{R}_+ \times (a, b)$. Note that in the Brownian case, the Kennedy martingale can be expressed in terms of $W^{(q)}$ and $Z^{(q)}$.

In the next example, we examine a special case when there exists an interesting function h which satisfies (16).

Example 4. We retain the assumptions of the present paragraph, that is X is spectrally negative and integrable.

Furthermore, we make the following hypothesis on the Laplace exponent of X. Let $\lambda_0 = \inf\{\lambda : \mathbb{E}[e^{\lambda X_1}] < \infty\} \leqslant 0$, and assume that $\lambda_0 < 0$. We also assume that $\beta^* = \lim_{\lambda \to \lambda_0+} \psi(\lambda) > 0$. Under these conditions, we can work out an interesting special case of Prop. 4.

The Laplace exponent of X, $\psi(\lambda) = \ln \mathbb{E}[e^{\lambda X_1}]$, is C^∞ and strictly convex on (λ_0, ∞). Then for each $0 < \beta < \beta^*$, there are two distinct solutions to the equation $\psi(\lambda) = \beta$; denote them for instance by $\gamma_- < 0 < \gamma_+$. The function $h(x) = Ae^{-\gamma_- x} + Be^{-\gamma_+ x}$ (where A and B are constants) verifies

$$-\beta h(x) - \mu h'(x) + \frac{\sigma^2}{2}h''(x) + \int_{-\infty}^{0} \big(h(x-z) - h(x) + zh'(x)\big)\, \mathrm{d}\nu(z) = 0$$

and the boundary condition of (14): $h'(0) = \alpha h(0)$ implies $-(A\gamma_- + B\gamma_+) = \alpha(A+B)$, that is $B = -A\frac{\alpha+\gamma_-}{\alpha+\gamma_+}$. Hence h is given by

$$h(x) = C\big((\alpha + \gamma_+)e^{-\gamma_- x} - (\alpha + \gamma_-)e^{-\gamma_+ x}\big)$$

for some constant C. It is not hard to check that h can be rewritten as

$$h(x) = e^{-\eta x}\big((\alpha + \eta)\sinh(\delta x) + \delta \cosh(\delta x)\big) \tag{19}$$

for a suitable choice of C, where

$$\eta = \eta(\beta) = \frac{1}{2}(\gamma_+ + \gamma_-), \qquad \delta = \delta(\beta) = \frac{1}{2}(\gamma_+ - \gamma_-).$$

It is quite remarkable that (19) is identical to (9), up to a change of parameters, and it would be interesting to develop further this analogy.

4 Some applications

This last section is dedicated to some examples of application of the martingales we have introduced in the previous sections: We examine two results that have already been obtained by using excursion theory, and derive them again by martingale arguments. We obtain more elementary, if less intuitive proofs.

4.1 The Wiener–Hopf factorization

In this paragraph, we show how our martingales can be used to recover the celebrated Wiener–Hopf factorization:

Theorem 1. Let θ be an exponential variable of parameter q, independent of X. Then S_θ and $X_\theta - S_\theta$ are independent, and hence:

$$\mathbb{E}[e^{iuX_\theta}] = \mathbb{E}[e^{iu(X_\theta - S_\theta)}]\, \mathbb{E}[e^{iuS_\theta}]. \tag{20}$$

The above theorem is equivalent to the existence, for every $q > 0$, of two functions ϕ_q^+ and ϕ_q^-, such that:

1. ϕ_q^+ and ϕ_q^- are the characteristic functions of infinitely divisible laws;
2. ϕ_q^+ is analytic in $\{\Im(z) \leqslant 0\}$ and ϕ_q^- is analytic in $\{\Im(z) \geqslant 0\}$;
3. For every $u \in \mathbb{R}$,

$$\frac{q}{q + \phi(u)} = \phi_q^+(u)\phi_q^-(u). \tag{21}$$

Standard arguments from complex analysis then entail that the functions ϕ_q^+ and ϕ_q^- are unique.

Recall the basic fact that if σ is the inverse local time of R at 0 and if we set $H = X \circ \sigma \equiv S \circ \sigma$, the process (σ, H) is a bivariate subordinator, called the ladder process of X. This follows from the fact that σ is the inverse of an additive functional and satisfies $(S - X)_{\sigma(t)} = 0$ for all t. This subordinator is killed at some rate c if $S_\infty < \infty$. We denote by κ its Laplace exponent: $\mathbb{E}[e^{-q\sigma_t - \lambda H_t}] = e^{-t\kappa(q,\lambda)}$.

Also, recall that the continuous part S^c of the supremum S is not trivial if and only if it is a version of the local time at 0 for the reflected process $S - X$. Indeed, it is clear first of all that the support of dS^c is included in $\overline{\{t : S_t = X_t\}}$; we have already used this fact above. Next, let T be an a.s. finite stopping time such that $S_T = X_T$ a.s. Then, we have

$$S_{T+t} = X_T + 0 \vee \sup\{X_{T+s} - X_T, s \leqslant t\}$$

so that

$$\left(R_{T+t}, S_{T+t}^c - S_T^c\right)_{t \geqslant 0} = \left(X_T - X_{T+t} + 0 \vee \sup\{X_{T+s} - X_T, s \leqslant t\},\right.$$
$$\left.0 \vee \sup\{X_{T+s} - X_T, s \leqslant t\} - \sum_{s \leqslant t} \Delta S_{T+s}\right)_{t \geqslant 0}.$$

By the Lévy property of X, we see that $(R_{T+t}, S_{T+t}^c - S_T^c)_{t \geqslant 0}$ is independent of $\mathcal{F}_T$ and has the same law as $(R_t, S_t^c)_{t \geqslant 0}$. Proposition IV.5 in [6] then entails that S^c is a version of the local time of R at 0.

Hence, when $S^c \not\equiv 0$, we choose it as the local time of $S - X$ at 0 and we then have $\sigma = (S^c)^{-1}$; if $S^c \equiv 0$, we take the same definition as in [9] for this local time. Note that it is possible that 0 be regular for itself relative to R even if $S^c \equiv 0$, so R can have a continuous Markov local time at 0 also in this case.

The ladder exponent has the following Lévy–Khintchine representation: for some b, c, $k \geqslant 0$,

$$\kappa(q, \lambda) = c + bq + k\lambda + \int_{(0,\infty)^2} \left(1 - e^{-qt - \lambda z}\right) l(dt, dz).$$

This corresponds to the Lévy–Itô path decomposition

$$\sigma_t = bt + \sum_{s \leqslant t} \Delta \sigma_s, \qquad t < \zeta$$

$$H_t = kt + \sum_{s \leqslant t} \Delta H_s, \qquad t < \zeta$$

where $\zeta = S_\infty$ is the lifetime. Note that if $S^c \equiv 0$, then $H = S \circ \sigma$ is a pure jump process and therefore $k = 0$. Otherwise, we have, for almost all ω

$$H_t(\omega) = S_{\sigma_t(\omega)}(\omega) = S^c_{\sigma_t(\omega)}(\omega) + S^d_{\sigma_t(\omega)}(\omega) = t + S^d_{\sigma_t(\omega)}(\omega)$$

and so $k = 1$. It is also known that $k > 0$ is equivalent to X creeping upwards with positive probability ([6, Th. VI.19]).

The following identity, which will be useful in the sequel, is an immediate consequence of the Lévy–Khintchine representation of κ:

$$\kappa(q, \lambda) = \kappa(q, 0) + k\lambda + \int_{(0,\infty)^2} e^{-qt} \left(1 - e^{-\lambda z}\right) l(\mathrm{d}t, \mathrm{d}z). \tag{22}$$

In the remainder of this paragraph, we shall work under the following assumptions:

- X does not drift to $-\infty$, that is $\limsup_{t \to \infty} X_t = +\infty$.
- The Lévy measure ν of X satisfies: $\int_0^1 x \, \nu(\mathrm{d}x) < \infty$.
- θ denotes an exponential variable with parameter $q > 0$, independent of X.

The first assumption implies in particular that the ladder process has an infinite lifetime, i.e. $c = 0$.

In order to prove the Wiener–Hopf factorization, we prepare a series of lemmas; these results are essentially already known, but we prove them here in the spirit of the present paper, using only elementary martingale arguments.

Lemma 2. *The process* $\left(M_t^{(u)}, t \geqslant 0\right)$, *where*

$$M_t^{(u)} = e^{iu(X_t - S_t)} - e^{-iuS_0} + \Psi(u) \int_0^t e^{iu(X_r - S_r)} \, \mathrm{d}r$$
$$+ iuS_t^c - \sum_{r \leqslant t} e^{iu(X_r - S_r)} \left(1 - e^{iu\Delta S_r}\right) \tag{23}$$

is a martingale.

Proof. We already know (see [13] or Example 1) that $M^{(u)}$ is a local martingale, and we now show that $\sup_{s \leqslant t} |M_s^{(u)}|$ is integrable for any t. First let us note that if S^c is not trivial, then S_θ^c has an exponential law since S^c is a local time. Let us now turn to the jump term. By the basic compensation lemma, the quantity

$$\mathbb{E}\left[\sum_{r \leqslant t} |1 - e^{iu\Delta S_r}|\right]$$

exists as soon as

$$\int_0^t dr \int_0^\infty \nu(dz)\, \mathbb{E}\big[1_{R_r<z}|1-e^{iu(z-R_r)}|\big] < \infty.$$

But since $|1-e^{iux}| = iux + O(x^2)$ as $x \to 0$, this latter condition is fulfilled thanks to our hypothesis on ν. The remaining terms are bounded.   □

The next lemma is proved in [12] by elementary methods essentially building on discrete time approximations. For the sake of completeness, we now present a proof in the spirit of the present paper.

Lemma 3. *The following equality holds for each $u \in \mathbb{R}$:*

$$\mathbb{E}\big[e^{iuS_\theta}\big] = \frac{\kappa(q,0)}{\kappa(q,-iu)}. \tag{24}$$

Proof. Using Itô's formula and a basic compensation, it is easy to see that the process $(\Sigma_t^{(u)}, t \geqslant 0)$ defined by

$$\Sigma_t^{(u)} = e^{iuS_t} + \lambda \int_0^t e^{iuS_v}\, dS_v^c + \int_0^t \left\{ e^{iuS_v} \int_{R_v}^\infty \big(1 - e^{iu(x-R_v)}\big)\,\nu(dx) \right\} dv,$$

is a martingale, hence $\mathbb{E}\big[\Sigma_\theta^{(u)}\big] = 1$ (we use the same argument as in the proof of the previous lemma as to the integrability of $\sup_{s\leqslant t}|\Sigma_s^{(u)}|$). Let us now compute each term of $\mathbb{E}\big[\Sigma_\theta^{(u)}\big]$. First, we claim that

$$\mathbb{E}\left[\int_0^\theta e^{iuS_v}\, dS_v^c\right] = \frac{k}{\kappa(q,-iu)}.$$

It is obvious if $S^c \equiv 0$, and if $S^c \not\equiv 0$, we have

$$\mathbb{E}\left[\int_0^\theta e^{iuS_v}\, dS_v^c\right] = \mathbb{E}\left[\int_0^\infty e^{-qv+iuS_v}\, dS_v^c\right] = \mathbb{E}\left[\int_0^\infty e^{-q\sigma_v+iuH_v}\, dv\right]$$
$$= \frac{1}{\kappa(q,-iu)}.$$

Next, notice the a.s. equality between sets

$$\{(v,S_{v-},\Delta S_v) : v > 0,\ \Delta S_v > 0\} = \{(\sigma_w, H_{w-}, \Delta H_w) : w > 0,\ \Delta H_w > 0\}.$$

Using the basic compensation lemma, it follows that

$$\mathbb{E}\left[\int_0^\theta \left\{ e^{iuS_v} \int_{R_v}^\infty \left(1 - e^{iu(x-R_v)}\right) \nu(dx) \right\} dv \right]$$

$$= \mathbb{E}\left[\int_0^\infty \left\{ e^{-qv+iuS_v} \int_{R_v}^\infty \left(1 - e^{iu(x-R_v)}\right) \nu(dx) \right\} dv \right]$$

$$= \mathbb{E}\left[\sum_{v>0} e^{-qv+iuS_{v-}} \left(1 - e^{iu\Delta S_v}\right) 1_{\Delta S_v > 0} \right]$$

$$= \mathbb{E}\left[\sum_{v>0} e^{-q\sigma_v+iuH_{v-}} \left(1 - e^{iu\Delta H_v}\right) \right]$$

$$= \mathbb{E}\left[\int_0^\infty e^{-q\sigma_v+iuH_v} \, dv \int_{(0,\infty)^2} e^{-qs} \left(1 - e^{iuh}\right) l(ds, dh) \right]$$

$$= \frac{1}{\kappa(q,-iu)} \int_{(0,\infty)^2} e^{-qs} \left(1 - e^{iuh}\right) l(ds, dh).$$

Putting pieces together, we obtain

$$\mathbb{E}\left[e^{iuS_\theta}\right] = \frac{1}{\kappa(q,-iu)}\left(\kappa(q,-iu) + kiu - \int_{(0,\infty)^2} e^{-qs}\left(1 - e^{iuh}\right) l(ds, dh)\right).$$

Now, from the Lévy–Khintchine representation (22) of κ, we recognize that the term in brackets in the above formula is nothing else but $\kappa(q,0)$. □

Remark 5. Formula (24) shows that S_θ has the same law as $\widetilde{Z}_{\tilde{\theta}}$ where $\widetilde{Z}$ is a subordinator with Laplace exponent $\phi(\lambda) = \kappa(q,\lambda) - \kappa(q,0)$ and $\tilde{\theta}$ is an exponential variable with parameter $\kappa(q,0)$, independent of $\widetilde{Z}$. More precisely, (24) yields, for every $\lambda \geq 0$

$$\mathbb{E}\left[e^{-\lambda S_\theta}\right] = \widetilde{\mathbb{E}}\left[e^{-\lambda H_{\tilde{\theta}}}\right]$$

where $\widetilde{\mathbb{P}}$ is the Esscher transform

$$\widetilde{\mathbb{P}}|_{\mathcal{G}_t} = e^{-q\sigma_t + t\kappa(q,0)} \mathbb{P}|_{\mathcal{G}_t}$$

with $\mathcal{G}_t = \mathcal{F}_{\sigma_t}$. It would be interesting to know more about the relationships (if any) between fluctuation theory and this Esscher transform.

The following result is a slightly different form of [19, Formula (13)], which is proved using excursion theory. We provide here a martingale proof.

Lemma 4. *For all $u \in \mathbb{R}$,*

$$\mathbb{E}\left[e^{iuS_\theta}\right] = \left(1 - \frac{iu}{\kappa(q,0)} + \mathbb{E}\left[\int_0^\theta dr \int_{R_r}^\infty \left(1 - e^{iu(x-R_r)}\right) \nu(dx)\right]\right)^{-1}. \quad (25)$$

Proof. Let us set, for $q > 0$ and $u \in \mathbb{R}$:

$$A = \mathbb{E}\left[\sum_{v>0} e^{-q\sigma_v}\left(1 - e^{iu\Delta H_v}\right)\right].$$

By the basic compensation lemma, we have

$$A = \mathbb{E}\left[\sum_{v>0} e^{-q\sigma_{v-}}e^{-q\Delta\sigma_v}\left(1 - e^{iu\Delta H_v}\right)\right]$$

$$= \mathbb{E}\left[\int_0^\infty e^{-q\sigma_v}\,dv \int_{(0,\infty)^2} e^{-qs}\left(1 - e^{iuh}\right)l(ds,dh)\right]$$

where l is the Lévy measure of the ladder process (σ, H). The foregoing formula can be written as

$$A = \left(\int_0^\infty \mathbb{E}\left[e^{-q\sigma_v}\right]dv\right)\left(\int_{(0,\infty)^2} e^{-qs}\left(1 - e^{iuh}\right)l(ds,dh)\right)$$

$$= \frac{1}{\kappa(q,0)}\int_{(0,\infty)^2} e^{-qs}\left(1 - e^{iuh}\right)l(ds,dh).$$

Now, observe that we have the a.s. equality between the sets

$$\{(\sigma_s, \Delta H_s) : s > 0\} = \{(v, X_v - S_{v-}) : v > 0, X_v > S_{v-}\};$$

this follows from the very definition of (σ, H). Hence, we have:

$$A = \mathbb{E}\left[\sum_{v>0} e^{-qv}\left(1 - e^{iu(X_v - S_{v-})}\right)1_{X_v>S_{v-}}\right]$$

$$= \mathbb{E}\left[\sum_{v>0} e^{-qv}\left(1 - e^{iu(\Delta X_v + X_{v-} - S_{v-})}\right)1_{\Delta X_v > S_{v-} - X_{v-}}\right]$$

$$= \mathbb{E}\left[\int_0^\infty e^{-qv}\,dv \int_{S_v-X_v}^\infty \left(1 - e^{iu(x+X_v-S_v)}\right)\nu(dx)\right]$$

as follows once more from Lemma 1. Then, it is not hard to see that

$$A = \frac{1}{q}\mathbb{E}\left[\int_{R_\theta}^\infty \left(1 - e^{iu(x-R_\theta)}\right)\nu(dx)\right]$$

$$= \mathbb{E}\left[\int_0^\theta dr \int_{R_r}^\infty \left(1 - e^{iu(x-R_r)}\right)\nu(dx)\right].$$

By what we have just shown above, we obtain:

$$\kappa(q,\lambda) = \kappa(q,0) - iu + \frac{\kappa(q,0)}{q}\int_0^\infty \mathbb{P}[R_\theta \in dy]\int_y^\infty \left(1 - e^{iu(x-y)}\right)\nu(dx). \quad (26)$$

(25) now follows easily from combining (26) and (24). $\square$

Proof of Theorem 1. By Lemma 2, we have $\mathbb{E}\big[M_\theta^{(u)}\big] = 0$. Let us compute the expectation of each term on the right-hand side of (23). First, we have

$$\mathbb{E}\left[\int_0^\theta e^{iu(X_r - S_r)}\, dr\right] = \int_0^\infty e^{-qr}\, \mathbb{E}\big[e^{iu(X_r - S_r)}\big]\, dr = \frac{1}{q}\, \mathbb{E}\big[e^{iu(X_\theta - S_\theta)}\big].$$

Next, we have

$$\mathbb{E}[S_\theta^c] = \frac{k}{\kappa(q,0)}\,.$$

This is clear if $S^c \equiv 0$, while if $S^c \not\equiv 0$, an immediate change of variables gives

$$\mathbb{E}[S_\theta^c] = \mathbb{E}\left[\int_0^\infty dS_r^c e^{-qr}\right] = \int_0^\infty \mathbb{E}[e^{-q\sigma_t}]\, dt = \frac{1}{\kappa(q,0)}\,.$$

Let us now turn to the last term of (23). Since the summand vanishes unless $\Delta S_r \neq 0$, which is equivalent to $\Delta X_r > R_{r-}$, we have

$$\mathbb{E}\left[\sum_{r \leqslant \theta} e^{iu(X_r - S_r)}\big(1 - e^{iu\Delta S_r}\big)\right] = \mathbb{E}\left[\sum_{r \leqslant \theta} 1_{\Delta X_r > R_{r-}}\big(1 - e^{iu(\Delta X_r - R_{r-})}\big)\right]$$

$$= \mathbb{E}\left[\int_0^\theta dr \int_{R_r}^\infty \big(1 - e^{iu(x - R_r)}\big)\, \nu(dx)\right]$$

where we recall that ν is the Lévy measure of X and the second equality follows from the basic compensation lemma. Putting pieces together, we obtain:

$$\mathbb{E}\big[e^{iu(X_\theta - S_\theta)}\big] = \frac{q}{q + \Psi(u)}\left(1 - \frac{iu}{\kappa(q,0)} + \mathbb{E}\left[\int_0^\theta dr \int_{R_r}^\infty \big(1 - e^{iu(x - R_r)}\big)\, \nu(dx)\right]\right)$$

$$= \mathbb{E}\big[e^{iuX_\theta}\big]\left(1 - \frac{iu}{\kappa(q,0)} + \mathbb{E}\left[\int_0^\theta dr \int_{R_r}^\infty \big(1 - e^{iu(xR_r)}\big)\, \nu(dx)\right]\right)$$

and we conclude thanks to (25). $\square$

Remark 6. In the spectrally negative case, the argument above is due to [16]; it is then very simple, since there is no sum of jumps in $M^{(u)}$ and $S_\theta \equiv S_\theta^c$ simply has an exponential distribution.

Let us consider the formula:

$$\mathbb{E}\big[e^{iu(X_\theta - S_\theta)}\big] = \frac{q}{q + \Psi(u)}\left(1 - \frac{iu}{\kappa(q,0)} + \mathbb{E}\left[\int_0^\theta dr \int_{R_r}^\infty \big(1 - e^{iu(x - R_r)}\big)\, \nu(dx)\right]\right)$$

in the case of a spectrally negative process X. Then, ν is supported by $\mathbb{R}_-$ and $S = S^c$ is a version of the local time of R at 0; the inverse local time is thus $\sigma = S^{-1}$. A simple martingale argument shows that this subordinator has Laplace exponent Φ (see [6]). Recall on the other hand that the Laplace exponent of σ is given by $\kappa(q,0)$. As a consequence, the above formula takes the form

$$\mathbb{E}\left[e^{iu(X_\theta - S_\theta)}\right] = \frac{q}{q + \Psi(u)}\left(1 - \frac{iu}{\Phi(q)}\right) = \frac{q}{\Phi(q)}\frac{\Phi(q) - iu}{q + \Psi(u)}.$$

Now by the well-known duality lemma, I_θ has the same law as $X_\theta - S_\theta$ and the foregoing equality is valid with any λ such that $\Re(\lambda) \geqslant 0$ instead of iu, so that we finally obtain (see also [6, Chap. VII Formula (3)]):

$$\mathbb{E}\left[e^{\lambda I_\theta}\right] = \frac{q}{\Phi(q)}\frac{\Phi(q) + \lambda}{q - \psi(\lambda)}, \qquad \Re(\lambda) \geqslant 0. \tag{27}$$

4.2 The exit problem for the reflected process (spectrally negative case)

In this paragraph, we assume that X is spectrally negative, and we are interested in the joint law of (τ_k, R_{τ_k}) where $\tau_k = \inf\{t : R_t > k\}$. The knowledge of this joint law, given by (29) below, has applications in finance ([2, 4, 18]), and was originally obtained in [4] using excursion theory (see also [11] for a different approach). We also assume for simplicity that the Lévy measure of X has no atom, so that $W^{(q)}$ is a C^1 function, as shown in [11] (our results still hold in the general case, provided we interpret $W^{(q)\prime}$ as the right derivative of $W^{(q)}$).

Recall the scale functions $W^{(q)}$ and $Z^{(q)}$ give rise naturally to some martingales related to the two-sided exit problem for X. They can also be used to construct martingales related to R. Before we proceed, let us introduce the Esscher transform of $\mathbb{P}$ with parameter v, i.e. the probability measure defined by

$$\mathbb{P}^v|_{\mathcal{F}_t} = e^{vX_t - t\psi(v)}\,\mathbb{P}|_{\mathcal{F}_t},$$

and recall that under $\mathbb{P}^v$, X is a spectrally negative Lévy process with Laplace exponent $\psi_v(\lambda) = \psi(v + \lambda) - \psi(v)$. The scale functions relative to $\mathbb{P}^v$ are denoted $W_v^{(q)}$ and $Z_v^{(q)}$.

We can now state:

Proposition 5. *Fix $0 < x < k$, and set*

$$M_t = e^{-vR_t - (q + \psi(v)t)} \times$$

$$\times \left(Z_v^{(q)}(k - R_t) - \frac{vZ_v^{(q)}(k) + qW_v^{(q)}(k)}{vW_v^{(q)}(k) + W_v^{(q)\prime}(k)} W_v^{(q)}(k - R_t) \right). \tag{28}$$

Then $(M_{t \wedge \tau_k}, t \geqslant 0)$ is a $\mathbb{P}_{R_0 = x}$-martingale.

Proof. Set $T = \inf\{t : X_t \notin (-k, 0)\}$ and

$$N_t = e^{-qt}\left(Z_v^{(q)}(X_t + k) + \alpha W^{(q)}(X_t + k)\right), \qquad \alpha \in \mathbb{R}.$$

Then by Example 3, $(N_{t \wedge T}, t \geqslant 0)$ is a $\mathbb{P}_{-y}^v$-martingale for all $y \in (0, k)$, so that if

$$\widetilde{N}_t = e^{vX_t - (q + \psi(v))t} \big(Z_v^{(q)}(X_t + k) + \alpha W_v^{(q)}(X_t + k) \big),$$

$\big(\widetilde{N}_{t \wedge T}, t \geqslant 0 \big)$ is a $\mathbb{P}_{-y}$-martingale, for all $y \in (0, k)$. By Corollary 1, if we set

$$\widetilde{M}_t = e^{-vR_t - (q + \psi(v))t} \big(Z_v^{(q)}(k - R_t) + \alpha W_v^{(q)}(k - R_t) \big)$$
$$+ \int_0^t e^{-(q + \psi(v))s} \big[v \big(Z_v^{(q)}(k) + \alpha W_v^{(q)}(k) \big) + q W_v^{(q)}(k) + \alpha W_v^{(q)}(k) \big] \, dS_s$$

then $\big(\widetilde{M}_{t \wedge \tau_{k,\epsilon}}, t \geqslant 0 \big)$ is a $\mathbb{P}_{R_0 = x}$-martingale, where $\epsilon \in (0, x)$ and $\tau_{k,\epsilon} = \inf\{t : R_t \notin (\epsilon, k)\}$. By choosing

$$\alpha = - \frac{v Z_v^{(q)}(k) + q W_v^{(q)}(k)}{v W_v^{(q)}(k) + W_v^{(q)'}(k)}$$

we obtain that $(M_{t \wedge \tau_{k,\epsilon}}, t \geqslant 0)$ is a $\mathbb{P}_{R_0 = x}$-martingale. Now $\tau_{k,\epsilon} \uparrow \tau_k$ as $\epsilon \to 0$ so that $R_{\tau_{k,\epsilon}} \to R_{\tau_k}$ by quasi left continuity. It is then clear that $M_{t \wedge \tau_{k,\epsilon}} \to M_{t \wedge \tau_k}$ in L^1 as $\epsilon \to 0$. $\square$

From the preceding martingale property, we deduce the joint Laplace transform of (τ_k, R_{τ_k}):

Corollary 3. *For all $v \geqslant 0$ and $q \geqslant 0$, we have:*

$$\mathbb{E}_{R_0 = x} \big[e^{-vR_{\tau_k} - (q + \psi(v))\tau_k} \big]$$
$$= e^{-vx} \left(Z_v^{(q)}(k - x) - \frac{v Z_v^{(q)}(k) + q W_v^{(q)}(k)}{v W_v^{(q)}(k) + W_v^{(q)'}(k)} \, W_v^{(q)}(k - x) \right). \quad (29)$$

Formula (29) was derived in [4] by means of excursion theory.

To complete the description of the process at time τ_k, we give the following trivariate law, which follows immediately from (29) after an Esscher transform:

Corollary 4. *For u, v, $q \geqslant 0$, we have:*

$$\mathbb{E} \big[e^{uX_{\tau_k} - vR_{\tau_k} - (q + \psi(u+v))\tau_k} \big]$$
$$= Z_{u+v}^{(q)}(k) - \frac{v Z_{u+v}^{(q)}(k) + q W_{u+v}^{(q)}(k)}{v W_{u+v}^{(q)}(k) + W_{u+v}^{(q)'}(k)} \, W_{u+v}^{(q)}(k). \quad (30)$$

Remark 7. It may also be of interest to study the first hitting time $\tau_a' = \inf\{t : S_t - X_t = a\}$. Note that $\tau_a' \geqslant \tau_a$ and that $S_{\tau_a'} = S_{\tau_a}$ a.s.. Applying the Markov property at time τ_a and using (30), one can then show that

$$\mathbb{E} \left[e^{-\alpha S_{\tau_a'} - q\tau_a'} \right] = e^{\Phi(q)a} \, \frac{W^{(q)'}(a) - \Phi(q)W^{(q)}(a)}{W^{(q)'}(a) + \alpha W^{(q)}(a)}.$$

4.3 A partial extension

We now generalize the previous discussion to more exit problems related to the reflected process. In this paragraph, we still assume that X is spectrally negative, and furthermore that it is a martingale. Let Θ be a left-continuous, decreasing function defined on $\mathbb{R}_+$, such that $\Theta(0+) = a \leqslant +\infty$ and $\Theta(+\infty) = 0$. As shown in [5], one can associate to Θ a centered probability measure m by the formula:

$$m\big([x, \infty)\big) = \exp\left\{-\int_{-\infty}^{x} \frac{\mathrm{d}\Xi^c(s)}{\Xi(s) - s}\right\} \times \prod_{s < x} \frac{\Xi(s) - s}{\Xi(s+) - s}$$

where the function Ξ is defined by $\Theta(x) = x - \Xi^{-1}(x)$. The function Ξ is an increasing, left-continuous function such that $\Xi(-\infty) = 0$ and $\Xi(y) = y \Rightarrow \Xi(x) = x$ for all $x \geqslant y$. Moreover, $\sup\{x : \Xi(x) = 0\} = -a$ and $\inf\{x : \Xi(x) = x\} = \inf\{x : \Theta(x) = 0\}$.

It is shown in [5] that Ξ is the Hardy–Littlewood transform of m. Conversely, to a probability measure m, one can associate its Hardy–Littlewood transform

$$\Xi_m(x) = \begin{cases} \dfrac{1}{m([x, \infty))} \displaystyle\int_{-\infty}^{x} t\, m(\mathrm{d}t) & \text{if } m([x, \infty)) > 0 \\[2ex] x & \text{otherwise,} \end{cases}$$

and a function $\Theta_m(x) = x - \Xi_m^{-1}(x)$ that has the properties of the function Θ.

Our goal here is to study the stopping time $T = \inf\{t : S_t - X_t \geqslant \Theta(S_t)\}$. We first prepare an intermediate result.

Lemma 5. *Assume that X has bounded jumps. Then, we have*

$$\mathbb{E}[X_T \mid S_T = s] = s - \frac{W(\Theta(s))}{W'(\Theta(s))}. \tag{31}$$

Proof. Set $g_T = \sup\{t < T : S_t = X_t\}$. The law of $(X_t, g_T < t < T)$ conditional on $\mathcal{F}_{g_T}$ is the same as the law of $(X_t, t < T^-(s - \Theta(s)))$, started at s, conditioned to stay below s. That is, we have:

$$\mathbb{E}[X_T \mid \mathcal{F}_{g_T}, S_T = s] = \lim_{\epsilon \downarrow 0} \mathbb{E}_{s - \epsilon}\big[X_{T^-(s - \Theta(s))} \,\big|\, T^-\big(s - \Theta(s)\big) < T^+(s)\big]$$

$$= \lim_{\epsilon \downarrow 0} \mathbb{E}\big[s - \epsilon + X_{T^-(\epsilon - \Theta(s))} \,\big|\, T^-\big(\epsilon - \Theta(s)\big) < T^+(\epsilon)\big].$$

On the one hand, we have from Example 3:

$$\mathbb{P}\big[T^-\big(\epsilon - \Theta(s)\big) < T^+(\epsilon)\big] = 1 - \frac{W(\Theta(s) - \epsilon)}{W(\Theta(s))}.$$

On the other hand, for a, $b > 0$, let $\tau = T^-(-a) \wedge T^+(b)$; since the jumps of X are bounded, $(X_{t \wedge \tau}, t \geq 0)$ is a uniformly integrable martingale, so that by the optional stopping theorem:

$$\mathbb{E}[X_\tau] = 0.$$

The foregoing can be rewritten

$$\mathbb{E}[X_\tau 1_{\tau = T^-(-a)}] + b\,\mathbb{P}[\tau = T^+(b)] = 0.$$

We deduce that

$$\mathbb{E}[X_{T^-(\epsilon - \Theta(s))} 1_{T^-(\epsilon - \Theta(s)) < T^+(\epsilon)}] = -\epsilon\,\frac{W(\Theta(s) - \epsilon)}{W(\Theta(s))}$$

and then

$$\mathbb{E}[X_T \,|\, \mathcal{F}_{g_T}, S_T = s] = \lim_{\epsilon \downarrow 0} \left(s - \epsilon - \epsilon\,\frac{W(\Theta(s) - \epsilon)}{W(\Theta(s))} \left(1 - \frac{W(\Theta(s) - \epsilon)}{W(\Theta(s))}\right)^{-1} \right)$$

$$= s - \frac{W(\Theta(s))}{W'(\Theta(s))}. \qquad \square$$

Next, we obtain the law of S_T.

Proposition 6. *The law of S_T is given by*

$$\mathbb{P}[S_T > x] = \exp\left(-\int_0^x \frac{W'(\Theta(y))}{W(\Theta(y))}\,dy\right). \tag{32}$$

Proof. Let us first assume that the jumps of X are bounded. Plainly, for every continuous function g with compact support, the local martingale $(M_t^g, t \leq T)$ (see (15)) given by

$$M_t^g = G(S_t) - (S_t - X_t)g(S_t)$$

where $G(x) = \int_0^x g(y)\,dy$, is uniformly integrable. Hence, for all such g

$$\mathbb{E}[G(S_T)] = \mathbb{E}[(S_T - X_T)g(S_T)].$$

Inserting the expression of $\mathbb{E}[X_T \,|\, S_T]$ given in the preceding lemma, we get:

$$\mathbb{E}[G(S_T)] = \mathbb{E}\left[g(S_T)\frac{W(\Theta(S_T))}{W'(\Theta(S_T))}\right].$$

We can solve this integral equation for the law of S_T and this gives the announced result.

The general case follows by approximation. $\qquad \square$

Remark 8. Applying standard excursion techniques, it is possible to obtain (a characterization of) the law of X_T. However the expression thus obtained seems to be quite complex, so that it would be difficult to exploit such a result.

The previous result enables us to show that the construction given in [5] of a solution to Skorokhod's embedding problem is specific to Brownian motion. Indeed let us consider a centered probability measure m. We associate to m its Hardy–Littlewood transform Ξ_m, as recalled at the beginning of this paragraph. Going along the line of [5], it is natural to consider the following stopping time as a candidate for the solution to Skorokhod's problem:

$$T'_m = \inf\{t : S_t - X_t = \Theta_m(S_t)\},$$

where $\Theta_m(x) = x - \Xi_m^{-1}(x)$ $(x \geqslant 0)$. The following result shows that the method of [5] generally fails except for Brownian motion.

Proposition 7. *Assume that $\Theta_m(\mathbb{R}_+) = \mathbb{R}_+$ and that for all continuous functions g with compact support, the local martingale $(M_t^g, t \leqslant T'_m)$ is uniformly integrable. Then X is Brownian motion.*

Proof. Under the assumption of the proposition, we have for all continuous functions g with compact support:

$$\mathbb{E}\big[G\big(S_{T'_m}\big)\big] = \mathbb{E}\big[g\big(S_{T'_m}\big)\Theta_m\big(S_{T'_m}\big)\big],$$

and as in the proof of Prop. 6, we deduce that the law of $S_{T'_m}$ is given by

$$\mathbb{P}\big[S_{T'_m} > x\big] = \exp\bigg(-\int_0^x \frac{dy}{\Theta_m(y)}\bigg).$$

However if $T_m = \inf\{t : S_t - X_t \geqslant \Theta_m(S_t)\}$, we have (32):

$$\mathbb{P}\big[S_{T_m} > x\big] = \exp\bigg(-\int_0^x \frac{W'(\Theta_m(y))}{W(\Theta_m(y))}\,dy\bigg).$$

Now, we have plainly $S_{T_m} = S_{T'_m}$, and comparing the two formulas above we obtain that

$$W(x) = xW'(x)$$

for all $x \in \Theta_m(\mathbb{R}_+) = \mathbb{R}_+$. This implies $W(x) = x$ and hence, $\psi(\lambda) = \lambda^2$, i.e. X is Brownian motion (more precisely, $X_t = \sqrt{2}\,B_t$, where B is a standard Brownian motion).   □

Remarks 9.

1. The condition $\Theta_m(\mathbb{R}_+) = \mathbb{R}_+$ is fulfilled if and only if the support of m is connected and is not bounded below.
2. A solution to Skorokhod's problem, in a general framework which embeds the case of Lévy processes, is developed in [8] using excursion theory; the stopping time obtained has an explicit, but not so simple expression.

The following example shows that even if the support of m is bounded below, the method of [5] is likely to work only for Brownian motion.

Example 5. For $a > 0$, set

$$m_a(\mathrm{d}x) = \frac{\mathrm{d}x}{a} \, 1_{x > -a} \, \mathrm{e}^{-\frac{1}{a}(x+a)}$$

so that $\Theta_a(x) := \Theta_{m_a}(x) = a$ for all x, and we have $T_{m_a} = \tau_a$, $T'_{m_a} = \tau'_a$. Remark 7 shows that the law of $S_{\tau_a} = S_{\tau'_a}$ is given by

$$\mathbb{P}\big[S_{\tau_a} > x\big] = \exp\left(-x \, \frac{W'(a)}{W(a)}\right).$$

If the martingales $(M^g_t, \, t \leqslant \tau'_a)$ are uniformly integrable (where $a > 0$ and g runs through continuous functions with compact support) we also have

$$\mathbb{P}\big[S_{\tau'_a} > x\big] = \mathrm{e}^{-x/a}.$$

We conclude again that $W(x) = x$ for all $x > 0$, hence X is Brownian motion.

Acknowledgement. Some of the developments of this paper were inspired by the thesis [18] of M. Pistorius. We also thank an anonymous referee whose remarks helped improve the presentation of the present paper.

References

1. S. Asmussen: Applied Probability and Queues. 2nd edition, Springer, Berlin Heidelberg New-York (2003).
2. S. Asmussen, F. Avram and M. Pistorius: Russian and American put options under exponential phase-type Lévy models. *Stoch. Proc. Appl.*, **109**, 79–112 (2004).
3. S. Asmussen and O. Kella: A multi-dimensional martingale for Markov additive processes and its applications. *Adv. Appl. Probab.*, **32**, 376–393 (2000).
4. F. Avram, A. Kyprianou and M. Pistorius: Exit problems for spectrally negative Lévy processes and applications to Russian, American and Canadized options. *Ann. Appl. Probab.*, **14**, 215–238 (2004).
5. J. Azéma and M. Yor: Une solution simple au problème de Skorokhod. In *Sém. Probab. XIII*, Lect. Notes Math. **721**, pages 91–115. Springer, Berlin Heidelberg New York (1979).
6. J. Bertoin: Lévy processes. Cambridge University Press, Cambridge (1996).
7. J. Bertoin: On the first exit-time of a completely asymmetric stable process from a finite interval. *Bull. London Math. Soc.*, **5**, 514–520 (1996).
8. J. Bertoin and Y. Le Jan. Representation of measures by balayage from a regular recurrent point. *Ann. Probab.*, **20**(1), 538–548 (1992).
9. N. Bingham. Fluctuation theory in continuous time. *Adv. Appl. Probab.*, **7**, 705–766 (1975).
10. L. Chaumont and M. Yor: Exercises in probability: a guided tour from measure theory to random processes, via conditioning. Cambridge University Press, Cambridge (2003).

11. R. A. Doney: Some excursion calculations for spectrally one-sided Lévy processes. In *Sém. Probab. XXXVIII*. Springer, Berlin Heidelberg New York (2004).
12. B. Fristedt: Sample functions of stochastic processes with stationary, independent increments. *Adv. Probab. Relat. Top.*, **3**, 241–396 (1974).
13. O. Kella and W. Whitt: Useful martingales for stochastic storage with Lévy input. *J. Appl. Probab.*, **29**, 396–403 (1992).
14. D. Kennedy. Some martingales related to cumulative sum tests and single-server queues. *Stoch. Proc. Appl.*, **4**, 261–269 (1976).
15. U. Küchler and S. Lauritzen: Exponential families, extreme point models and minimal space-time invariant functions for stochastic processes with stationary and independent increments. *Scand. J. Statist.*, **16**, 237–261 (1989).
16. A. Kyprianou and Z. Palmowski: Fluctuations of spectrally negative Markov additive processes (2003).
17. L. Nguyen-Ngoc: On the excursions of reflected Lévy processes. Prépublication 852, Laboratoire de Probabilités et Modèles Aléatoires, Universités Paris 6 et 7 (2003).
18. M. Pistorius: Exit problems of Lévy processes with applications in finance. PhD Thesis, Utrecht University (2003).
19. L. C. G. Rogers: A new identity for real Lévy processes. *Ann. Inst. Henri Poincaré, Probab. Stat.*, **20**, 21–34 (1984).
20. H. M. Taylor: A stopped Brownian motion formula. *Ann. Probab.*, **3**, 234–246 (1975).
21. D. Williams: On a stopped Brownian formula of H. M. Taylor. In *Sém. Probab. X*, Lect. Notes Math. **511**, pages 235–239. Springer, Berlin Heidelberg New York (1976).
22. V. M. Zolotarev. One-dimensional stable distributions. Number 65 in Translations of mathematical monographs. American Mathematical Society, Providence (1986).

Generalised Ornstein–Uhlenbeck Processes and the Convergence of Lévy Integrals*

K. Bruce Erickson and Ross A. Maller

Department of Mathematics,
University of Washington,
Seattle, WA, 98195, USA
e-mail: erickson@math.washington.edu
and
Centre for Financial Mathematics, MSI,
and School of Finance & Applied Statistics,
Australian National University, Canberra, Australia
e-mail: Ross.Maller@anu.edu.au

Summary. Exponential functionals of the form $\int_0^t e^{-\xi_{s-}} \, d\eta_s$ constructed from a two dimensional Lévy process (ξ, η) are of interest and application in many areas. In particular, the question of the convergence of the integral $\int_0^\infty e^{-\xi_{t-}} \, d\eta_t$ arises in recent investigations such as those of Barndorff–Nielsen and Shephard [3] in financial econometrics, and in those of Carmona, Petit and Yor [9], and Yor [40], [41], where it is related among other things to the existence of an invariant measure for a generalised Ornstein–Uhlenbeck process. We give a complete solution to the convergence question for integrals of the form $\int_0^\infty g(\xi_{t-}) \, d\eta_t$, when $g(t) = e^{-t}$ and η_t is general, or $g(\cdot)$ is a nonincreasing function and $d\eta_t = dt$, and some other related results. The necessary and sufficient conditions for convergence are stated in terms of the *canonical characteristics* of the Lévy process. Some applications in various areas (compound Poisson processes, subordinated perpetuities, the Doléans-Dade exponential) are also outlined.

2000 MSC Subject Classifications: primary: 60H05, 60H30, 60J30
secondary: 60J15, 60F15, 60K05

Key words: stochastic integral, Lévy process, subordinated perpetuity, compound Poisson process, Ornstein–Uhlenbeck-type process, Lamperti transformation.

1 Introduction and Results.

Stochastic integrals with functionals of Lévy processes as integrand and/or integrator occur in many areas. For a good introduction and background to the

---

* This work is partially supported by ARC Grant DP0210572.

kinds of problem we consider, see Carmona, Petit and Yor [9] and their references; some further examples and applications are given below. The authors of
[9] express great interest in exponential functionals of the form $\int_0^t e^{-\xi_s-} \, d\eta_s$,
where (ξ, η) is a two dimensional Lévy process. These are related to generalised
Ornstein–Uhlenbeck processes which have found application in mathematical
finance (option pricing [39], insurance and perpetuities [13], [33], risk theory
[32]) mathematical physics, and random dynamical systems. In particular, [9]
and [13] give results concerning the properties of the improper stochastic inte-
gral ("stochastic perpetuity") $\int_0^\infty e^{-\xi_{t-}} \, d\eta_t$, in some special cases, *under the
assumption that the integral converges*. See, e.g., Theorem 3.1 of [9], where it
is shown that the distribution of this integral (when it exists and is a.s. finite,
and when ξ and η are independent) is the unique invariant measure of the
generalised Ornstein–Uhlenbeck process having the distribution of

$$X_t = e^{-\xi_t} \left(x + \int_0^t e^{\xi_{s-}} \, d\eta_s \right), \qquad t \geqslant 0, \, x \in \mathbb{R}.$$

In this paper we give a complete solution to the question of convergence for
integrals of the type $\int_0^\infty e^{-\xi_{t-}} \, d\eta_t$, in the form of necessary and sufficient conditions for convergence stated in terms of the *canonical characteristics* of the
Lévy process, rather than in terms of the difficult-to-access one-dimensional
distributions of the process. Combined with Theorem 3.1 of [9], this completes
the existence problem for the invariant measure of the generalised Ornstein–
Uhlenbeck process in a very explicit way, in the situation of [9]. We go on to
consider some related integrals and other applications.

   To introduce and illustrate the methods and ideas to be used, we start
with a simpler, one-dimensional, setup. Our first theorem gives necessary and
sufficient conditions for convergence of the integral $\int_0^\infty g(\xi_t^+) \, dt$, for a positive
non-increasing function $g(\cdot)$. (Throughout, $x^+ = x \vee 0$ is the positive part of a
number x.) We assume ξ_t is a Lévy process with canonical triplet $(\gamma_\xi, \sigma_\xi^2, \Pi_\xi)$
(see below for the formal definitions and setting). Define, for $x > 0$,

$$\overline{\Pi}_\xi^+(x) = \Pi_\xi((x, \infty)) \quad \text{and} \quad \overline{\Pi}_\xi^-(x) = \Pi_\xi((-\infty, -x)), \qquad (1.1)$$

for the tails of Π, and the real function

$$A_\xi(x) = \gamma_\xi + \overline{\Pi}_\xi^+(1) + \int_1^x \overline{\Pi}_\xi^+(y) \, dy, \qquad x > 0. \qquad (1.2)$$

$A_\xi(x)$ is a kind of truncated mean related to the positive Lévy measure of
ξ, which occurs when estimating the orders of magnitudes of certain renewal
functions, as in Erickson [15], p. 377. When ξ drifts to $+\infty$ a.s., $A_\xi(x)$ is
positive for large enough $x > 0$, as we discuss below.

Theorem 1. *Suppose* $\lim_{t \to \infty} \xi_t = +\infty$ *a.s., so that* $A_\xi(x) > 0$ *for all* x
larger than some $a > 0$, *and that* $g(\cdot)$ *is a finite positive non-constant non-
increasing function on* $[0, \infty)$. *Then the integral*

$$\int_0^\infty g(\xi_t^+)\,\mathrm{d}t \tag{1.3}$$

is finite, or infinite, a.s., according as the integral

$$\int_{(a,\infty)} \left(\frac{x}{A_\xi(x)}\right) |\mathrm{d}g(x)| \tag{1.4}$$

converges or diverges.

The precursor of Theorem 1 is a result of Erickson [16] for renewal processes (increasing random walks), which we apply together with an asymptotic estimate for the rate of growth of the renewal measure of ξ_t in terms of $x/A_\xi(x)$ (Bertoin [4], p. 74) to prove Theorem 1. Bertoin's estimate in turn is a generalisation of a corresponding estimate for random walks due to Erickson [15]. These estimates can further be used to give easily stated necessary and sufficient conditions for $\lim_{t\to\infty} \xi_t = +\infty$ a.s; Doney and Maller [12] show that this is equivalent to

$$\int_1^\infty \overline{\Pi}_\xi^+(y)\,\mathrm{d}y = \infty \text{ and } J_- < \infty, \quad \text{or} \quad 0 < \mathbf{E}\,\xi_1 \leqslant \mathbf{E}\,|\xi_1| < \infty, \tag{1.5}$$

where

$$J_- = \int_{(1,\infty)} \left(\frac{x}{1 + \int_1^x \overline{\Pi}_\xi^+(y)\,\mathrm{d}y}\right) |\overline{\Pi}_\xi^-(\mathrm{d}x)|. \tag{1.6}$$

Analogous conditions characterise drift to minus infinity ($\lim_{t\to\infty} \xi_t = -\infty$ a.s.) or $\{\xi_t\}$ *oscillates* (a.s., $\limsup_{t\to\infty} \xi_t = +\infty$ and $\liminf_{t\to\infty} \xi_t = -\infty$). See also Sato [37]. When $\lim_{t\to\infty} \xi_t = +\infty$ a.s., the limit $\lim_{x\to\infty} A_\xi(x)$ exists and is positive; in fact, it is $+\infty$ or no smaller than $\mathbf{E}\,\xi_1 > 0$ according as the first or second condition in (1.5) holds. Bertoin [4], p. 100, also has a version of Erickson's [16] result, for subordinators. While a direct application of these arguments (estimating the order of growth of the renewal function when $\xi_t \to \infty$ a.s.) does not seem to work in the general case, nevertheless, we can prove Theorem 1 using the above results, and one of Pruitt [35].

The cases $\lim_{t\to\infty} \xi_t = -\infty$ a.s. or $\{\xi_t\}$ oscillates lead trivially to the divergence of the integral in (1.3), because our requirement that $g(0) > 0$ implies that the range of the integral will contain arbitrarily large intervals of t with $g(\xi_t^+) = g(0) > 0$, in these cases. Thus we have a complete description of the a.s. convergence of the integral in (1.3) in terms of the characteristics $(\gamma_\xi, \sigma_\xi^2, \Pi_\xi)$ of the Lévy process ξ_t. With these ideas in mind we can try to broaden the class of integrals which can be treated by similar methods. In particular, especially for the purposes of the applications already mentioned, we wish to allow integration with respect to a further Lévy process, η_t, say, which is not necessarily independent of ξ_t. We achieve complete generality in this direction when $g(t) = \mathrm{e}^{-t}$ (Theorem 2), and, to some extent, when $g(t)$ is a nonanticipating function that does not grow too fast (Theorem 3).

Before stating these results we need to set out our framework in a little more detail. We shall use a canonical setup $(\Omega, \mathcal{F}, \mathbf{P})$ for a bivariate Lévy process $W = (\xi, \eta)$. For Ω we take the space of càdlàg (right continuous with left limits, also called ricowill) paths $\omega : [0, \infty) \mapsto \mathbb{R}^2$, such that $\omega(0) = (0,0)$. Then $W(t, \omega) = \omega(t) = (\xi(t, \omega), \eta(t, \omega))$. We let $\mathcal{F}_t^0$ be the natural filtration generated by the random variables $\{W_s; 0 \leqslant s \leqslant t\}$ and $\mathcal{F}^0$ the σ-field generated by $\{W_s, 0 \leqslant s < \infty\}$. Also $\mathcal{F}$ is the completion of $\mathcal{F}^0$ and $\{\mathcal{F}_t\}$ is the usual augmentation of $\{\mathcal{F}_t^0\}$ making it a filtration satisfying the "usual hypotheses". The process W is an infinite lifetime Lévy process with respect to the probability measure $\mathbf{P}$. Its characteristic exponent, $\psi(\theta) := -(1/t) \log \mathbf{E} \exp(\mathrm{i}\langle \theta, W_t \rangle)$, will be written in the form:

$$\psi(\theta) = -\mathrm{i}\langle \tilde{\gamma}, \theta \rangle + \tfrac{1}{2}\langle \theta, \Sigma\theta \rangle + \iint_{|w| \leqslant 1} \left(1 - e^{\mathrm{i}\langle w, \theta \rangle} + \mathrm{i}\langle w, \theta \rangle \right) \Pi\{dw\}$$

$$+ \iint_{|w| > 1} \left(1 - e^{\mathrm{i}\langle w, \theta \rangle}\right) \Pi\{dw\}, \quad \text{for } \theta \in \mathbb{R}^2. \quad (1.7)$$

In (1.7), the $\langle \cdot, \cdot \rangle$ denotes inner product in $\mathbb{R}^2$, $\tilde{\gamma} = (\tilde{\gamma}_1, \tilde{\gamma}_2)$ is a nonstochastic vector in $\mathbb{R}^2$, and $\Sigma = (\sigma_{rs})$, $r, s = 1, 2$, is a nonstochastic 2×2 non-negative definite matrix. Finally, Π, the Lévy measure, is a measure on $\mathbb{R}^2 \setminus \{0\}$. It is uniquely determined by the process and satisfies $\iint \min(|w|^2, 1)\, \Pi\{dw\} < \infty$. See Bertoin [4], p. 3, Sato [37], Ch. 2.

Corresponding to (1.7) is the representation of W as a sum of four mutually independent processes:

$$W_t = (\xi_t, \eta_t) = (\gamma_\xi t, \gamma_\eta t) + (B_t, C_t) + \left(X_t^{(1)}, Y_t^{(1)}\right) + \left(X_t^{(2)}, Y_t^{(2)}\right). \quad (1.8)$$

Each of these four processes is itself a bivariate Lévy process, adapted to the same filtration as $\{W_t\}$. In detail:

(i) $\{(\gamma_\xi t, \gamma_\eta t)\}$, where γ_ξ and γ_η (not usually the same as $\tilde{\gamma}_1, \tilde{\gamma}_2$, see (1.11) below) are (non-stochastic) constants, is the drift term.

(ii) $\{(B_t, C_t)\}$ is a Brownian motion on $\mathbb{R}^2$ with mean $(0, 0)$ and covariance matrix $t\Sigma$. We denote the individual variances by $\sigma_\xi^2 t$ and $\sigma_\eta^2 t$, respectively.

(iii) $\{(X_t^{(1)}, Y_t^{(1)})\}$ is a discontinuous process with jumps of magnitude not ever exceeding the value 1. It may have infinitely many jumps in every time interval. It has finite moments of all orders (indeed, each component has a finite exponential moment; see the end of the proof of Protter [34], Theorem 34, Ch. I, p. 24), and $\mathbf{E}\, X_t^{(1)} = \mathbf{E}\, Y_t^{(1)} \equiv 0$.

(iv) $\{(X_t^{(2)}, Y_t^{(2)})\}$ is a discontinuous jump process with jumps of magnitude always exceeding the value 1. The sample functions of this process are ordinary step functions; it is a bivariate compound Poisson process. The component processes, too, are compound Poisson processes, not necessarily independent of each other. Their jump measures need not be concentrated on $(-\infty, -1] \cup [1, \infty)$ but they do have finite total mass. These processes may have infinite expectations.

We will denote the sum of the two pure jump Lévy processes as $\{(X_t, Y_t)\}$. Thus

$$\Delta W_t = (\Delta X_t, \Delta Y_t) = (X_t - X_{t-}, Y_t - Y_{t-}), \qquad t > 0,$$

denotes the jump process of W. If A is a Borel subset of $\mathbb{R}^2 \setminus \{(0,0)\}$, then $\Pi\{A\}$ equals the expected number of jumps of W of (vector) magnitude in A occuring during any unit time interval, i.e.

$$\Pi\{A\} = \mathbf{E} \sum_{t < s \leqslant t+1} 1_{\{\Delta W_s \in A\}}. \tag{1.9}$$

(The expectation does not depend on t.)

The component Lévy processes ξ_t and η_t have canonical triplets given by $(\gamma_\xi, \sigma_\xi^2, \Pi_\xi)$ and $(\gamma_\eta, \sigma_\eta^2, \Pi_\eta)$, where

$$\Pi_\xi\{B\} := \int_{\mathbb{R}} \Pi\{B, dy\} \quad \text{and} \quad \Pi_\eta\{B\} := \int_{\mathbb{R}} \Pi\{dx, B\}, \tag{1.10}$$

for B a Borel subset of $\mathbb{R} \setminus \{0\}$, and

$$\gamma_\xi := \tilde{\gamma}_1 + \int_{|x| \leqslant 1} x \int_{|y| > \sqrt{1-x^2}} \Pi\{dx, dy\}, \tag{1.11}$$

and similarly for γ_η. In addition to the $\overline{\Pi}_\xi^+$, $\overline{\Pi}_\xi^-$ introduced in (1.1), we will write, for $x > 0$,

$$\overline{\Pi}_\xi(x) = \overline{\Pi}_\xi^+(x) + \overline{\Pi}_\xi^-(x)$$

for the tailsum of Π_ξ. Similar notations are defined for Π_η. Recall the definition (1.2) of $A_\xi(x)$. Note that the functions $x \mapsto A_\xi(x)$ and $x \mapsto x/A_\xi(x)$ do not decrease, for large enough x, when (1.5) holds.

Our next theorem examines the convergence of $\int_0^\infty e^{-\xi_{t-}} d\eta_t$. For stochastic integrals like this, or, more generally, $\int_A^B g(\xi_{s-}) d\eta_s$, where $g(\cdot)$ is a measurable function, we will take the definition(s) as found in Protter [34]. Throughout, we will assume that neither ξ_t nor η_t degenerate at 0.

Theorem 2. *Suppose $A_\xi(x) > 0$ for all x larger than some $a > 0$. If*

$$\lim_{t \to \infty} \xi_t = +\infty \text{ a.s., and } I_{\xi,\eta} := \int_{(e^a, \infty)} \left(\frac{\log y}{A_\xi(\log y)} \right) |\overline{\Pi}_\eta\{dy\}| < \infty, \tag{1.12}$$

then

$$\mathbf{P}\left(\lim_{t \to \infty} \int_0^t e^{-\xi_{s-}} d\eta_s \text{ exists and is finite} \right) = 1. \tag{1.13}$$

Conversely, if (1.12) fails, then (1.13) fails.

In greater detail we have: if $\lim_{t\to\infty} \xi_t = +\infty$ *a.s. but* $I_{\xi,\eta} = \infty$, *then*

$$\left| \int_0^t e^{-\xi_{s-}}\, d\eta_s \right| \xrightarrow{P} \infty \text{ as } t \to \infty. \tag{1.14}$$

If on the other hand ξ_t *does not tend to* $+\infty$ *a.s. as* t *tends to infinity, then either (1.14) holds or there exists a constant* $k \in \mathbb{R} \setminus \{0\}$ *such that*

$$\mathbf{P}\left(\int_0^t e^{-\xi_{s-}}\, d\eta_s = k(1 - e^{-\xi_t}) \,\forall\, t > 0 \right) = 1, \tag{1.15}$$

and, a.s., the integral $\int_0^t e^{-\xi_{s-}}\, d\eta_s$ *again fails to converge as* $t \to \infty$.

Some brief comments are in order.

As observed following (1.5), $\lim_{t\to\infty} \xi_t = +\infty$ a.s. implies $\lim_{x\to\infty} A_\xi(x)$ exists and is not smaller than $\mathbf{E}\,\xi_1 \in (0,\infty]$, and thus the condition $A_\xi(x) > 0$ for large enough x is automatically satisfied in the case of interest. In the converse, this condition is not in fact needed if the denominator in $I_{\xi,\eta}$ in (1.12) is replaced by, e.g., $1 + \int_a^{\log y} \overline{\Pi}_\xi^+(y) dy$. Further, if $\mathbf{E}\,|\xi_1| < \infty$, the denominator is actually irrelevant to the convergence of the integral, and $I_{\xi,\eta}$ then converges if and only if $\int_1^\infty \log y\, |\overline{\Pi}_\eta\{dy\}|$ converges. This is thus a sufficient but not in general necessary condition for convergence of $\int_0^\infty e^{-\xi_{t-}}\, d\eta_t$.

The converse in Theorem 2 can be clarified somewhat as follows. First, (1.12) can fail in three ways: (i) $\xi_t \to \infty$, a.s., but the integral $I_{\xi,\eta}$ diverges; (ii) $\xi_t \to -\infty$ a.s.; (iii) $\{\xi_t\}$ oscillates. In the first two cases the theorem tells us, via (1.15) in the second case, that (1.14) holds. In the third case it says only that (1.14) holds or (1.15) holds. These last two possibilities are not mutually exclusive; in fact (1.15) implies (1.14) when (and only when) $\xi_t \to -\infty$ in probability. (A non-trivial example for which (1.13) and (1.14) both fail is in Case 2 below.)

The divergence of the integral can be further analysed as follows.

Proposition 1. *Suppose* $\{\xi_t\}$ *oscillates and (1.15) holds. Then:*
(i) *we have*

$$0 \leqslant \liminf_{t\to\infty} \left| \int_0^t e^{-\xi_{s-}}\, d\eta_s \right| < \limsup_{t\to\infty} \left| \int_0^t e^{-\xi_{s-}}\, d\eta_s \right| = \infty \text{ a.s.}; \tag{1.16}$$

(ii) *if, in addition,* $\Pi_\eta = 0$, *we have* $\xi_t = B_t$ *and* $\eta_t = k(B_t - \sigma_\xi^2 t/2)$ *for some* $k \neq 0$, *where* $\sigma_\xi > 0$, *and (1.16) holds;*
(iii) *if, instead,* $\Pi \neq 0$, *the support of* Π *must lie in a curve of the form*

$$\{(x,y) : y + ke^{-x} = k\}, \quad \text{for some } k \in \mathbb{R}. \tag{1.17}$$

The situations in (1.15) and (1.17) are kinds of degenerate cases, but they can be important, and they do not preclude (1.14) in general. Suppose for example that $\{\xi_t\}$ oscillates, and the support of Π degenerates with positive mass only at a point (x_0, y_0) with $|x_0| > 1$, $|y_0| > 1$. Then (1.17) holds with $k = y_0/(1 - e^{-x_0})$. To simplify further, suppose the Brownian components are absent and $\gamma_\eta = 0$. Then $(\xi_t, \eta_t) = (\gamma_\xi t + x_0 N_t, y_0 N_t)$, where N_t is a Poisson process of rate λ, say, with jumps at times $T_1 < T_2 < \cdots$. For $\{\xi_t\}$ to oscillate we must have $\gamma_\xi + x_0\lambda = 0$, so $\gamma_\xi \neq 0$. Now, with $T_0 = 0$ and $\Delta T_j = T_j - T_{j-1}$,

$$
\begin{aligned}
\int_0^t e^{-\xi_{s-}} \, d\eta_s &= \int_0^t e^{-\gamma_\xi s + \gamma_\xi N_{s-}/\lambda} d(y_0 N_s) \\
&= y_0 \sum_{i=1}^{N_t} \exp\left(-\gamma_\xi \sum_{j=1}^i \Delta T_j + \gamma_\xi(i-1)/\lambda\right) \\
&= y_0 e^{-\gamma_\xi/\lambda} \sum_{i=1}^{N_t} \left(\prod_{j=1}^i e^{-\gamma_\xi(\Delta T_j - 1/\lambda)}\right).
\end{aligned}
\tag{1.18}
$$

This can be shown to satisfy (1.14) by virtue of the divergence part of Theorem 2.1 of Goldie and Maller [19], which applies to discrete time perpetuities. To see this, apply that result to the i.i.d. sequence with typical member $(Q_1, M_1) = (1, e^{-\gamma_\xi(T_1 - 1/\lambda)})$; the condition $\mathbf{P}(Q_1 + M_1 c = c) = 1$ in that theorem does not hold for any $c \in \mathbb{R}$, despite the fact that (1.17) holds. This example shows that the continuous and discrete time results can have distinctly different features. (See also the example in Case 3 below.)

In view of (1.18), it is tempting at this point to suggest deriving Theorem 2 in general by a direct application of the discrete time perpetuity results in [19] to a discretisation of the integral $Z_t := \int_0^t e^{-\xi_{s-}} \, d\eta_s$ (the discrete skeleton $\{Z_n\}_{n=1,2,\ldots}$, for example), and then "filling in the gaps" between the discrete points in some way. There are at least two difficulties with this approach. In the first place, "filling in the gaps" to infer the a.s. convergence of Z_t (or divergence in probability of $|Z_t|$ to infinity) from that of the subsequence Z_n is not trivial. But more importantly, the discrete time perpetuity would be constructed from components (Q_i, M_i) distributed as $\left(e^{-\xi_{1-}}, \int_0^1 e^{-\xi_{s-}} \, d\eta_s\right)$, or similar, so direct application of the results of [19] would give uncheckable conditions phrased in terms of the marginal distributions of these components, which are quite inaccessible. We do in fact reduce the problem to that of [19] eventually, but not by a direct discretisation. The approach adopted herein seems to be the most efficient way to get conditions phrased in terms of the characteristics of the Lévy process.

We now analyse some other special cases of interest.

Case 1. *Suppose Π_ξ is degenerate at 0*, i.e. Π has support in the y-axis. Then $\xi_t = \gamma_\xi t + B_t$ has no jump component – it is Brownian motion with drift. It can diverge to ∞ a.s. only if $\gamma_\xi > 0$. By (1.2), $A_\xi(x) = \gamma_\xi$, so Theorem 2 gives: $\gamma_\xi > 0$ and $\int_{(1,\infty)} \log y \, |\overline{\Pi}_\eta\{dy\}| < \infty$ implies $\int_0^\infty e^{-\gamma_\xi t - B_t} \, d\eta_t$ exists and is finite a.s. In particular, B_t may be absent here (by taking $\sigma_\xi = 0$), in which case the a.s. existence of $\int_0^\infty e^{-\gamma_\xi t} d\eta_t$ when $\gamma_\xi > 0$ and $\int_{(1,\infty)} \log y \, |\overline{\Pi}_\eta\{dy\}| < \infty$ follows.

The support of Π, contained in $\{(x, y) : x = 0\}$, is not of the form in (1.17) for any $k \in \mathbb{R}$, as long as Π_η is non-trivial. The converse part of Theorem 2 together with Proposition 1 thus tells us that if $\gamma_\xi > 0$ and $\int_{(1,\infty)} \log y \, |\overline{\Pi}_\eta\{dy\}| = \infty$, or if $\gamma_\xi \leqslant 0$ (thus, ξ_t does not tend to ∞ a.s.) and $\Pi_\eta \neq 0$, then, as $t \to \infty$, $\left| \int_0^t e^{-\gamma_\xi s - B_s} d\eta_s \right| \xrightarrow{P} \infty$. In this example the failure of (1.17) rules out (1.15) and so establishes (1.14).

Case 2. *Suppose Π_η is degenerate at 0.* The forward direction of Theorem 2 then gives that $\lim_{t \to \infty} \xi_t = +\infty$, a.s., implies the convergence of

$$\int_0^\infty e^{-\xi_t} d\eta_t = \int_0^\infty e^{-\xi_t} (\gamma_\eta \, dt + dC_t).$$

Since we may have $\gamma_\eta = 0$ or C_t absent here, we have the a.s. convergence of $\int_0^\infty e^{-\xi_t} dt$ and $\int_0^\infty e^{-\xi_t} dC_t$, when $\xi_t \to \infty$ a.s. In particular, $\int_0^\infty e^{-\gamma_\xi t - B_t} dt$ and $\int_0^\infty e^{-\gamma_\xi t - B_t} dC_t$ converge a.s. when $\gamma_\xi > 0$. This covers subordinated perpetuities, constructed from Brownian motions, where it is often assumed in addition (unnecessarily, for purposes of convergence) that B_t and C_t are independent. The distribution of the first of these integrals, $\int_0^\infty e^{-\gamma_\xi t - B_t} dt$, when $\gamma_\xi > 0$, is known quite explicitly; it is a multiple of the reciprocal of a Gamma random variable [13]. The second integral, $\int_0^\infty e^{-\gamma_\xi t - B_t} dC_t$, has a Pearson Type IV distribution when B and C are independent ([9], Prop. 3.2).

The converse part of Theorem 2 tells us in this case (Π_η degenerate at 0) that both $\int_0^\infty e^{-\gamma_\xi t - B_t} dt$ and $\int_0^\infty e^{-\gamma_\xi t - B_t} dC_t$ diverge in case $\gamma_\xi \leqslant 0$, as we would expect. In this case ($\gamma_\xi \leqslant 0$), we have, as $t \to \infty$, $\left| \int_0^t e^{-\gamma_\xi s - B_s} ds \right| \xrightarrow{P} \infty$, but not necessarily $\left| \int_0^t e^{-\xi_s} dC_s \right| \xrightarrow{P} \infty$; we could have the situation in Proposition 1, with $\xi_t = B_t$ and $\eta_t = -\sigma_\xi^2 t/2 + B_t$, and then (1.15) holds with $k = 1$, but not (1.14). The support of Π for this case is the x–axis, $\{(x, y) : y = 0\}$, which satisfies (1.17) for $k = 0$.

Case 3. A *bivariate compound Poisson process* is constituted from i.i.d. 2-vectors (Q_i, M_i) added together at the jump times of an independent Poisson process of rate λ, say. Suppose $\mathbf{P}\{Q_1 = 0\} < 1$ and $\mathbf{P}\{M_1 = 0\} = 0$. Let N_t be the number of jumps of the Poisson process in the time interval $(0, t]$, with $N_0 = 0$. The process

$$W_t = (\xi_t, \eta_t) = \sum_{i=1}^{N_t} (-\log|M_i|, Q_i), \qquad t \geqslant 0,$$

(with $W_0 = 0$) is a Lévy process with canonical triplet $(0, 0, \Pi)$, where Π is λ times a probability measure: $\Pi(A) = \lambda \mathbf{P}\{(-\log|M_1|, Q_1) \in A\}$ for Borel $A \subseteq \mathbb{R}^2$. So $\Pi_\xi(dm)$ and $\Pi_\eta(dq)$ are proportional to (in fact, also λ

times) the probability measures $\mathbf{P}\{-\log|M_1| \in \mathrm{d}m\}$ and $\mathbf{P}\{Q_1 \in \mathrm{d}q\}$. The corresponding stochastic integral is

$$Z_t = \int_0^t \mathrm{e}^{-\xi_{s-}}\,\mathrm{d}\eta_s = \sum_{i=1}^{N_t}\left(\prod_{j=1}^{i-1}|M_j|\right)Q_i, \qquad t \geqslant 0, \qquad (1.19)$$

and is the value of the discrete time perpetuity

$$\tilde{Z}_n := \sum_{i=1}^{n}\left(\prod_{j=1}^{i-1}|M_j|\right)Q_i, \qquad n = 1,2,\ldots$$

at the random time N_t. It follows from Theorem 2 that, if

$$\xi_t = -\log\left(\prod_{j=1}^{N_t}|M_j|\right) \to \infty \quad \text{a.s. as } t \to \infty, \qquad (1.20)$$

or, equivalently, if $\lim_{n\to\infty}\prod_{j=1}^{n}|M_j| = 0$ a.s., and the expression $\mathbf{E}_Q :=$ $\mathbf{E}\left(\log^+|Q_1|/A_\xi(\log^+|Q_1|)\right)$ is finite, where

$$A_\xi(x) = \lambda\,\mathbf{P}\{-\log|M_1| > 1\} + \lambda\int_1^x \mathbf{P}\{-\log|M_1| > y\}\,\mathrm{d}y,$$

then $Z_t \to Z_\infty$ a.s. as $t \to \infty$ for an a.s. finite Z_∞; while if $\mathbf{P}\{Q_1 + k|M_1| = k\} < 1$ for all $k \in \mathbb{R}$, so that (1.17) does not hold, and (1.20) fails, or else (1.20) holds and $\mathbf{E}_Q = \infty$, then $|Z_t| \overset{P}{\to} \infty$ as $t \to \infty$. This is an analogue of the discrete time theorem of [19]. Our proof of Theorem 2 employs at one stage a reduction to this case.

Comparing this example to the one following Proposition 1 highlights the differences that the continuous time setting can introduce. The "interesting" aspect of that example arises from the drift coefficient, γ_ξ. The continuous time analysis also opens the way to investigation of further expressions obtained from integration by parts, etc.

The analogy with [19] is not complete in one respect; notice that we have $|M_j|$ rather than M_j in (1.19). Theorem 2 as it stands does not allow for a signed integrand. However, we have the following.

Theorem 3. *Let* $f = \{f_t\}$ *be a locally predictable square integrable functional of the Lévy process* $W = (\xi,\eta)$ *satisfying* $\lim_{t\to\infty}\mathrm{e}^{-\varepsilon t}\,\mathbf{E}\,f_t^2 = 0$ *and* $\lim_{t\to\infty}\mathrm{e}^{-\varepsilon t}f_t = 0$ *a.s. for each* $\varepsilon > 0$. *Then, under* (1.12), *the integral*

$$\int_0^\infty f_t\,\mathrm{e}^{-\xi_{t-}}\,\mathrm{d}\eta_t \qquad (1.21)$$

exists and is finite a.s.

Remarks. In the case of *convergence* – when the integral $Z_\infty := \int_0^\infty e^{-\xi_{t-}} d\eta_t$ exists finitely – we can ask for its properties. [9], [13], [39], [40], [41] and others investigate this in particular when the integrator and/or integrand are Brownian, giving detailed distributional information (or even an explicit expression for the distribution). For more general Lévy integrators and/or integrands, we would not expect to obtain such specific information, but the tail behaviour, at least, of Z_∞ can be investigated by the methods of Kesten [26] and Goldie [18]. From their results we would conjecture a heavy tailed (power law) behaviour of Z_∞ under some assumptions. See also Grincevičius [22, 23, 24]. In the discrete case, other kinds of properties (e.g., continuity of the distribution of the limit) have been studied by Grincevičius [20], [22] and Vervaat [38].

In the case of *divergence* – we can ask how quickly the integral Z_t diverges, in some sense, as $t \to \infty$. In the discrete case, Grincevičius [21] shows how to norm Z so as to obtain a finite, nonzero limit. One can use his results and discretisation to obtain some information on the magnitude of the stochastic integral in the case of divergence, but further refined results along these lines would require a deeper analysis. For another approach in discrete time, see Babillot, Bougerol and Elie [1] and Brofferio [8].

2 Applications

1. Dominance of a Lévy process over its jumps

For any Lévy process ξ_t, an extension of Exercise 6(a) on page 100 of Bertoin [4] shows that, for an increasing function $f : [0, \infty) \mapsto [1, \infty)$, we have

$$\mathbf{P}\{\Delta\xi_t > f(\xi_{t-}) \text{ i.o. as } t \to \infty\} = 0 \tag{2.1}$$

iff

$$\int_0^\infty \overline{\Pi}_\xi^+(f(\xi_t)) \, dt < \infty \quad \text{a.s.} \tag{2.2}$$

Suppose in addition that $\xi_t \to \infty$ a.s. as $t \to \infty$. Applying Theorem 1 with $g(x) = \overline{\Pi}_\xi^+(f(x))$ shows that these are equivalent to

$$\int_1^\infty \left(\frac{x}{A_\xi(x)}\right) |d\overline{\Pi}_\xi^+(f(x))| < \infty, \tag{2.3}$$

or to

$$\int_e^\infty \left(\frac{\log x}{A_\xi(\log x)}\right) d\Pi_\xi(f(\log x)) < \infty. \tag{2.4}$$

This is of the form in Theorem 2 if we define η_t by

$$\eta_t = \sum_{0 < s \leqslant t} e^{f^-(\Delta\xi_s)} 1_{\{\Delta\xi_s > f(0)\}},$$

where

$$f^{\leftarrow}(x) = \sup\{z : f(z) \leqslant x\}, \qquad x \geqslant 1,$$

is a generalised inverse of f. Then η_t is a Lévy process with $\Pi_\eta(x) = \Pi_\xi(f(\log^+ x))$, and (2.4) shows that (2.3) holds iff

$$\int_e^\infty \left(\frac{\log x}{A_\xi(\log x)} \right) \Pi_\eta\{dx\} < \infty. \tag{2.5}$$

The process $W_t = (\xi_t, \eta_t)$ is a bivariate Lévy process with marginal measures Π_ξ, Π_η. Assuming $\Pi_\xi \neq 0$, Theorem 2 applies to show that (2.5) and hence (2.1) holds if and only if $\int_0^\infty e^{-\xi_{t-}} d\eta_t$ converges a.s.

When f grows at most algebraically, or, more generally, if f is nondecreasing and $\limsup_{x \to \infty} f(\lambda x)/f(x) < \infty$ for each $\lambda > 1$, then (2.1) is equivalent to

$$\limsup_{t \to \infty} \left(\frac{\Delta \xi_t}{f(\xi_{t-})} \right) \leqslant 0 \quad \text{a.s.},$$

and this is then equivalent to (2.5). In particular, when $f(x) = x \vee 1$, then (2.3) is equivalent to $A_\xi(\infty) < \infty$, thus to $\int_1^\infty \overline{\Pi}_\xi^+(x) \, dx < \infty$, which is thus equivalent to $\limsup_{t \to \infty} (\Delta \xi_t/\xi_{t-}) \leqslant 0$ a.s. Other results like this are easily worked out.

2. Integrating the Doléans-Dade Exponential

The Doléans-Dade Exponential ([11]) of a Lévy process L_t which has canonical triplet $(\gamma_L, \sigma_L^2, \Pi_L)$, is

$$\mathcal{E}(L)_t = e^{L_t - \frac{1}{2}[L,L]_t} \prod_{0 < s \leqslant t} (1 + \Delta L_s) e^{-\Delta L_s + \frac{1}{2}(\Delta L_s)^2},$$

where

$$[L, L]_t = \sigma_L^2 t + \sum_{0 < s \leqslant t} (\Delta L_s)^2$$

is the quadratic variation process. Suppose that $\Pi_L(\cdot)$ attributes no mass to $(-\infty, -1]$. Then we can write $\mathcal{E}(L)_t = e^{-\xi_t}$, where

$$\xi_t = -L_t + (\sigma_L^2/2)t - \sum_{0 < s \leqslant t} (\log(1 + \Delta L_s) - \Delta L_s)$$

is a Lévy process. It follows that $\Delta \xi_t = -\log(1 + \Delta L_t)$, so if L_t is coupled with η_t in a bivariate Lévy process, whose Lévy measure is $\Pi_{L,\eta}$, then (ξ, η) is a bivariate Lévy process with measure satisfying (for $x > 0$, $y > 0$)

$$\Pi_{\xi,\eta}\{(x,\infty) \times (y,\infty)\} = \mathbf{E} \sum_{0 < s \leqslant t} 1_{\{\Delta \xi_s > x, \Delta \eta_s > y\}}$$

$$= \Pi_{L,\eta}\{(-\infty, e^{-x} - 1) \times (y, \infty)\}. \tag{2.6}$$

Thus $\overline{\Pi}_\xi^+(x) = \overline{\Pi}_L^-(1 - e^{-x})$, and, similarly, $\overline{\Pi}_\xi^-(x) = \overline{\Pi}_L^+(e^x - 1)$, for $x > 0$. The support of $\Pi_{\xi,\eta}$ lies on the curve $y + ke^{-x} = k$ iff $\Delta\eta_t + ke^{-\Delta\xi_t} = k$, iff $\Delta\eta_t + k\Delta L_t = 0$, or, equivalently, the support of $\Pi_{L,\eta}$ lies on the curve $y = -kx$, which we assume not to be the case for any $k \in \mathbb{R}$. Corresponding to (1.2), define

$$A_\xi(x) = -\gamma_L + \sigma_L^2/2 + \overline{\Pi}_\xi^+(1) + \int_1^x \overline{\Pi}_\xi^+(y)\,\mathrm{d}y, \qquad x > 0.$$

Suppose also that $\xi_t \to \infty$ a.s. as $t \to \infty$. Then we have

$$\int_0^\infty e^{-\xi_{t-}}\,\mathrm{d}\eta_t = \int_0^\infty \mathcal{E}(L)_{t-}\,\mathrm{d}\eta_t$$

converges a.s. iff

$$\int_e^\infty \left(\frac{\log x}{A_\xi(\log x)}\right)|\overline{\Pi}_\eta\{\mathrm{d}x\}| < \infty. \tag{2.7}$$

The expressions in (1.5) and (1.6), which characterise $\lim_{t\to\infty}\xi_t = \infty$ a.s., and in (2.7) are easily written in terms of the Lévy measure of $\Pi_{L,\eta}$, using (2.6).

3. Other applications and extensions

As mentioned, subordinated perpetuities occur in various applications, especially in Finance (e.g., [3], [9], [40], [41]) and in time series modelling (e.g., Brockwell [6], [7]). Convergence of the integral in the latter case corresponds to the existence of a stationary version of the series which is of importance in modelling applications. In the former case, the distribution of the integral, when it converges, is of interest. As remarked at the end of Section 1, in the general Lévy process situation little is known in this direction. For an application of the discrete time results to branching random walks, see Iksanov [25] and his references.

For another kind of extension, representations of fractional Brownian motion, ν_t, say, in terms of integrals of certain kernel functions against ordinary Brownian motions can be used to examine the convergence of integrals like $\int_0^\infty g(\xi_t)\,\mathrm{d}\nu_t$. (These can be defined even though fractional Brownian motion is not a semimartingale; see [14], [29], [31].) Theorem 3 may be of use here.

For many other applications of Theorem 2 in a variety of areas, see [2], and also Klüppelberg et al. [28]. Theorem 3 has applications in the insurance area (e.g., [30]).

3 Proofs

Proof of Theorem 1. Suppose first that ξ_t is a subordinator (so that $\lim_{t\to\infty}\xi_t = +\infty$ a.s.). Obviously $\mathbf{E}\int_0^\infty g(\xi_t)\,\mathrm{d}t < \infty$ implies that the integral $\int_0^\infty g(\xi_t)\,\mathrm{d}t$ is finite a.s. Conversely, assume $\int_0^\infty g(\xi_t)\,\mathrm{d}t$ is finite with positive probability. Writing

$$\xi_t = S_{\lfloor t \rfloor} + (\xi_t - \xi_{\lfloor t \rfloor}) = S_{\lfloor t \rfloor + 1} - (\xi_{\lfloor t \rfloor + 1} - \xi_t),$$

where $\lfloor t \rfloor$ is the integer part of t and $S_n = \sum_1^n (\xi_i - \xi_{i-1})$ is a renewal process, we have $S_{\lfloor t \rfloor} \leqslant \xi_t \leqslant S_{\lfloor t \rfloor + 1}$, so we see that $\int_0^\infty g(S_{\lfloor t \rfloor}) \, dt$ and hence $\sum_{n \geqslant 1} g(S_n)$ are finite with positive probability. Applying Prop. 1 of Erickson [16] gives $\mathbf{E} \sum_{n \geqslant 1} g(S_n)$ finite and thus $\mathbf{E} \int_0^\infty g(\xi_t) \, dt$ finite. Hence $\mathbf{P}\big(\int_0^\infty g(\xi_t) \, dt < \infty\big)$ is 0 or 1 according as $\mathbf{E} \int_0^\infty g(\xi_t) \, dt$ is finite or infinite. But

$$\mathbf{E} \int_0^\infty g(\xi_t) \, dt = \int_0^\infty \int_0^\infty g(y) \, \mathbf{P}(\xi_t \in dy) \, dt = \int_0^\infty g(y) \, U(dy) \qquad (3.1)$$

where $U(dy) = \int_0^\infty \mathbf{P}(\xi_t \in dy) \, dt$. From Bertoin [4], p. 74,

$$U(x) \asymp \frac{x}{\int_1^x \overline{\Pi}_\xi(y) \, dy} \asymp \frac{x}{A_\xi(x)} \qquad (\text{as } x \to \infty), \qquad (3.2)$$

where "$\asymp$" means that the ratio of the two sides is bounded away from 0 and ∞ as $x \to \infty$. So the integral in (1.4) converges if and only if the integral in (3.1) does, as we see after integrating by parts (and using the monotonicity properties of $g(x)$ and $U(x)$). Thus Theorem 1 is true for subordinators.

Next, for a general ξ we assume $\lim_{t \to \infty} \xi_t = +\infty$ a.s. Consider two cases. If $\mathbf{E} |\xi_1| < \infty$ (and so $\mu := \mathbf{E} \xi_1 > 0$) then $\xi_t \sim t\mu$ a.s. as $t \to \infty$ (that is, $\lim_{t \to \infty} \xi_t / t = \mu$ a.s.). Let $\mathbf{P}\big(\int_0^\infty g(\xi_t^+) \, dt < \infty\big) > 0$, and $t_1 := \sup\{t : \xi_t > 2\mu t\}$. Then $t_1 < \infty$ a.s. and $\mathbf{P}\big(\int_{t_1}^\infty g(2\mu t) \, dt < \infty\big) > 0$. Since $\int_0^{t_1} g(2\mu t) \, dt < \infty$ a.s., we have $\int_0^\infty g(2\mu t) \, dt < \infty$ and this is equivalent to (1.4) in this case since $\lim_{x \to \infty} A_\xi(x) = A_\xi(\infty) \in [\mu, \infty)$. A similar argument using $t_2 = \sup\{t : \xi_t \leqslant \mu t / 2\}$, shows that (1.4) finite implies (1.3) is finite with probability equal to 1. So the result is true when $\mathbf{E} |\xi_1| < \infty$.

Next assume $\mathbf{E} |\xi_1| = \infty$. In a similar way to (1.8), we can write

$$\xi_t = \sum_{i=1}^{N_t} J_i + O(t)$$

where $O(t)/t$ is bounded a.s. as $t \to \infty$, the J_i are i.i.d with

$$\mathbf{P}(J_1 \in dx) = \Pi_\xi(dx) 1_{\{x : |x| > 1\}} / \Pi_\xi(\{x : |x| > 1\}),$$

and N_t is a Poisson process of rate $c := \Pi_\xi(\{x : |x| > 1\})$, independent of the J_i. Then $\xi_t \to \infty$ a.s. implies (1.5), and thus, by a corresponding random walk version (Kesten and Maller [27]), we have $\sum_1^n J_i \to \infty$ a.s. as $n \to \infty$. We also have $\mathbf{E} |J_1| = \infty$, so by Pruitt ([35], Lemma 8.1), $\sum_1^n J_i^- = o\big(\sum_1^n J_i^+\big)$ a.s., and thus, almost surely, $\sum_1^n J_i \sim \sum_1^n J_i^+$ as $n \to \infty$, in the sense that the ratio of the two sides tends to 1 a.s. as $n \to \infty$. Thus, almost surely, $\sum_1^{N_t} J_i \sim \sum_1^{N_t} J_i^+$ as $t \to \infty$. It follows that, when $\xi_t \to \infty$ a.s.,

$$\xi_t \sim Y_t := \sum_1^{N_t} J_i^+, \quad \text{almost surely, as } t \to \infty.$$

Now suppose $\mathbf{P}\left(\int_0^\infty g(\xi_t^+)\,dt < \infty\right) > 0$. Define $t_1 := \sup\{t : \xi_t > 2Y_t\}$, so $t_1 < \infty$ a.s. and $\mathbf{P}\left(\int_{t_1}^\infty g(2Y_t)\,dt < \infty\right) > 0$. Consequently we have $\mathbf{P}\left(\int_0^\infty g(2Y_t)\,dt < \infty\right) > 0$. Apply the result for subordinators to Y_t to get

$$\int_1^\infty \left(\frac{x}{A_Y(x)}\right)|dg(x)| < \infty$$

where $A_Y(x) = \overline{\varPi}_Y^+(1) + \int_1^x \overline{\varPi}_Y^+(y)\,dy$ (in an obvious notation). This is equivalent to (1.4) in this case since

$$A_Y(x) \asymp \int_0^x \overline{\varPi}_Y^+(y)\,dy \asymp \int_1^x \overline{\varPi}_\xi^+(y)\,dy \asymp A_\xi(x).$$

A similar argument, using $t_2 = \sup\{t : \xi_t \leqslant Y_t/2\}$, works in the other direction. □

Proof of Theorem 2, Sufficiency. This and the other proofs make continual use of the decomposition into "small" and "big" jumps, i.e., the last two processes in (1.8). From (1.8) we can write

$$(\xi_t, \eta_t) = \left(\gamma_\xi t + B_t + X_t^{(1)} + X_t^{(2)}, \ \gamma_\eta t + C_t + Y_t^{(1)} + Y_t^{(2)}\right). \tag{3.3}$$

This decomposition gives rise to a similar partitioning of the integral:

$$\int_0^t e^{-\xi_{s-}}\,d\eta_s = \int_0^t e^{-\xi_{s-}}\,d\varGamma_s + \int_0^t e^{-\xi_{s-}}\,dY_s^{(2)}, \tag{3.4}$$

where, for convenience, we write

$$\varGamma_t = \gamma_\eta t + C_t + Y_t^{(1)}, \tag{3.5}$$

and we analyze each integral as $t \to \infty$. In this analysis we will show that the first integral on the righthand side of (3.4) always converges as $t \to \infty$ when $\lim_{t \to \infty} \xi_t = \infty$ a.s. The convergence of the integral in (1.12) becomes of equal importance to the convergence of the second integral on the righthand side of (3.4).

Following a common convention, a stochastic integral of the form $\int_A^B$ means $\int_{(A,B]}$ and the distinction matters – sometimes we will write $\int_{A+}^B$ to emphasise this. Because the Lévy process η is a semimartingale and because the integrand is of class **L** (adapted, left continuous with right limits), it follows from the theory in Chapter II of Protter [34] (see in particular Theorem 9 and its corollary in [34], page 48, the definitions on pages 49 and 51, and Theorem 19, page 55) that these integrals are finite a.s., and are defined for all A

and B, with $0 \leqslant A < B < \infty$, simultaneously and for all trajectories outside of a null set (not depending on A and B). This implies that A and B can be any finite positive random variables satisfying $A \leqslant B$, a.s. Moreover the process $t \mapsto \int_0^t e^{-\xi_{s-}} \, d\eta_s$ is càdlàg and a semimartingale.

We shall need the following lemma.

Lemma 1. *Suppose* $\lim_{t\to\infty} \xi_t = \infty$ *a.s. There exists a strictly positive* μ, *which may be infinite, such that, almost surely,*

$$\mu = \lim_{t\to\infty} \frac{\xi_t}{t} = \gamma_\xi + \lim_{t\to\infty} \frac{X_t^{(1)} + X_t^{(2)}}{t} = \gamma_\xi + \lim_{t\to\infty} \frac{X_t^{(2)}}{t}. \tag{3.6}$$

Proof. First, $\lim_{t\to\infty}(1/t)B_t = \lim_{t\to\infty}(1/t)X_t^{(1)} = 0$, a.s., since both of these processes have mean 0 and the strong law applies. Thus we can safely ignore these components.

Next, if $\mathbf{E}\,|\xi_1| < \infty$, then the strong law again applies to give us the desired conclusion and in this case we can take

$$\mu = \mathbf{E}\,\xi_1 = \gamma_\xi + \mathbf{E}\,X_1^{(2)} > 0. \tag{3.7}$$

That $\mu > 0$ is forced by the assumption that $\lim_{t\to\infty} \xi_t = \infty$ a.s., because $\{\xi_t\}$ oscillates or drifts to $-\infty$ a.s. as $t \to \infty$ when $\mu \leqslant 0$, see Bertoin [4], p. 183.

However, if $\mathbf{E}\,|\xi_1| = \infty$, then the existence of the limit of ξ_t/t in (3.6) and that it also equals $+\infty$ when $\lim_{t\to\infty} \xi_t = \infty$ a.s. is somewhat more delicate. Consult [12], [37], [5] and [15] for the proof.   □

The integral $\int_0^\infty e^{-\xi_{s-}} \, d\Gamma_s$.

To deal with the first integral on the righthand side of (3.4), it suffices to show that the integral over (L, ∞) converges for some (random but finite) $L \geqslant 0$. Fix $c \in (0, \mu)$, where μ is defined in (3.6). Then there exists $L \geqslant 0$ (which may be random) such that $\xi_{s-} \geqslant sc$ for all $s > L$. (In fact we may take $L = \sup\{s > 0 : \xi_{s-} - cs \leqslant 0\}$, if the set of the sup is non-empty, and $L = 0$ otherwise. This is a random variable because ξ is càdlàg, so separable. Note that L is *not* a stopping time.) Clearly,

$$\int_L^\infty e^{-\xi_{s-}} |\gamma_\eta| \, ds \leqslant |\gamma_\eta| \int_L^\infty e^{-cs} \, ds < \infty,$$

so there is no harm in supposing that $\gamma_\eta = 0$. But on removing the term $\gamma_\eta t$ from Γ_t we find that the diminished process $\{\Gamma_t\}$ becomes a square-integrable, mean 0, martingale with quadratic variation kt for some positive constant k.

To continue, the random variable $\exp\{-\xi_{s-}\}$ may well have infinite expectation. To get around this define $\lambda_t := \max\{ct, \xi_t\}$, so that $\lambda_{s-} \geqslant cs$ for all $s \geqslant 0$, and $\lambda_{s-} = \xi_{s-}$ for all $s > L$. Hence, with probability 1,

$$\lim_{t\to\infty} \int_L^{L\vee t} e^{-\xi_{s-}} \, d\Gamma_s = \lim_{t\to\infty} \int_L^{L\vee t} e^{-\lambda_{s-}} \, d\Gamma_s.$$

On the other hand

$$\mathbf{E}\left(\int_0^t e^{-\lambda_s-}\,d\Gamma_s\right)^2 = k\int_0^t \mathbf{E}\{e^{-2\lambda_s-}\}\,ds \leqslant k\int_0^\infty e^{-2cs}\,ds < \infty. \qquad (3.8)$$

This establishes that the martingale $t \mapsto \int_0^t e^{-\lambda_s-}\,d\Gamma_s$ has bounded (actually converging) second moments. In this situation it must therefore converge with probability 1 (and in mean) as $t \to \infty$. Consequently $\int_0^\infty e^{-\xi_t-}\,d\Gamma_t$ converges with probability equal to 1 (though not necessarily in mean).

The integral $\int_0^\infty e^{-\xi_t-}\,dY_t^{(2)}$

We now show that the second integral on the righthand side of (3.4) converges a.s. as $t \to \infty$, under the stated conditions, namely, assume that both conditions of (1.12) hold. We can suppose in addition that $Y^{(2)} \neq 0$ a.s., for otherwise the result is trivial. Thus $\Pi_n(\cdot)$ is not degenerate at 0, so $\Pi(\cdot)$ also is not degenerate at 0, and consequently Π assigns positive mass to a region of the form $\{w : |w| > r > 0\}$. Only the positivity of r is relevant to our proof and there is no loss in generality in supposing that $r = 1$. Thus $\alpha \equiv \Pi\{(w : |w| > 1)\} > 0$. Similarly there is no loss in generality in supposing that $a = 1$ in the integral of (1.12).

Let N_t be the number of jumps of W falling in $\{|w| > 1\}$ during $(0,t]$, and let $0 = T_0 < T_1 < \cdots$ be the times at which these occur. The increments $\{T_i - T_{i-1}\}$ are i.i.d. exponentially distributed of rate α. Define

$$M_j = e^{-[\xi(T_{j+1}-)-\xi(T_j-)]},$$

and

$$Q_i = Y^{(2)}(T_i) - Y^{(2)}(T_i-) = Y^{(2)}(T_i) - Y^{(2)}(T_{i-1}).$$

(Note that $Y^{(2)}$ is constant between successive T_j.) The integral may be re-cast thus:

$$\int_0^t e^{-\xi_s-}\,dY_s^{(2)} = \sum_{i=1}^{N_t} e^{-\xi_{T_i}-}\,\Delta Y_{T_i}^{(2)} = \sum_{i=1}^{N_t}\left(\prod_{j=1}^{i-1} M_j\right)Q_i. \qquad (3.9)$$

The sequence of random 2-vectors $\{(M_i, Q_i)\}$ is i.i.d. with common distribution the same as that of $(e^{-[\xi(T_2-)-\xi(T_1-)]}, Y_{T_1}^{(2)})$, and is independent of the sequence of jump times $\{T_i\}$. (The independence of (M_i, Q_i) and $\{(M_j, Q_j)\}_{j \geqslant i+2}$ is fairly obvious. That each (M_i, Q_i) is independent of (M_{i+1}, Q_{i+1}) is also clear if one uses the first expression above for Q_i: $Q_i = Y^{(2)}(T_i) - Y^{(2)}(T_i-)$. Naturally, the total independence of the jump times with the values of the jumps is also crucial.)

The process $\xi^* = \{\xi(T_1 + t) - \xi(T_1); 0 \leqslant t < T_2 - T_1\}$ has the same law as that of $\xi^{**} = \{\xi(t); 0 \leqslant t < T_1\}$ and they are independent of each other. Also $\xi(T_2) - \xi(T_2-)$ and $\xi(T_1) - \xi(T_1-)$ have the same distribution and are independent of each other and of the two processes ξ^* and ξ^{**}. These considerations lead to the following:

$$\text{Law}\{\xi(T_2-) - \xi(T_1-); \mathbf{P}\} = \text{Law}\{\xi(T_1); \mathbf{P}\}. \tag{3.10}$$

We may now apply Theorem 2.1 of [19]. Note that $\mathbf{P}\{M_1 = 0\} = 0$, and also $\mathbf{P}\{Q_1 = 0\} < 1$, as required to apply that theorem (if $Q_1 = Y_{T_1}^{(2)} = 0$ a.s. then $\text{Var}(Y_{T_1}^{(2)}) = 0$, so $Y_1^{(2)} = 0$ a.s., which we assumed not to be the case).

Thus, suppose $\prod_{i=1}^{n} M_i \to 0$ a.s. as $n \to \infty$, and

$$I_{M,Q} := \int_{(1,\infty)} \left(\frac{\log q}{A_M(\log q)} \right) \mathbf{P}\{|Q_1| \in dq\} < \infty, \tag{3.11}$$

where $A_M(x) = \int_0^x \mathbf{P}\{-\log M_1 > u\}\, du$, for $x \geq 0$. Then the random sequence

$$\tilde{Z}_n := \sum_{i=1}^{n} \left(\prod_{j=1}^{i-1} M_j \right) Q_i, \qquad n = 1, 2, \ldots \tag{3.12}$$

converges a.s. to a finite limit, $\tilde{Z}_\infty$, say, as $n \to \infty$. This implies that the righthand side of (3.9) converges a.s. to $\tilde{Z}_\infty$ as $t \to \infty$, since $N_t \to \infty$ a.s. as $t \to \infty$ (again recall $\Pi\{w : |w| > 1\} > 0$). Now $T_n \to \infty$ a.s., as $n \to \infty$, and since $\lim_{t \to \infty} \xi_t = \infty$ a.s., it follows that

$$\prod_{i=1}^{n} M_i = e^{-\xi(T_n-)} \longrightarrow 0 \quad \text{a.s. as } n \to \infty.$$

Thus for $\tilde{Z}_\infty \equiv Z_\infty = \int_0^\infty e^{-\xi_t-}\, dY_t^{(2)}$ to exist a.s. it only remains to show that the convergence of $I_{\xi,\eta}$ in (1.12) implies the convergence of $I_{M,Q}$ in (3.11), when $\xi_t \to \infty$ a.s.

For this we need expressions for the distributions of the M_i and Q_i. By (1.7) the compound Poisson process

$$\left(X_t^{(2)}, Y_t^{(2)} \right) = \sum_{i=1}^{N_t} (\Delta X_{T_i}^{(2)}, \Delta Y_{T_i}^{(2)})$$

has Lévy exponent $\psi^{(2)}(\theta)$ satisfying

$$\psi^{(2)}(\theta) = -(1/t) \log \mathbf{E}\, e^{i\langle \theta, (X_t^{(2)}, Y_t^{(2)}) \rangle} = \iint_{|w|>1} \{1 - e^{i\langle \theta, w \rangle}\}\, \Pi\{dw\}, \tag{3.13}$$

for $\theta \in \mathbb{R}^2$. So $(\Delta X_{T_1}^{(2)}, \Delta Y_{T_1}^{(2)}) = (X_{T_1}^{(2)}, Y_{T_1}^{(2)})$ has joint distribution

$$\mathbf{P}\{X_{T_1}^{(2)} \in dx, Y_{T_1}^{(2)} \in dy\} = 1_{\{x^2+y^2>1\}}\Pi\{dx, dy\}/\alpha,$$

with $\alpha = \Pi(\{(x,y) : |(x,y)| > 1\}) > 0$. Thus

$$\mathbf{P}\{Q_1 \in dq\} = \mathbf{P}\{Y_{T_1}^{(2)} \in dq\}$$

$$= 1_{\{|q|\leq 1\}} \int_{|x| > \sqrt{1-q^2}} \Pi\{dx, dq\}/\alpha + 1_{\{|q|>1\}} \int_{x \in \mathbb{R}} \Pi\{dx, dq\}/\alpha. \tag{3.14}$$

If $|q| > 1$ this implies that

$$\mathbf{P}\{Q_1 \in dq\} = \Pi_\eta(dq)/\alpha. \tag{3.15}$$

Next, by (3.10) we have

$$A_M(x) = \int_0^x \mathbf{P}\{-\log M_1 > u\}\, du = \int_0^x \mathbf{P}\{\xi_{T_1} > u\}\, du,$$

and, under (1.12) we have (1.5). Suppose $0 < \mathbf{E}\,\xi_1 \leqslant \mathbf{E}\,|\xi_1| < \infty$. Now $\{\xi(T_n-)\}$ is a sequence of partial sums of i.i.d. random variables each of which has the same distribution as $\xi(T_1)$, by (3.10). By Wald's Equation we get that

$$\mathbf{E}\,\xi(T_1) = \mathbf{E}\,\xi(1)\,\mathbf{E}\,T_1 = \mathbf{E}\,\xi(1)/\alpha.$$

The right hand side is finite and strictly positive. Hence

$$\lim_{x \to \infty} A_M(x) = \int_0^\infty \mathbf{P}\{\xi(T_1) > u\}\, du = \mathbf{E}\,\xi^+(T_1) \geqslant \mathbf{E}\,\xi(T_1) > 0.$$

As this limit is positive and finite, the denominator in $I_{M,Q}$ (see (3.11)) can be ignored. The denominator in $I_{\xi,\eta}$ is (see (1.2))

$$\gamma_\xi + \overline{\Pi}_\xi^+(1) + \int_1^{\log y} \overline{\Pi}_\xi^+(u)\, du \longrightarrow \gamma_\xi + \int_{(1,\infty)} u\, \Pi_\xi\{du\}, \quad \text{as } y \to \infty. \tag{3.16}$$

Now

$$\mathbf{E}\,\xi_1 = \gamma_\xi + \int_{|u|>1} u\, \Pi_\xi\{du\},$$

so the righthand side of (3.16) is not smaller than $\mathbf{E}\,\xi_1$, which is positive, as we assumed, so the denominator in $I_{\xi,\eta}$ can be ignored also. Alternatively, suppose the first condition in (1.5) holds. Now by a similar calculation to (3.14), if $u > 1$,

$$\mathbf{P}\{X_{T_1}^{(2)} > u\} = \int_{(u,\infty)} \Pi_\xi\{dx\}/\alpha = \overline{\Pi}_\xi^+(u)/\alpha, \tag{3.17}$$

so we have $\mathbf{E}(X_{T_1}^{(2)})^+ = \infty$. For $x > 1$

$$A_M(x) = \int_0^x \mathbf{P}\{\xi_{T_1} > u\}\, du = \int_0^x \mathbf{P}\{\gamma_\xi T_1 + B_{T_1} + X_{T_1}^{(1)} + X_{T_1}^{(2)} > u\}\, du$$

$$\geqslant \int_0^x \mathbf{P}\{X_{T_1}^{(2)} > 2u\}\, du - \int_0^x \mathbf{P}\{\gamma_\xi T_1 + B_{T_1} + X_{T_1}^{(1)} \leqslant -u\}\, du \tag{3.18}$$

The last integral here has a finite limit as $x \to \infty$, because the random variable involved has a finite mean. The first integral tends to infinity because $\mathbf{E}(X_{T_1}^{(2)})^+ = \infty$. Thus for some finite constant $C > 0$

$$A_M(x) \geqslant C \int_0^x \mathbf{P}\{X_{T_1}^{(2)} > 2u\} \, \mathrm{d}u$$

for all large x. However the latter integral is bounded below by one half of $\int_1^x \mathbf{P}\{X_{T_1}^{(2)} > u\} \, \mathrm{d}u$ (which also tends to ∞ because of the infinite expectation). Hence from (3.17),

$$A_M(x) \geqslant (C/2\alpha) \int_1^x \overline{\Pi}_\xi^+(u) \, \mathrm{d}u. \tag{3.19}$$

Together (3.15) and (3.19) show that $I_{\xi,\eta} < \infty$ implies $I_{M,Q} < \infty$.

To this point we have established: (1.12) *implies $\int_0^\infty \mathrm{e}^{-\xi_{t-}} \, \mathrm{d}\eta_t$ exists and is finite a.s.*

Proof of Theorem 2, the Converse. Let us first suppose (1.12) fails by virtue of $I_{\xi,\eta} = \infty$, but $\xi_t \to \infty$ a.s. as $t \to \infty$. Then by the sufficiency part of the proof, the infinite integrals of $\mathrm{e}^{-\xi_{t-}}$ with respect to $\mathrm{d}t$, $\mathrm{d}C_t$ and $\mathrm{d}Y_t^{(1)}$ are all finite, a.s. Suppose $|\tilde{Z}_{N_t}| \xrightarrow{P} \infty$, where $\tilde{Z}_n$ is the random variable defined in (3.12). Then (3.9) shows that the integral with respect to $\mathrm{d}Y_t^{(2)}$ diverges, in fact

$$\left| \int_0^t \mathrm{e}^{-\xi_{s-}} \, \mathrm{d}Y_s^{(2)} \right| \xrightarrow{P} \infty, \qquad \text{as } t \to \infty.$$

Thus (1.14) will be the case. Now $|\tilde{Z}_n| \xrightarrow{P} \infty \ (n \to \infty)$ implies $|\tilde{Z}_{N_t}| \xrightarrow{P} \infty$, because for $z > 0$, $\varepsilon > 0$,

$$\mathbf{P}\{|\tilde{Z}_{N_t}| \leqslant z\} \leqslant \sum_{n > n_0} \mathbf{P}\{|\tilde{Z}_n| \leqslant z\} \mathbf{P}\{N_t = n\} + \mathbf{P}\{N_t \leqslant n_0\}$$

$$\leqslant \varepsilon + \mathbf{P}\{N_t \leqslant n_0\} \tag{3.20}$$

once $n_0 = n_0(\varepsilon, z)$ is so large that $\mathbf{P}\{|\tilde{Z}_n| \leqslant z\} \leqslant \varepsilon$ for $n > n_0$, and $N_t \to \infty$ a.s. But $|\tilde{Z}_n| \xrightarrow{P} \infty \ (n \to \infty)$ is implied by $I_{M,Q} = \infty$, according to Lemma 5.5 of [19]. So for this part of the converse it suffices to show that $I_{\xi,\eta} = \infty$ implies $I_{M,Q} = \infty$, when $\xi_t \to \infty$ a.s. We showed (following (3.15) above) that this is the case when $\mathbf{E}|\xi_1| < \infty$, since the denominators in $I_{M,Q}$ and $I_{\xi,\eta}$ can be ignored then.

Alternatively, suppose the first condition in (1.5) holds. We need a reverse inequality to (3.19). From the second equality in (3.18) we get the inequality

$$A_M(x) \leqslant \int_0^x \mathbf{P}\{X_{T_1}^{(2)} > u/2\} \, \mathrm{d}u + \int_0^x \mathbf{P}\{\gamma_\xi T_1 + B_{T_1} + X_{T_1}^{(1)} > u/2\} \, \mathrm{d}u$$

$$\leqslant 2 \int_1^{x/2} \mathbf{P}\{X_{T_1}^{(2)} > v\} \, \mathrm{d}v + K \tag{3.21}$$

where

$$K = \int_0^2 \mathbf{P}\{X_{T_1}^{(2)} > u/2\}\, du + \int_0^\infty \mathbf{P}\{\gamma_\xi T_1 + B_{T_1} + X_{T_1}^{(1)} > u/2\}\, du < \infty.$$

Again the integral on the right in (3.21) diverges as $x \to \infty$ because $\mathbf{E}(X_{T_1}^{(2)})^+ = \infty$. It follows that for some finite constant $C' > 0$

$$A_M(x) \leqslant 2C' \int_1^{x/2} \mathbf{P}\{X_{T_1}^{(2)} > v\}\, dv \leqslant (2C'/\alpha) \int_1^x \overline{\Pi}_\xi^+(u)\, du$$

for large enough x. This inequality shows that $I_{\xi,\eta} = \infty$ implies $I_{M,Q} = \infty$.

To complete the proof of the converse we have to consider the case when ξ_t does not tend to ∞ a.s. as $t \to \infty$, so assume this. Let

$$(Q^t, M^t) = \left(\int_{0+}^t e^{-\xi_{s-}}\, d\eta_s,\ e^{-\xi_t} \right). \tag{3.22}$$

The first step is to show:

Lemma 2. *There is a $t > 0$ such that for every constant $k \in \mathbb{R}$,*

$$\mathbf{P}\{Q^t + kM^t = k\} < 1, \tag{3.23}$$

or else (1.15) *holds.*

Proof. Suppose, contrary to (3.23), that for each $t > 0$ there is indeed a constant $k_t = k(t) \in \mathbb{R}$ such that

$$\int_0^t e^{-\xi_{s-}}\, d\eta_s + k_t e^{-\xi_t} = k_t \quad \text{a.s.}$$

Take $t = t_2$ and $t = t_1$ $(0 < t_1 < t_2)$, in this formula and subtract to get

$$e^{-\xi(t_1)} \int_{t_1}^{t_2} e^{-[\xi(s-)-\xi(t_1)]}\, d\eta_s = k(t_2)\big(1 - e^{-\xi(t_2)}\big) - k(t_1)\big(1 - e^{-\xi(t_1)}\big),$$

or

$$\int_{t_1}^{t_2} e^{-[\xi(s-)-\xi(t_1)]}\, d\eta_s + k(t_2)e^{-[\xi(t_2)-\xi(t_1)]} = [k(t_2) - k(t_1)]e^{\xi(t_1)} + k(t_1).$$

Here, the lefthand side is independent of the righthand side, so both sides are degenerate rvs and this implies that $k(t_2) = k(t_1)$, i.e., $k_t = k$ is independent of t. Thus

$$\int_0^t e^{-\xi_{s-}}\, d\eta_s = k(1 - e^{-\xi_t}) \quad \text{a.s,} \tag{3.24}$$

which is (1.15), except that the null set may depend on t. But we can get (3.24) to hold for a countable dense set H of t outside of a null set of paths, as both sides of (3.24) are well defined càdlàg processes. Hence a passage to the limit $t \downarrow s$, $t \in H$, shows that we may assume that (3.24) holds for all t simultaneously outside of a single null set of paths. Finally, $k = 0$ in (3.24) would imply $\Delta \eta_t = 0$, hence $\Pi_\eta = 0$, and $\gamma_\eta t + C_t$ degenerate at 0, hence $\gamma_\eta = \sigma_\eta = 0$. Thus η would be degenerate at 0, and we have excluded this case. □

Now (still assuming that ξ_t does not tend to ∞ a.s.) (3.23) implies (1.14), as follows. Given (3.23), choose a value u such that

$$\mathbf{P}\{Q^u + kM^u = k\} < 1 \quad \text{for all } k \in \mathbb{R}. \tag{3.25}$$

For this u define

$$Z_n^u = \int_{0+}^{nu} e^{-\xi_{s-}} \, d\eta_s, \qquad n = 1, 2, \ldots$$

Then

$$Z_{n+1}^u = Z_n^u + e^{-\xi_{nu}} \int_{nu+}^{(n+1)u} e^{-(\xi_{s-} - \xi_{nu})} \, d\eta_s = Z_n^u + \Pi_n^u Q_{n+1}^u, \quad \text{say}, \tag{3.26}$$

where

$$\Pi_n^u = e^{-\xi_{nu}} = \prod_{i=1}^{n} e^{-(\xi_{iu} - \xi_{(i-1)u})} = \prod_{i=1}^{n} M_i^u, \quad \text{say}.$$

Notice that

$$(Q_i^u, M_i^u) = \left(\int_{(i-1)u+}^{iu} e^{-(\xi_{s-} - \xi_{(i-1)u})} \, d\eta_s, \ e^{-(\xi_{iu} - \xi_{(i-1)u})} \right)$$

are i.i.d., each with the distribution of (Q^u, M^u) in (3.22), while, from (3.26),

$$Z_n^u = \sum_{i=1}^{n} \Pi_{i-1}^u Q_i^u, \qquad n = 1, 2, \ldots$$

(with $\Pi_0^u = 1$). Because of (3.25), Theorem 2.1 of [19] applies to give that $|Z_n^u| \xrightarrow{P} \infty$ as $n \to \infty$ provided that Π_n^u does not tend to 0 a.s. as $n \to \infty$, or, equivalently, that ξ_{nu} does not tend to ∞ a.s. as $n \to \infty$. Suppose for the moment this is the case and take any $t > 0$. Let $n_t = n(u, t) = \lfloor t/u \rfloor$, so $n_t u \leqslant t < (n_t + 1)u$. Let $r_t = r(u, t) = t - n_t u$, and write

$$Z_t = \int_{0+}^{t} e^{-\xi_{s-}} \, d\eta_s = \int_{0+}^{r_t} e^{-\xi_{s-}} \, d\eta_s + e^{-\xi_{r_t}} \int_{r_t+}^{n_t u + r_t} e^{-(\xi_{s-} - \xi_{r_t})} \, d\eta_s$$

$$= Q^{r_t} + M^{r_t} Z_{n_t}(u, r_t), \quad \text{say}, \tag{3.27}$$

where

$$Z_{n_t}(u, r_t) = \int_{r_t+}^{n_t u + r_t} e^{-(\xi_{s-} - \xi_{r_t})} \, d\eta_s \overset{D}{=} \int_{0+}^{n_t u} e^{-\xi_{s-}} \, d\eta_s = Z_{n_t}^u.$$

Since $|Z_n^u| \overset{P}{\to} \infty$ as $n \to \infty$ we have $|Z_{n_t}^u| \overset{P}{\to} \infty$ as $n_t \to \infty$, and because Q^{r_t} and M^{r_t} in (3.27) are bounded away from ∞ and 0, a.s., respectively, we have $|Z_t| \overset{P}{\to} \infty$ as $t \to \infty$, as required.

It remains to show that if ξ_t does not tend to ∞ a.s., then ξ_{nu} does not tend to ∞ a.s. as $n \to \infty$. We need only prove this for $u = 1$. But then the result is true by the following lemma.

Lemma 3. *For a Lévy process* ξ_t, $\xi_n \to \infty$ *a.s. as* $n \to \infty$ *implies* $\xi_t \to \infty$ *a.s. as* $t \to \infty$.

Proof. This is proved in [17]. A different proof was provided by Chaumont [10]. We omit further details here.   □

Returning to the proof of the converse assertion of Theorem 2, we have shown that if $\lim_{t \to \infty} \xi_t = \infty$ a.s. and $I_{\xi,\eta} = \infty$, or if ξ_t does not tend to ∞ as $t \to \infty$, then (1.14) or (1.15) holds and (1.13) fails. This concludes the proof.   □

Proof of Proposition 1. Assume $\{\xi_t\}$ oscillates and (1.15) holds.

(i) That (1.16) holds is immediate.

(ii) Assume in addition that $\Pi_\eta = 0$. Then (1.15) gives

$$\int_0^t e^{-\xi_{s-}} \left(\gamma_\eta \, ds + dC_s\right) = k(1 - e^{-\xi_t}), \quad t \geqslant 0, \quad \text{for some } k \in \mathbb{R} \setminus \{0\}. \quad (3.28)$$

(3.28) then implies that ξ_t has no jump component, so $\xi_t = \gamma_\xi t + B_t$, and, since $\{\xi_t\}$ oscillates, $\gamma_\xi = 0$ and $\sigma_\xi \neq 0$. Thus (3.28) reduces to

$$\int_0^t e^{-B_s} \left(\gamma_\eta \, ds + dC_s\right) = k(1 - e^{-B_t}), \quad t \geqslant 0, \quad \text{for some } k \neq 0. \quad (3.29)$$

Differentiate using Ito's lemma to get

$$dC_t = k \, dB_t - \left(k\sigma_\xi^2/2 + \gamma_\eta\right) dt,$$

which is only possible if $\gamma_\eta = -k\sigma_\xi^2/2$ and $C_t = kB_t$. We cannot have $\sigma_\eta = 0$ otherwise $\sigma_\xi = 0$ and then $\gamma_\eta = 0$, and again η_t degenerates to 0. Thus $\sigma_\eta > 0$ and we conclude that $\eta_t = k(B_t - \sigma_\xi^2 t/2) = k(\xi_t - \sigma_\xi^2 t/2)$ in this case.

(iii) Next assume that $\Pi \neq 0$. From (1.15) we deduce that

$$e^{-\xi_{t-}} \Delta\eta_t = -k\left(e^{-\xi_t} - e^{-\xi_{t-}}\right),$$

so $\Delta\eta_t = -k(e^{-\Delta\xi_t} - 1)$. From this and (1.9), one sees that the support of Π must lie on the curve $\{(x, y) : y = -k(e^{-x} - 1)\}$, as in (1.17).   □

Proof of Theorem 3. Let (1.12) hold and suppose f satisfies the given conditions. Instead of (3.8) we can write

$$\mathbf{E}\left(\int_0^\infty f_s e^{-\lambda_{s-}} \, d\Gamma_s\right)^2 = k \int_0^\infty \mathbf{E}\{|f_s|^2 \, e^{-2\lambda_s}\} \, ds$$

$$\leqslant k\beta \int_0^\infty e^{-2(c-\varepsilon)s} \, ds,$$

for some $\beta > 0$, with c as in (3.8). The last integral is finite if ε is chosen smaller than c. The process $t \mapsto \int_0^t f_s e^{-\lambda_{s-}} \, d\Gamma_s$ is also a martingale (recall that the process $s \mapsto f_s$ is assumed non-anticipating), and by the preceding calculation has bounded second moments. Therefore it converges a.s. as $t \to \infty$ (and in mean). On the other hand $\int_L^{L \vee t} f_s \exp(-\xi_{s-}) \, d\Gamma_s \equiv \int_L^{L \vee t} f_s \exp(-\lambda_{s-}) \, d\Gamma_s$, so $\int_0^\infty f_s \exp(-\xi_{s-}) \, d\Gamma_s$ also converges a.s.

It remains to consider

$$\int_0^\infty f_s e^{-\xi_{s-}} \, dY_s^{(2)}.$$

If $\{T_i\}$ is the increasing sequence of jump times of $(X^{(2)}, Y^{(2)})$, and if

$$Q_i = Y_{T_i}^{(2)} - Y_{T_i-}^{(2)} = Y_{T_i}^{(2)} - Y_{T_{i-1}}^{(2)}, \qquad M_{i-1} = e^{-\{\xi(T_i-)-\xi(T_{i-1}-)\}}$$

for $i \geqslant 1$, then

$$\left|\int_{0+}^\infty f_s e^{-\xi_{s-}} \, dY_s^{(2)}\right| \leqslant \sum_{i=1}^\infty \left(\prod_{j=0}^{i-1} M_j\right) |f_{T_i}| \, |Q_i|.$$

By assumption, $|f_{T_i}| \leqslant e^{\varepsilon T_i}$ a.s for large enough i. Under (1.12), the righthand side of (3.12) converges a.s., so by Lemma 5.2 of [19], $e^{an}(\prod_{i=1}^n |M_i|)|Q_n| \to 0$ a.s. as $n \to \infty$, for some $a > 0$. The T_i are Gamma(i, α) rvs, so if $\varepsilon/\alpha < a$, the series $\sum e^{\varepsilon T_i - ai}$ converges a.s. and we can complete the proof. $\square$

Remark. Strictly speaking, in the proof of Theorem 2, we only proved existence of the integral $\int_{0+}^\infty e^{-\xi_{t-}} \, d\eta_t$ as an improper integral, that is, as the a.s. limit as $t \to +\infty$ of the processes $\int_{0+}^t e^{-\xi_{s-}} \, d\eta_s$. But the working in the proof of Theorem 3, together with dominated convergence ([34], Theorem 32, Ch. IV, p. 145) shows that the integral in fact exists in the sense of [34], that is, as the ucp limit (uniformly on compacts in probability) of simple predictable functions, in the terminology of Protter [34], Ch. II.

Acknowledgements. Sincere thanks to Profs Ron Doney, Charles Goldie, Harry Kesten, Claudia Klüppelberg, and Zhen-Qing Chen for their helpful interest in this project; in particular to Charles for drawing our attention to these kinds of problem in the first place. Thanks also to Alex Lindner for pointing out some needed corrections. The second author is also grateful for the hospitality of the Department of Mathematics, University of Washington.

References

1. Babillot, M., Bougerol, P. and Elie, L. (1997) The random difference equation $X_n = A_n X_{n-1} + B_n$ in the critical case. *Ann. Probab.*, **25**, 478–493.
2. Barndorff-Nielsen, O.E. (Ed.) (2002) *Proceedings of the Second MaPhysto Conference*, Centre for Math. Phys. & Stochastics Publications, Aarhus, no. 22; http://www.maphysto.dk/
3. Barndorff-Nielsen, O.E. and Shephard, N. (2001) Modelling by Lévy processes for financial econometrics. In: *Lévy processes, Theory and Applications*, pp. 283–318, O. E. Barndorff-Nielsen, T. Mikosch, S. Resnick, Eds, Birkhäuser, Boston.
4. Bertoin, J. (1996) *Lévy Processes.* Cambridge University Press.
5. Bertoin, J. (1997) Regularity of the half-line for Lévy processes. *Bull. Sci. Math.* **121**, 345–354.
6. Brockwell, P.J. (2001) Lévy driven CARMA processes, *Ann. Inst. Statist. Math.*, **52**, 1–18.
7. Brockwell, P.J. (2002) Continuous time ARMA processes, preprint.
8. Brofferio, S. (2003) How a centered random walk on the affine group goes to infinity. *Ann. Inst. H. Poincaré*, **39**, 371–384.
9. Carmona, P., Petit, F. and Marc Yor, (2001) Exponential Functionals of Lévy Processes. In: *Lévy processes, Theory and Applications*, pp. 41–55, O. E. Barndorff-Nielsen, T. Mikosch, S. Resnick, Eds, Birkhäuser, Boston.
10. Chaumont, A. (2003) Personal communication.
11. Doléans-Dade, C. (1970) Quelques applications de la formule de changement de variables pour les semimartingales. *Z. Wahrsch. Verw. Gebiete* **16**, 181–194.
12. Doney, R. and Maller, R.A. (2002) Stability and attraction to normality for Lévy processes at zero and at infinity. *J. Theoret. Prob.*, **15** (n° 3), 751–792.
13. Dufresne, D. (1990) The distribution of a perpetuity, with application to risk theory and pension funding, *Scand. Actuar. J.*, **9**, 39–79.
14. Dzhaparidze, K. and van Zanten, H. (2002) A series expansion of fractional Brownian Motion, CWI report 2002-14.
15. Erickson, K.B. (1973) The strong law of large numbers when the mean is undefined. *Trans. Amer. Math. Soc.* **185**, 371–381.
16. Erickson, K. B. (1981) A limit theorem for renewal sequences with an application to local time. *Z. Wahrsch. Verw. Gebiete* **57**, 535–558.
17. Erickson, K.B. and Maller, R.A. (2003) The strong law for subordinated random walks and Lévy processes, to appear *J. Theoret. Prob.*
18. Goldie, C.M. (1991) Implicit renewal theory and the tails of solutions of random equations. *Ann. Appl. Probab.* **1**, 126–166.
19. Goldie, C.M. and Maller, R.A. (2000) Stability of perpetuities. *Ann. Probab.* **28**, 1195–1218.
20. Grincevičius, A. K. (1974) On the continuity of the distribution of a sum of dependent variables connected with independent walks on lines. *Theor. Probab. Appl.* **19**, 163–168.
21. Grincevičius, A. K. (1975a) Limit theorems for products of random linear transformations on the line. *Lithuanian Math. J.* **15**, 568–579.
22. Grincevičius, A. K. (1975b) One limit distribution for a random walk on the line. *Lithuanian Math. J.* **15**, 580–589.
23. Grincevičius, A. K. (1980) Products of random affine transformations. *Lithuanian Math. J.* **20**, 279–282.

24. Grincevičius, A. K. (1981) A random difference equation. *Lithuanian Math. J.*, **21**, 302–306.
25. Iksanov, A.M (2003) Elementary fixed points of the BRW smoothing transforms with infinite number of summands. Preprint.
26. Kesten, Harry (1973) Random difference equations and renewal theory for products of random matrices. *Acta Math* **131**, 242–250.
27. Kesten, Harry and Maller, R.A. (1996) Two renewal theorems for random walks tending to infinity, *Prob. Theor. Rel. Fields*, **106**, 1–38.
28. Klüppelberg, C., Lindner, A. and Maller, R.A. (2004) A continuous time GARCH process driven by a Lévy process: stationarity and second order behaviour, *J. Appl. Prob.*, to appear.
29. Mandelbrot, B.B. and Van Ness, J.W. (1968) Fractional Brownian motions, fractional noises and applications. *SIAM Rev.* **10**, 422–437.
30. Milevsky, M.A. (1999) Martingales, scale functions and stochastic life annuities: a note. *Insurance: Mathematics and Economics*, **24**, 149–154.
31. Norros, I., Valkeila, E. and Virtamo, J. (1999) An elementary approach to a Girsanov formula and other analytical results on fractional Brownian motions. *Bernoulli* **5**, 571–587.
32. Paulsen, J. (1993) Risk theory in a stochastic economic environment. *Stoch. Proc. and Appl.*, **46**, 327–361.
33. Paulsen, J. and Hove, A. (1999) Markov chain Monte Carlo simulation of the distribution of some perpetuities. *Adv. Appl Prob.*, **31**, 112–134.
34. Protter, P. (1990) *Stochastic Integration and Differential Equations, A New Approach.* Springer-Verlag.
35. Pruitt, W.E. (1981) General one-sided laws of the iterated logarithm. *Ann. Probab.* **9**, 1–48.
36. Sato, K. (1999) *Lévy Processes and Infinitely Divisible Distributions.* Cambridge Studies in Advanced Mathematics, 68, Cambridge University Press, Cambridge.
37. Sato, K. (2001) Basic results on Lévy processes. In: *Lévy processes, Theory and Applications*, pp. 3–38, O. E. Barndorff-Nielsen, T. Mikosch, S. Resnick, Eds, Birkhäuser, Boston.
38. Vervaat, W. (1979) On a stochastic difference equation and a representation of non-negative infinitely divisible random variables. *Adv. Appl. Prob.*, **11**, 750–783.
39. Yor, Marc (1992) Sur certaines fonctionnelles du mouvement brownien réel, *J. Appl Prob.* **29**, 202–208.
40. Yor, Marc (2001) Interpretations in terms of Brownian and Bessel meanders of the distribution of a subordinated perpetuity. In: *Lévy Processes, Theory and Applications*, pp. 361–375, O. E. Barndorff-Nielsen, T. Mikosch, S. Resnick, Eds, Birkhäuser, Boston.
41. Yor, Marc (2001) *Exponential Functionals of Brownian Motion and Related Processes.* Springer, New York, Berlin.

Spectral Gap for log-Concave Probability Measures on the Real Line

Pierre Fougères

Laboratoire de Statistique et Probabilités,[*]
Université Paul Sabatier, Toulouse, France,
& Department of Mathematics,[**]
Imperial College London, United Kingdom.
e-mail: p.fougeres@imperial.ac.uk

Summary. We propose a simple parameter to describe the exact order of the Poincaré constant (or the inverse of the spectral gap) for a log-concave probability measure on the real line. This parameter is the square of the mean value of the distance to the median. Bobkov recently derived a similar result in terms of the variance of the measure. His approach was based on the study of the Cheeger constant. Our viewpoint is quite different and makes use of the Muckenhoupt functional and of a variational computation in the set of convex functions.

1 Introduction

A log-concave measure $\mu(dx)$ on the real line is an absolutely continuous measure with respect to the LEBESGUE measure, with density $\exp(-\Phi)$, where Φ is a convex function. We assume that μ is a finite measure, or equivalently that Φ goes to infinity as $|x| \to +\infty$, and we normalize μ to a probability. As we shall see later, such a measure always satisfies a POINCARÉ inequality, that is, there exists a constant $C_2 < \infty$ such that, for any smooth function with compact support $f : \mathbb{R} \to \mathbb{R}$,

$$\mathbf{Var}_\mu(f) = \int f^2 \, d\mu - \left(\int f \, d\mu \right)^2 \leqslant C_2 \int f'^2 \, d\mu.$$

When Φ is smooth enough, this well known inequality is strongly related to the natural operator associated with μ which acts on a smooth function f by

$$\mathbf{L}f = f'' - \Phi' f'.$$

Clearly, **L** satisfies the fundamental integration by parts formula: for any smooth compactly supported function f,

---

[*] Address during the work.
[**] Current address.

$$\forall g, \qquad \int \mathbf{L} f\, g \,\mathrm{d}\mu = -\int f'\, g'\, d\mu.$$

Hence, this unbounded self-adjoint operator on $\mathbf{L}^2(\mu)$ is negative. Moreover, the first non-zero eigenvalue λ_1 of $-\mathbf{L}$, the so called *spectral gap*, is given by the inverse of the best constant C_2 in (1). From now on, C_2 will stand for this best constant. As a consequence, the constant C_2 determines the ergodic behaviour of the semigroup generated by $\mathbf{L}$. The latter is a Markovian ergodic semigroup of operators $(\mathbf{P}_t)_{t \geqslant 0}$ on $\mathbf{L}^2(\mu)$ and inequality (1) is equivalent to the exponential rate convergence

$$\|\mathbf{P}_t f - \mu(f)\|_{2,\mu}^2 \leqslant e^{-2\,t/C_2}\,\mathbf{Var}_\mu(f),$$

where $\mu(f) = \int f \,\mathrm{d}\mu$ is the mean of f under the measure μ.

These well known results are valid in a much more general setting and can be found in many earlier papers (see for instance [Bak94, Roy99, ABC$^+$00]). They emphasize the importance of the constant C_2 and justify the following question: how does this constant depend on the measure μ?

A complete answer, based on MUCKENHOUPT's paper on HARDY inequalities [Muc72], has been recently given for measures on a one-dimensional space. This is due to MICLO for measures on $\mathbb{Z}$ ([Mic99a, Mic99b]) and to BOBKOV and GÖTZE for general measures on the real line (see [BG99]). They get a functional of the measure μ which describes the exact order of the constant C_2. Let us now briefly present this result in our framework.

Let Φ be a function on $\mathbb{R}$ (not necessarily convex) and

$$\mu(\mathrm{d}x) = Z^{-1} \exp\big(-\Phi(x)\big) \,\mathrm{d}x \qquad (1)$$

the associated BOLTZMANN measure, supposed to be finite and normalized to a probability. Let m be the (unique) median of μ. Define

$$B_+ = \sup_{x \geqslant m} \left(\int_m^x e^{\Phi(t)}\,\mathrm{d}t \int_x^{+\infty} e^{-\Phi(t)}\,\mathrm{d}t \right)$$

and

$$B_- = \sup_{x \leqslant m} \left(\int_x^m e^{\Phi(t)}\,\mathrm{d}t \int_{-\infty}^x e^{-\Phi(t)}\,\mathrm{d}t \right).$$

Then call $B(\Phi) \overset{\text{def.}}{=} \max(B_+, B_-)$ the MUCKENHOUPT functional. Now, there exists a POINCARÉ inequality for μ if and only if $B(\Phi)$ is finite, and then one has

$$\frac{1}{2}\,B(\Phi) \leqslant C_2 \leqslant 4\,B(\Phi).$$

See [ABC$^+$00] for instance for more details. Notice that BOBKOV and GÖTZE get a similar result with an appropriate functional to characterize the logarithmic SOBOLEV constant of the measure μ.

Now, suppose again that Φ is convex. It is not so easy to be convinced that $B(\Phi)$ is finite, even if it is true. The aim of this paper is to find a simpler and clearer functional of Φ which still describes the order of the POINCARÉ constant of the measure μ. The following theorem answers this question.

Theorem 1.1. *Let Φ be a convex function on $\mathbb{R}$. Then the probability measure*

$$\mu(\mathrm{d}x) = Z^{-1} \exp\bigl(-\Phi(x)\bigr)\,\mathrm{d}x$$

satisfies a POINCARÉ *inequality*

$$\mathbf{Var}_\mu(f) = \int f^2\,\mathrm{d}\mu - \left(\int f\,\mathrm{d}\mu\right)^2 \leqslant C_2 \int f'^2\,\mathrm{d}\mu \qquad (2)$$

and there exists a universal constant D (not depending on Φ) such that, if $C_2(\Phi)$ stands for the best constant in (2), then

$$\left(\int |x - m|\,\mu(\mathrm{d}x)\right)^2 \leqslant C_2(\Phi) \leqslant D \left(\int |x - m|\,\mu(\mathrm{d}x)\right)^2.$$

In other words, up to universal constants, the POINCARÉ constant of a log-concave measure is nothing but the square of the mean value of the distance to the median.

Remark 1.2. We shall see that the constant D can be chosen equal to 16. This precise value will not be derived from our computations but will appear as a consequence of a closer analysis of the method described in [Bob99].

Remark 1.3. The first result which could indicate the importance of the distance function for the estimate of the POINCARÉ (or rather log-SOBOLEV) constant was WANG's theorem (see [Wan97]). Applied to log-concave measures on Euclidean spaces, it says that such a measure satisfies a log-SOBOLEV inequality if and only if the square of the distance function is exponentially integrable. The log-SOBOLEV constant obtained by WANG depends in a non trivial manner on this exponential integral.

The following lemma is used in the proof of theorem 1.1 and is interesting in itself. It says that a convex function cannot be much smaller than its value at the median of the associated log-concave measure.

Lemma 1.4. *Let Φ be a convex function on $\mathbb{R}$ which goes to infinity at infinity, and let m be the median of the associated log-concave probability measure. Then*

$$\Phi \geqslant \Phi(m) - \log 2.$$

As will be seen later, S. BOBKOV proved that the best constant in the CHEEGER inequality satisfied by a one-dimensional log-concave probability measure is $e^{\Phi(m)}/2$. Hence, the previous lemma says that the inverse of the CHEEGER constant is an upper bound of the density of such a measure. Lemma 1.4 will be useful in section 2.6 (and will be proved there) to solve explicitly the variational problem on a simple enough class of functions.

After having obtained theorem 1.1, we discovered a paper of S. BOBKOV (see [Bob99] for details) which deals in particular with the same question. The functional which BOBKOV considers is the variance of the measure and he gets

$$\mathbf{Var}(\mu) \leqslant C_2(\Phi) \leqslant 12\,\mathbf{Var}(\mu),$$

where $\mathbf{Var}(\mu)$ is defined by

$$\mathbf{Var}(\mu) = \int x^2\,\mu(\mathrm{d}x) - \left(\int x\,\mu(\mathrm{d}x)\right)^2.$$

His approach is in some sense more complete than ours. Namely, he obtains a similar control for the log-SOBOLEV constant of a log-concave measure on the real line in terms of the ORLICZ norm

$$\|x - \mathbf{E}_\mu(x)\|_\Psi^2$$

associated with the YOUNG function $\Psi(s) = \exp(s^2) - 1$. Moreover, he extends his result to log-concave measures on the Euclidean space of dimension n: he gets an upper control of the constant C_2 in terms of $\|x\|_{2,\mu}^2$ but which doesn't remain of the good order in general.

Nevertheless, we think that our approach, based on totally different tools, is worth being explained. Whereas BOBKOV's point of view makes use of a control of the best constant in the CHEEGER inequality, we perform a variation computation in the set of convex functions, by means of the MUCKENHOUPT functional. This method could potentially be adapted to provide extensions to the non-convex setting (with a second derivative bounded from below) or to study other functionals.

Moreover, one of the indirect but striking consequence of our study is the universal comparison between two simple parameters for any log-concave measure on the real line: the variance and the (square of the) mean value of the distance to the median. Precisely, one has

$$\left(\int |x - m|\,\mu(\mathrm{d}x)\right)^2 \leqslant \mathbf{Var}(\mu) \leqslant 16\left(\int |x - m|\,\mu(\mathrm{d}x)\right)^2,$$

where m is the median of μ. A similar estimate can be obtained as well as an application of BOREL's exponential inequality on tails of distributions of norms under log-concave probability measures (see [Bor74]). More details are given at the end of the paper.

Section 2 is devoted to the variational proof of theorem 1.1. Afterwards, we turn in section 3 to a brief adaptation of BOBKOV's point of view to find again our result by a totally different way. A short and formal argument due to D. BAKRY to get a POINCARÉ inequality is given in appendix 3.3. This argument is essentially self-contained, readable at this stage and, although slightly more theoretical than the rest of the paper, it might convince quickly the reader that convex functions do satisfy a POINCARÉ inequality.

2 Variational proof of the main result

From now on, the median m of the measure μ is also called the median of Φ. Proving theorem 1.1 amounts to find a constant D such that, for any convex function Φ with median m,

$$B(\Phi) \leqslant D \left(\int |x - m| \, \mu(\mathrm{d}x) \right)^2.$$

The other inequality is immediate indeed since, by the $\mathbf{L}^1$ variational characterization of the median, the CAUCHY–SCHWARZ inequality and the LIPSCHITZ property of the identical function, one has

$$\int |x - m| \, \mu(\mathrm{d}x) \leqslant \int |x - \mathbf{E}(\mu)| \, \mu(\mathrm{d}x) \leqslant \sqrt{\mathbf{Var}(\mu)} \leqslant \sqrt{C_2},$$

where $\mathbf{E}(\mu) = \int x \, \mu(\mathrm{d}x)$ is the expectation (or the mean) of μ.

2.1 General strategy of the proof

One may easily reduce the number of parameters by using simple transformations on convex functions. First, it is enough to show that

$$B_+(\Phi) \leqslant D \left(\int_m^\infty |x - m| \, \mu(\mathrm{d}x) \right)^2.$$

By homogeneity, one may assume that $\Phi(m) = 0$. Now, by changing Φ into $\Phi(\cdot - m)$, one may suppose that $m = 0$ and the condition becomes, for any convex function Φ with median 0 which vanishes at 0 and any $x > 0$,

$$\int_0^x e^{\Phi(t)} \, \mathrm{d}t \int_x^\infty e^{-\Phi(t)} \, \mathrm{d}t \preccurlyeq \left(\int_0^\infty t \, e^{-\Phi(t)} \, \mathrm{d}t \right)^2 \left(\int_0^\infty e^{-\Phi(t)} \, \mathrm{d}t \right)^{-2},$$

where $\preccurlyeq$ stands for $\leqslant$ up to a universal constant. Now, remark that changing Φ into $\Phi(\cdot/x)$ allows to consider $x = 1$. Hence, the problem is reduced to minimize the functional

$$\frac{1}{Z_\infty(\Phi) \, E_c(\Phi)} \left(\frac{M_c(\Phi) + M_\infty(\Phi)}{Z_c(\Phi) + Z_\infty(\Phi)} \right)^2$$

on the set of convex functions with median 0 which vanish at 0. Here, we put

$$Z_c(\Phi) = \int_0^1 e^{-\Phi(t)} \, \mathrm{d}t, \qquad M_c(\Phi) = \int_0^1 t \, e^{-\Phi(t)} \, \mathrm{d}t, \qquad E_c(\Phi) = \int_0^1 e^{\Phi(t)} \, \mathrm{d}t,$$

$$Z_\infty(\Phi) = \int_1^\infty e^{-\Phi(t)} \, \mathrm{d}t, \qquad M_\infty(\Phi) = \int_1^\infty t \, e^{-\Phi(t)} \, \mathrm{d}t.$$

We split the integrals in order to dissociate the two components of Φ on the *compact* $[0, 1]$ and the infinite interval $[1, \infty)$ respectively.

Definition 2.1. *The functionals Z will be called the* mass *functionals, M the* mean *functionals and E_c the* inverse mass *functional.*

As we are only concerned with the restriction of Φ to the half line $[0, \infty)$, it would be interesting to characterize the convex functions on this interval which have a convex extension with median 0 on the whole line. This question is settled by the following lemma.

Lemma 2.2. *Let Φ be a convex function on $\mathbb{R}^+$ which goes to infinity at $+\infty$. Then Φ is the restriction of a convex function on $\mathbb{R}$ which goes to infinity at infinity as well and with median 0, if and only if,*

$$\Phi'(0) \geqslant \frac{-1}{Z_c(\Phi) + Z_\infty(\Phi)}. \tag{3}$$

Proof of lemma 2.2. This is a simple argument. The case of interest is when $\Phi'(0_+) < 0$. If Φ has a convex extension, say $\overline{\Phi}$, then $\overline{\Phi}'(0_-) \leqslant \Phi'(0_+)$ and $\overline{\Phi}(t) \geqslant L(t)$ for any $t \in \mathbb{R}$, where L is the linear function vanishing at 0 and with slope $\overline{\Phi}(0_-)$. Let Z_- be the mass functional of a function on $\mathbb{R}^-$. One has $Z_-(\overline{\Phi}) = Z_c(\Phi) + Z_\infty(\Phi)$ as the median is 0. And $Z_-(\overline{\Phi}) \leqslant Z_-(L)$. It remains to compute explicitly the latter. Conversely, the condition makes sure that the linear function L with slope $\Phi'(0_+)$ at 0 is a convenient extension. $\square$

As a consequence, define the class $\mathcal{C}$ of convex functions on $\mathbb{R}^+$ going to infinity at $+\infty$, vanishing at 0 and such that

$$\Phi'(0) \geqslant \frac{-1}{Z_c(\Phi) + Z_\infty(\Phi)}.$$

We now fix some denominations which will be used throughout the paper.

Definition 2.3. *We shall consider continuous piecewise linear functions on $\mathbb{R}^+$. Among them, the k-piecewise linear functions are those with k linear pieces. The $k - 1$ knots of such a function are the points where the derivative is broken.*

Let $\mathcal{T}$ be the class of 3-piecewise linear convex functions vanishing at 0 on a floating compact interval $[0, \beta]$, for some $\beta \geqslant 1$. We furthermore require these functions to have a knot at point 1 and the other inside the interval $[0, 1]$. We consider the functions in $\mathcal{T}$ as functions defined on the entire half-line by extending them by ∞ on $(\beta, +\infty)$. The proof of theorem 1.1 can be split into the following two propositions.

Proposition 2.4 (Restriction to a simple class of convex functions). *Let*

$$K(\Phi) = \frac{1}{Z_\infty(\Phi)\, E_c(\Phi)} \left(\frac{M_c(\Phi) + M_\infty(\Phi)}{Z_c(\Phi) + Z_\infty(\Phi)} \right)^2$$

be the functional of interest on the class $\mathcal{C}$. Then, the 3-piecewise linear functions in $\mathcal{T} \cap \mathcal{C}$ are almost extremals of K. Precisely,

$$\min_{\Phi \in \mathcal{C}} K(\Phi) \geqslant \frac{1}{2} \min_{\Phi \in \mathcal{T} \cap \mathcal{C}} K(\Phi).$$

Proposition 2.5. *The functional $K(\Phi)$ on the class $\mathcal{T} \cap \mathcal{C}$ introduced before is bounded from below by a positive constant.*

Remark 2.6. We shall sometimes call the functions of $\mathcal{T} \cap \mathcal{C}$ *minimizing functions* of the variational problem even though they are not exactly so.

Let us sketch the strategy we chose to investigate this minimization problem. The complexity of the involved functional K can be weakened by fixing some appropriate parameters. By this way, we may focus on a simpler functional

$$K_M(\Phi) = \frac{\left(M_c(\Phi) + M\right)^2}{E_c(\Phi)},$$

$M \geqslant 0$, acting on convex functions defined on the compact interval $[0, 1]$. By approximation in the uniform norm, attention may be restricted to piecewise linear functions. A precise study of the functionals M_c and E_c then allows to reduce gradually the number of linear pieces by an induction argument. This leads to proposition 2.4. As for proposition 2.5, the proof is based on classical minimization under constraints (on a domain of $\mathbb{R}^5$) and asymptotic expansions.

In what follows, we have tried to present these arguments in a short way in order to emphasize the key points. Some routine proofs are left to the reader who may refer to [Fou02] for complete details.

2.2 Simplifying the problem by fixing parameters

The restrictions to $[0, 1]$ and $[1, \infty)$ of a function Φ in the class $\mathcal{C}$ will be called respectively its *finite* (or *compact*) and *infinite* components. Let $\mathcal{C}_c$ and $\mathcal{C}_\infty$ be the sets of all these possible components for different $\Phi \in \mathcal{C}$.

Independence conditionally to fixed parameters. The class $\mathcal{C}$ is not the product of $\mathcal{C}_c$ and $\mathcal{C}_\infty$ or, in other words, the two components of Φ are not independent. But the point is that, conditionally to three well-chosen parameters, they are. These parameters are the value x at 1, the slope b at 1_+ and the mass Z_∞ on the infinite interval.

Here are some more details. Note that the necessary and sufficient condition for x and b to correspond to a function in $\mathcal{C}$, that is, $x = \Phi(1)$ and $b = \Phi'(1_+)$ for some $\Phi \in \mathcal{C}$, is

$$b \geqslant x > -\log 2. \tag{4}$$

This lower bound appears by considering the limit function L_x linear with slope x between 0 and 1 and infinite afterwards. Condition (3) for L_x is exactly $x > -\log 2$. Let us here emphasize that the previous notation L_x to describe this linear function will be used throughout the paper.

Let $\mathcal{C}_\infty(x, b)$ be the set of all infinite components with the corresponding parameters (satisfying (4)). Fix $Z_\infty \in Z_\infty(\mathcal{C}_\infty(x, b))$ and extend the previous notation in a natural way to define $\mathcal{C}(x, b, Z_\infty)$, $\mathcal{C}_\infty(x, b, Z_\infty)$ and $\mathcal{C}_c(x, b, Z_\infty)$ respectively. Then one has

$$\mathcal{C}(x, b, Z_\infty) \simeq \mathcal{C}_c(x, b, Z_\infty) \times \mathcal{C}_\infty(x, b, Z_\infty).$$

Moreover, $\mathcal{C}_c(x, b, Z_\infty)$ is the set of convex functions φ on $[0, 1]$ such that

(i) $\varphi(0) = 0$ and $\varphi(1) = x$,
(ii) $\varphi'(1) \leqslant b$ and $\varphi'(0) \geqslant -1/(Z_c(\varphi) + Z_\infty)$.

Minimizing the mass functional on the infinite interval. On $\mathcal{C}_\infty(x, b, Z_\infty)$, the mean functional $M_\infty(\cdot)$ is minimal when the mass Z_∞ is as close to 1 as possible, i.e., when the infinite component is the linear function

$$l_{x, b, Z_\infty} = \big(x + b(\cdot - 1)\big)\mathbf{1}_{[1, \alpha)} + \infty \mathbf{1}_{[\alpha, \infty)}.$$

The threshold α is fixed so that the mass is Z_∞. In short,

$$\min_{\psi \in \mathcal{C}_\infty(x, b, Z_\infty)} M_\infty(\psi) \overset{\text{def.}}{=} M_\infty(l_{x, b, Z_\infty}) \overset{\text{def.}}{=} M_\infty(x, b, Z_\infty).$$

Simplified minimization problem. The previous paragraph ensures that a minimizing function of the functional K on $\mathcal{C}$ introduced in proposition 2.4 is necessarily linear on an interval $[1, \beta]$ and infinite after β.

The next step of simplification consists in fixing the mass on the compact interval $[0, 1]$. The minimization problem can thus be reduced to the following

Simplified problem. *There exists a positive ϵ such that, for any real numbers x and b such that $b \geqslant x > -\log(2)$, any masses $Z_\infty \in Z_\infty(\mathcal{C}_\infty(x, b))$ and $Z_c \in Z_c(\mathcal{C}_c(x, b, Z_\infty))$, and any function $\Phi \in \mathcal{C}_c(x, b, Z_c, Z_\infty)$,*

$$\epsilon Z_\infty \left(Z_c + Z_\infty\right)^2 \leqslant \frac{\big(M_c(\Phi) + M_\infty(x, b, Z_\infty)\big)^2}{E_c(\Phi)}. \tag{5}$$

Here, $\mathcal{C}_c(x, b, Z_c, Z_\infty)$ must be understood as the subset of functions in $\mathcal{C}_c(x, b, Z_\infty)$ with fixed mass Z_c.

Remark 2.7. It is shown in [Fou02] that $Z_\infty(\mathcal{C}_\infty(x, b)) = (0, \mathcal{Z}_\infty(x, b))$ and $Z_c(\mathcal{C}_c(x, b, Z_\infty)) = (Z_c(L_x), \mathcal{Z}_c(x, b, Z_\infty))$ for some explicit functions $\mathcal{Z}_c$ and $\mathcal{Z}_\infty$.

2.3 Approximation by piecewise linear functions

The previous simplified problem (5) leads to minimize a functional of the form

$$K_M(\Phi) \stackrel{\text{def.}}{=} \frac{\left(M_c(\Phi) + M\right)^2}{E_c(\Phi)}, \qquad M \geqslant 0, \tag{6}$$

for Φ in $\mathcal{C}_c(x, b, Z_c, Z_\infty)$.

The first step consists in approximating Φ by a sequence $(\Phi_n)_n$ still in $\mathcal{C}_c(x, b, Z_c, Z_\infty)$ of piecewise linear functions which converges uniformly to Φ. Such an approximation is obtained as follows. Given a partition σ of $[0, 1]$, let the functions

$$\overline{\Phi}(\sigma) \geqslant \Phi \geqslant \underline{\Phi}(\sigma),$$

which are 1 or 2-piecewise linear between two points of the partition, be defined as follows. $\overline{\Phi}(\sigma)$ is the secant line of Φ between these two points, $\underline{\Phi}(\sigma)$ is the supremum of the two tangent lines of Φ at these points. These two functions converge uniformly to Φ as the partition size goes to 0.

Now, zooming in between two consecutive points σ_i and σ_{i+1} of the partition, the situation is described by figure 1. In the family of 2-piecewise linear

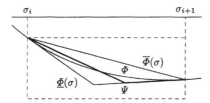

Fig. 1. Local approximation by piecewise linear functions

functions on $[\sigma_i, \sigma_{i+1}]$ interpolating between $\underline{\Phi}(\sigma)$ and $\overline{\Phi}(\sigma)$, there is exactly one function Ψ with the same mass as Φ on the considered interval. Globally, one gets a piecewise linear Ψ in $\mathcal{C}_c(x, b, Z_c, Z_\infty)$ such that

$$\|\Psi - \Phi\|_\infty \leqslant \max\left(\left\|\overline{\Phi} - \Phi\right\|_\infty, \left\|\underline{\Phi} - \Phi\right\|_\infty\right).$$

As the partition size goes to 0, one gets the expected sequence $(\Phi_n)_n$. As $K_M(\Phi_n)$ converges to $K_M(\Phi)$ as $n \to +\infty$, we may assume in the sequel that the considered function $\Phi \in \mathcal{C}_c(x, b, Z_c, Z_\infty)$ is piecewise linear.

2.4 Precise study on 2-piecewise linear functions

As will be seen, we are not able in general to determine the exact minimizing functions of the functional $K_M(\Phi)$ introduced in (6). Nevertheless, up to the constant $1/2$ of proposition 2.4, these are linear with two pieces. To prove this, an induction argument will be used. The main step consists in establishing that, on the set of piecewise linear convex functions with less than three pieces, the extremal values of the functionals M_c and E_c are achieved by some 2-piecewise linear functions.

We now enter the particularity of the Muckenhoupt functional by computing explicitly the involved functionals on linear functions.

Functionals on linear functions

Let Φ be the linear function on the interval $[\alpha, \beta]$ such that $\Phi(\alpha) = u$ and $\Phi(\beta) = v$. One has

$$Z(\Phi) = (\beta - \alpha) S(-u, -v),$$
$$E(\Phi) = (\beta - \alpha) S(u, v),$$
$$M(\Phi) = \frac{(\beta - \alpha)^2}{v - u} \left(S(-u, -v) - e^{-v} \right) + \alpha (\beta - \alpha) S(-u, -v). \tag{7}$$

The function $S(u, v)$ is the slope of the secant line of the exponential function between the two points u and v:

$$S(u, v) \stackrel{\text{def.}}{=} \frac{e^u - e^v}{u - v}.$$

Note that this function satisfies the following differential equation (which will be helpful later)

$$\partial_1 S(u, v) = \partial_2 S(v, u) = \frac{1}{u - v} \left(e^u - S(u, v) \right). \tag{8}$$

2-piecewise linear functions with fixed mass

We focus here on the set of convex functions on $[0, 1]$ with fixed boundary values and fixed mass functional

$$\overline{C}_c(x, Z_c) = \{\Phi : \Phi(0) = 0, \Phi(1) = x, Z_c(\Phi) = Z_c\}$$

and more particularly on its subset of 2-piecewise linear functions $\overline{C}_c^2(x, Z_c)$. Note that, from remark 2.7, $Z_c \geqslant Z_c(L_x) = S(0, -x)$.

A 2-piecewise linear function Φ on $[0, 1]$ with fixed boundary values is specified by two parameters: its knot α and the value u at this point α. Then, the constraint

$$Z_c(\Phi) = \alpha S(0, -u) + (1 - \alpha) S(-x, -u) = Z_c$$

determines a curve. This curve is globally parametrized by

$$\alpha = \alpha(u) = \frac{Z_c - S(-x, -u)}{S(0, -u) - S(-x, -u)}, \tag{9}$$

for u in an interval U. Precisely, $U = [u_0, u_1]$ if $x > 0$, $[u_1, u_0]$ if $x < 0$, where u_0 and u_1 are given by $Z_c = S(0, -u_1) = S(-x, -u_0)$ or equivalently $\alpha(u_0) = 0$ and $\alpha(u_1) = 1$. This condition is nothing but the fixed mass condition for the corresponding functions Φ_0 and Φ_1 to be introduced later in section 2.4 (see also figure 2). Remark that such a 2-piecewise linear function is either

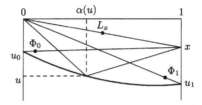

Fig. 2. 2-piecewise linear functions with fixed mass

convex or concave. Nevertheless, the convexity is automatically given by the necessary condition $Z_c \geqslant S(0, -x)$, which then entails that

$$\alpha x \geqslant u. \tag{10}$$

The derivative

$$\alpha'(u) = -\frac{\partial Z_c}{\partial u} \bigg/ \frac{\partial Z_c}{\partial \alpha} = \frac{\alpha \, \partial_2 S(0, -u) + (1 - \alpha) \, \partial_2 S(-x, -u)}{S(0, -u) - S(-x, -u)} \tag{11}$$

is strictly of the same sign as x, since $\partial_2 S \geqslant 0$.

Mean functional

A geometric argument points out the extremals of the mean functional $M_c(\cdot)$ on $\overline{C}_c(x, Z_c)$. The identical function $t \in \mathbb{R} \mapsto t$ being increasing, if the mass of Φ is fixed to $Z_c(\Phi) = Z_c$,

$$M_c(\Phi) = \int_0^1 t \, e^{-\Phi(t)} \, dt$$

is minimal as this mass is as close to 0 as possible, i.e., if Φ is as small as possible near the point 0. Hence, the minimizing function Φ is the linear function Φ_0 with boundary values u_0 at 0 and x at 1. This function may be considered as a 2-piecewise linear function in $\overline{C}_c^2(x, Z_c)$ with infinite slope at 0 (even if $\Phi_0(0) \neq 0$).

Inverse mass functional

Thanks to the parametrization (9) of the set $\overline{C}_c^2(x, Z_c)$, one is led to study the variations of the function

$$E(u) = \alpha(u) \, S(0, u) + \big(1 - \alpha(u)\big) \, S(x, u). \tag{12}$$

Proposition 2.8. *The previous function E is decreasing.*

Proof. Here is a way to compute the derivative of E, splitting it into different terms whose sign is easily determined. By (12), (9) and (11), the derivative of E may be expressed as

$$E'(u) = \alpha \, A(u, x) + (1 - \alpha) \, B(u, x),$$

where we put

$$A(u, x) = \frac{S(0, u) - S(x, u)}{S(0, -u) - S(-x, -u)} \, \partial_2 S(0, -u) + \partial_2 S(0, u)$$

and

$$B(u, x) = \frac{S(0, u) - S(x, u)}{S(0, -u) - S(-x, -u)} \, \partial_2 S(-x, -u) + \partial_2 S(x, u).$$

An interesting *symmetry* property satisfied by these two functions makes the choice of the previous notation clearer. That is

$$B(u, x) = e^x \, A(u - x, -x). \tag{13}$$

To check it, come back to the definition of S, use the differential equation (8) and perform a development (with a little bit patience). Now, note (after another tedious computation) that $A(u, x)$ can be rewritten as

$$A(u, x) = \frac{2}{u^2 \, (u - x) \, \big(S(0, -u) - S(-x, -u)\big)} \, \psi_x(u).$$

The involved function $\psi_x(u)$ is defined by

$$\psi_x(u) = x + \sinh(x) + \sinh(u - x) - \sinh(u) + (u - x) \cosh(u) - u \cosh(u - x)$$

and has the same sign as x. And, thanks to (13), we get

$$B(u, x) = \frac{2 \, e^x}{(u - x)^2 \, u \, \big(S(0, x - u) - S(x, x - u)\big)} \, \psi_{-x}(u - x).$$

Well,

$$\psi_{-x}(u - x) = -\psi_x(u)$$

and

$$e^{-x}\big(S(0, x - u) - S(x, x - u)\big) = S(-x, -u) - S(0, -u).$$

So that

$$E'(u) = \frac{2\,\psi_x(u)}{u^2\,(u-x)^2\,(S(0,-u) - S(-x,-u))}\,(u - \alpha\,x).$$

But $u \leqslant \alpha\,x$ is exactly the convexity constraint (10). Finally, as ψ_x has the same sign as x, the ratio

$$\frac{\psi_x(u)}{S(0,-u) - S(-x,-u)}$$

is always non negative. □

Extremal functions

The minimizing function of the mean functional M_c was shown in section 2.4 to be the linear function Φ_0 with boundary values u_0 and x. Now, let Φ_1 be the symmetric linear function with boundary values 0 and u_1 obtained at the limit when α goes to 1 (see figure 2). Following the previous proposition, the maximizing function of the inverse mass functional E_c on $\overline{C}_c^2(x, Z_c)$ is either Φ_0 (if $x > 0$) or Φ_1 (if $x < 0$).

2.5 Induction and *almost* minimizing functions

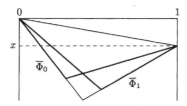

Fig. 3. Extremal functions under constraints

Thanks to the preceding study of the mean and inverse mass functionals on 2-piecewise linear functions, we will be able to find the extremal functions of these functionals on the whole set $C_c(x, b, Z_c, Z_\infty)$. We have to replace the functions Φ_0 and Φ_1 introduced before by functions $\overline{\Phi}_0$ and $\overline{\Phi}_1$ satisfying the corresponding constraints. Those are the ones with mass Z_c as close as possible to 0 and 1 respectively (see figure 3).

Lemma 2.9. *For any function* $\Phi \in C_c(x, b, Z_c, Z_\infty)$,

$$M_c(\overline{\Phi}_0) \leqslant M_c(\Phi) \leqslant M_c(\overline{\Phi}_1).$$

Moreover,

$$E_c(\overleftarrow{\Phi}_1) \leqslant E_c(\Phi) \leqslant E_c(\overline{\Phi}_0), \qquad if \qquad x \geqslant 0,$$
$$E_c(\overline{\Phi}_0) \leqslant E_c(\Phi) \leqslant E_c(\overline{\Phi}_1), \qquad if \qquad -\log(2) < x \leqslant 0.$$

Proof. As seen in section 2.3, one may assume Φ to be n-piecewise linear for some n. Let $I = [0, \gamma]$ be an interval on which Φ is 3-piecewise linear. We now describe some $(n-1)$-piecewise linear functions $\overleftarrow{\Phi}_{n-1}$ and $\overrightarrow{\Phi}_{n-1}$ in $\mathcal{C}_c(x, b, Z_c, Z_\infty)$ which coincide with Φ outside I. On I, $\overleftarrow{\Phi}_{n-1}$ (resp. $\overrightarrow{\Phi}_{n-1}$) is the 2-piecewise linear function with the same mass $Z_c(\Phi)$ as Φ constructed by extension of the slope of Φ at its first (resp. second) knot α (resp. β). We focus on the comparison of Φ and $\overleftarrow{\Phi}_{n-1}$. They coincide outside $[\alpha, \gamma]$ and

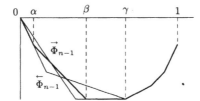

Fig. 4. Induction argument

their restrictions to this interval are 2-piecewise linear functions. One can then make use of the results seen in subsection 2.4 and 2.4 to compare the values of the mean and inverse mass functionals on these two functions. The same thing occurs for Φ and $\overrightarrow{\Phi}_{n-1}$ on $[0, \beta]$. On the one side,

$$M_c(\overleftarrow{\Phi}_{n-1}) \leqslant M_c(\Phi) \leqslant M_c(\overrightarrow{\Phi}_{n-1}). \tag{14}$$

On the other side, according to the respective values of Φ at α, β and γ,

$$E_c(\overleftarrow{\Phi}_{n-1}) \leqslant E_c(\Phi) \leqslant E_c(\overrightarrow{\Phi}_{n-1}), \tag{15}$$

or

$$E_c(\overrightarrow{\Phi}_{n-1}) \leqslant E_c(\Phi) \leqslant E_c(\overleftarrow{\Phi}_{n-1}). \tag{16}$$

By induction, we end up with 2-piecewise linear functions $\overleftarrow{\Phi}_2$ and $\overrightarrow{\Phi}_2$ on $[0, 1]$ which satisfy inequalities (14) and (15) or (16). Now, for these 2-piecewise linear functions, the expected comparisons of lemma 2.9 come from the results proven in subsection 2.4. $\qquad \square$

Hence, the extremals of the mean and inverse mass functionals are completely determined. For the functional K_M needed in proposition 2.4, it is a bit more complicated. When $x \geqslant 0$, everything goes off well. Indeed, M_c achieves its minimum as E_c achieves its maximum. So, $\overline{\Phi}_0$ is the minimizing function of $K_M(\Phi)$ on $\mathcal{C}_c(x, b, Z_c, Z_\infty)$. For $-\log(2) < x < 0$, however, a closer look is necessary. Thanks to lemma 2.9,

$$K_M(\Phi) \geqslant \frac{E_c(\overline{\Phi}_0)}{E_c(\overline{\Phi}_1)} K_M(\overline{\Phi}_0) \geqslant \frac{E_c(\Phi_0)}{E_c(\Phi_1)} K_M(\overline{\Phi}_0).$$

The reader may check that $E_c(\Phi_0)/E_c(\Phi_1)$ equals $e^x G(e^x Z_c)/G(Z_c)$, where the function $G(z) \overset{\text{def}}{=} S(0, -S^{-1}(0, \cdot)(z))$ is decreasing. Here, $S^{-1}(0, \cdot)$ stands for the inverse function of $S(0, \cdot)$. Hence, this ratio is bounded from below by $1/2$ uniformly in the parameters $-\log(2) < x < 0$ and $Z_c \geqslant 0$. Finally, one gets $K_M(\Phi) \geqslant 1/2 K_M(\overline{\Phi}_0)$ and the proof of proposition 2.4 is complete.

2.6 Minimization for *almost* extremals

A classical (but a bit strenuous) study shows that the functional K is bounded from below by a positive constant on $\mathcal{T} \cap \mathcal{C}$ as claimed in proposition 2.5. We won't give precise details to avoid lenghty (and non crucial) computations. Nevertheless, we now point out some guidelines to make the task easier for the interested reader.

A domain in $\mathbb{R}^5$

A 3-piecewise linear function in $\mathcal{T}$ is described by five parameters: the knot α inside $(0, 1)$, the value u of the function at this point, the value x and the slope r at point 1_+ and the threshold β after which the function is infinite. Furthermore, let p and q be the slopes at 0 and 1_- (see figure 5).

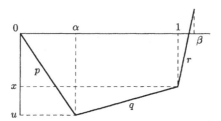

Fig. 5. 3-piecewise linear functions

The convexity constraints are

$$p = \frac{u}{\alpha} \leqslant x \qquad \text{and} \qquad q = \frac{x - u}{1 - \alpha} \leqslant r. \tag{17}$$

For the function to belong to $\mathcal{C}$ (it must have a convex extension with median 0 as in lemma 2.2), an extra constraint must be satisfied:

$$p\,Z \geqslant -1. \tag{18}$$

Here, Z stands for $Z_c + Z_\infty$. We let the reader express the mass, inverse mass and mean functionals (acting on functions in $\mathcal{T}$) as functions of the five chosen parameters (use paragraph (2.4)). The set $\mathcal{T} \cap \mathcal{C}$ may hence be viewed as a domain in $\mathbb{R}^5$ determined by the constraints (17) and (18).

Bounded values on the compact interval

Lemma 1.4 presented in the introduction claims that the density of a log-concave probability measure is bounded from above by twice its value at the median. This property highly simplifies the functional K of interest when the value x at 1 is less than a fixed constant $A \geqslant 1$. It will be made more precise later. Meanwhile, we focus on the proof of the lemma.

Proof of lemma 1.4. Let Φ be a convex function on $\mathbb{R}$ going to infinity at infinity. Recall we want to establish the estimate

$$\Phi \geqslant \Phi(m) - \log(2),$$

where m is the median of Φ. One may assume that $m = 0$ and that $\Phi(0) = 0$. Then the restriction of Φ on $[0, \infty)$ satisfies the extension condition of lemma 2.2 and, for any $t > 0$, the convex function

$$l(s) = \frac{\Phi(t)}{t}\, s\, \mathbf{1}_{[0,t]}(s) + (+\infty)\, \mathbf{1}_{(t,+\infty)}(s)$$

is still in class $\mathcal{C}$ as $l'(0) \geqslant \Phi'(0)$ and $Z(l) \leqslant Z(\Phi)$. Consequently,

$$l'(0)\, Z(l) = 1 - e^{-\Phi(t)} \geqslant -1 \qquad \text{or} \qquad \Phi(t) \geqslant -\log(2).$$

One gets the lower bound for the restriction of Φ on $(-\infty, 0]$ by symmetry. $\square$

As a consequence, when $x \leqslant A$, the functions Z_c, M_c and E_c are of order 1. And one is led to the simpler function

$$H(r, \beta) = \frac{1}{\sqrt{\overline{Z}_\infty}}\, \frac{1 + \overline{M}_\infty}{1 + \overline{Z}_\infty},$$

where $\overline{Z}_\infty$ and $\overline{M}_\infty$ stand for $e^x Z_\infty$ and $e^x M_\infty$ which do not depend on x any more. The domain on which $H(r, \beta)$ has to be minimized may be shown to be

$$-\log(2) \leqslant r \qquad \text{and} \qquad 1 \leqslant \beta \leqslant \beta_{\max}(r)$$

where

$$\beta_{\max}(r) = \begin{cases} -\log(2)/r & \text{for } -\log(2) \leqslant r < 0, \\ +\infty & \text{for } r \geqslant 0. \end{cases}$$

The EULER–LAGRANGE equations for fixed mass Z_∞ show that $H(r,\beta)$ has no extremum in the interior of the previous domain. And, on its boundary, one gets

$$H(r,1) = H(+\infty,\beta) = \lim_{r \to 0_-} H\big(r,\beta_{\max}(r)\big) = \lim_{r \to -\log 2} H\big(r,\beta_{\max}(r)\big) = +\infty$$

while $H(r,\infty) \approx (1+r)/\sqrt{r}$ is positively bounded from below. Note that the symbol $f \approx g$ stands for

$$\exists B > 0 \quad \text{such that} \quad B^{-1} f \leqslant g \leqslant B f.$$

Large values and behaviour at the limit

The previous simplification is no longer valid when x goes to infinity. To deal with these large values, the analysis is dependent upon the sign of the derivative p at 0.

Negative derivative

This is the case where constraint (18) becomes effective. Define $v = -u > 0$ (to deal with positive quantities). We split the domain D given by (18) into subdomain D_γ, $0 < \gamma \leqslant 1$, defined by

$$p\,Z = v \left[S(-x,v) - S(0,v) - \frac{S(-x,v) + Z_\infty}{\alpha} \right] = -\gamma. \tag{19}$$

Note that $p\,Z$ is maximal for $\alpha = 1$, so that there exists an α satisfying (19) if and only if

$$v\left(-Z_\infty - F(-v)\right) \geqslant -\gamma.$$

It leads to the parametrization of D_γ

$$\forall 0 < v \leqslant v_{\max}(Z_\infty), \qquad \alpha = \alpha(v,x) = \frac{S(-x,v) + Z_\infty}{S(-x,v) + z_\gamma(v)}$$

for some $z_\gamma(v) \geqslant Z_\infty$ and $v_{\max}(Z_\infty)$ to be specified (see [Fou02]). Remarking that $Z_\infty \leqslant \mathrm{e}^{-x}/x$, the reader may check that

$$\alpha \approx \frac{1}{1 + \lambda x} \tag{20}$$

where the new parameter $\lambda = z_\gamma(v) - Z_\infty \geqslant 0$ is introduced. A further asymptotic study shows that

$$Z \approx Z_c \approx \frac{1+\lambda}{1+\lambda x} \quad \text{and} \quad E_c \approx \frac{1+\lambda\,\mathrm{e}^x}{1+\lambda x}. \tag{21}$$

Here, we used $Z_\infty = o(1/x) = o(Z_c)$.

Now, equation (7) leads to the following expression for the mean functional

$$M_c = \alpha^2 \frac{1 + (v-1)e^v}{v^2} + \frac{(1-\alpha)^2}{x+v} \left(S(v, -x) - e^{-x}\right) + \alpha\,(1-\alpha)\,S(v, -x).$$

As each term of this sum is positive, it suffices to find their orders separately. One gets $M_c \approx \alpha^2 + (1-\alpha)^2/x^2 + \alpha\,(1-\alpha)/x$ and so

$$M_c \approx \frac{1 + \lambda^2}{1 + \lambda^2\,x^2}. \tag{22}$$

From (21) and (22), it follows that $(M_c/Z)^2 \approx M_c$. Using $Z_\infty \leqslant e^{-x}/x$ once more, one has

$$K = \frac{1}{E_c\,Z_\infty} \left(\frac{M_c + M_\infty}{Z}\right)^2 \succcurlyeq x\,e^x \frac{1 + \lambda x}{1 + \lambda\,e^x} \frac{1 + \lambda^2}{1 + \lambda^2\,x^2} \succcurlyeq 1,$$

where the above bound is uniform in $\lambda \geqslant 0$ and $x \geqslant A$ for some value of the threshold $A \geqslant 1$. Recall that $\succcurlyeq$ stands for $\geqslant$ up to a universal constant. At the end of the day, note that the parameter γ was hidden up to now in $\lambda = z_\gamma(v) - Z_\infty$. Hence, the positive bound we got does not depend on γ and is valid on the whole domain D.

Non negative derivative

In that case, $0 \leqslant u \leqslant \alpha\,x$ and

$$S(0, -x) = Z_c(L_x) \leqslant Z_c \leqslant Z_c(0) = 1. \tag{23}$$

Hence $Z_c \succcurlyeq 1/x$ so that $Z_\infty = o(Z_c)$ and

$$K \succcurlyeq \frac{x\,e^x}{E_c} \left(\frac{M_c}{Z_c}\right)^2. \tag{24}$$

Now, following paragraph 2.4, for a 2-piecewise linear function on $[0,1]$ with value x at 1 and fixed mass Z_c, one has $\alpha = \alpha(u)$, where u lies in $U = [u_0, u_1]$ (as $x \geqslant 0$) with $u_0 = -S^{-1}(-x, \cdot)(Z_c)$ and $u_1 = -S^{-1}(0, \cdot)(Z_c)$. According to (23), $u_0 \leqslant 0 \leqslant u_1$, so that $0 \in U$. It was shown that, in case when $x \geqslant 0$, $M_c{}^2/E_c$ is an increasing function of parameter u. Hence, it achieves its minimum (when the slope p is non negative at 0) at $u = 0$. And we may restrict ourselves to the case $u = 0$ in (24). For the remaining function $I(\alpha, x)$, one gets

$$I(\alpha, x) \approx \frac{1}{\alpha\,x\,e^{-x} + (1-\alpha)} \left(\frac{\alpha^2\,x^2 + (1-\alpha)^2 + \alpha\,(1-\alpha)\,x}{\alpha\,x + 1}\right)^2 \succcurlyeq 1,$$

uniformly in $\alpha \in (0,1)$ and $x \geqslant A$. The proof of proposition 2.5 is complete.

3 Bobkov's argument via the Cheeger constant

In this section, we present another proof of theorem 1.1 thanks to a closer analysis of BOBKOV's argument (see [Bob99]). This approach gives rise to an exact estimate of the constant D, which may be chosen equal to 16. We start with some generalities on the CHEEGER inequality and then come to the proof of the result.

3.1 Preliminaries on the Cheeger inequality

Let μ be a probability measure on the Euclidean space $\mathbb{R}^n$. For any measurable set A, the surface measure of A is

$$\mu_s(\partial A) = \liminf_{\varepsilon \to 0} \frac{\mu(A^\varepsilon) - \mu(A)}{\varepsilon} \,,$$

where

$$A^\varepsilon = \{x \in \mathbb{R}^n \ : \ |x - a| < \varepsilon, \text{ for some } a \in A\}$$

is called the ε-neighbourhood of A. The measure μ satisfies a CHEEGER inequality if there exists a constant $C_1 > 0$ such that, for any measurable set,

$$\min\big(\mu(A), \mu(A^c)\big) \leqslant C_1 \, \mu_s(\partial A). \tag{25}$$

This inequality is nothing but a uniform control of the measure of any BOREL set such that $\mu(A) \leqslant 1/2$ by its surface measure. It was introduced by CHEEGER to get an estimate of the spectral gap of the LAPLACE–BELTRAMI operator on a Riemannian manifold (see [Che70]). As the Gaussian isoperimetric inequality, the CHEEGER inequality has a functional equivalent version. This functional CHEEGER inequality says that, for any compactly supported smooth function f on $\mathbb{R}^n$ with median m,

$$\int |f - m| \, d\mu \leqslant C_1 \int |\nabla f| \, d\mu. \tag{26}$$

The median m is precisely the median of the law of f under measure μ. This formulation shows that the CHEEGER inequality is a $\mathbf{L}^1$ or isoperimetric version of the POINCARÉ inequality. This is the reason of the indices 1 and 2 of the constants.

Formally, we get the equivalence between (25) and (26) as follows. First, if $\mu(A) \leqslant 1/2$ then 0 is a median of $\mathbf{1}_A$, so that (26) applied to such indicator functions is exactly (25). Conversely, if g is smooth enough with median 0, using the co-area formula on the manifold $\{g > 0\}$, one gets

$$\int_{\{g>0\}} g \, d\mu = \int_0^{+\infty} \mu(\{g > t\}) \, dt$$

$$\leqslant C_1 \int_0^{+\infty} \mu_s(\{g = t\}) \, dt = C_1 \int_{\{g>0\}} |\nabla g| \, d\mu,$$

as $\mu(\{g > t\}) \leqslant 1/2$. Then, (26) follows by applying the previous inequality to both $f - m$ and $m - f$.

Isoperimetric function for measures on the real line

Assume

$$\mu(\mathrm{d}x) = Z^{-1} \exp\bigl(-\varPhi(x)\bigr)\,\mathrm{d}x$$

is a BOLTZMANN probability measure on $\mathbb{R}$ whose phase function $\varPhi$ is continuous. Following S. BOBKOV, note $F(t) = \mu((-\infty, t])$ the distribution function of μ (which is continuously differentiable and strictly increasing) and define the *isoperimetric function* of μ by

$$\iota_\mu(p) = F' \circ F^{-1}(p), \qquad p \in [0,1].$$

For the exponential measure $\epsilon(\mathrm{d}x) = \exp(-|x|)\,\mathrm{d}x/2$, the function $\iota(p)$ is the function $\min(p, 1-p)$ appearing in the CHEEGER inequality, whereas for the Gaussian distribution it is the function $\mathcal{U}(p)$ of the Gaussian isoperimetric inequality (see [Bob96b], [Led96], [Led99] or [Fou00]).

Why this terminology *isoperimetric function*? Trivially, if H is a half-line, one has $\mu_s(\partial H) = \iota_\mu(\mu(H))$. Rigorously, the isoperimetric function of the measure μ is

$$I_\mu(p) = \inf_{\mu(A)=p} \mu_s(\partial A).$$

By definition, I_μ is the greatest function I such that, for any BOREL set A,

$$I\bigl(\mu(A)\bigr) \leqslant \mu_s(\partial A). \tag{27}$$

So, requiring ι_μ to equal I_μ is exactly saying that inequality (27) for $I = I_\mu$ has extremal sets with any fixed measure p and that these sets are half-lines. According to [Bob96b] (see also [Bob96a]), it is the case when μ is the Gaussian measure. It is known that the same thing occurs for the exponential measure. Hence, the isoperimetric inequality of $\epsilon(\mathrm{d}x)$ is exactly

$$\min\bigl(\epsilon(A), 1 - \epsilon(A)\bigr) \leqslant \epsilon_s(\partial A)$$

with half-lines as extremal sets (see [Tal91]). So the best constant in the CHEEGER inequality for $\epsilon(\mathrm{d}x)$ is 1.

Another consequence is that the CHEEGER inequality appears like a comparison with the typical model of the exponential measure. This situation is very similar to the one of Gaussian measure with respect to the BOBKOV isoperimetric inequality. In [BL96], BAKRY and LEDOUX deduce from this Gaussian isoperimetric inequality of constant C_1 (of $\mathbf{L}^1$ type) the corresponding $\mathbf{L}^2$ type inequality, i.e., logarithmic SOBOLEV inequality of constant $C_2 = 2\,C_1^2$. The constant 2 comes from the logarithmic SOBOLEV constant of the Gaussian model. The same argument may be applied to the CHEEGER and POINCARÉ inequalities. Whereas this result is well known, we now sketch the proof copied in this new context in order to emphasize the unity of methods to compare inequalities of $\mathbf{L}^1$ and $\mathbf{L}^2$ types (in the case of diffusion operators).

Cheeger and Poincaré inequalities

It is easy to be convinced that the POINCARÉ constant of the exponential measure $\epsilon(\mathrm{d}x)$ is $C_2(\epsilon) = 4$. Trivially, the MUCKENHOUPT functional takes value 1 in this example. Hence, by (1), $C_2(\epsilon) \leqslant 4$. Applying the POINCARÉ inequality to the monomials x^k for odd k, one gets $C_2(\epsilon) \geqslant 4$ as k goes to infinity. Remark that it shows optimality of the constant 4 in the MUCKENHOUPT estimate.

The following argument roughly comes from [BL96]. In this paragraph, μ is a probability measure on $\mathbb{R}^n$. Let f be a fixed smooth function. We may assume that the law of f has a positive density G'. Remark that, in the case $r \leqslant m$, 0 is a median of $g \overset{\mathrm{def.}}{=} 1_{(-\infty, r]} \circ f$. In the other case $r \geqslant m$, 1 is a median. Applying the functional CHEEGER inequality to g, one gets the differential inequality

$$\min\big(G(r), 1 - G(r)\big) \leqslant C_1 \theta(r) G'(r)$$

where G is the distribution function of the law of f under μ and

$$\theta(r) = \mathbf{E}_\mu\big(|\nabla f| \, / f = r\big)$$

is a version of the conditional expectation of $|\nabla f|$ given f. Call F_ϵ the distribution function of $\epsilon(\mathrm{d}x)$. In terms of the function $k = G^{-1} \circ F_\epsilon$ which has the same law under ϵ as f under μ, the previous inequality says exactly that

$$k' \leqslant C_1 \, \theta \circ k. \tag{28}$$

The POINCARÉ inequality for ϵ applied to k, conjugate with (28), gives rise to the POINCARÉ inequality

$$\mathbf{Var}_\mu(f) \leqslant 4 \, C_1^2 \int |\nabla f|^2 \, \mathrm{d}\mu.$$

3.2 The one-dimensional Cheeger inequality

We have seen before that the CHEEGER inequality may be considered as a comparison with the typical model of the exponential measure. As this model is also the typical model of log-concave measures, it is not surprising that these measures satisfy a CHEEGER inequality. As for the POINCARÉ inequality, to determine the exact value of the best constant $C_1(\mu)$ for a log-concave measure is of great interest. This highly non trivial question is still open in dimension more than 2. In dimension 1, the answer was given by BOBKOV and HOUDRÉ (see [BH97, Bob99]). We briefly present their results below.

Cheeger constant for Boltzmann measures on $\mathbb{R}$

Let
$$\mu(\mathrm{d}x) = Z^{-1} \exp\bigl(-\Phi(x)\bigr) \, \mathrm{d}x$$
be a BOLTZMANN probability measure on $\mathbb{R}$ with Φ continuous, and F be its distribution function. Then

Theorem 3.1 (Bobkov–Houdré, [BH97]). *The* CHEEGER *constant is given by*

$$C_1 \stackrel{def.}{=} \sup_A \frac{\min\bigl(\mu(A), 1 - \mu(A)\bigr)}{\mu_s(\partial A)} = \sup_{x \in \mathbb{R}} \frac{\min\bigl(F(x), 1 - F(x)\bigr)}{F'(x)} \, .$$

In other terms, one may consider only half-lines in the supremum.

Sketch of proof. Taking $A = (-\infty, x]$, one gets

$$C_1 \geqslant K_\mu \stackrel{def.}{=} \sup_{x \in \mathbb{R}} \frac{\min\bigl(F(x), 1 - F(x)\bigr)}{F'(x)} \, .$$

Hence, one may assume that K_μ is finite. The function $k = F^{-1} \circ F_\epsilon$ has the same law under ϵ as the identity function $x \mapsto x$ under μ. Moreover, one easily checks the LIPSCHITZ bound
$$k' \leqslant K_\mu.$$

Now, if f is of median m under μ, $g = f \circ k$ has median m under ϵ and $g' \leqslant K_\mu f' \circ k$. Applying the CHEEGER inequality with constant 1 for ϵ to the function k, one gets

$$\int |f - m| \, \mathrm{d}\mu = \int |g - m| \, \mathrm{d}\epsilon \leqslant \int |g'| \, \mathrm{d}\epsilon \leqslant K_\mu \int |f' \circ k| \, \mathrm{d}\epsilon = K_\mu \int |f'| \, \mathrm{d}\mu.$$

Finally, $C_1 \leqslant K_\mu$. $\qquad\square$

Cheeger constant for log-concave measure on $\mathbb{R}$

In the case Φ convex, theorem 3.1 may be stated with much more precision. This is due to BOBKOV's characterization of BOLTZMANN and log-concave measures in terms of their isoperimetric functions ι. In few words, there is a one-to-one correspondence between the family of BOLTZMANN probability measures μ (up to translations) and the family of positive continuous functions ι on $]0, 1[$ vanishing on the boundary. For a measure μ with median 0, it is given by
$$\iota = F' \circ F^{-1}$$
and

$$F^{-1}(u) = \int_{1/2}^u \frac{\mathrm{d}p}{\iota(p)} \, , \qquad u \in [0, 1]. \tag{29}$$

If we assume as previously that the function Φ we consider is finite (i.e., the domain of μ is the whole line), then the function $1/\iota$ has to be non integrable at 0 and 1. The fundamental trick is that this correspondence maps the family of log-concave measures into the family of concave isoperimetric functions ι.

Theorem 3.2 (Bobkov, [Bob99]). *Let* $\mu(\mathrm{d}x) = Z^{-1} \exp\!\big(-\Phi(x)\big)\,\mathrm{d}x$ *be a log-concave probability measure of median* m *on the real line. Then, the* CHEEGER *constant is*

$$C_1 = \frac{1}{2}\,\exp\!\big(\Phi(m)\big).$$

In other words, the CHEEGER *inequality admits the two half-lines determined by* m *as extremal sets of measure* $1/2$.

Sketch of proof. The key argument is the later characterization (in terms of isoperimetric functions) of log-concave measures. Namely,

$$C_1 = \sup_{x \in \mathbb{R}} \frac{\min\!\big(F(x), 1 - F(x)\big)}{F'(x)} = \sup_{p \in (0,1)} \frac{\min(p, 1 - p)}{\iota(p)}.$$

As ι is concave and coincides with the secant 2-piecewise linear function

$$2\,\iota(1/2)\min(p, 1 - p)$$

at points 0, $1/2$ and 1, the latter lies below ι everywhere. Consequently,

$$C_1 \leqslant \big(2\,\iota(1/2)\big)^{-1} = 1/2 \, \exp\!\big(\Phi(m)\big). \qquad \square$$

A new look at lemma 1.4 shows that if μ is log-concave on the real line, then its density is bounded from above by the inverse of its CHEEGER constant.

3.3 An alternative proof of Theorem 1.1 via the Cheeger inequality

In this section, we present a new proof of our main result via Bobkov's argument. More precisely, Bobkov proved in [Bob99] that

$$C_1^2 \leqslant 3\,\mathbf{Var}(\mu)$$

and then ended up with the estimate

$$C_2 \leqslant 12\,\mathbf{Var}(\mu).$$

But we may use his method as well to get a similar bound in terms of the functional we studied. The proof is even simpler.

Theorem 3.3. *Let* $\mu(\mathrm{d}x) = Z^{-1} \exp\bigl(-\Phi(x)\bigr)\,\mathrm{d}x$ *be a log-concave probability measure of median* m *on the real line. Then*

$$C_1 \leqslant 2 \int |x - m|\,\mu(\mathrm{d}x)$$

and consequently

$$C_2 \leqslant 16 \left(\int |x - m|\,\mu(\mathrm{d}x) \right)^2.$$

Proof. We may assume $m = 0$. According to (29), one has

$$\int_{-\infty}^{+\infty} |x|\,F'(x)\,\mathrm{d}x = \int_0^1 |F^{-1}(u)|\,\mathrm{d}u = \int_0^1 \left| \int_{1/2}^u \frac{\mathrm{d}p}{\iota(p)} \right|\,\mathrm{d}u.$$

Define $\iota_\alpha(p) = \iota(1/2) + \alpha\,(p - 1/2)$, the linear line with slope α which coincides with ι at point $1/2$. As ι is concave (and $\mathcal{C}^1$), it lies below its tangent line at point $1/2$. Let $\overline{\alpha} = \iota'(1/2)$. One gets $\iota \leqslant \iota_{\overline{\alpha}}$ and $\int |x|\,\mu(\mathrm{d}x) \geqslant H(\overline{\alpha})$ where we put

$$H(\alpha) \stackrel{\text{def.}}{=} \int_0^1 \left| \int_{1/2}^u \frac{\mathrm{d}p}{\iota_\alpha(p)} \right|\,\mathrm{d}u, \qquad \alpha \in \bigl(-2\,\iota(1/2), 2\,\iota(1/2)\bigr).$$

Note that this interval contains $\overline{\alpha}$ by the concavity of ι and that $H(\alpha)$ is well defined on it.

As a function of α, $1/\iota_\alpha(p)$ is convex, and consequently $H(\alpha)$ is convex as well. Moreover, thanks to a change of variable, notice that $H(\alpha) = H(-\alpha)$, so that $H(\alpha) \geqslant H(0)$. Remarking that $\iota_0 \equiv \iota(1/2)$, one gets

$$\int |x|\,\mu(\mathrm{d}x) \geqslant \frac{1}{\iota(1/2)} \int_0^1 |u - 1/2|\,\mathrm{d}u \geqslant \exp\bigl(\Phi(m)\bigr)/4 = C_1/2.$$

This is the claimed assertion.   □

As mentioned in the introduction, a direct consequence of this result is the universal comparison

$$\left(\int |x - m|\,\mu(\mathrm{d}x) \right)^2 \leqslant \mathbf{Var}(\mu) \leqslant 16 \left(\int |x - m|\,\mu(\mathrm{d}x) \right)^2 \qquad (30)$$

between the variance and the (square of) the average distance to the median for log-concave probability measures. The constant 16 is probably not optimal. Moreover, as pointed out by the referee, the previous estimate could have been obtained as a corollary of BORELL's exponential decay of tails of log-concave probability measures (see [Bor74] lemma 3.1 for a general statement). We briefly present a naive derivation of that result by this way. It gives rise to a worse constant (even though it could certainly be refined).

As far as we are concerned, a direct consequence of BORELL's inequality is the following. Let a and b be real numbers ($b > 0$). Define

$$G_{\mu,a}(b) = \mu(|x - a| \leq b) \qquad \text{and} \qquad b(\mu, a) = G_{\mu,a}^{-1}(3/4)$$

(3/4 is here chosen for simplicity). Then, for any log-concave probability measure μ,

$$\forall t \geq 0, \qquad \mu(|x - a| \geq t\, b(\mu, a)) \leq 3^{(1-t)/2}. \tag{31}$$

Let $\mathbf{E}(\mu) = \int x\, \mu(\mathrm{d}x)$ be the expectation of μ. Thanks to the above inequality and formulae like

$$\int |x - \mathbf{E}(\mu)|^2\, \mu(\mathrm{d}x) = 2 \int_0^{+\infty} t\mu(|x - \mathbf{E}(\mu)| \geq t)\, \mathrm{d}t$$

and

$$\int |x - \mathbf{E}(\mu)|\, \mu(\mathrm{d}x) \geq b\, \mu(|x - \mathbf{E}(\mu)| \geq b),$$

one gets

$$\int |x - \mathbf{E}(\mu)|^2\, \mu(\mathrm{d}x) \approx b^2(\mu, \mathbf{E}(\mu)) \approx \left(\int |x - \mathbf{E}(\mu)|\, \mu(\mathrm{d}x) \right)^2.$$

It remains to replace the mean by the median in the latter; this follows from

$$|\mathbf{E}(\mu) - m| \leq \int |x - m|\, \mu(\mathrm{d}x)$$

and the triangle inequality.

Finally, one obtains inequality (30), with the constant $2^9 \sqrt{3}/(\log 3)^2$ instead of 16.

Appendix: Simple criterion for the Poincaré inequality

The present argument due to D. BAKRY is valid in the general setting of Markov diffusion operators. However, as such a generality is not useful at all in the rest of the paper and to make things simple, we shall limit ourselves to the case of second order elliptic differential operators on manifolds (refer to [Bak94], [Led00] or [ABC$^+$00] for the description of abstract diffusion operators).

Let M be an n-dimensional smooth connected manifold and consider a second order elliptic differential operator $\mathbf{L}$ on it (in non divergence form). In any local coordinates system, $\mathbf{L}$ acts on smooth functions by

$$\mathbf{L}f(x) = \sum_{i,j=1}^n g^{ij}(x)\, \frac{\partial^2 f}{\partial x_i \partial x_j}(x) + \sum_{i=1}^n b^i(x)\, \frac{\partial f}{\partial x_i}(x),$$

where $(g^{ij}(x))$ is a symmetric definite positive matrix with smooth coefficients with respect to x, that is a (co)metric. Note that $b^i(x)$ is smooth as well. Let Γ be the square field operator associated to $\mathbf{L}$ by the formula

$$\Gamma(f,h)(x) = \frac{1}{2}\big(\mathbf{L}(fh) - f\,\mathbf{L}h - h\,\mathbf{L}f\big) = \sum_{i,j} g^{ij}(x)\,\frac{\partial f}{\partial x_i}(x)\,\frac{\partial h}{\partial x_j}(x).$$

We write $\Gamma(f)$ for $\Gamma(f,f)$. It is nothing but the square of the length of the gradient w.r.t. the Riemannian metric associated to the second-order part of $\mathbf{L}$ (the inverse matrix $(g_{ij}(x))$ of $(g^{ij}(x))$). Thanks to the LAPLACE–BELTRAMI operator given by this metric, $\mathbf{L}$ may be rewritten in a more concise way as $\mathbf{L} = \Delta + X$, for some vector field X. The square field operator of such diffusion operators satisfies the following derivation property

$$\Gamma\big(\Psi(f)\big) = \big(\Psi'(f)\big)^2\,\Gamma(f), \tag{32}$$

for any smooth Ψ.

Equip M with a probability measure μ_Φ absolutely continuous w.r.t. the Riemannian measure and with density $Z_\Phi^{-1}\exp(-\Phi)$. The operator $\mathbf{L}$ is said to be reversible w.r.t. the measure μ_Φ if, for any smooth functions f and h,

$$\int \mathbf{L}f\,h\,\mathrm{d}\mu_\Phi = -\int \Gamma(f,h)\,\mathrm{d}\mu_\Phi.$$

It occurs provided $X = -\nabla\Phi$ and then

$$\mathbf{L} = \Delta - \Gamma(\Phi,\cdot).$$

A LIPSCHITZ continuous function u (w.r.t. the Riemannian distance) is called a contraction if $\Gamma(u) \leqslant 1$. Note that, for non smooth functions, it occurs almost everywhere thanks to RADEMACHER's theorem. We are now able to state the

Theorem A.1[3]. *Let $\mathbf{L}$, Γ and μ_Φ as before. Suppose there exists a contraction u such that, for some real numbers $\alpha > 0$ and β, and some point $x_0 \in M$,*

$$(\mathbf{L}u + \alpha)\,\mathrm{d}\mu_\Phi \leqslant \beta\,\delta_{x_0} \tag{33}$$

as measures. Then the measure μ_Φ satisfies the following POINCARÉ inequality: for any compactly supported smooth function f on M,

$$\mathbf{Var}_{\mu_\Phi}(f) \leqslant \frac{4}{\alpha^2}\int \Gamma(f)\,\mathrm{d}\mu_\Phi. \qquad\qquad PI\,(4/\alpha^2)$$

Remark A.2. To make the following formal computations more rigorous, we should construct a suitable approximation of the contraction u by smooth functions. We limit ourselves to this formal level to emphasize the key points of the proof.

Remark A.3. Considering two finite (signed) measures μ and ν on M, $\mu \leqslant \nu$ means here that, for any smooth bounded non negative function f,

---

[3] I give D. BAKRY my sincere thanks for allowing me to write this argument here.

$$\int f \, d\mu \leqslant \int f \, d\nu.$$

Note that, in the following proof, this inequality is only required for non negative smooth bounded functions whose derivatives go to 0 at infinity.

Proof. As $d\mu_\Phi \leqslant \alpha^{-1} (\delta_{x_0} - \mathbf{L}u \, d\mu_\Phi)$, one has

$$\int \left(f - f(x_0)\right)^2 d\mu_\Phi \leqslant -\alpha^{-1} \int \left(f - f(x_0)\right)^2 \mathbf{L}u \, d\mu_\Phi$$

$$\leqslant \alpha^{-1} \int \mathbf{\Gamma}\left(\left(f - f(x_0)\right)^2, u\right) d\mu_\Phi.$$

Thanks to the diffusion property (32), the latter may be rewritten as

$$2\,\alpha^{-1} \int \left(f - f(x_0)\right) \mathbf{\Gamma}(f, u) \, d\mu_\Phi.$$

By the CAUCHY–SCHWARZ inequality for the bilinear form $\mathbf{\Gamma}$,

$$\mathbf{\Gamma}(f, u) \leqslant \sqrt{\mathbf{\Gamma}(f)}$$

since u is a contraction. Another use of CAUCHY–SCHWARZ inequality now in $\mathbf{L}^2(\mu_\Phi)$ leads to

$$\int \left(f - f(x_0)\right)^2 d\mu_\Phi \leqslant 2\,\alpha^{-1} \left(\int \left(f - f(x_0)\right)^2 d\mu_\Phi\right)^{1/2} \left(\int \mathbf{\Gamma}(f) \, d\mu_\Phi\right)^{1/2}.$$

Finally,

$$\mathbf{Var}_{\mu_\Phi}(f) \leqslant \int \left(f - f(x_0)\right)^2 d\mu_\Phi \leqslant \frac{4}{\alpha^2} \int \mathbf{\Gamma}(f) \, d\mu_\Phi.$$

This is the desired claim.   □

We now focus on BOLTZMANN measures μ_Φ on $\mathbb{R}$ as introduced in (1). Assume that Φ is smooth except possibly at some point $x_0 \in \mathbb{R}$ where it has right and left derivatives. Suppose furthermore that its derivative is positively bounded from below (in absolute value) or more precisely that, for any $x \in \mathbb{R}\backslash\{x_0\}$,

$$\Phi'(x) \, \text{sign}(x - x_0) \geqslant \alpha, \tag{34}$$

for some $\alpha > 0$. Let $u(x) \stackrel{\text{def.}}{=} |x - x_0|$ be the distance function from x_0. Then, evaluating $\mathbf{L}u \, d\mu_\Phi$ on test functions f via the integration by parts formula

$$\int f \, \mathbf{L}u \, d\mu_\Phi = \int \mathbf{L}f \, u \, d\mu_\Phi,$$

one gets

$$\mathbf{L}u \, d\mu_\Phi = 2\,Z_\Phi^{-1} \, e^{-\Phi(x_0)} \, \delta_{x_0} - \Phi' \, \text{sign}(\cdot - x_0) \, d\mu_\Phi.$$

The appropriate test functions are bounded, smooth and with derivatives vanishing at infinity. One must also assume a very weak additional assumption on Φ for the boundary terms to vanish. That is, $|\Phi'(x)| e^{-\Phi(x)} = o(1/|x|)$ as $|x|$ goes to ∞. Under this condition and by (34), u satisfies (33) and $PI\left(4/\alpha^2\right)$ holds.

Now, if Φ is convex, one may always perturb Φ by a bounded function and use the above argument for the remaining function. Namely, considering two real numbers s and t such that $\Phi'(s) < 0$ and $\Phi'(t) > 0$, replace Φ on (s, t) by the maximum of its tangent lines at s and t (Φ is left unchanged outside this interval). The modified function satisfies (34) for appropriate α and x_0. The POINCARÉ inequality hence obtained for the measure associated to the perturbed function leads to a POINCARÉ inequality for μ_Φ. This is a direct consequence of the following stability property under bounded perturbation (see [ABC$^+$00] for instance for a proof).

Theorem A.4 (Bounded perturbation for the Poincaré inequality).
Let μ_Φ be as before. Consider a bounded measurable function V on M and assume that the probability $\mu_{\Phi+V}$ with density $Z_{\Phi+V}^{-1} \exp(-(\Phi + V))$ w.r.t. the Riemannian measure satisfies a POINCARÉ inequality $PI(C_2)$. Then μ_Φ satisfies the POINCARÉ inequality $PI(e^{2\operatorname{osc}(V)} C_2)$, where $\operatorname{osc}(V) = \sup V - \inf V$.

Acknowledgement. I am very grateful to Dominique Bakry for initiating my investigation of optimal constants by means of simple parameters, and also for many enlightning discussions during the work. Pierre Mathieu's careful reading and sincere remarks were very helpful. I thank the referee for suggesting many improvements in the presentation of the paper and for mentioning useful references as well.

References

[ABC$^+$00] C. ANÉ, S. BLACHÈRE, D. CHAFAÏ, P. FOUGÈRES, I. GENTIL, F. MALRIEU, C. ROBERTO and G. SCHEFFER, *Sur les inégalités de Sobolev logarithmiques*, Panoramas et Synthèses, no. 10, Société Mathématique de France, Paris, 2000.

[Bak94] D. BAKRY, "L'hypercontractivité et son utilisation en théorie des semigroupes", Lectures on probability theory (École d'été de Saint-Flour, 1992), Springer, Berlin, 1994, p. 1–114.

[BG99] S. G. BOBKOV and F. GÖTZE, "Exponential integrability and transportation cost related to logarithmic Sobolev inequalities", *J. Funct. Anal.* **163** (1999), no. 1, p. 1–28.

[BH97] S. G. BOBKOV and C. HOUDRÉ, "Isoperimetric constants for product probability measures", *Ann. Probab.* **25** (1997), no. 1, p. 184–205.

[BL96] D. BAKRY and M. LEDOUX, "Lévy-Gromov's isoperimetric inequality for an infinite-dimensional diffusion generator", *Invent. Math.* **123** (1996), no. 2, p. 259–281.

[Bob96a] S. G. BOBKOV, "Extremal properties of half-spaces for log-concave distributions", *Ann. Probab.* **24** (1996), no. 1, p. 35–48.

[Bob96b] _____, "A functional form of the isoperimetric inequality for the Gaussian measure", *J. Funct. Anal.* **135** (1996), no. 1, p. 39–49.

[Bob99] _____, "Isoperimetric and analytic inequalities for log-concave probability measures", *Ann. Probab.* **27** (1999), no. 4, p. 1903–1921.

[Bor74] C. BORELL, "Convex measures on locally convex spaces", *Ark. Mat.* **12** (1974), p. 239–252.

[Che70] J. CHEEGER, "A lower bound for the smallest eigenvalue of the Laplacian", Problems in analysis (Papers dedicated to Salomon Bochner, 1969), Princeton Univ. Press, Princeton, N. J., 1970, p. 195–199.

[Fou00] P. FOUGÈRES, "Hypercontractivité et isopérimétrie gaussienne. Applications aux systèmes de spins", *Ann. Inst. H. Poincaré Probab. Statist.* **36** (2000), no. 5, p. 647–689.

[Fou02] _____, "Inégalités fonctionnelles liées aux formes de Dirichlet. De l'isopérimétrie aux inégalités de Sobolev", Thèse soutenue sous la direction de D. Bakry à l'université Paul Sabatier, 2002.

[Led96] M. LEDOUX, "Isoperimetry and Gaussian analysis", Lectures on probability theory and statistics (École d'été de Saint-Flour, 1994), Springer, Berlin, 1996, p. 165–294.

[Led99] _____, "Concentration of measure and logarithmic Sobolev inequalities", Séminaire de Probabilités, XXXIII, Springer, Berlin, 1999, p. 120–216.

[Led00] _____, "The geometry of Markov diffusion generators", *Ann. Fac. Sci. Toulouse* **IX** (2000), p. 305–366.

[Mic99a] L. MICLO, "An example of application of discrete Hardy's inequalities", *Markov Process. Related Fields* **5** (1999), p. 319–330.

[Mic99b] _____, "Relations entre isopérimetrie et trou spectral pour les chaînes de Markov finies", *Probab. Theor. Relat. Fields* **114** (1999), p. 431–485.

[Muc72] B. MUCKENHOUPT, "Hardy's inequality with weights", *Studia Math.* **44** (1972), p. 31–38, collection of articles honoring the completion by Antoni Zygmund of 50 years of scientific activity, I.

[Roy99] G. ROYER, *Une initiation aux inégalités de Sobolev logarithmiques*, Cours Spécialisés, no. 5, Société Mathématique de France, Paris, 1999.

[Tal91] M. TALAGRAND, "A new isoperimetric inequality and the concentration of measure phenomenon", Geometric aspects of functional analysis (1989–90), Springer, Berlin, 1991, p. 94–124.

[Wan97] F.-Y. WANG, "Logarithmic Sobolev inequalities on noncompact Riemannian manifolds", *Probab. Theor. Relat. Fields* **109** (1997), no. 3, p. 417–424.

Propriété de Choquet–Deny et fonctions harmoniques sur les hypergroupes commutatifs

Laurent Godefroy

Université de Tours, Laboratoire de Mathématiques
et Physique Théorique, Parc de Grandmont, 37200 Tours.
e-mail: godefroy@univ-tours.fr

Résumé. Dans cet article, on prouve un résultat de type Choquet–Deny sur les hypergroupes commutatifs : les seules fonctions harmoniques continues bornées des marches aléatoires irréductibles sont les fonctions constantes. On en déduit la description de la frontière de Martin par une méthode de relativisation.

Introduction

La description du cône des fonctions harmoniques des marches aléatoires sur un groupe a fait l'objet de nombreuses études. On renvoie le lecteur aux travaux de Choquet et Deny [4], Furstenberg [9], Conze et Guivarc'h [5], Derriennic [6] et Raugi [22] qui ont respectivement étudié le cas des groupes abéliens, semi-simples, nilpotents, libre et résolubles connexes. Choquet et Deny ont par exemple montré que sur un groupe abélien, les fonctions harmoniques d'une marche aléatoire adaptée s'écrivent comme une moyenne d'exponentielles et que les seules fonctions harmoniques bornées sont les constantes.

Notre propos ici est d'étendre ce type de résultat aux marches aléatoires sur des hypergroupes commutatifs. De telles chaînes de Markov ont montré leur intérêt, par exemple dans la modélisation de marches réfléchies sur $\mathbb{N}^d$ et $\mathbb{R}^d_+$, et ont de nombreuses applications [10].

Rappelons que sur un hypergroupe commutatif $(X, *)$, dont $(T_y, \ y \in X)$ désignent les opérateurs de translation, une marche aléatoire de loi μ est une chaîne de Markov homogène de noyau de transition $P(x, \mathrm{d}y) = \delta_x * \mu(\mathrm{d}y)$. Les fonctions harmoniques associées sont alors les fonctions positives vérifiant $h(x) = \int_X T_y h(x) \, \mu^-(\mathrm{d}y)$ pour tout x dans X.

Après une première partie composée de rappels et notations, nous énonçons et démontrons dans la deuxième section une propriété de type Choquet–Deny. Plus précisément, nous prouvons que les seules fonctions harmoniques continues bornées d'une marche aléatoire irréductible sur un hypergroupe commutatif sont les fonctions constantes. Pour ce faire, nous adaptons à notre contexte une méthode de Raugi [21].

Dans la troisième et dernière partie, nous caractérisons les génératrices extrémales du cône des fonctions harmoniques et appliquons le théorème de représentation intégrale de Choquet [3]. Notre résultat est le suivant. Soit $(X, *)$ un hypergroupe commutatif et involutif. On dit qu'une fonction χ est multiplicative si $T_y\chi(x) = \chi(x)\chi(y)$ pour tout $x, y \in X$. Pour une marche aléatoire de loi μ sur $(X, *)$ on note

$$\mathcal{S}_\mu = \left\{ \chi \text{ multiplicatives et positives sur } X \text{ telles que } \int_X \chi(x)\,\mu(\mathrm{d}x) = 1 \right\}.$$

Alors, toute fonction μ-harmonique h admet une unique représentation intégrale

$$h(\,.\,) = \int_{\mathcal{S}_\mu} \chi(\,.\,)\,\nu(\mathrm{d}\chi),$$

où ν est une mesure borélienne positive sur $\mathcal{S}_\mu$. La preuve repose sur une méthode de relativisation et sur le résultat de la Section 2.

1 Hypergroupes et marches aléatoires

1.1 Généralités sur les hypergroupes

Soit X un espace topologique localement compact séparé. On notera par $M(X)$ l'espace des mesures de Radon bornées sur X et par $M^1(X)$ le sous-ensemble de $M(X)$ constitué des mesures de probabilités. Pour $\mu \in M(X)$, $\mathrm{supp}(\mu)$ désignera le support de μ et pour $x \in X$, δ_x sera la masse de Dirac au point x. On notera également par $C_c(X)$ l'espace des fonctions continues à support compact sur X.

Définition 1. *On dit que $(X, *)$ est un hypergroupe si $*$ est une opération bilinéaire et associative sur $M(X)$ vérifiant les conditions suivantes [1]:*

(i) *$\delta_x * \delta_y$ appartient à $M^1(X)$ et est à support compact pour tous x, y dans X.*

(ii) *L'application $X \times X \to M^1(X)$, $(x, y) \mapsto \delta_x * \delta_y$ est continue pour la topologie vague.*

(iii) *L'application $(x, y) \mapsto \mathrm{supp}(\delta_x * \delta_y)$ de $X \times X$ dans l'espace des parties compactes de X muni de la topologie de Michael, est continue.*

(iv) *Il existe un élément e dans X, appelé unité, tel que $\delta_x * \delta_e = \delta_e * \delta_x = \delta_x$ pour tout x dans X.*

(v) *Il existe un homéomorphisme involutif de X, $x \mapsto x^-$, tel que e appartienne à $\mathrm{supp}(\delta_x * \delta_y)$ si et seulement si $y = x^-$; et dont le prolongement naturel à $M(X)$ vérifie $(\delta_x * \delta_y)^- = \delta_{y^-} * \delta_{x^-}$.*

À noter que les groupes sont exactement les hypergroupes pour lesquels $\delta_x * \delta_y$ est une masse de Dirac quels que soient x et y dans X.

Le produit de convolution de deux mesures μ, $\nu \in M(X)$ est alors la mesure $\mu * \nu$ donnée par :

$$\langle \mu * \nu, f \rangle = \int_X \int_X \langle \delta_x * \delta_y, f \rangle \, \mu(\mathrm{d}x)\,\nu(\mathrm{d}y) \qquad (f \in C_c(X)).$$

Pour deux parties A et B de X, on notera par $A * B$ l'ensemble :

$$A * B = \bigcup_{x \in A,\, y \in B} \mathrm{supp}(\delta_x * \delta_y).$$

Pour $y \in X$ et pour $\mu \in M(X)$, on définit les opérateurs de translation T_y et T_μ sur $C_c(X)$ par les formules :

$$T_y f(x) = \langle \delta_x * \delta_{y^-}, f \rangle \qquad (x \in X),$$

$$T_\mu f(x) = \langle \delta_x * \mu^-, f \rangle = \int_X T_y f(x)\,\mu(\mathrm{d}y) \qquad (y \in X).$$

Hypothèse. Dans la suite, on ne considérera que des hypergroupes commutatifs, c'est-à-dire des hypergroupes pour lesquels l'opération $*$ est commutative.

De tels hypergroupes possèdent une mesure de Haar, unique à une constante multiplicative près, c'est-à-dire une mesure de Radon positive m qui vérifie :
$$\langle m, T_y f \rangle = \langle m, f \rangle \qquad (f \in C_c(X),\ y \in X).$$
Il faut noter que $\mathrm{supp}(m) = X$.

On peut également définir le produit de convolution de deux fonctions f, $g \in C_c(X)$ comme étant la fonction $f * g$ donnée par :

$$f * g(x) = \int_X f(y) T_y g(x)\,m(\mathrm{d}y) \qquad (x \in X).$$

Il convient de remarquer que $f * g$ est la densité par rapport à m de la mesure $\mu * \nu$ où $\mu = fm$ et $\nu = gm$.

Enfin, pour clore ces rappels, introduisons la notion de fonction multiplicative.

Définition 2. *Soit χ une fonction continue sur X. On dit que χ est multiplicative si $\chi(e) = 1$ et si pour tout x, $y \in X$*

$$T_y \chi(x) = \chi(x)\chi(y).$$

1.2 Marches aléatoires sur les hypergroupes

Soit $(X, *)$ un hypergroupe. Nous dirons qu'une chaîne de Markov homogène sur X est une marche aléatoire, si son noyau de transition P commute avec les translations de $(X, *)$, c'est-à-dire si pour tout $x \in X$, $PT_x = T_x P$. On peut montrer (cf. [11]) qu'une telle condition implique que P soit de la forme T_{μ^-}, pour une mesure de probabilité μ sur X que l'on appelle loi de la marche aléatoire. Les transitions de la marche sont alors données par $P(x, A) = \delta_x * \mu(A)$, pour x dans X et A borélien de X.

Comme dans le cas des groupes, voir [23] à ce sujet, on introduit les notions de marche adaptée ou irréductible. Pour μ une mesure de probabilité sur X, on notera dans la suite par X_μ (resp. Y_μ) la plus petite partie fermée H de X contenant supp(μ) telle que $H * H \subset H$ et $H^- = H$ (resp. $H * H \subset H$).

Définition 3. *Une marche aléatoire de loi μ sera dite adaptée si $X_\mu = X$ et irréductible si $Y_\mu = X$.*

On peut donner (cf. [11]) une description précise de X_μ et Y_μ :

Proposition 1. *Soit $\mu \in M^1(X)$. On a :*

$$X_\mu = \overline{\bigcup_{p+q \geqslant 1} \operatorname{supp}(\mu^{*p} * (\mu^-)^{*q})}$$

$$Y_\mu = \overline{\bigcup_{p \geqslant 1} \operatorname{supp}(\mu^{*p})},$$

où l'on a posé par convention $\mu^{*0} = \delta_e$.

2 La propriété de Choquet–Deny

Nous allons ici prouver une propriété de type Choquet–Deny (cf [4]), à savoir que les seules fonctions harmoniques continues bornées d'une marche aléatoire irréductible sont les fonctions constantes. Pour ce faire, nous adaptons au cas des hypergroupes une méthode élémentaire et élégante due à Albert Raugi [21]. Voir aussi à ce sujet, sous une présentation différente, l'article de Derriennic [7].

Théorème 1. *Soit $\mu \in M^1(X)$. Une fonction borélienne bornée h vérifie $T_{\mu^-} h = h$ si et seulement si, pour tout $k \geqslant 1$ et pour $(\mu^-)^{*k}$ presque tout $y \in X$, $T_y h = h$.*

Preuve. Soit h une fonction borélienne bornée vérifiant $T_{\mu^-} h = h$.

On définit une suite de fonctions $(H_n)_{n \in \mathbb{N}}$ comme suit :

$$H_0(x) = \int_X \big(T_y h(x) - h(x)\big)^2 \mu^-(\mathrm{d}y), \qquad x \in X \tag{1}$$

$$H_n = T_\mu^n {}_- H_0, \qquad n \geqslant 1. \tag{2}$$

On obtient facilement la relation

$$H_0(x) = \int_X \big(T_y h(x)\big)^2 \mu^-(\mathrm{d}y) - h^2(x).$$

De plus, d'après l'inégalité de Jensen, on a pour tout x, y dans X :

$$\big(T_y h(x)\big)^2 = \langle \delta_x * \delta_{y^-}, h\rangle^2 \leqslant \langle \delta_x * \delta_{y^-}, h^2\rangle = T_y(h^2)(x).$$

Ainsi, on a $H_0 \leqslant T_{\mu^-}(h^2) - h^2$. On en déduit que pour tout entier n

$$\sum_{k=0}^n H_k \leqslant T_{\mu^-}^{n+1}(h^2) - h^2 \leqslant \|h\|_\infty^2.$$

La fonction h étant bornée, ceci prouve que la série de fonctions de terme général positif H_k est convergente.

D'autre part, pour tout x appartenant à X, on a :

$$
\begin{aligned}
T_{\mu^-} H_0(x) &= \int_X T_y H_0(x) \mu^-(\mathrm{d}y) \\
&= \int_X \int_X H_0(z)\, \delta_x * \delta_{y^-}(\mathrm{d}z)\, \mu^-(\mathrm{d}y) \\
&= \int_X \int_X \int_X \big(T_\omega h(z) - h(z)\big)^2 \mu^-(\mathrm{d}\omega)\, \delta_x * \delta_{y^-}(\mathrm{d}z)\, \mu^-(\mathrm{d}y) \\
&= \int_X \int_X \int_X \big(T_\omega h(z) - h(z)\big)^2 \delta_x * \delta_{y^-}(\mathrm{d}z)\, \mu^-(\mathrm{d}y)\, \mu^-(\mathrm{d}\omega) \\
&\geqslant \int_X \int_X \left(\int_X \big(T_\omega h(z) - h(z)\big) \delta_x * \delta_{y^-}(\mathrm{d}z) \right)^2 \mu^-(\mathrm{d}y)\, \mu^-(\mathrm{d}\omega) \\
&\geqslant \int_X \int_X \big(T_y T_\omega h(x) - T_y h(x)\big)^2 \mu^-(\mathrm{d}y)\, \mu^-(\mathrm{d}\omega) \\
&\geqslant \int_X \left(\int_X \big(T_y T_\omega h(x) - T_y h(x)\big) \mu^-(\mathrm{d}y) \right)^2 \mu^-(\mathrm{d}\omega) \\
&\geqslant \int_X \big(T_\omega h(x) - h(x)\big)^2 \mu^-(\mathrm{d}\omega) \\
&\geqslant H_0(x).
\end{aligned}
$$

Pour démontrer cette suite d'inégalités, on a successivement utilisé le théorème de Fubini, l'inégalité de Jensen et le caractère harmonique de la fonction h. On en déduit que la suite de fonctions $(H_n)_{n \in \mathbb{N}}$ est croissante.

Les fonctions H_n, $n \in \mathbb{N}$, sont donc nécessairement nulles. En particulier, la nullité de H_0 implique que pour tout $x \in X$ et pour μ^- presque tout $y \in X$, $T_y h(x) = h(x)$.

On montre de même que pour tout $k \geqslant 1$, pour tout $x \in X$ et pour $(\mu^-)^{*k}$ presque tout $y \in X$, on a $T_y h(x) = h(x)$.   □

Corollaire 1 (Propriété de Choquet–Deny). *Soit* $\mu \in M^1(X)$ *telle que* $Y_\mu = X$. *Alors, les solutions continues bornées de l'équation* $T_{\mu^-} h = h$ *sont les fonctions constantes.*

Preuve. On démontre ce résultat en utilisant le théorème précédent, l'hypothèse d'irréductibilité et le fait que $\mathrm{supp}(\mu^-) = \left(\mathrm{supp}(\mu)\right)^-$.   □

Remarque 1. Sur les groupes abéliens, cette propriété est vraie pour une loi seulement supposée adaptée. Avec cette hypothèse, la question reste ouverte pour les hypergroupes.

Signalons cependant que Gebuhrer l'a résolue dans [12] pour certains hypergroupes à croissance polynomiale.

3 Frontière de Martin des marches aléatoires

Hypothèses. Dans cette partie, on ne considérera que des hypergroupes involutifs, c'est-à-dire des hypergroupes pour lesquels l'involution décrite par (v) dans la Définition 1 est l'identité. De tels hypergroupes sont nécessairement commutatifs. De plus, les marches aléatoires seront supposées irréductibles sur $(X, *)$, et leurs lois seront à support compact et à densité continue par rapport à la mesure de Haar de X.

Définition 4. *On notera par* $\mathcal{C}_\mu$ *l'ensemble des fonctions harmoniques continues positives relatives à la marche aléatoire de loi* μ, *c'est-à-dire l'ensemble des fonctions continues positives* h *sur* X *vérifiant* $T_\mu h = h$. *On notera par* $\mathcal{B}_\mu$ *le sous-ensemble de* $\mathcal{C}_\mu$ *constitué des fonctions* h *vérifiant l'équation* $h(e) = 1$.

Commençons par donner quelques propriétés de $\mathcal{C}_\mu$.

3.1 Propriétés du cône des fonctions harmoniques

Lemme 1. *Soit* $h \in \mathcal{C}_\mu$. *Si* h *n'est pas la fonction nulle, alors* h *ne s'annule pas sur* H.

Preuve. Il s'agit d'une application directe de l'hypothèse d'irréductibilité.   □

Proposition 2. *L'ensemble* $\mathcal{C}_\mu$ *est un cône convexe réticulé pour son ordre propre, dont* $\mathcal{B}_\mu$ *est une base compacte pour la topologie de la convergence uniforme sur les compacts.*

Preuve. Le cône $\mathcal{C}_\mu$ est réticulé pour son ordre propre, puisque si h et h' sont deux fonctions appartenant à $\mathcal{C}_\mu$, la formule suivante définit la borne supérieure dans $\mathcal{C}_\mu$ de h et h' (voir [18]) :

$$h \vee h' = \lim_{n \to \infty} \uparrow T_{\mu^{*n}}\big(\max(h, h')\big).$$

Soir r la densité de μ par rapport à m. Par hypothèse, r est continue et à support compact.

On montre facilement par récurrence que pour $n \geqslant 1$, $\|r^{*\,n}\|_\infty \leqslant \|r\|_\infty$. Cette inégalité prouve que la série de fonctions $\sum_{n \geqslant 1} 2^{-n} r^{*\,n}$ converge normalement sur X, et que sa somme S est continue. De plus, S est strictement positive sur X. En effet, puisque $Y_\mu = X$ et que $\mu^{*n} = r^{*\,n} m$, pour tout $x \in X$, il existe $n \geqslant 1$ tel que $r^{*\,n}(x) > 0$.

Montrons maintenant que pour $h \in \mathcal{C}_\mu$ et x, $y \in X$, on a

$$|h(x) - h(y)| \leqslant h(e)\varepsilon(x, y), \tag{3}$$

où

$$\varepsilon(x, y) = \sup_{z \in X} \frac{|T_x r(z) - T_y r(z)|}{S(z)}.$$

On a en effet

$$
\begin{aligned}
h(x) - h(y) &= T_\mu h(x) - T_\mu h(y) \\
&= \int_X T_x h(z) r(z)\, m(\mathrm{d}z) - \int_X T_y h(z) r(z)\, m(\mathrm{d}z) \\
&= \int_X h(z) T_x r(z) m(\mathrm{d}z) - \int_X h(z) T_y r(z)\, m(\mathrm{d}z) \\
&= \int_X h(z) \frac{T_x r(z) - T_y r(z)}{S(z)} S(z)\, m(\mathrm{d}z).
\end{aligned}
$$

Ainsi,

$$|h(x) - h(y)| \leqslant \left(\int_X h(z) S(z)\, m(\mathrm{d}z) \right) \varepsilon(x, y).$$

Enfin,

$$
\begin{aligned}
\int_X h(z) S(z)\, m(\mathrm{d}z) &= \sum_{n \geqslant 1} 2^{-n} \int_X h(z) r^{*\,n}\, m(\mathrm{d}z) \\
&= \sum_{n \geqslant 1} 2^{-n} T_{\mu^{*\,n}} h(e) \\
&= \sum_{n \geqslant 1} 2^{-n} h(e) = h(e).
\end{aligned}
$$

L'inégalité (3) et le théorème d'Ascoli permettent de prouver que les restrictions des éléments de $\mathcal{B}_\mu$ à tout compact de X forment une famille relativement compacte pour la topologie de la convergence uniforme. L'espace X étant localement compact, on en déduit que $\mathcal{B}_\mu$ est relativement compacte pour la topologie de la convergence uniforme sur les compacts. On conclut en remarquant que $\mathcal{B}_\mu$ est fermé dans $\mathcal{C}_\mu$. □

Il est facile de voir que $\mathcal{C}_\mu$ contient les fonctions constantes et les fonctions multiplicatives positives χ de X vérifiant $\int_X \chi(x)\,\mu(\mathrm{d}x) = 1$. Avant d'expliciter les points extrémaux de $\mathcal{B}_\mu$, on va étudier la marche aléatoire relativisée par une telle fonction.

3.2 Marche aléatoire relativisée

Soit χ une fonction multiplicative positive de $(X, *)$ vérifiant $\int_X \chi(x)\,\mu(\mathrm{d}x) = 1$. Elle est donc harmonique pour la marche aléatoire de loi μ.

Commençons par rappeler un résultat de Voit [24] :

Proposition 3. *Pour x, $y \in X$, on pose*

$$\delta_x \circ \delta_y(\mathrm{d}z) = \frac{1}{\chi(x)\chi(y)}\,\chi(z)\,\delta_x * \delta_y(\mathrm{d}z).$$

Alors, $(X, \circ)$ est un hypergroupe involutif d'élément unité e. La convolée de deux mesures μ, $\nu \in M(X)$ est donnée par

$$\mu \circ \nu(\mathrm{d}z) = \chi(z)\left(\frac{\mu}{\chi} * \frac{\nu}{\chi}\right)(\mathrm{d}z).$$

D'autre part, étant donné un noyau markovien P, à toute fonction P-harmonique positive h on peut associer un nouveau noyau markovien P^h en posant

$$P^h(x, \mathrm{d}y) = \frac{1}{h(x)}\,P(x, \mathrm{d}y)h(y).$$

On dit que P^h est le noyau relativisé de P par la fonction harmonique h [23].

La proposition suivante montre que si P est le noyau de la marche aléatoire de loi μ sur $(X, *)$, alors P^χ est aussi le noyau d'une marche aléatoire mais sur $(X, \circ)$.

Proposition 4. *Soit μ^χ la mesure de probabilité sur X définie par $\mu^\chi(\mathrm{d}y) = \chi(y)\,\mu(\mathrm{d}y)$. Alors, P^χ est le noyau de la marche aléatoire sur $(X, \circ)$ de loi μ^χ.*

Preuve. Il s'agit d'une simple vérification. On a pour tout x dans X :

$$\begin{aligned}
\delta_x \circ \mu^\chi(\mathrm{d}t) &= \frac{\chi(t)}{\chi(x)} \int_X \frac{\delta_x * \delta_y(\mathrm{d}t)}{\chi(y)}\,\mu^\chi(\mathrm{d}y) \\
&= \frac{\chi(t)}{\chi(x)} \int_X \delta_x * \delta_y(\mathrm{d}t)\,\mu(\mathrm{d}y) \\
&= \frac{\chi(t)}{\chi(x)}\,\delta_x * \mu(\mathrm{d}t) \\
&= P^\chi(x, \mathrm{d}t).
\end{aligned}$$

□

3.3 Points extrémaux du cône des fonctions harmoniques et formule de représentation intégrale

Théorème 2. *Les points extrémaux du convexe $\mathcal{B}_\mu$ sont les fonctions multiplicatives positives χ sur X vérifiant*

$$\int_X \chi(x)\,\mu(\mathrm{d}x) = 1.$$

Preuve. Soit χ un point extrémal du convexe $\mathcal{B}_\mu$. La fonction χ est harmonique donc

$$\chi(\,\cdot\,) = \int_X T_y\chi(\,\cdot\,)\,\mu(\mathrm{d}y),$$

et, puisque d'après le Lemme 1 χ ne s'annule pas, on a

$$\chi(\,\cdot\,) = \int_X \frac{T_y\chi(\,\cdot\,)}{\chi(y)}\chi(y)\,\mu(\mathrm{d}y). \tag{4}$$

L'hypergroupe étant commutatif, pour tout $y \in X$, $T_y\chi(\,\cdot\,)$ est une fonction harmonique et ainsi $T_y\chi(\,\cdot\,)/\chi(y) \in \mathcal{B}_\mu$. De plus, $\int_X \chi(y)\mu(\mathrm{d}y) = T_\mu\chi(e) = \chi(e) = 1$, donc l'équation (4) implique, puisque χ est extrémal, que pour tout $y \in X$, $T_y\chi(\,\cdot\,)/\chi(y) = \chi$. Ce résultat de convexité assez intuitif est par exemple démontré dans [2, Chap. IV, §7, Proposition 3]. La fonction χ est donc de la forme annoncée.

Réciproquement, considérons une fonction multiplicative positive χ telle que $\int_X \chi(x)\,\mu(\mathrm{d}x) = 1$. On a déjà remarqué qu'une telle condition implique que $\chi \in \mathcal{B}_\mu$. Posons $P(x,\mathrm{d}y) = \delta_x * \mu(\mathrm{d}y)$ et utilisons les notations de la Section 3.2. Il est facile de voir qu'une fonction h est harmonique pour P si et seulement si $\frac{h}{\chi}$ est harmonique pour P^χ. Ainsi, χ sera harmonique extrémale pour P si et seulement si la fonction constante égale à $\mathbf{1}$ est harmonique extrémale pour P^χ. Cette dernière assertion résulte de la Proposition 4 et du Corollaire 1, dont les hypothèses sont bien vérifiées : la loi μ^χ est irréductible puisque $(\mu^\chi)^{\circ n} = \chi\mu^{*n}$ et $\chi > 0$. Ainsi, le résultat est démontré. $\square$

Notons

$$\mathcal{S}_\mu = \left\{ \chi \text{ multiplicatives et positives sur } X \text{ telles que } \int_X \chi(x)\,\mu(\mathrm{d}x) = 1 \right\}.$$

Une simple application du théorème de Choquet [3], nous donne alors la formule de représentation intégrale suivante :

Corollaire 2. *Il y a une correspondance biunivoque entre le cône des fonctions harmoniques $\mathcal{C}_\mu$ et l'ensemble des mesures boréliennes, positives, bornées sur $\mathcal{S}_\mu$. Cette correspondance est donnée par la formule :*

$$h(\,\cdot\,) = \int_{\mathcal{S}_\mu} \chi(\,\cdot\,)\,\nu(\mathrm{d}\chi).$$

En théorie, une telle formule s'obtient également en étudiant le comportement asymptotique des noyaux de Green

$$G(x, \mathrm{d}y) = \frac{\delta_x * \sum_{n \geqslant 0} \mu^{*n}(\mathrm{d}y)}{\sum_{n \geqslant 0} \mu^{*n}(\mathrm{d}y)}.$$

Le lecteur intéressé consultera à ce sujet la théorie de Martin [17] adaptée au cadre des chaînes de Markov [8, 16, 15]. En pratique, cette dernière méthode est délicate et très technique, voir par exemple [19] pour son application au cas des marches aléatoires sur $\mathbb{Z}^d$. Ici, $\mathcal{S}_\mu$ s'identifie à la partie extrémale de la frontière de Martin. La détermination explicite de $\mathcal{S}_\mu$ pour certains hypergroupes d-dimensionnels a été effectuée par l'auteur dans [13] et [14].

Une autre possibilité pour obtenir la formule de représentation intégrale du Corollaire 2, est d'adapter la méthode employée par Raugi dans [20]. Basée sur des résultats classiques de la théorie des martingales, elle évite l'usage du théorème de Choquet.

Références

1. BLOOM W. et HEYER H. *Harmonic analysis of probability measures on hypergroups*, volume 20 of *de Gruyter Studies in Mathematics*. Walter de Gruyter & Co., Berlin, 1995.
2. BOURBAKI N. *Éléments de mathématique. Fasc. XIII. Livre VI: Intégration. Chapitres 1, 2, 3 et 4 : Inégalités de convexité, Espaces de Riesz, Mesures sur les espaces localement compacts, Prolongement d'une mesure, Espaces L^p*. Deuxième édition revue et augmentée. Actualités Scientifiques et Industrielles, No. 1175. Hermann, Paris, 1965.
3. CHOQUET G. Existence et unicité des représentations intégrales au moyen des points extrémaux dans les cônes convexes. In *Séminaire Bourbaki, Vol. 4*, pages Exp. No. 139, 33–47. Soc. Math. France, Paris, 1995.
4. CHOQUET G. et DENY J. Sur l'équation de convolution $\mu = \mu \star \sigma$. *C. R. Acad. Sci. Paris, série I*, 250 : 799–801, 1960.
5. CONZE J.-P. et GUIVARC'H Y. Propriété de droite fixe et fonctions propres des opérateurs de convolution. In *Théorie du potentiel et analyse harmonique*, number 404 in Lecture Note in Math., pages 126–133. Springer-Verlag, 1974.
6. DERRIENNIC Y. Marche aléatoire sur le groupe libre et frontière de Martin. *Z. Wahrscheinlichkeitstheorie und Verw. Gebiete*, 32 (4) : 261–276, 1975.
7. DERRIENNIC Y. Sur le théorème de point fixe de Brunel et le théorème de Choquet-Deny. *Ann. Sci. Univ. Clermont-Ferrand II Probab. Appl.*, (4) : 107–111, 1985.
8. DOOB J. L. Discrete potential theory and boundaries. *J. Math. Mech.*, 8 : 433–458; erratum 993, 1959.
9. FURSTENBERG H. Translation-invariant cones of functions on semi-simple Lie groups. *Bull. Amer. Math. Soc.*, 71 : 271–326, 1965.
10. GALLARDO L. A multidimensional central limit theorem for random walks on hypergroups. *Stoch. Stoch. Rep.*, 73 (1–2) : 1–23, 2002.

11. GALLARDO L. et GEBUHRER O. Marches aléatoires et hypergroupes. *Exposition. Math.*, 5 (1) : 41–73, 1987.

12. GEBUHRER M. O. Algèbres de Banach commutatives de mesures de Radon bornées et théorèmes ergodiques de type L^1. *C. R. Acad. Sci. Paris Sér. I Math.*, 306 (2) : 67–70, 1988.

13. GODEFROY L. Frontière de Martin des marches aléatoires sur certains hypergroupes d-dimensionnels. *C. R. Acad. Sci. Paris Sér. I Math.*, 334 (11) : 1029–1034, 2002.

14. GODEFROY L. Martin boundary of polynomial random walks on the d-dimensional lattice of non-negative integers. *J. Theor. Probab.*, 2003. To appear.

15. GUIVARC'H Y., JI L. et TAYLOR J. C. *Compactifications of symmetric spaces*, volume 156 of *Progress in Mathematics*. Birkhäuser Boston Inc., Boston, MA, 1998.

16. HUNT G. A. Markoff chains and Martin boundaries. *Illinois J. Math.*, 4 : 313–340, 1960.

17. MARTIN R. S. Minimal positive harmonic functions. *Trans. Amer. Math. Soc.*, 49 : 137–172, 1941.

18. NEVEU J. Chaînes de Markov et théorie du potentiel. *Ann. Fac. Sci. Univ. Clermont-Ferrand*, (24) : 37–89, 1964.

19. NEY P. et SPITZER F. The Martin boundary for random walk. *Trans. Amer. Math. Soc.*, 121 : 116–132, 1966.

20. RAUGI A. Une démonstration d'un théorème de Choquet-Deny par les martingales. *Ann. Inst. H. Poincaré Sect. B (N.S.)*, 19 (1) : 101–109, 1983.

21. RAUGI A. Sur une équation de convolution. *Gaz. Math.*, (55) : 93–94, 1993.

22. RAUGI A. Fonctions harmoniques positives sur certains groupes de Lie résolubles connexes. *Bull. Soc. Math. France*, 124 (4) : 649–684, 1996.

23. REVUZ D. *Markov chains*, volume 11 of *North-Holland Mathematical Library*. North-Holland Publishing Co., Amsterdam, second edition, 1984.

24. VOIT M. Positive characters on commutative hypergroups and some applications. *Math. Z.*, 198 (3) : 405–421, 1988.

Exponential Decay Parameters Associated with Excessive Measures

Mioara Buiculescu

Romanian Academy
Institute of Mathematical Statistics
and Applied Mathematics
Calea 13 Septembrie nr.13
RO-76100 Bucharest, Romania
e-mail: bmioara@csm.ro

Summary. Let X be a Markov process with semigroup (P_t) and m an excessive measure of X. With m we associate the spectral radius $\lambda_r^{(p)}(m)$ of (P_t) on $L^p(m)$ ($1 \leqslant p \leqslant \infty$) and the exit parameter $\lambda_e^{\mathcal{C}}(m)$ defined for an m-nest $\mathcal{C} = (C_n)$ in terms of the corresponding first exit times (τ_n). We discuss the impact of these parameters as well as their connection with other parameters of interest for the process.

Mathematics subject classifications (2000): 60J25, 60J45.

Key words: Markov process, excessive measure, decay parameter.

1 Introduction

When X is a transient Markov process it is sometimes possible to associate with it a new process $\widetilde{X}$ endowed with a conservative measure $\widetilde{m}$. Limit theorems obtained for $\widetilde{X}$ relative to $\widetilde{m}$ provide information on the long term behaviour of the initial process X. The process $\widetilde{X}$ and the measure $\widetilde{m}$ are obtained by means of a γ-subinvariant function and a γ-subinvariant measure (to be precisely defined in the sequel), whence the interest in the classes of γ-subinvariant functions and γ-subinvariant measures and the parameters associated with them.

In case of Harris irreducible processes this is a well established theory, sometimes called λ-*theory* (see [Ber97], [NN86], [TT79]). Related results for processes that are not necessarily irreducible are given in [Glo88] and [Str82].

The present paper is concerned with this kind of problems and they are considered in the context of the theory of excessive measures ([DMM92] and [Get90]). Unless otherwise mentioned the process X is assumed to be Borel right with state space $(E, \mathcal{E})$, semigroup (P_t), resolvent (U^q) and lifetime ζ.

For any $\gamma \geqslant 0$ we consider:

$\mathbf{M}^\gamma(X) := \{\eta : \sigma\text{-finite measure on } (E, \mathcal{E}) \text{ such that } e^{\gamma t}\eta P_t \leqslant \eta, \forall t \geqslant 0\}$,

$\mathbf{F}^\gamma(X) := \{f \in \mathcal{E}^* : f \geqslant 0 \text{ on } E \text{ such that } e^{\gamma t}P_t f \leqslant f, \forall t \geqslant 0\}$.

An element of $\mathbf{M}^\gamma(X)$ (resp. $\mathbf{F}^\gamma(X)$) is called a γ-*subinvariant mesure* (resp. a γ-*subinvariant function*). They are called γ-invariant measures (resp. γ-invariant functions) when all the corresponding inequalities become equalities. Standard candidates as members of $\mathbf{F}^\gamma(X)$ are

$$\Phi^\gamma f(x) := \int_0^\infty e^{\gamma t}P_t f(x)\, dt, \qquad f \in \mathcal{E}^*, \, f \geqslant 0,$$

and as members of $\mathbf{M}^\gamma(X)$ the measures $\mu\Phi^\gamma$ provided they are σ-finite.

The following global parameters are of interest and were considered in various contexts ([Glo88], [NN86], [Str82], [TT79]):

$\lambda_\pi := \sup\{\gamma \geqslant 0 : \mathbf{M}^\gamma(X) \neq \{0\}\}$;

$\lambda_\varphi := \sup\{\gamma \geqslant 0 : \exists f \in \mathbf{F}^\gamma(X), \, f > 0 \text{ on } E, \, f \text{ not identically } \infty\}$.

When $\gamma = 0$ instead of $\mathbf{M}^\gamma(X)$ we write as usual $\text{Exc}(X)$ and consider its well known important subclasses:

$\text{Pur}(X) := \{m \in \text{Exc}(X) : mP_t(h) \to 0 \text{ when } t \to \infty, \forall h > 0, m(h) < \infty\}$;

$\text{Inv}(X) := \{m \in \text{Exc}(X) : mP_t = m, \forall t \geqslant 0\}$;

$\text{Con}(X) := \{m \in \text{Exc}(X) : m(Uh < \infty) = 0, \forall h > 0, \, m(h) < \infty\}$;

$\text{Dis}(X) := \{m \in \text{Exc}(X) : m(Uh = \infty) = 0, \forall h > 0, \, m(h) < \infty\}$.

Whenever $m \in \text{Exc}(X)$, each P_t, $t > 0$, and each qU^q, $q > 0$, may be thought as a contraction from $L^p(m)$ to $L^p(m)$ for $1 \leqslant p \leqslant \infty$; also, the semigroup (P_t) is strongly continuous on $L^p(m)$, $1 \leqslant p \leqslant \infty$ (these facts are discussed in a more general setting in [Get99]). As usual let $\lambda_r^{(p)}(m)$, the *spectral radius* of (P_t) on $L^p(m)$, $1 \leqslant p \leqslant \infty$, be defined as

$$\lambda_r^{(p)}(m) := \lim_{t \to \infty}\{-t^{-1}\ln\|P_t\|_{L^p(m)}\}.$$

The second section is devoted to the impact of $\lambda_r^{(p)}(m)$, $1 \leqslant p \leqslant \infty$, as decay parameters. First the connection of $\lambda_r^{(2)}(m)$ with λ_π is discussed. Then under certain restrictions on the process (imposed by the application of a very powerful result of Takeda [Tak00]) one gets that $\lambda_r^{(p)}(m)$ are independent of p and one identifies this common value with the decay parameter associated with X as irreducible process. The remaining part of section 2 is concerned exclusively with properties of $\lambda_r^{(1)}(m)$ having in view especially the connection with $\lambda_\pi, \lambda_\varphi$. An expression of $\lambda_r^{(1)}(m)$ in terms of the Kuznetsov measure Q_m associated with m is given. This allows to distinguish those measures m

for which $\lambda_r^{(1)}(m) > 0$ among the purely excessive ones. Finally, the underlying construction in the classical context of quasi-stationary distributions for irreducible processes is retrieved in the general case emphasizing the role of $\lambda_r^{(1)}(m)$.

Section 3 introduces the exit parameter associated with an m-nest, when m is a dissipative measure. These are similar to some parameters considered in [Str82] in a more specific context.

Notation. As is standard, we denote by $\mathcal{E}^e$ the σ-algebra generated by excessive functions and for any $B \in \mathcal{E}^e$, $T_B := \inf\{t > 0 : X_t \in B\}$, $\tau_B := T_{E \setminus B}$, $P_B f(x) := P^x\{f(X_{T_B}); T_B < \infty\}$.

2 Spectral radius as decay parameter

We start by revealing the special role played by the spectral radius in case of some irreducible processes. Recall that the process X is said to be μ-*irreducible*, with μ a σ-finite measure, if:

$$\mu\text{-(I)} \qquad\qquad \mu(B) > 0 \implies \forall x \in E, \quad U^1(x, B) > 0.$$

The following theorem introduces the *decay parameter* associated with the irreducible process X.

Theorem 1 ([TT79]). *For any Markov process X satisfying μ-(I) there exist a μ-polar set Γ, an increasing sequence of sets $(B_n) \subseteq \mathcal{E}^e$ with $E = \bigcup_n B_n$ and a parameter $\lambda \in [0, \infty[$ such that:*

(i)   *For any $\gamma < \lambda$ we have $\Phi^\gamma(x, B_n) < \infty$, $\forall x \notin \Gamma$, $n \in \mathbb{N}$.*
(ii) *For any $\gamma > \lambda$ we have $\Phi^\gamma(x, B) \equiv \infty$, $\forall B \in \mathcal{E}$, $\mu(B) > 0$.*
(iii) *The process is either λ-transient, i.e. (i) holds for $\gamma = \lambda$, or it is λ-recurrent, i.e. (ii) holds for $\gamma = \lambda$.*
(iv) $\lambda = \lambda_\pi = \lambda_\varphi$.

We recall also that the whole theory of irreducible processes is based on the existence of a remarkable class of sets, namely:

$$\mathcal{L}(\mu) := \{B \in \mathcal{E}^e : \mu(B) > 0 \text{ for which there exists a measure } \nu_B \neq 0$$
$$\text{such that } U^1(x, \,.\,) \geqslant \nu_B(\,.\,), \forall x \in B\}.$$

The impact of $\mathcal{L}(\mu)$ comes from the fact that whenever B is in $\mathcal{L}(\mu)$ and $\Phi^\gamma(x, B) = \infty$ for some $x \in E$, one has $\Phi^\gamma(x, A) = \infty$ for any $A \in \mathcal{E}$ such that $\mu(A) > 0$.

The next result is concerned with the connection between $\lambda_r^{(2)}(m)$ and λ_π.

Proposition 1. (i) *For any excessive measure m we have $\lambda_r^{(2)}(m) \leqslant \lambda_\pi$.*

(ii) *Suppose that the state space is locally compact with countable base and that X is a Feller process, m-symmetric with respect to the Radon measure m. Suppose further that m-(I) holds and that the support of m has non-empty interior. Then $\lambda_r^{(2)}(m) = \lambda$.*

Proof. (i) We shall actually show that given $h \in \mathcal{E}$ such that $0 < h \leqslant 1$ and $m(h) < \infty$, the parameter

$$\lambda(m;h) := \liminf_{t \to \infty}\left\{ -t^{-1}\ln\frac{(h,P_t h)_m}{\|h\|_{L^{(2)}(m)}} \right\}$$

satisfies $\lambda(m;h) \leqslant \lambda_\pi$. This will be enough to prove (i) since $\lambda_r^{(2)}(m) \leqslant \lambda(m;h)$. To this end consider $\gamma < \lambda(m;h)$, $\gamma' \in\ ]\gamma, \lambda(m;h)[$ and let $t_{\gamma'}$ be such that $(h,P_t h)_m \leqslant e^{-\gamma' t}\|h\|_{L^{(2)}(m)}$, $\forall t \geqslant t_{\gamma'}$. Then the measure $\nu_\gamma(g) := (\gamma' - \gamma)\int_{t_{\gamma'}}^{\infty} e^{t\gamma} m(hP_t g)\,dt$, $g \in \mathcal{E}$, $g \geqslant 0$, is in $\mathbf{M}^\gamma$.

(ii) By (i) and m-(I) we have $\lambda_r^{(2)}(m) \leqslant \lambda_\pi = \lambda$. We will next prove that for any $\gamma > \lambda_r^{(2)}(m)$ there exists a compact K such that $m(K) > 0$ and $\Phi^\gamma 1_K \equiv \infty$. According to Proposition 6.3.8 (ii) in [MT96], whose hypotheses are the ones in (ii) but for the m-symmetry, any compact K such that $m(K) > 0$ is in $\mathcal{L}(m)$. This will be enough to ensure that $\gamma \geqslant \lambda$ and thus $\lambda \leqslant \lambda_r^{(2)}(m)$. Let $(E_z)_{z \in \mathbb{R}}$ be the resolution of the generator of $\{P_t;\ t > 0\}$ on $L^2(m)$ and let $\gamma' \in\ ]\lambda_r^{(2)}(m), \gamma[$. There exists a function φ, continuous and with compact support such that $E_{-\gamma'}\varphi \neq E_{-\lambda_r^{(2)}(m)}\varphi$. Then φ may be written as $\varphi = \int_{-\gamma'}^{-\lambda_r^{(2)}(m)} dE_z \varphi$ and

$$(|\varphi|, P_t|\varphi|)_m \geqslant (\varphi, P_t\varphi)_m = \int_{-\gamma'}^{-\lambda_r^{(2)}(m)} e^{zt}\,d(\varphi, E_z\varphi) \geqslant e^{-\gamma' t}\Big(\varphi, E_{-\lambda_r^{(2)}(m)}\varphi\Big)_m.$$

Whence $(|\varphi|, \Phi^\gamma|\varphi|)_m = \infty$, which in turn implies $(1_{K_\varphi}, \Phi^\gamma 1_{K_\varphi})_m = \infty$, K_φ being the compact support of φ.

By m-(I) we have only two possibilities: either $\Phi^\gamma 1_{K_\varphi}$ is finite up to an m-polar set, or $\Phi^\gamma 1_{K_\varphi} \equiv \infty$. Assuming the first possibility we get an $M > 0$ such that $m(\Phi^\gamma 1_{K_\varphi} \leqslant M) > 0$ and therefore there exists a compact K such that $K \subseteq \{\Phi^\gamma 1_{K_\varphi} \leqslant M\}$. This implies $m(1_K \Phi^\gamma 1_{K_\varphi}) \leqslant Mm(K) < \infty$. On the other hand from $K_\varphi \in \mathcal{L}(m)$ and from the classical formula $\Phi^\gamma 1_K = \sum_{n=1}^{\infty}(1+\gamma)^{n-1}U^{1(n)}1_K$ we have $\Phi^\gamma 1_K \geqslant \nu_{K_\varphi}(K)(1+\gamma)\Phi^\gamma 1_{K_\varphi}$. Using this and m-symmetry we get

$$(1_K, \Phi^\gamma 1_{K_\varphi})_m = (1_{K_\varphi}, \Phi^\gamma 1_K)_m \geqslant \nu_{K_\varphi}(K)(1+\gamma)(1_{K_\varphi}, \Phi^\gamma 1_{K_\varphi})_m = \infty.$$

The obtained contradiction rules out the possibility that $\Phi^\gamma 1_{K_\varphi}$ is finite up to an m-polar set and thus $\Phi^\gamma 1_{K_\varphi} \equiv \infty$, the property which was to be proved. $\square$

We now briefly discuss two other forms of irreducibility that will be involved in the sequel. We begin with

(I)   for any finely open, non-empty set Γ we have $P_\Gamma 1(x) > 0$, $\forall x \in E$.

Condition (I) amounts to the property that all states communicate in the sense of [ADR66]. Immediate consequences of it are : the fact that all excessive measures are equivalent and the property that any $\xi \in \mathrm{Exc}(X)$ satisfies ξ-(I).

The next condition of irreducibility is the one imposed in [Tak00]. Let m be a σ-finite measure on $(E, \mathcal{E})$ and $\mathcal{I}(m) := \{A \in \mathcal{E} : 1_A P_t f = P_t 1_A f \text{ a.e.-} m$ for any bounded $f \in L^2(m)\}$. The condition is the following:

m-(I)'   for any $A \in \mathcal{I}(m)$ we have either $m(A) = 0$ or $m(E\backslash A) = 0$.

In what follows we consider conditions under which it is possible to relate the three forms of irreducibility.

Lemma 1. *Let m belong to $\mathrm{Exc}(X)$.*

(i) *If m is Radon and m-(I) holds, then m-(I)' also holds.*
(ii) *If the process is m-symmetric, if m is a reference measure and if m-(I)' holds, then (I) also holds.*

Proof. To get (i) let $A \in \mathcal{I}(m)$ be such that $m(A) > 0$. For any compact $K \subseteq E\backslash A$ we have $m(1_A U^1 1_K) = 0$, implying by m-(I) that $m(K) = 0$, whence $m(E\backslash A) = 0$.

For (ii) let us note that by m-symmetry any absorbing set A is in $\mathcal{I}(m)$ and then apply this to the set $A = \{U^1 1_\Gamma = 0\}$ with Γ finely open, non-empty. Taking into account that m is a reference measure, the possibility that $m(E\backslash A) = 0$ is ruled out and from $m(A) = 0$ we get $P_\Gamma 1(x) > 0$, $\forall x \in E$. $\square$

We now turn to the very special case, indicated in the introduction, when the p-independence of $\lambda_r^{(p)}(m)$, $1 \leqslant p \leqslant \infty$, occurs.

Theorem 2. *Assume that the state space $(E, \mathcal{E})$ is a locally compact metric space with countable base and that X is an m-symmetric Markov process, with m a Radon measure. Assume also that the following conditions are satisfied:*

(i) *m-(I)' holds.*
(ii) *For each $t > 0$ and $x \in E$, $P_t(x, \,.\,) \ll m$.*
(iii) *For any $t > 0$, $P_t f \in \mathcal{C}_0(E)$ whenever $f \in \mathcal{C}_0(E)$, where $\mathcal{C}_0(E)$ denotes the space of all continuous functions vanishing at infinity.*
(iv) *$U^1 1 \in \mathcal{C}_0(E)$.*

Then $\lambda_r^{(p)}(m)$, $1 \leqslant p \leqslant \infty$, is independent of $p \in [1, \infty]$ and the common value $\lambda_r(m)$ coincides with the decay parameter λ associated with X as irreducible process.

Proof. The imposed conditions are precisely those in Theorem 2.3 in [Tak00] which provides the p-independence of $\lambda_r^{(p)}(m)$, $1 \leqslant p \leqslant \infty$.

The process is subject to (I) by Lemma 1 (ii) and the equality $\lambda_r(m) = \lambda$ follows from Proposition 1 (ii), since under the present assumptions the support of m is E.   □

Remarks. 1. The conditions in Theorem 2 are met by the process (X, τ_D) equal to the d-dimensional Brownian motion killed at τ_D, where D is a regular domain satisfying $m(D) < \infty$. In this remarkable particular case we actually have a much stronger property, namely the corresponding transition operator P_t^D, $t > 0$, is a compact operator and has the same eigenvalues $\{\exp(-\lambda_k); \; k = 1, 2, \dots\}$ with $0 < \lambda_1 \leqslant \lambda_2 \leqslant \cdots < \infty$, in all *appropriate spaces* $L^{(p)}(m_D)$, $1 \leqslant p < \infty$, and $\mathcal{C}_0(D)$ ([CZ95], Theorem 2.7). The decay parameter λ coincides in this case with λ_1.

2. In [Sat85], Sato studies the impact of λ_r^∞ as decay parameter assuming (iii), a condition weaker than (v) in Theorem 2 and the following condition of irreducibility: for any open set $G \in \mathcal{E}$ one has $P_G 1(x) > 0$, $\forall x \in E$. Symmetry and absolute continuity are not assumed.

3. In [Str82], Stroock proposed for a Radon measure m

$$\lambda_\sigma(m) := \sup_{\substack{\varphi \in C_c(E) \\ \varphi \geqslant 0}} \left\{ \limsup_{t \to \infty} \{ t^{-1} \ln (\varphi, P_t \varphi)_m \} \right\}$$

as decay parameter and shows that—under his conditions and m-symmetry—it coincides with the right end point of the spectrum of the generator on $L^2(m)$ (actually $-\lambda_r^{(2)}(m)$ in that case). Here $\mathcal{C}_c(E)$ denotes the space of all real valued continuous functions on E with compact support.

In what follows we turn to specific properties of $\lambda_r^{(1)}(m)$, $m \in \text{Exc}(X)$.

Let Q_m and Y be the Kuznetsov measure and process associated with m. Let also α be the birthtime of Y and $\mathcal{H}(m) := \{ h \in \mathcal{E} : h \geqslant 0, \, 0 < m(h) < \infty \}$.

Proposition 2. *For any $m \in \text{Exc}(X)$ we have*

$$\lambda_r^{(1)}(m) = \sup \left\{ \gamma \geqslant 0 : \lim_{t \to \infty} e^{\gamma t} Q_m \big(h(Y_0); \, \alpha < -t \big) = 0, \, \forall h \in \mathcal{H}(m) \right\}.$$

If $\lambda_r^{(1)}(m) > 0$, then $m \in \text{Pur}(X)$.

Proof. First we note that by the Markov property of Y under Q_m and the stationarity of Q_m we have for any $f \in \mathcal{E}$, $f \geqslant 0$

$$m(P_t f) = Q_m \big(P^{Y_0} \big(f(X_t) \big); \, \alpha < 0 \big) = Q_m \big(f(Y_t); \, \alpha < 0 \big)$$
$$= Q_m \big(f(Y_0); \, \alpha < -t \big).$$

Recall now the following well known property of the spectral radius: for any $a < \lambda_r^{(1)}(m)$ there exists $M_a \geqslant 1$ such that $e^{at} \| P_t \|_{L^1(m)} \leqslant M_a$. Then put $\tilde{\lambda}(m) := \sup\{\gamma \geqslant 0 : \lim_{t\to\infty} e^{\gamma t} m P_t(h) = 0, \forall h \in \mathcal{H}(m)\}$. To show that $\lambda_r^{(1)}(m) \leqslant \tilde{\lambda}(m)$, let $\gamma < \lambda_r^{(1)}(m)$ and $\gamma' \in \,]\gamma, \lambda_r^{(1)}(m)[\,$; then there exists $M_{\gamma'} \geqslant 1$ such that $e^{\gamma' t} m P_t(h) \leqslant M_{\gamma'} m(h)$, $\forall h \in \mathcal{H}(m)$, and thus $e^{\gamma t} m P_t(h) \to 0$ as $t \to \infty$, $\forall h \in \mathcal{H}(m)$. The converse inequality follows observing that for any $\gamma < \tilde{\lambda}(m)$, the family of bounded operators $T_t := e^{\gamma t} P_t$, $t \geqslant 0$, is such that $\sup_t \|T_t f\| < \infty$, $\forall f \in L^1(m)$. By the uniform boundedness principle we have $\sup_t \|T_t\| < \infty$, implying $\gamma \leqslant \lambda_r^{(1)}(m)$.

Finally from the obtained formula we get $Q_m(\alpha = -\infty) = 0$ when $\lambda_r^{(1)}(m) > 0$, which is well known to be equivalent to $m \in \mathrm{Pur}(X)$. $\square$

We give now further results on $\lambda_r^{(1)}(m)$, $m \in \mathrm{Exc}(X)$, that will be of interest in connection with $\lambda_\pi, \lambda_\varphi$.

Proposition 3. (i) *If* $m \in \mathbf{M}^\gamma(X)$, *then* $\gamma \leqslant \lambda_r^{(1)}(m)$. *If* m *is* γ-*invariant then* $\lambda_r^{(1)}(m) = \gamma$.
(ii) *For any* $\gamma < \lambda_r^{(1)}(m)$ *there exist* $\nu_\gamma \in \mathbf{M}^\gamma(X)$ *and* $f_\gamma \in \mathbf{F}^\gamma(X)$, $f_\gamma > 0$, *such that* $\nu_\gamma(f_\gamma) < \infty$.
(iii) *For any* $t > 0$, $e^{-\lambda_r^{(1)}(m)t}$ *is in the spectrum of* P_t *on* $L^1(m)$.

Proof. Property (i) is checked by direct verification.

For (ii) let $\gamma < \lambda_r^{(1)}(m)$, $\gamma' \in \,]\gamma, \lambda_r^{(1)}(m)[\,$ and $h \in \mathcal{E}$, $h > 0$, $m(h) < \infty$. Let also $M_{\gamma'} \geqslant 1$ be such that $e^{\gamma t} m(P_t f) \leqslant M_{\gamma'} m(f)$, $\forall t > 0$, $\forall f \in \mathcal{E}$, $0 < m(f) < \infty$. We then set:

$$\nu_\gamma := m\Phi^\gamma \qquad \text{and} \qquad f_\gamma := \Phi^\gamma h$$

and these are the required elements in $\mathbf{M}^\gamma(X)$, respectively $\mathbf{F}^\gamma(X)$ because we can successively check that $m(f_\gamma) \leqslant M_{\gamma'}(\gamma' - \gamma)^{-1} m(h)$ and $\nu_\gamma(f_\gamma) \leqslant [M_{\gamma'}(\gamma' - \gamma)^{-1}]^2 m(h)$.

Property (iii) is a consequence of a very powerful result (Theorem 7.7 in [Dav81]) applied to the positive (in the sense that it applies non-negative functions from $L^1(m)$ into functions of the same kind) operator P_t, $t > 0$. (Unfortunately this does not ensure that any of the corresponding eigenfunctions is non-negative). $\square$

Corollary 1. (i) $\lambda_\pi = \sup_{m\in\mathrm{Exc}(X)}\{\lambda_r^{(1)}(m)\} = \sup_{m\in\mathrm{Pur}(X)}\{\lambda_r^{(1)}(m)\}$.
(ii) $\lambda_\pi = \lambda_\varphi = \lambda_{\pi,\varphi}$ *where*

$$\lambda_{\pi,\varphi} := \sup\{\gamma \geqslant 0 : \exists f \in \mathbf{F}^\gamma(X), \, f > 0 \text{ on } E$$
$$\text{and } \nu \in \mathbf{M}^\gamma(X) \text{ such that } \nu(f) < \infty\}.$$

To further emphasize the role played by $\lambda_r^{(1)}(m)$ we end up this section with the construction of the process $\tilde{X}$ and the measure $\tilde{m}$ alluded to in the introduction. This amounts to considering $m \in \mathbf{M}^\gamma(X)$ and $h \in \mathbf{F}^\gamma(X)$ such that $m(E \backslash E_h) = 0$, where $E_h := \{0 < h < \infty\}$. With h we associate

$\tilde{h} := \uparrow \lim_{t \downarrow 0} e^{\gamma t} P_t h$ so that $\tilde{h} \in \mathbf{F}^\gamma$, $\tilde{h} \leqslant h$, $\{\tilde{h} < h\}$ is a set of zero potential and $\tilde{h}$ is an excessive function.

The supermartingale multiplicative functional

$$M_t := e^{\gamma t}\left[\tilde{h}(X_0)\right]^{-1}\tilde{h}(X_t)1_{\{0<\tilde{h}(X_0)<\infty\}}1_{\{t<T_{\{h=0\}}\}}$$

defines a subprocess $\tilde{X}$ with state space $\tilde{E} := \{0 < \tilde{h} < \infty\}$ and semigroup

$$\tilde{P}_t f(x) := e^{\gamma t}\left[\tilde{h}(x)\right]^{-1}P_t\left(\tilde{h}f\right)(x), \qquad x \in \tilde{E}, \, f \in \mathcal{E}_{|\tilde{E}}, \, f \geqslant 0.$$

The process $\tilde{X}$ is in turn a right Markov process ([Sha88], § 62) and $\tilde{m} := \tilde{h}m$ belongs to $\mathrm{Exc}(\tilde{X})$; $\tilde{m} \in \mathrm{Inv}(\tilde{X})$ when m is γ-invariant.

A necessary and sufficient condition for $\tilde{m}$ to be in $\mathrm{Con}(\tilde{X})$ (which is a precondition for developing an ergodic theory with respect to $\tilde{m}$) is the following: for any $f \in \mathcal{E}$, $f > 0$ such that $m(f1_{\tilde{E}}) < \infty$ we have $m(\Phi^\gamma f1_{\tilde{E}} < \infty) = 0$. Note that when this condition is fulfilled we necessarily have $\gamma \geqslant \lambda_r^{(1)}(m)$; since m was taken from $\mathbf{M}^\gamma(X)$ we must have in fact in this case $\lambda_r^{(1)}(m) = \gamma$.

3 Exit parameters

Theorem 1 suggests that λ_π is (at least in the irreducible case) related to the amount of time spent by the process in *small sets*. The parameter λ_π may be also characterized in terms of escape from such sets and we are going to provide conditions for this. An alternative set of conditions are imposed in [Str82] in order to obtain Radon instead of σ-finite measures.

Recall from [FG96] that an *m-nest* associated with $m \in \mathrm{Dis}(X)$ is defined as an increasing sequence of finely open sets $\mathcal{C} = (C_n) \subseteq \mathcal{E}^e$ such that $P^m(\lim_n \tau_n < \infty) = 0$, where $\tau_n := \tau_{C_n}$.

For each $n \in \mathbb{N}$ let $(P_{t,n})$, (U_n^q) denote the semigroup and resolvent associated with the killed process (X, τ_n) and

$$\Phi_n^\gamma := \int_0^\infty e^{\gamma t} P_{t,n}\, dt = \sum_{p=1}^\infty (1+\gamma)^{p-1}U_n^{1(p)}f.$$

The m-nest (C_n) of interest for our problem will be assumed to have the following additional property:

$(*)$ there exists $D \in \mathcal{E}^e$, $D \subseteq C_1$, such that $U(x, B) > 0$, $\forall x \in E$ and $U_1^1(x, \,.\,) \geqslant \nu(\,.\,)$, $\forall x \in D$, where $\nu(\Gamma) := m_D(\Gamma)[m(D)]^{-1}$.

With the m-nest $\mathcal{C}$ having property $(*)$ we associate

$$\lambda_e^{\mathcal{C}}(m) := \sup\{\gamma \geqslant 0 : P^\nu(e^{\gamma \tau_n}) < \infty, \, \forall n \in \mathbb{N}\}.$$

Proposition 4. *Let C be an m-nest satisfying condition $(*)$. Then $\lambda_e^C(m) \leqslant \lambda_\pi$.*

Proof. Let $\gamma < \lambda_e^C(m)$. In order to construct a measure in $\mathbf{M}^\gamma(X)$ we start by considering the measures $\nu_n := \nu \Phi_n^\gamma$, $n \in \mathbb{N}$. For each $n \in \mathbb{N}$, $\nu_n \in \mathbf{M}^\gamma(X, \tau_n)$ and it is finite because $\nu_n(P^\cdot(\mathrm{e}^{\gamma'\tau_n})) < \infty$ for $\gamma' \in \,]\gamma, \lambda_e^C(m)[$ as is easily checked. Also, for each $n \in \mathbb{N}$, $\nu_n(D) > 0$, $P^{\nu_n}(\lim \tau_n < \zeta) = 0$ and, due to condition $(*)$, $\nu_n \geqslant (1 + \gamma)\nu_n(D)\nu$.

Next let $\mu_n := [\nu_n(D)]^{-1}\nu_n$, $n \in \mathbb{N}$. For each $n \in \mathbb{N}$ the measure μ_n is in turn finite and such that $\mu_n \in \mathbf{M}^\gamma(X, \tau_n)$, $\mu_n(D) = 1$, $\mu_n \geqslant (1 + \gamma)\nu$ and $P^{\mu_n}(\lim \tau_n < \zeta) = 0$.

Let further $\eta_n := \inf_{p \geqslant n} \mu_p$, $n \in \mathbb{N}$, define an increasing sequence of measures such that $\eta_n \in \mathbf{M}^\gamma(X, \tau_n)$, $\forall n \in \mathbb{N}$. For each $n \geqslant 1$, one has $\eta_n(D) \leqslant 1$, $\eta_n \geqslant (1 + \gamma)$ and

$$\eta_n(U^1 1_D) = \lim_{k \to \infty} P^{\eta_n}\left(\int_0^\infty \mathrm{e}^{-u} 1_D(X_u)\, \mathrm{d}u\right)$$
$$\leqslant \lim_{k \to \infty} \eta_k\big(U_k^1(D)\big) \leqslant \lim_{k \to \infty} \eta_k(D) \leqslant 1.$$

Finally, let $\eta := \uparrow\lim_{n \to \infty} \eta_n$. Obviously $\eta \geqslant (1 + \gamma)\nu$ and it is σ-finite because $\eta(U^1 1_D) \leqslant 1$ and $U^1 1_D(x) > 0$, $\forall x \in E$. It remains to show that $\eta \in \mathbf{M}^\gamma(X)$; this follows from the fact that for any $n \in \mathbb{N}$

$$\mathrm{e}^{\gamma t} \eta_n(P_t f) = \mathrm{e}^{\gamma t} \lim_{k \to \infty} P^{\eta_n}\big(f(X_t); t < \tau_k\big) \leqslant \lim_{k \to \infty} \eta_n(f) = \eta(f)$$

for each $f \in \mathcal{E}$, $f \geqslant 0$. $\qquad\square$

It is perhaps worth mentioning that while there exist a number of m-nests associated with $m \in \mathrm{Dis}(X)$ (see [FG96] in his respect), condition $(*)$ is quite restrictive. Among other things the very existence of a set D with the properties involved in $(*)$ entails the ν-irreducibility of the process X.

Acknowledgment. The results of this paper were partially obtained while the author benefited of a kind invitation from INSA-Rouen and they were presented at the Séminaire de Probabilités de Rouen.

References

[ADR66] Azéma, J., Kaplan-Duflo, M., Revuz, D.: Récurrence fine des processus de Markov. *Ann. Inst. Henri Poincaré* **2**, 185–220 (1966).

[Ber97] Bertoin, J.: Exponential decay and ergodicity of completely asymmetric Lévy processes in a finite interval. *Ann. Appl. Probab.* **7**, 156–169 (1997).

[CZ95] Chung, K.L., Zhao, Z.: *From Brownian Motion to Schrödinger's Equation.* Springer, Berlin Heidelberg New York (1995).

[Dav81] Davies, E.B.: *One parameter semigroups.* Oxford University Press, Oxford (1981).

[DMM92] Dellacherie, C., Maisonneuve, B., Meyer, P.A.: *Probabilités et Potentiel.* Ch. XVII–XXIV, Hermann, Paris (1992).

[FG96]   Fitzsimmons, P.J., Getoor, R.K.: Smooth measures and continuous additive functionals of right Markov processes. In: *Ito's Stochastic Calculus and Probability Theory.* Springer, Berlin Heidelberg New York (1996).

[Get90] Getoor, R.K.: *Excessive Measures.* Birkhäuser, Boston (1990).

[Get99] Getoor, R.K.: Measure perturbations of Markovian semigroups. *Potential Analysis* 11, 101–133 (1999).

[Glo88] Glover, J.: Quasi-stationary distributions, eigenmeasures and eigenfunctions of Markov processes. In: *Seminar on Stochastic Processes 1984.* Birkhäuser, Boston (1986).

[MT96]   Meyn, S.P., Tweedie, R.L.: *Markov Chains and Sochastic Stability.* Springer, Berlin Heidelberg New York (1996).

[NN86]   Niemi, S., Nummelin, E.: On non-singular renewal kernels with applications to a semigroup of transition kernels. *Stochastic Process. Appl.* 22, 177–202 (1986).

[Sat85] Sato, S.: An inequality for the spectral radius of Markov processes. *Kodai Math. J.* 8, 5–13 (1985).

[Sha88] Sharpe, M.J.: *General Theory of Markov Processes.* Academic Press, San Diego (1988).

[Str82] Stroock, D.: On the spectrum of Markov semigoups and the existence of invariant measures. In: *Functionl Analysis in Markov Processes.* Springer, Berlin Heidelberg New York (1982).

[Tak00] Takeda, M.: L^p-independence of the spectral radius of symmetric Markov semigroups. *Canadian Math. Soc., Conference Proceedings* 20, 613–623 (2000).

[TT79]   Tuominen, P., Tweedie, R.J.: Exponential decay of general Markov processes and their discrete skeletons. *Adv. Appl. Probab.* 11, 784–802 (1979).

[Yos80] Yosida, K.: *Functional Analysis, 6th ed.* Springer, Berlin Heidelberg New York (1980).

Positive Bilinear Mappings
Associated with Stochastic Processes

Valentin Grecea

Institute of Mathematics "Simion Stoilow"
of the Romanian Academy
P.O.Box 1-764
RO 70700 Bucharest
Romania
e-mail: Valentin.Grecea@imar.ro

Summary. On some vector spaces of adapted stochastic processes, we define increasing families of positive bilinear forms, which generalize the usual square brackets $[X, Y]$ and angle brackets $\langle X, Y \rangle$. We study the corresponding Hardy spaces especially for $p = 1$ or 2, and extend to this abstract framework results of Fefferman type from martingale theory.

1 Introduction

The role of the square bracket in martingale theory cannot be overestimated. The starting point of our study was Fefferman's theorem establishing the duality between H^1 and BMO by a positive bilinear form naturally associated to the square bracket. But there also exist other similar forms acting on different spaces leading to similar results (roughly speaking, of Fefferman type); see Pratelli [3], Stein [4] and Yor [5].

Our purpose is to unify all these results of Fefferman type in a common framework. In particular, our main result, Theorem 3.6, simultaneously extends (essentially) Fefferman's theorem and the similar results from [4] and [5]; the relationship with Pratelli's result is slightly different (and simpler).

2 Description of the framework; preliminaries

Throughout this paper $(\Omega, \mathcal{F}, \mathcal{F}_t, P)$ is a complete probability space endowed with a filtration $\mathcal{F}_t$ satisfying the usual conditions: it is right continuous and $\mathcal{F}_0$ contains all negligible sets of $\mathcal{F}$. We put $\mathcal{F}_{0-} = \mathcal{F}_0$.

On the vector space of all real valued, adapted processes on $R_+ \times \Omega$ we consider the equivalence relation: $X \sim Y$ iff X and Y are indistinguishable, that is, iff the set $\{\omega \in \Omega : \exists t \geqslant 0$ such that $X_t(\omega) \neq Y_t(\omega)\}$ is negligible. We denote by $\mathcal{A}$ the vector space (with induced operations) of equivalence classes with respect to this relation. When no confusion is possible, we identify a process with its equivalence class.

In the sequel we consider a vector subspace $\mathcal{S}$ of $\mathcal{A}$, and a symmetric bilinear mapping $[.\,,.]$ from $\mathcal{S} \times \mathcal{S}$ to $\mathcal{A}$, satisfying the following property: for any $X \in \mathcal{S}$, the process $[X, X]$ is *positive, increasing* and *right continuous*. We say that $[.\,,.]$ is a *positive bilinear mapping*. By polarization, it follows that for any $X, Y \in \mathcal{S}$, the process $[X, Y]$ is adapted, right continuous with finite variation on each trajectory.

Example 1. For any local martingales X, Y we consider their square bracket, that is, the unique adapted right continuous process with finite variation, denoted by $[X, Y]$, such that:

1) $XY - [X, Y]$ is a local martingale;
2) $\Delta[X, Y]_t = (\Delta X_t)(\Delta Y_t)$ for any $t > 0$;
3) $[X, Y]_0 = X_0 Y_0$.

It is known that the square bracket extends to semimartingales as a positive bilinear mapping still possessing 2) and 3) (see [1, VII, 44]).

Example 2. For any locally square integrable local martingales X and Y we consider their angle bracket $\langle X, Y \rangle$, the unique predictable right continuous process with finite variation such that $XY - \langle X, Y \rangle$ is a local martingale null at 0. It is in fact the predictable compensator of the square bracket $[X, Y]$ (see [1, VII, 39]).

Example 3. We denote by Λ_0 the space of thin ("minces" in French) optional processes X such that the increasing process

$$A_t^X = \sum_{0 \leqslant s \leqslant t} X_s^2$$

is finite for finite t. We call $\{X, X\}_t$ this process and we define $\{X, Y\}$ for any X and Y in Λ_0 by polarization: $\{X, Y\}_t = \sum_{0 \leqslant s \leqslant t} X_s Y_s$ (see [5]).

Example 4. Given a discrete filtration $\mathcal{F}_n$, we consider the space of sequences of (finite) random variables $(X_n)_{n \geqslant 0}$ such that X_n is $\mathcal{F}_n$-measurable for any $n \in N$. For any such sequences X and Y we define the sequence of random variables (see [4])

$$\{X, Y\}_n = \sum_{m=0}^{n} X_m Y_m.$$

Of course this situation may be "imbedded" in the above by considering the filtration $\mathcal{F}_t = \mathcal{F}_n$ for $n \leqslant t < n + 1$.

The following simple fact (suggested by the end of the proof of [1, VII, 53]) is fundamental for the sequel. We write shortly $[X, Y]_s^t = [X, Y]_t - [X, Y]_s$ for $s < t$, even for $s = 0_-$, which means $[X, Y]_t$.

Proposition 2.1. *For any fixed stopping times S, T such that $S \leqslant T$, and X, $Y \in \mathcal{S}$, we have*

$$(2.1) \qquad \left| [X, Y]_S^T \right| \leqslant \left([X, X]_S^T \right)^{\frac{1}{2}} \left([Y, Y]_S^T \right)^{\frac{1}{2}} \qquad \text{a.s.}$$

Proof. For any fixed $r \in \mathbb{R}$, we have

$$(2.2) \qquad 0 \leqslant [X+rY, X+rY]_S^T = [X, X]_S^T + 2r[X, Y]_S^T + r^2[Y, Y]_S^T \qquad \text{a.s.}$$

The exceptional set on which this inequality does not hold for some rational r is then negligible and hence (2.2) holds on the complement of this set for any $r \in \mathbb{R}$, by continuity; this obviously implies (2.1). $\qquad \square$

The proof of the next extension to our framework of the classical inequality of H. Kunita and S. Watanabe is now an easy adaptation of the proof given in [1, VII, 53].

Theorem 2.2. *Let X, $Y \in \mathcal{S}$ and H, K be measurable processes (not necessarily adapted). We have then*

$$(2.3) \qquad \int_0^\infty |H_s| |K_s| \, \bigl| \mathrm{d}[X, Y]_s \bigr| \leqslant \left(\int_0^\infty H_s^2 \, \mathrm{d}[X, X]_s \right)^{\frac{1}{2}} \left(\int_0^\infty K_s^2 \, \mathrm{d}[Y, Y]_s \right)^{\frac{1}{2}} \qquad \text{a.s.}$$

Proof. We obviously may reduce to the case where H, K are bounded and supported on some interval $[0, N]$. Also, we may replace the left side of (2.3) by

$$\left| \int_0^\infty H_s K_s \, \mathrm{d}[X, Y]_s \right|.$$

Now, using twice the monotone class theorem, we are reduced to the case where

$$H = H_0 I_{\{0\}} + H_1 I_{]0, s_1]} + \cdots + H_m I_{]s_{m-1}, s_m]}$$
$$K = K_0 I_{\{0\}} + K_1 I_{]0, t_1]} + \cdots + K_n I_{]t_{n-1}, t_n]}$$

with the H_i and K_j measurable and bounded. We obviously may assume then that $m = n$, $s_i = t_i$ for $i = 1, \ldots, n$. Putting $s_0 = 0$ and using (2.1), we now get by addition

$$\left| \int_0^\infty H_s K_s \, \mathrm{d}[X, Y]_s \right| \leqslant |H_0 K_0 [X, Y]_0| + \sum_{i=1}^n |H_i K_i [X, Y]_{s_{i-1}}^{s_i}|$$

$$\leqslant |H_0| \left([X, X]_0 \right)^{\frac{1}{2}} |K_0| \left([Y, Y]_0 \right)^{\frac{1}{2}} + \sum_{i=1}^n \left(H_i^2 [X, X]_{s_{i-1}}^{s_i} \right)^{\frac{1}{2}} \left(K_i^2 [Y, Y]_{s_{i-1}}^{s_i} \right)^{\frac{1}{2}} \qquad \text{a.s.}$$

and (2.3) follows by the Schwarz inequality. $\qquad \square$

We are now able to extend the Fefferman inequality to our setting; the proof is the same as in [1, VII, 86], which is itself an adaptation of the proof given by C. Herz in discrete time.

Theorem 2.3. *Let* $X, Y \in S$ *and* H, K *be optional processes. Let* $c \in [0, \infty]$ *be such that*

$$(2.4) \qquad \mathbb{E}\left[\int_{[T,\infty]} K_s^2 \, d[Y,Y]_s \,\middle|\, \mathcal{F}_T\right] \leqslant c^2 \qquad \text{a.s.}$$

for all stopping times T. *Then we have*

$$(2.5) \qquad \mathbb{E}\left[\int_{[0,\infty]} |H_s||K_s| \, |d[X,Y]_s|\right] \leqslant c\sqrt{2}\, \mathbb{E}\left[\left(\int_{[0,\infty]} H_s^2 \, d[X,X]_s\right)^{\frac{1}{2}}\right].$$

Proof. We may of course suppose $c < \infty$, and $H, K \geqslant 0$. We consider the (positive) increasing processes

$$\alpha_t = \int_{[0,t]} H_s^2 \, d[X,X]_s \qquad \text{and} \qquad \beta_t = \int_{[0,t]} K_s^2 \, d[Y,Y]_s,$$

to which we associate the positive optional processes U and V defined by the relations

$$U_s^2 = \begin{cases} \dfrac{H_s^2}{\sqrt{\alpha_s} + \sqrt{\alpha_{s-}}} & \text{for } \alpha_s > 0, \\ 0 & \text{for } \alpha_s = 0; \end{cases} \qquad V_s^2 = K_s^2 \sqrt{\alpha_s}.$$

The processes U and V have the following three properties:

$$(2.6) \qquad\qquad H_s K_s \leqslant \sqrt{2}\, U_s V_s$$

almost surely with respect to the measure $\big|d[X,Y]_s(\omega)\big|$ for almost all $\omega \in \Omega$,

$$(2.7) \qquad \mathbb{E}\left[\int_{[0,\infty]} U_s^2 \, d[X,X]_s\right] = \mathbb{E}\left[\sqrt{\alpha_\infty}\right],$$

$$(2.8) \qquad \mathbb{E}\left[\int_{[0,\infty]} V_s^2 \, d[Y,Y]_s\right] \leqslant c^2 \, \mathbb{E}\left[\sqrt{\alpha_\infty}\right].$$

Indeed, one first checks that (2.6) holds almost surely with respect to the measure $d[X,X]_s(\omega)$ for each ω (an elementary measure-theoretic exercise on the line). One then uses Theorem 2.2 (taking for H the indicator of the optional set $\{HK > \sqrt{2}\,UV\}$ and $K = I_{[0,n]\times\Omega}$, and letting n tend to infinity) to justify relation (2.6). Relation (2.7) follows from the stronger relation

$$\int_{[0,t]} U_s^2 \, d[X,X]_s = \sqrt{\alpha_t} \qquad \forall t \geqslant 0,$$

which stems from the choice of U and from [1, VI, 91 c)]. As to (2.8), we have

$$
(2.9) \quad
\begin{aligned}
\mathbb{E}\left[\int V_s^2 \, d[Y,Y]_s\right] &= \mathbb{E}\left[\int K_s^2 \sqrt{\alpha_s} \, d[Y,Y]_s\right] \\
&= \mathbb{E}\left[\int \sqrt{\alpha_s} \, d\beta_s\right] = \mathbb{E}\left[\int (\beta_\infty - \beta_{s-}) \, d\sqrt{\alpha_s}\right]
\end{aligned}
$$

from the integration by parts formula [1, VI, 90].

Since α_s, and hence $\sqrt{\alpha_s}$ too, is optional, we may replace the process $(\beta_\infty - \beta_{s-})_s$ by its optional projection, which we know is dominated by c^2 by hypothesis. Finally, using (2.6), (2.3) (applied to U and V), the Hölder inequality, (2.7) and (2.8), we have

$$
\begin{aligned}
\mathbb{E}\left[\int H_s K_s \, |d[X,Y]_s|\right] &\leqslant \sqrt{2}\,\mathbb{E}\left[\int U_s V_s \, |d[X,Y]_s|\right] \\
&\leqslant \sqrt{2}\left(\mathbb{E}\left[\int U_s^2 \, d[X,X]_s\right]\right)^{\frac{1}{2}}\left(\mathbb{E}\left[\int V_s^2 \, d[Y,Y]_s\right]\right)^{\frac{1}{2}} \leqslant \sqrt{2}\,c\,\mathbb{E}[\sqrt{\alpha_\infty}]
\end{aligned}
$$

and the proof is over.   □

Remark 2.4. Suppose in addition that the processes $[X,X]$ and H are predictable and condition (2.4) is replaced by the weaker condition

$$
(2.4') \qquad \mathbb{E}\left[\int_{[T,\infty]} K_s^2 \, d[Y,Y]_s \;\middle|\; \mathcal{F}_{T-}\right] \leqslant c^2 \qquad \text{a.s.}
$$

for all predictable stopping times T. Then the same conclusion (2.5) holds.

The proof is the same, except that one passes from (2.9) to (2.8) by considering the predictable projection of the process $(\beta_\infty - \beta_{s-})$, the measure $d(\sqrt{\alpha_s})$ now being *predictable*.

For example, (2.4') is implied by the condition

$$
K_0^2 \, [Y,Y]_0 + \mathbb{E}\left[\int_{(T,\infty]} K_s^2 \, d[Y,Y]_s \;\middle|\; \mathcal{F}_T\right] \leqslant c^2 \qquad \text{a.s.}
$$

for all stopping times T. Conversely (2.4') implies that:

a) $K_0^2 \, [Y,Y]_0 \leqslant c^2$;

b) $\mathbb{E}\left[\int_{(T,\infty]} K_s^2 \, d[Y,Y]_s \mid \mathcal{F}_T\right] \leqslant c^2$ a.s. for all stopping times T. (Approximate T by the predictable stopping time $T + 1/n$.)

3 The main result

First, remark that (2.1) leads immediately to the Minkowski-type inequality: for all $X, Y \in \mathcal{S}$ and $s < t$, one has

(3.1) $$\left([X+Y,X+Y]_s^t\right)^{\frac{1}{2}} \leqslant \left([X,X]_s^t\right)^{\frac{1}{2}} + \left([Y,Y]_s^t\right)^{\frac{1}{2}};$$

by taking $s = 0_-$, $t = \infty$, this enables us to consider for any $1 \leqslant p \leqslant \infty$ the seminorms on $\mathcal{S}$:

$$\|X\|_{H^p} = \left(\mathbb{E}\left[\left([X,X]_\infty^{\frac{p}{2}}\right)\right]\right)^{\frac{1}{p}} = \left\|[X,X]_\infty^{\frac{1}{2}}\right\|_{L^p}$$

and to define $H^p = \{X \in \mathcal{S} : \|X\|_{H^p} < \infty\}$ as subspaces of $\mathcal{S}$.

Throughout the rest of the paper, we assume that the following implication holds for all $X \in \mathcal{S}$:

(S) $$[X,X]_\infty \equiv 0 \quad \Longrightarrow \quad X \equiv 0$$

(that is, all above seminorms are in fact norms on the corresponding spaces), and in addition H^2 is *complete* with respect to $\|\cdot\|_{H^2}$ (that is, H^2 is a Hilbert space). On the vector space H^2 we also consider the restriction of the norm $\|\cdot\|_{H^1}$, and we denote it by $(H^2, \|\cdot\|_{H^1})$. Finally, we consider on $\mathcal{S}$ the norm $\|\cdot\|_{BMO}$ so defined: $\|X\|_{BMO}$ is the smallest (possibly infinite) constant $c \geqslant 0$ such that, for all stopping times T, one has $\mathbb{E}\left[[X,X]_\infty - [X,X]_{T_-} \mid \mathcal{F}_T\right] \leqslant c^2$. Homogeneity of $\|\cdot\|_{BMO}$ is obvious. In order to check its subadditivity, we note the inequality

(3.2)
$$\left|[X,Y]_\infty - [X,Y]_{T_-}\right| \leqslant \left([X,X]_\infty - [X,X]_{T_-}\right)^{\frac{1}{2}}\left([Y,Y]_\infty - [Y,Y]_{T_-}\right)^{\frac{1}{2}} \quad \text{a.s.}$$

which can be proved exactly as (2.1), or deduced from it. Let now $X, Y \in \mathcal{S}$ and $c_1, c_2 \in [0, \infty]$ be such that

$$\mathbb{E}\left[[X,X]_\infty - [X,X]_{T_-} \mid \mathcal{F}_T\right] \leqslant c_1^2 \quad \text{a.s.}$$
$$\mathbb{E}\left[[Y,Y]_\infty - [Y,Y]_{T_-} \mid \mathcal{F}_T\right] \leqslant c_2^2 \quad \text{a.s.}$$

for some fixed stopping time T. We have

$$\mathbb{E}\left[[X+Y,X+Y]_\infty - [X+Y,X+Y]_{T_-} \mid \mathcal{F}_T\right]$$
$$\leqslant c_1^2 + c_2^2 + 2\,\mathbb{E}\left[[X,Y]_\infty - [X,Y]_{T_-} \mid \mathcal{F}_T\right]$$
$$\leqslant c_1^2 + c_2^2 + 2\,\mathbb{E}\left[\left([X,X]_\infty - [X,X]_{T_-}\right)^{\frac{1}{2}}\left([Y,Y]_\infty - [Y,Y]_{T_-}\right)^{\frac{1}{2}} \mid \mathcal{F}_T\right] \quad \text{a.s.}$$

For any $A \in \mathcal{F}_T$ the Hölder inequality gives

$$\int_A \left([X,X]_\infty - [X,X]_{T_-}\right)^{\frac{1}{2}}\left([Y,Y]_\infty - [Y,Y]_{T_-}\right)^{\frac{1}{2}} d\mathbb{P}$$
$$\leqslant \left(\int_A \left([X,X]_\infty - [X,X]_{T_-}\right) d\mathbb{P}\right)^{\frac{1}{2}}\left(\int_A \left([Y,Y]_\infty - [Y,Y]_{T_-}\right) d\mathbb{P}\right)^{\frac{1}{2}}$$
$$\leqslant c_1 c_2\, \mathbb{P}[A]$$

and we conclude that

$$\mathbb{E}\big[[X{+}Y, X{+}Y]_\infty - [X{+}Y, X{+}Y]_{T-}\big] \mid \mathcal{F}_T\big] \leqslant (c_1 + c_2)^2 \qquad \text{a.s.}$$

We can now define $BMO = \{X \in \mathcal{S} : \|X\|_{BMO} < \infty\}$ as a subspace of $\mathcal{S}$; taking $T = 0$ in its definition we see that $BMO \subset H^2$.

Using the above language, we remark that th. 2.3 (with $H \equiv K \equiv 1$) implies that the mapping

$$BMO \ni Y \mapsto \mathbb{E}\big[[\,.\,, Y]_\infty\big] \in (H^2, \|\,.\,\|_{H^1})^*$$

is a linear continuous injection with norm $\leqslant \sqrt{2}$; in fact we have more, namely, $\mathbb{E}\big[[\,.\,, Y]_\infty\big]$ defines an element of $(H^1)^*$ with norm at most $\sqrt{2}\,\|Y\|_{BMO}$). The aim of this section is to show that this mapping admits a continuous inverse (as a consequence BMO is a Banach space). To this end we impose some additional conditions on $\mathcal{S}$ and $[\,.\,,.\,]$, following the intuitive idea that the r.v. $[X, X]_t$ represents the accumulation up to time t of the information about some property which depends quadratically upon the behaviour of X on $[0, t]$.

A1. For all $X \in \mathcal{S}$ and for all $t \in [0, \infty)$ and $H \in \mathcal{F}_t$, the following implication holds: $X = 0$ on $([0, t] \times \Omega) \cup ([t, \infty) \times H) \Rightarrow [X, X] = 0$ on $([0, t] \times \Omega) \cup ([t, \infty) \times H)$. (This expresses the fact that $[X, X]$ marks nothing as long as X remains identically null, and still continues to mark nothing from the last moment when X is identically null, on some subset of Ω on which X continues to be null up to ∞.)

A1'. For all $X \in \mathcal{S}$, $t \in [0, \infty)$, and $H \in \mathcal{F}_t$, if $X = 0$ on $[0, t] \times \Omega$, the process $I_H X$ belongs to $\mathcal{S}$.

A2. For all $X \in \mathcal{S}$ and $t \in [0, \infty)$, there exists some $\widetilde{X} \in \mathcal{S}$ such that $\widetilde{X} = X$ on $[0, t] \times \Omega$ and $\big[\widetilde{X}, \widetilde{X}\big]_\infty = [X, X]_t$. (Note that $\big[\widetilde{X}, \widetilde{X}\big] = [X, X]$ on $[0, t] \times \Omega$, from A1 and prop. 2.1. Roughly speaking, we can modify X from any moment such that the property marked by $[\,.\,,.\,]$ is stopped.)

Definition 3.1. Given a fixed stopping time T, an element $X \in H^2$ is called a T-*atom* if $[X, X]_\infty = \Delta[X, X]_T$. (We put $[X, X]_\infty = [X, X]_{\infty-} = \lim_{t \to \infty}[X, X]_t$, so that $\Delta[X, X]_T = 0$ on the set $\{T = \infty\}$, which may be big.)

Proposition 3.2. *Let X be a T-atom and let $Y \in \mathcal{S}$ be arbitrary. Then the process $[X, Y]$ verifies $[X, Y]_t = \Delta[X, Y]_T I_{\{T \leqslant t\}}$; it is null on $[0, T)$ and pathwise constant on $[T, \infty)$.*

Proof. For $Y = X$ this follows directly from above definition. For arbitrary Y, use prop. 2.1 and conclude by the optional section theorem. □

Definition 3.3. The positive bilinear mapping $[\,.\,,.\,]$ is said to have *square linear jumps* if for all stopping times T there exists a linear application $\Psi_T : H^2 \to L^2$ such that

(3.5) $$\Delta[X, X]_T = \big(\Psi_T(X)\big)^2 \qquad \text{a.s.}$$

for any $X \in H^2$.

By polarization and linearity of Ψ_T we then have $\Delta[X,Y]_T = \Psi_T(X)\Psi_T(Y)$ a.s. for all $X, Y \in H^2$. Note also that $\Psi_T(X) = 0$ a.s. on $\{T = \infty\}$, so that Ψ_T may be considered as taking values in $L^2(\{T < \infty\})$.

Fix now a family $\mathcal{T}$ of stopping times such that for any stopping time T, the graph of T is contained in a countable union of graphs of stopping times belonging to $\mathcal{T}$.

Denote by $\mathcal{A}_T$ the set of T-atoms, which is a linear subspace of H^2 by prop. 3.2.

Definition 3.4. We say that the set of atoms is $\mathcal{T}$-*full* if for any $T \in \mathcal{T}$ we have $\Psi_T(H^2) = \Psi_T(\mathcal{A}_T)$ as identical subspaces of L^2.

A3. The set of atoms is $\mathcal{T}$-full. One can see that $\mathcal{A}_T$ is closed in H^2 (complete), and hence $\Psi_T(H^2)$ is closed in L^2 for any $T \in \mathcal{T}$ by isometry.

Proposition 3.5. *Suppose that* $[.,.]$ *has square linear jumps, and A3 holds. Then for any* $U \in L^2$ *and* $T \in \mathcal{T}$*, there exists a unique atom* $X \in \mathcal{A}_T$ *that satisfies for all* $Y \in H^2$ *the following relations:*

$$(3.6) \qquad \mathbb{E}\big[[X,Y]_\infty\big] = \mathbb{E}\big[\Delta[X,Y]_T\big] = \mathbb{E}[\Psi_T(X)\Psi_T(Y)] = \mathbb{E}[U\Psi_T(Y)] .$$

Proof. Denote by U' the orthogonal projection of U onto the closed space $\Psi_T(H^2)$, and pick $X \in \mathcal{A}_T$ such that $U' = \Psi_T(X)$. The first equality follows from prop. 3.2, the other ones are obvious now; uniqueness of X is also clear. $\square$

In the sequel we refer to the correspondence $U \mapsto X$ as the *atomic map* for fixed $T \in \mathcal{T}$, and we denote this map by a_T.

Theorem 3.6. a) *Suppose that A1 and A2 hold, and let* $\varphi \in (H^2, \|.\|_{H^1})^*$. *Then there exists a unique* $Y \in H^2$ *such that* $\varphi(.) = \mathbb{E}\big[[.,Y]_\infty\big]$. *Moreover*

$$(3.7) \qquad\qquad \mathbb{E}\big[[Y,Y]_\infty - [Y,Y]_T \mid \mathcal{F}_T\big] \leqslant \|\varphi\|^2$$

for all stopping times T.

b) *If in addition* $[.,.]$ *has square linear jumps, if A3 holds too and if the family of atomic maps* $(a_T)_{T \in \mathcal{T}}$ *is uniformly bounded in norms from* $(L^2, \|.\|_{L^1})$ *to* $(H^2, \|.\|_{H^1})$ *by some* $M > 0$*, then* $Y \in BMO$ *and moreover* $\|Y\|_{BMO} \leqslant (M^2 + 1)^{\frac{1}{2}}\|\varphi\|$.

Proof. a) Remark first that T may be replaced by t (a constant stopping time) in (3.7); this follows from the right continuity of $[.,.]$ and from a classical approximation of T by a decreasing sequence of discrete stopping times.

Since the norm $\|.\|_{H^1}$ is obviously dominated by $\|.\|_{H^2}$, the existence and uniqueness of $Y \in H^2$ representing φ as desired is assured by the Riesz representation theorem. To prove (3.7), fix $t \in [0, \infty)$ and $H \in \mathcal{F}_t$, and consider the element X of $\mathcal{S}$ defined by $X = I_H(Y - \tilde{Y})$ (use A2 and A1$'$). Let us show that

(3.8) $[X,Y]_\infty = [X,X]_\infty = \big([Y,Y]_\infty - [Y,Y]_t\big) I_H.$

We have for any $s \geqslant t$ the relations

(3.9) $[X,Y]_s = \big[I_H(Y-\widetilde{Y}),Y\big]_s = \big[I_H(Y-\widetilde{Y}),Y-\widetilde{Y}\big]_s + \big[I_H(Y-\widetilde{Y}),\widetilde{Y}\big]_s.$

But since $\big[\widetilde{Y},\widetilde{Y}\big]_s = \big[\widetilde{Y},\widetilde{Y}\big]_t$ (by A2), it follows from prop. 2.1 that

(3.10) $\big[I_H(Y-\widetilde{Y}),\widetilde{Y}\big]_s = \big[I_H(Y-\widetilde{Y}),\widetilde{Y}\big]_t = 0$ a.s.,

because $I_H(Y-\widetilde{Y})$ is null on $[0,t] \times \Omega$, hence $\big[I_H(Y-\widetilde{Y}),I_H(Y-\widetilde{Y})\big]$ is null on $[0,t] \times \Omega$ by A1, and one more application of prop. 2.1 suffices.

Putting $Z = Y - \widetilde{Y}$, to show that $[X,Y]_s = [X,X]_s$ (and s will then go to infinity), it now suffices to check that

(3.11) $[I_H Z, Z]_s = [I_H Z, I_H Z]_s = I_H [Z,Z]_s$ a.s.

The first equality is equivalent to $[I_H Z, I_{H^c} Z]_s = 0$, which is a consequence of prop. 2.1, by using A1.

Since obviously $[I_H Z, Z] + [I_{H^c} Z, Z] = [Z,Z]$, it suffices for the second one to see that

$$[I_H Z, Z]_s = 0 \qquad \text{a.s. on } H^c$$
$$[I_{H^c} Z, Z]_s = 0 \qquad \text{a.s. on } H,$$

one more application of prop 2.1, by using A1.

To show now the second half of (3.8), take $s \geqslant t$. We have from (3.11)

$$[X,X]_s = [I_H Z, I_H Z]_s = I_H [Z,Z]_s \qquad \text{a.s.}$$

and therefore we may suppose that $H = \Omega$, $X = Z$. We have finally

$$[X,X]_s = [Y-\widetilde{Y}, Y-\widetilde{Y}]_s = [Y,Y]_s - 2[Y,\widetilde{Y}]_s + [\widetilde{Y},\widetilde{Y}]_s,$$

and the desired relation follows from (3.10) by letting s tend to infinity.

We can now write

(3.12)
$$\mathbb{E}\big[[X,Y]_\infty\big] = \varphi(X) \leqslant \|\varphi\| \, \mathbb{E}\big[[X,X]_\infty^{\frac{1}{2}}\big]$$
$$= \|\varphi\| \, \mathbb{E}\big[I_H\big([Y,Y]_\infty - [Y,Y]_t\big)^{\frac{1}{2}}\big].$$

On the other hand it follows from the Hölder inequality that

(3.13) $\mathbb{E}\big[I_H\big([Y,Y]_\infty - [Y,Y]_t\big)^{\frac{1}{2}}\big] \leqslant \mathbb{P}[H]^{\frac{1}{2}} \big(\mathbb{E}\big[I_H\big([Y,Y]_\infty - [Y,Y]_t\big)\big]\big)^{\frac{1}{2}}.$

From (3.8), (3.12) and (3.13) we conclude that

$$\mathbb{E}\big[I_H\big([Y,Y]_\infty - [Y,Y]_t\big)\big] \leqslant \|\varphi\| \, \mathbb{P}[H]^{\frac{1}{2}} \big(\mathbb{E}\big[I_H\big([Y,Y]_\infty - [Y,Y]_t\big)\big]\big)^{\frac{1}{2}}$$

and this implies (3.7) since H is an arbitrary element of $\mathcal{F}_t$.

b) For any any $T \in \mathcal{T}$ and $U \in L^2$, putting $X = a_T(U)$ and using prop. 3.5, we have

$$(3.14) \qquad \mathbb{E}[U\Psi_T(Y)] = E\big[[X,Y]_\infty\big] \leqslant \|\varphi\| \, \|a_T(U)\|_{H^1} \leqslant M \|\varphi\| \, \|U\|_{L^1}.$$

As L^2 is a dense subspace of L^1 it follows:

$$(3.15) \qquad \big\|(\Delta[Y,Y]_T)^{\frac{1}{2}}\big\|_{L^\infty} = \|\Psi_T(Y)\|_{L^\infty} \leqslant M \, \|\varphi\|.$$

Since the graph of any stopping time is contained in a countable union of graphs of stopping times from $\mathcal{T}$, (3.15) holds in fact for all stopping times T.

Finally, summing (3.7) and the square of (3.15) we get

$$\mathbb{E}\big[[Y,Y]_\infty - [Y,Y]_{T-} \mid \mathcal{F}_T\big] \leqslant (M^2 + 1) \|\varphi\|^2 \qquad \text{a.s.};$$

as T is arbitrary, the desired conclusion follows:

$$\|Y\|_{BMO} \leqslant (M^2 + 1)^{\frac{1}{2}} \|\varphi\|. \qquad \qquad \square$$

Remark 3.7. The uniform boundedness of the family of atomic maps from $(L^2, \|\cdot\|_{L^1})$ to $(H^2, \|\cdot\|_{H^1})$ may seem strange and strong. However, it is also necessary for the validity of b) (supposing that all other conditions hold). Indeed, suppose that for some constant $C > 0$, every $\varphi \in (H^2, \|\cdot\|_{H^1})^*$ is represented by a (unique) $Y \in BMO$ such that $\|Y\|_{BMO} \leqslant C\|\varphi\|$. Then, by a well known consequence of the Hahn–Banach theorem and by (3.6), for any $T \in \mathcal{T}$ and for any $U \in L^2$, puting $X = a_T(U)$, one has

$$\|X\|_{H^1} \leqslant \sup_{\|Y\|_{BMO} \leqslant C} \big|\mathbb{E}\big[[X,Y]_\infty\big]\big| = C \sup_{\|Y\|_{BMO} \leqslant 1} \big|\mathbb{E}\big[[X,Y]_\infty\big]\big|$$

$$= C \sup_{\|Y\|_{BMO} \leqslant 1} \big|E[U\Psi_T(Y)]\big| \leqslant C \sup_{\Delta[Y,Y]_T \leqslant 1} \big|E[U\Psi_T(Y)]\big| \leqslant C \, \|U\|_{L^1},$$

since $|\Psi_T(Y)| = (\Delta[Y,Y]_T)^{\frac{1}{2}}$ and obviously $\Delta[Y,Y]_T \leqslant \|Y\|_{BMO}^2$. a.s.

Remark 3.8. For $X \in H^2$ and fixed $t \geqslant 0$, consider the element $\widetilde{X} \in S$ given by A2. Looking at relation (3.10) (with X instead of Y, and $H = \Omega$), $\widetilde{X}$ appears as the orthogonal projection of X onto the, orthogonal complement (in H^2) of the linear space $F = \{X \in H^2 : X = 0 \text{ on } [0,t] \times \Omega\}$. It suffices to check that F is closed in H^2; this follows from the fact that X belongs to F if and only if X is in H^2 and $[X,X]_t = 0$ (consequence of A1 and A2). Indeed, if $X^n \in F$, $X \in H^2$ and $X^n \to X$ in H^2, then taking $T = t$ and $S = 0_-$ in (2.1), one has

$$\dot{\mathbb{E}}\big[[X,X]_t\big] = \mathbb{E}\big[[X-X^n, X-X^n]_t\big] \leqslant \mathbb{E}\big[[X-X^n, X-X^n]_\infty\big] \longrightarrow 0.$$

4 Applications

We now illustrate the theory of section 3 by applying it to the four examples listed in section 2.

Example 1 (square bracket). The implication $[X, X]_\infty \equiv 0 \Rightarrow X \equiv 0$ is well known; see for example [1, VII, 52]. In fact we need this, and the axioms Ai, for $X \in H^2$ only. To check Ai, recall three basic properties of $[.,.]$:

a) $[X^T, Y] = [X, Y]^T$ a.s. for any stopping time T, where $X_t^T = X_{T \wedge t}$.

b) Let T be an (arbitrary) fixed stopping time, and let X be a local martingale. Then the process

$$X_s' = (X_{T+s} - X_T) \qquad (s \in [0, \infty))$$

is a local martingale with respect to the filtration $\mathcal{F}_s' = \mathcal{F}_{T+s}$, and we have $[X', X']_s = [X, X]_{T+s} - [X, X]_T$ for any $s \geqslant 0$. (We may consider as well X_{T-} (resp. $[X, X]_{T-}$) instead of X_T (resp. $[X, X]_T$).

c) If X is a local martingale and $H \in \mathcal{F}_0$, then $I_H X$ is a local martingale, and $[I_H X, I_H X] = I_H[X, X]$.

These three properties easily imply A1 and A2 (take $\widetilde{X} = X^t$). Axiom A1', which is also an extension of the first part of *c)*, is satisfied because of the definition of local martingales.

We now pass to the second set of conditions, dealing with the jumps of $[.,.]$. Of course we take $\Psi_T(X) = \Delta X_T$ in definition 3.3. Further, we take $\mathcal{T} = \{T : T$ is predictable or totally inaccessible$\}$; it is well known that the graph of any stopping time can be covered by a countable union of graphs of predictable s.t. and the graph of a totally inaccessible s.t. (see [1, IV, 81]). We then have $\Psi_T(H^2) = \Psi_T(\mathcal{A}_T) = L^2(\mathcal{F}_T | \{T < \infty\})$ for T totally inaccessible and $\Psi_T(H^2) = \Psi_T(\mathcal{A}_T) = \{U - \mathbb{E}[U \,|\, \mathcal{F}_{T-}] : U \in L^2(\mathcal{F}_T | \{T < \infty\})\} = \{U \in L^2(\mathcal{F}_T | \{T < \infty\}) : \mathbb{E}[U \,|\, \mathcal{F}_{T-}] = 0\}$ for T predictable. To justify these pleasant relations, we invoque (with slight modifications) the discussion from the proof of [1, VII, 74] (due to Lépingle): for an arbitrary stopping time T and $U \in L^2(\mathcal{F}_T)$, consider the process $A_t = U I_{\{T \leqslant t\}}$; it has *finite variation.* This A is obviously integrable, and if $\widetilde{A}$ denotes its dual predictable projection (compensator), consider $X = A - \widetilde{A}$, which is a T-atom such that $\Delta X_T = U$ on $\{T < \infty\}$ (hence $\Delta[X, X]_T = U^2$ on $\{T < \infty\}$) if T is totally inaccessible and $\Delta X_T = U - \mathbb{E}[U \,|\, \mathcal{F}_{T-}]$ on $\{T < \infty\}$ (hence $\Delta[X, X]_T = (U - \mathbb{E}[U \,|\, \mathcal{F}_{T-}])^2$ on $\{T < \infty\}$) if T is predictable.

Therefore A3 holds ($\Delta X_T = \mathbb{E}[X_\infty \,|\, \mathcal{F}_T] - \mathbb{E}[X_\infty \,|\, \mathcal{F}_{T-}]$ if T is predictable, generally $X_T \in L^2(\mathcal{F}_T)$ if T is arbitrary), and moreover one can see that the family of atomic maps $(a_T)_{T \in \mathcal{T}}$ is uniformly bounded in norm from $(L^2, \|.\|_{L^1})$ to $(H^2, \|.\|_{H^1})$ by the constant $M = 2$. Summing up, we see that the hypotheses of th. 3.6. *a)* and *b)* hold, so that our result extends Fefferman's theorem ([1, VII, 88]) *up to the assumption that H^2 is dense in H^1,* which makes it possible to identify $(H^1)^*$ with $(H^2, \|\,.\,\|_{H^1})^*$. Looking at the proof of this fact in [1, VII, 85], we see that it has nothing in common with Fefferman's inequality; this explains why we consider here $(H^2, \|\,.\,\|_{H^1})$ instead of H^1, and we prefer not to assume that H^2 be dense in H^1 and that $(H^2, \|\,.\,\|_{H^1})^*$ be equal to $(H^1)^*$.

Example 2 (angle bracket). Remark first that in this case H^2 is dense in H^1 by definition of S (to make possible the definition of the angle bracket $\langle\,.\,,\,.\,\rangle$).

In this case, following Pratelli [3], one defines $\|X\|_{BMO_2}$ as the smallest (possibly infinite) constant c such that

$$\mathbb{E}\big[\langle X, X\rangle_\infty - \langle X, X\rangle_T \mid \mathcal{F}_T\big] \leqslant c^2 \qquad \text{for all stopping times } T,$$

so that we are interested only in the statement of th. 3.6. a), whereas b) is uninteresting because of remark 2.4.

The validity of A1, A1′, A2 are again consequences of properties a), b) and c), which hold for the angle bracket too (of course we must take care to consider only locally square integrable martingales, for which $\langle\,.\,,\,.\,\rangle$ exists).

To complete the description of the dual of H^1 (suggested by remark 2.4) in this example, remark the following simple fact: if $\varphi \in (H^2, \|\,.\,\|_{H^1})^*$ and $Y \in H^2$ are such that $\varphi(\,.\,) = \mathbb{E}[\langle\,.\,, Y\rangle_\infty]$, then it follows (in addition to (3.7)) that $\|Y_0\|_{L^\infty} \leqslant \|\varphi\|$. Indeed, if X_0 denotes the constant process equal to the r.v. $X_0 \in L^2(\mathcal{F}_0)$, we may write

$$\mathbb{E}[X_0 Y_0] = \mathbb{E}[X_0 Y_\infty] = \mathbb{E}\big[\langle X_0, Y\rangle_\infty\big] \leqslant \|\varphi\|\,\|X_0\|_{H^1} = \|\varphi\|\,\|X_0\|_{L^1},$$

and since $L^2(\mathcal{F}^0)$ is dense in $L^1(\mathcal{F}^0)$, the desired relation follows. Summing up, we see that the dual of H^1 may in this case be identified with the linear space $\big\{X \in S : \|X\|_{BMO_2} < \infty,\ \|X_0\|_{L^\infty} < \infty\big\}$ endowed with the norm $\|X\| = \|X\|_{BMO_2} + \|X_0\|_{L^\infty}$.

Example 3. The matter is considerably simpler in this case. The validity of A1 and A1′ are obvious, and to A2 we take brutally $\widetilde{X} = X \cdot I_{[0,t]\times\Omega}$. The significance of the notion of T-atom has a strong intuitive support here. Naturally, we take $\Psi_T(X) = X_T \cdot I_{\{T<\infty\}}$ and $\mathcal{T}$ is the family of all stopping times; $\Psi_T(H^2) = \Psi_T(\mathcal{A}_T) = L^2(\mathcal{F}_T \mid \{T<\infty\})$ for any T, and the atomic maps are isometries from $(L^2(\mathcal{F}_T \mid \{T<\infty\}), \|\,.\,\|_{L^p})$ to $(H^2, \|\,.\,\|_{H^p})$ for all $1 \leqslant p \leqslant \infty$.

Example 4. It is quite similar to above; the discrete processes carry over to "mince" processes null outside the set $\mathbb{Z}_+ \times \Omega$, and $\{\,.\,,\,.\,\}$ extends in the obvious way to whole $\mathbb{R}_+$. Here we take $\Psi_T(X) = X_T \cdot I_{\{T\in\mathbb{Z}_+\}}$ and we have $\Psi_T(H^2) = \Psi_T(\mathcal{A}_T) = L^2(\mathcal{F}_T \mid \{T \in \mathbb{Z}_+\})$ for any stopping time T.

Remark. Axioms A1 and A1′ express some local properties of S and $[\,.\,,\,.\,]$. As to A1′, it refers only to S, whereas A1 suggests that the trajectory $[X, X](\omega)$ depends only on the trajectory $X(\omega)$.

For example, in the case of the square bracket, recall that for all $X \in H^2$, putting $t_i^n = i \cdot 2^{-n} t$, one has

$$[X, X]_t = \lim_n \sum_{i=0}^{2^n - 1} \left(X_{t_{i+1}^n} - X_{t_i^n} \right)^2$$

strongly in L^1 (see [2]), which implies a property stronger than A1.

References

1. Dellacherie, C. and Meyer, P. A. (1975, 1980). *Probabilités et Potentiel* (I–IV, V–VIII), Hermann, Paris.
2. Doléans-Dade, C. (1969). Variation quadratique des martingales continues à droite. *Ann. Math. Stat.* **40** 284–289.
3. Pratelli, M. (1976). Sur certains espaces de martingales localement de carré intégrable. *Séminaire de Probabilités X*, Lecture Notes 511, 401–413.
4. Stein, E.M. (1970). *Topics in harmonic analysis related to the Littlewood-Paley Theory*, Ann. Math. Studies **63**, Princeton.
5. Yor, M. (1978). Inégalités entre processus minces et applications. *C. R. Acad. Sci. Paris* **286**, 799–802.

An Almost Sure Approximation for the Predictable Process in the Doob–Meyer Decomposition Theorem

Adam Jakubowski[*]

Nicolaus Copernicus University,
Faculty of Mathematics and Computer Science,
ul. Chopina 12/18, 87-100 Toruń, Poland.
e-mail: adjakubo@mat.uni.torun.pl

Summary. We construct the Doob–Meyer decomposition of a submartingale as a pointwise superior limit of decompositions of discrete submartingales suitably built upon discretizations of the original process. This gives, in particular, a direct proof of predictability of the increasing process in the Doob–Meyer decomposition.

1 The Doob–Meyer Theorem

The Doob–Meyer decomposition theorem opened the way towards the theory of stochastic integration with respect to square integrable martingales and—consequently—semimartingales, as described in the seminal paper [7]. According to Kallenberg [4], this theorem is "the cornerstone of the modern probability theory". It is therefore not surprising that many proofs of it are known. To the author's knowledge, all the proofs heavily depend on a result due to Doléans-Dade [3], which identifies predictable increasing with "natural" increasing processes, as defined by Meyer [6].

In the present paper we develop ideas of another classical paper by K. Murali Rao [8] and construct a sequence of decompositions for which the superior limit is *pointwise* (in (t, ω)) equal to the desired one, and thus we obtain predictability in the easiest possible way.

Let $(\Omega, \mathcal{F}, \{\mathcal{F}_t\}_{t \in [0,T]}, P)$ be a stochastic basis, satisfying the "usual" conditions, i.e. the filtration $\{\mathcal{F}_t\}$ is right-continuous and $\mathcal{F}_0$ contains all P-null sets of $\mathcal{F}_T$. Let (D) denote the class of measurable processes $\{X_t\}_{t \in [0,T]}$ such that the family $\{X_\tau\}$ is uniformly integrable, where τ runs over all stopping times with respect to $\{\mathcal{F}_t\}_{t \in [0,T]}$. One of the variants of the Doob–Meyer theorem can be formulated as follows.

[*] Supported by Komitet Badań Naukowych under Grant No PB 0253/P03/2000/19 and completed while the author was visiting Université de Rouen. The author would like to thank people of Mathematics in Rouen for their hospitality.

Theorem 1. *Any submartingale of class (D) admits a unique decomposition $X_t = M_t + A_t$, where $\{M_t\}$ is a uniformly integrable martingale and $\{A_t\}$ is a predictable increasing process, with A_T integrable.*

In discrete time this theorem is trivial: if $\{X_k\}_{k=0,1,\ldots,k_0}$ is a submartingale with respect to $\{\mathcal{F}_k\}_{k=0,1,\ldots,k_0}$, we can set $A_0 = 0$ and

$$A_k = \sum_{j=1}^{k} E\big(X_j - X_{j-1} \mid \mathcal{F}_{j-1}\big), \qquad k = 1, 2, \ldots, k_0.$$

The appealing idea of Murali Rao [8] consists in approximating A_t by increasing processes defined by discretizations of X_t.

Let $\theta_n = \{0 = t_0^n < t_1^n < t_2^n < \ldots < t_{k_n}^n = T\}$, $n = 1, 2, \ldots$, be an increasing sequence of partitions of $[0, T]$, with

$$\max_{1 \leqslant k \leqslant k_n} t_k^n - t_{k-1}^n \longrightarrow 0, \qquad \text{as } n \to \infty.$$

By "discretizations" $\{X_t^n\}_{t \in \theta_n}$ of $\{X_t\}_{t \in [0,T]}$ we mean the processes defined by

$$X_t^n = X_{t_k^n} \quad \text{if} \quad t_k^n \leqslant t < t_{k+1}^n, \; X_T^n = X_T.$$

The process X^n is a submartingale with respect to the discrete filtration $\{\mathcal{F}_t\}_{t \in \theta_n}$ and by the above discrete scheme we obtain a sequence of *right continuous* representations

$$X_t^n = M_t^n + A_t^n,$$

where

$$A_t^n = 0 \quad \text{if} \quad 0 \leqslant t < t_1^n,$$

$$A_t^n = \sum_{j=1}^{k} E\big(X_{t_j^n} - X_{t_{j-1}^n} \mid \mathcal{F}_{t_{j-1}^n}\big) \quad \text{if} \quad t_k^n \leqslant t < t_{k+1}^n, \; k = 1, 2, \ldots, k_n - 1,$$

$$A_T^n = \sum_{j=1}^{k_n} E\big(X_{t_j^n} - X_{t_{j-1}^n} \mid \mathcal{F}_{t_{j-1}^n}\big).$$

Since A_t^n is $\mathcal{F}_{t_{k-1}^n}$-measurable for $t_k^n \leqslant t < t_{k+1}^n$, the processes A^n are predictable in the very intuitive manner.

The following facts can be extracted from [8].

Theorem 2. *If $\{A_t\}$ is continuous (equivalently: $\{X_t\}$ is "quasi-left continuous", or "regular" in the former terminology), then for $t \in \bigcup_{n=1}^{\infty} \theta_n$*

$$A_t^n \longrightarrow A_t \quad \text{in } L^1.$$

In the general case

$$A_t^n \longrightarrow A_t \quad \text{weakly in } L^1, \quad t \in \bigcup_{n=1}^{\infty} \theta_n.$$

The latter statement cannot be improved: by a counterexample due to Dellacherie and Doléans-Dade [2], there exists an increasing integrable process $\{X_t\}$ and a sequence θ_n of partitions of $[0, 1]$ such that the A_1^n's fail to converge in L^1 to A_1.

By a slight modification of the approximating sequence we can obtain convergence in the strong sense.

Theorem 3. *There exists a subsequence* $\{n_j\}$ *such that for* $t \in \bigcup_{n=1}^{\infty} \theta_n$ *and as* $J \to +\infty$

$$\frac{1}{J}\left(\sum_{j=1}^{J} A_t^{n_j}\right) \longrightarrow A_t, \qquad a.s. \text{ and in } L^1. \tag{1}$$

Remark 1.

1. In fact, in any subsequence we can find a further "good" subsequence with the property described in Theorem 3. In view of Komlós' Theorem 4 below, it is natural to say that the sequence $\{A^n\}$ is K-convergent to A.
2. We do not know whether the whole sequence converges in the Cesàro sense.

2 Proof of Theorem 3

In order to avoid repetitions of well-known computations, we choose the textbook [4] as a fixed reference and will refer to particular results therein.

The preparating steps are standard and are given on pages 413-4 in [4].

1. If X is a submartingale of class (D), then the family of all random variables of the form $\{A_{\tau_n}^n\}$, where τ_n is a stopping time taking values in θ_n, is uniformly integrable. In particular,

$$\sup_n E A_T^n < +\infty.$$

2. We can extract a subsequence $\{n_k\}$ such that $A_T^{n_k} \to \alpha$ weakly in L_1. We define

$$M_t = E(X_T - \alpha \mid \mathcal{F}_t), \qquad A_t = X_t - M_t.$$

Then we have also

$$A^{n_k}(t) \longrightarrow A_t \qquad \text{weakly in } L^1, \quad t \in \bigcup_{n=1}^{\infty} \theta_n.$$

In the main step of proof we use the famous theorem of Komlós [5] (see also [1] for the contemporary presentation related to exchangeability).

Theorem 4. *If* $\xi_1, \xi_2, \ldots$ *is a sequence of random variables for which*

$$\sup_n E|\xi_n| < +\infty,$$

then there exists a subsequence $\{n_j\}$ *and an integrable random variable* ξ_∞ *such that for every subsequence* $\{n_{j_k}\}$ *of* $\{n_j\}$ *we have with probability one, as* $K \to +\infty$,

$$\frac{\xi_{n_{j_1}} + \xi_{n_{j_2}} + \cdots + \xi_{n_{j_K}}}{K} \longrightarrow \xi_\infty.$$

By this theorem we can find a subsequence $\{n_{k_j}\} \subset \{n_k\}$ and a random variable α_T such that

$$\frac{1}{J}\left(\sum_{j=1}^{J} A_T^{n_{k_j}}\right) \longrightarrow \alpha_T, \qquad \text{a.s.}$$

Since $A_T^{n_k} \to \alpha = A_T$ weakly in L^1, the Cesàro means of any subsequence also converge weakly to the same limit and we have $\alpha_T = A_T$. Since the family $\{A_T^n\}$ is uniformly integrable, the above convergence holds in L^1 as well.

Now let us take any $t_0 \in \bigcup_{n=1}^\infty \theta_n$, $t_0 \neq T$. As before, one can find another subsequence $\{n_{k_{j_i}}\} \subset \{n_{k_j}\}$ such that

$$\frac{1}{I}\left(\sum_{i=1}^{I} A_{t_0}^{n_{k_{j_i}}}\right) \longrightarrow A_{t_0}, \qquad \text{a.s. and in } L^1.$$

By the exceptional "subsequence property" given in the Komlós theorem we can still claim that

$$\frac{1}{I}\left(\sum_{i=1}^{I} A_T^{n_{k_{j_i}}}\right) \longrightarrow A_T, \qquad \text{a.s. and in } L^1.$$

Repeating these steps for each $t_0 \in \bigcup_{n=1}^\infty \theta_n$ and then applying the diagonal procedure we find a subsequence fulfilling the requirements of Theorem 3.

It remains to identify the limit with the unique predictable increasing process given by the Doob–Meyer decomposition. This can be done using Rao's result, but given almost sure convergence everything can be done with bare hands:

3 Predictability—direct!

We shall provide a direct proof of predictability of the process A appearing as the limit in Theorem 3. For notational convenience we assume that (1) holds for the whole sequence A^n. We introduce two auxiliary sequences of stochastic processes given by the following formula.

$$\tilde{A}_0^n = 0,$$

$$\tilde{A}_t^n = \sum_{j=1}^{k} E\big(X_{t_j^n} - X_{t_{j-1}^n} \mid \mathcal{F}_{t_{j-1}^n}\big) \qquad \text{if } t_{k-1}^n < t \leqslant t_k^n, \ k = 1, 2, \dots, k_n,$$

$$\tilde{B}_t^n = \frac{1}{n} \sum_{j=1}^{n} \tilde{A}_t^j.$$

The processes $\tilde{A}^n$ are adapted to the filtration $\{\mathcal{F}_t\}_{t\in[0,T]}$ and their trajectories are *left continuous*, hence they are *predictable* by the very definition of the predictable σ-field. The same holds for the $\tilde{B}^n$.

It is sufficient to show that there exists a set E of probability zero such that for every $\omega \notin E$ and every $t \in [0, T]$

$$\limsup_{n\to\infty} \tilde{B}_t^n(\omega) = A_t(\omega). \tag{2}$$

We have for $t_0 \in \bigcup_{n=1}^{\infty} \theta_n$ and n large enough

$$\tilde{A}_{t_0}^n(\omega) = A_{t_0}^n(\omega),$$

hence outside of a set E' of probability zero

$$\tilde{B}_{t_0}^n(\omega) \to A_{t_0}(\omega).$$

Since $\bigcup_{n=1}^{\infty} \theta_n$ is dense in $[0, T]$, it follows that for $\omega \notin E'$, in every point of continuity of $A_{(\cdot)}(\omega)$ we have

$$\tilde{B}_t^n(\omega) \longrightarrow A_t(\omega).$$

Moreover, since A is right continuous we have always

$$\limsup_{n\to\infty} \tilde{B}_t^n(\omega) \leqslant A_t(\omega).$$

We conclude that (2) can be violated only in points of discontinuity of A.

We claim it suffices to prove that for each stopping time τ

$$\lim_{n\to\infty} E\tilde{A}_\tau^n \longrightarrow EA_\tau. \tag{3}$$

To see this let us observe that if (3) holds then

$$E \limsup_{n\to\infty} \tilde{B}_\tau^n \leqslant EA_\tau = \lim_{n\to\infty} E\tilde{B}_\tau^n \leqslant E\limsup_{n\to\infty} \tilde{B}_\tau^n,$$

where Fatou's lemma can be applied in the last inequality because

$$\tilde{B}_\tau^n \leqslant B_T^n \longrightarrow A_T \qquad \text{in } L^1.$$

In particular, for every stopping time τ, we have almost surely,

$$A_\tau = \limsup_{n \to \infty} \tilde{B}^n_\tau.$$

Now it is well known (and easy to prove in the case of increasing processes) that there exists a sequence $\{\tau_q\}$ of stopping times which exhaust all jumps of A, i.e. $P(\Delta A_\tau > 0) > 0$ implies $P(\tau = \tau_q) > 0$ for some q. For each q we have

$$A_{\tau_q} = \limsup_{n \to \infty} \tilde{B}^n_{\tau_q}, \qquad \text{a.s.}$$

Enlarging E' by a countable family of P-null sets (one for each τ_q), we obtain a set E of P-measure zero (belonging to $\mathcal{F}_0$ due to the "usual" condition) outside of which (2) holds.

In order to prove (3) let us observe that we can write

$$\tilde{A}^n_\tau = \sum_{k=1}^{k_n} A^n_{t^n_k} I(t^n_{k-1} < \tau \leqslant t^n_k).$$

Since τ is a stopping time, the event $\{t^n_{k-1} < \tau \leqslant t^n_k\}$ belongs to $\mathcal{F}_{t^n_k}$. If we define

$$\rho^n(\tau) = 0 \quad \text{if} \quad \tau = 0, \qquad \rho^n(\tau) = t^n_k \quad \text{if} \quad t^n_{k-1} < \tau \leqslant t^n_k,$$

then $\rho^n(\tau)$ is a stopping time with respect to the discrete filtration $\{\mathcal{F}_t\}_{t \in \theta_n}$, $\rho^n(\tau) \geqslant \tau$, $\rho^n(\tau) \searrow \tau$ and

$$\tilde{A}^n_\tau = A^n_{\rho^n(\tau)}.$$

By the properties of the (discrete) Doob–Meyer decomposition

$$EA^n_{\rho^n(\tau)} = EX^n_{\rho^n(\tau)} = EX_{\rho^n(\tau)} \searrow EX_\tau = EA_\tau.$$

We have proved that A is predictable. The proof of its uniqueness is standard (see e.g. Lemma 22.11 and Proposition 15.2 in [4]) and does not involve advanced tools.

References

1. Aldous, D.J. (1985): Exchangeablility and related topics. In: Aldous, D.J, Ibragimov, I.A., Jacod, J.,École d'été de probabilités de Saint-Flour, XIII - 1983, 1–198, Lecture Notes in Math., **1117**, Springer Berlin Heidelberg
2. Dellacherie, C., Doléans-Dade, C. (1971): Un contre-exemple au problème des laplaciens approchés. In: Séminaire Probab. V, 127–137, Lecture Notes in Math., **191**, Springer Berlin Heidelberg
3. Doléans, C. (1967): Processus croissants naturels et processus croissants très bien mesurables, C. R. Acad. Sci. Paris Sér. A-B, **264**, 874-876
4. Kallenberg, O. (1997): Foundations of Modern Probability, Springer New York Berlin Heidelberg

5. Komlós, J. (1967): A generalization of a problem of Steinhaus, Acta Math. Acad. Sci. Hungar., **18**, 217-229
6. Meyer, P.A. (1963): Decomposition of supermartingales: The uniqueness theorem, Illinois J. Math., **7**, 1–17
7. Meyer, P.A. (1976): Un cours sur les intégrales stochastiques. In: Séminaire Probab. X, 127–137, Lecture Notes in Math., **511**, Springer Berlin Heidelberg
8. Rao, K.M. (1969): On decomposition theorems of Meyer, Math. Scand., **24**, 66–78

On Stochastic Integrals up to Infinity
and Predictable Criteria for Integrability

Alexander Cherny and Albert Shiryaev

[1] Moscow State University,
    Faculty of Mechanics and Mathematics,
    Department of Probability Theory,
    119992 Moscow, Russia
    e-mail: cherny@mech.math.msu.su
[2] Steklov Mathematical Institute,
    Gubkin str., 8,
    119991 Moscow, Russia
    e-mail: albertsh@mi.ras.ru

Summary. The first goal of this paper is to give an adequate definition of the stochastic integral

$$\int_0^\infty H_s \, dX_s, \qquad (*)$$

where $H = (H_t)_{t \geqslant 0}$ is a predictable process and $X = (X_t)_{t \geqslant 0}$ is a semimartingale. We consider two different definitions of $(*)$: as a *stochastic integral up to infinity* and as an *improper stochastic integral*.

The second goal of the paper is to give necessary and sufficient conditions for the existence of the stochastic integral

$$\int_0^t H_s \, dX_s, \quad t \geqslant 0$$

and for the existence of the stochastic integral up to infinity $(*)$. These conditions are expressed in predictable terms, i.e. in terms of the predictable characteristics of X.

Moreover, we recall the notion of a *semimartingale up to infinity* (*martingale up to infinity*, etc.) and show its connection with the existence of the stochastic integral up to infinity. We also introduce the notion of γ-*localization*.

Key words: Characteristics of a semimartingale, Fundamental Theorems of Asset Pricing, γ-localization, improper stochastic integrals, Lévy processes, semimartingales up to infinity, stochastic integrals, stochastic integrals up to infinity.

1 Introduction

In classical analysis there are two approaches to defining the integral $\int_0^\infty h(s)\,ds$, where h is a Borel function. In the first approach, the improper integral $\int_0^\infty h(s)\,ds$ is defined as

$$\int_0^\infty h(s)\,ds := \lim_{t\to\infty} \int_0^t h(s)\,ds,$$

where $\int_0^t h(s)\,ds$ is the "usual" Lebesgue integral over $[0,t]$. In the second approach, the integral up to infinity $\int_0^\infty h(s)\,ds$ is defined as the Lebesgue integral over $[0,\infty)$. Obviously, the classes

$$L = \left\{ h : \forall t \geqslant 0, \ \int_0^t |h(s)|\,ds < \infty \right\}, \tag{1}$$

$$L_{\mathrm{imp}} = \left\{ h \in L : \exists \lim_{t\to\infty} \int_0^t h(s)\,ds \right\}, \tag{2}$$

$$L_{[0,\infty)} = \left\{ h : \int_0^\infty |h(s)|\,ds < \infty \right\} \tag{3}$$

satisfy the following strict inclusions: $L_{[0,\infty)} \subset L_{\mathrm{imp}} \subset L$.

This paper has two main goals. The first goal is to give the corresponding definitions of the *improper stochastic integral* $\int_0^\infty H_s\,dX_s$ and of the *stochastic integral up to infinity* $\int_0^\infty H_s\,dX_s$, where $H = H_t(\omega)$ is a predictable process and $X = X_t(\omega)$, $t \geqslant 0$, $\omega \in \Omega$ is a semimartingale. The second goal is to derive a criterion for the existence of the stochastic integral up to infinity $\int_0^\infty H_s\,dX_s$ given in "predictable" terms (see Theorem 4.5). For more information on "predictability", see the monograph [10] by J. Jacod and A.N. Shiryaev. The second edition of this monograph contains a predictable criterion for the existence of stochastic integrals $\int_0^t H_s\,dX_s$, $t \geqslant 0$ (see [10; Ch. III, Theorem 6.30]). In this paper, we also derive a predictable criterion for the existence of these integrals (see Theorem 3.2). Our method differs slightly from the one in [10]. As a result, we get a simpler criterion.

The notion of a stochastic integral up to infinity is closely connected with the notion of a *semimartingale up to infinity*. These processes as well as *martingales up to infinity*, etc. are considered in Section 2. C. Stricker [20] also considered "semimartingales jusqu'à l'infini". He used another definition, but it is equivalent to our definition. The notions of a *process with finite variation up to infinity* and *local martingale up to infinity* introduced in Section 2 are closely connected with the notion of γ-*localization* that is also introduced in Section 2.

In Section 3, we recall the definition of a stochastic integral (that is sometimes called the *vector stochastic integral*) and give the predictable criterion for integrability.

Section 4 contains several equivalent definitions of the stochastic integral up to infinity, the definition of the improper stochastic integral as well as the predictable criterion for integrability up to infinity. This criterion is then applied to stable Lévy processes.

In Section 5, we show how the stochastic integrals up to infinity can be used in mathematical finance (to be more precise, in the Fundamental Theorems of Asset Pricing).

Throughout the paper, a filtered probability space $(\Omega, \mathcal{F}, (\mathcal{F}_t)_{t \geq 0}, P)$ is supposed to be fixed. The filtration $(\mathcal{F}_t)$ is assumed to be right-continuous.

2 Semimartingales up to Infinity

1. Notations and definitions. In this section, we consider only one-dimensional processes. The extension to the multidimensional case is straightforward.

We will use the notations $\mathcal{A}_{\mathrm{loc}}$, $\mathcal{V}$, $\mathcal{M}$, $\mathcal{M}_{\mathrm{loc}}$, $\mathcal{S}_p$, and $\mathcal{S}$ for the classes of processes with locally integrable variation, processes with finite variation, martingales, local martingales, special semimartingales, and semimartingales, respectively.

Definition 2.1. We will call a process $Z = (Z_t)_{t \geq 0}$ a *process with locally integrable variation up to infinity* (resp: *process with finite variation up to infinity, martingale up to infinity, local martingale up to infinity, special semimartingale up to infinity, semimartingale up to infinity*) if there exists a process $\overline{Z} = (\overline{Z}_t)_{t \in [0,1]}$ such that

$$\overline{Z}_t = Z_{\frac{t}{1-t}}, \quad t < 1$$

and $\overline{Z}$ is a process with locally integrable variation (resp: process with finite variation, martingale, local martingale, special semimartingale, semimartingale) with respect to the filtration

$$\overline{\mathcal{F}}_t = \begin{cases} \mathcal{F}_{\frac{t}{1-t}}, & t < 1, \\ \mathcal{F}, & t = 1. \end{cases} \tag{4}$$

We will use the notations $\mathcal{A}_{\mathrm{loc}, [0,\infty)}$, $\mathcal{V}_{[0,\infty)}$, $\mathcal{M}_{[0,\infty)}$, $\mathcal{M}_{\mathrm{loc}, [0,\infty)}$, $\mathcal{S}_{p, [0,\infty)}$, $\mathcal{S}_{[0,\infty)}$ for these classes of processes.

Note that $\mathcal{A}_{\mathrm{loc}, [0,\infty)} \subset \mathcal{A}_{\mathrm{loc}}$, $\mathcal{V}_{[0,\infty)} \subset \mathcal{V}$, $\mathcal{M}_{[0,\infty)} \subset \mathcal{M}$, $\mathcal{M}_{\mathrm{loc}, [0,\infty)} \subset \mathcal{M}_{\mathrm{loc}}$, $\mathcal{S}_{p, [0,\infty)} \subset \mathcal{S}_p$, $\mathcal{S}_{[0,\infty)} \subset \mathcal{S}$, and all the inclusions are strict.

2. Basic properties. In stochastic analysis there exist two types of "localization" procedures: *localization* (see [10; Ch. I, §1d]) and *σ-localization* (see [10; Ch. III, §6e]). Let us introduce one more type.

Definition 2.2. Let $\mathcal{C}$ be a class of random processes. The γ-*localized* class $\mathcal{C}_\gamma$ consists of the processes X, for which there exists an increasing sequence of stopping times (τ_n) such that $\{\tau_n = \infty\} \uparrow \Omega$ a.s. and, for each n, the stopped process $X_t^{\tau_n} := X_{t \wedge \tau_n}$ belongs to $\mathcal{C}$.

Lemma 2.3. *We have* $\mathcal{A}_{\mathrm{loc},\, [0,\infty)} = \mathcal{A}_\gamma$, *where* $\mathcal{A}$ *is the class of processes with integrable variation.*

The proof is straightfforward.

Lemma 2.4. *The class* $\mathcal{M}_{[0,\infty)}$ *coincides with the class of uniformly integrable martingales.*

Proof. This statement follows from the fact that the class of uniformly integrable martingales coincides with the class of the Lévy martingales, i.e. processes Z of the form $Z_t = \mathsf{E}(Z_\infty \mid \mathcal{F}_t)$, $t \geq 0$.   □

Lemma 2.5. *The following conditions are equivalent:*
(i) $Z \in \mathcal{M}_{\mathrm{loc},\, [0,\infty)}$;
(ii) $Z \in (\mathcal{M}_{[0,\infty)})_\gamma$;
(iii) $Z \in \mathcal{M}_{\mathrm{loc}}$ *and* $[Z]^{1/2} \in \mathcal{A}\gamma$ ($[Z]$ *denotes the quadratic variation of* Z).

Proof. (i)$\Rightarrow$(iii) This implication follows from the fact that, for a local martingale M, $[M]^{1/2} \in \mathcal{A}_{\mathrm{loc}}$ (see [10; Ch. I, Corollary 4.55]).
(iii)$\Rightarrow$(ii) This implication follows from the Davis inequality (see [16; Ch. I, §9, Theorem 6]).
(ii)$\Rightarrow$(i) This implication is obvious.   □

The following statement characterizes the semimartingales as the "L^0-integrators". Recall that a collection of random variables $(\xi_\lambda)_{\lambda \in \Lambda}$ is *bounded in probability* if for any $\varepsilon > 0$, there exists $M > 0$ such that $\mathsf{P}(|\xi_\lambda| > M) < \varepsilon$ for any $\lambda \in \Lambda$. Recall that a stopping time is called *simple* if it takes only a finite number of values, all of which are finite.

Proposition 2.6. *Let* Z *be a càdlàg* $(\mathcal{F}_t)$-*adapted process. Then* $Z \in \mathcal{S}$ *if and only if for any* $t \geq 0$, *the collection*

$$\left\{ \int_0^t H_s \, dZ_s : H \text{ has the form } \sum_{i=1}^n h_i I_{]\!]S_i, T_i]\!]}, \text{ where } S_1 \leq T_1 \leq \cdots \leq S_n \leq T_n \right.$$

$$\left. \text{are simple } (\mathcal{F}_t)\text{-stopping times and } h_i \in [-1, 1] \right\}$$

is bounded in probability. (*Note that* $\int_0^t H_s \, dZ_s$ *here is actually a finite sum.*)

For the proof, see [2; Theorem 7.6].

The next statement characterizes the semimartingales up to infinity as the "L^0-integrators up to infinity". Recall that the space $\mathcal{H}^p$ consists of semimartingales Z, for which there exists a decomposition $Z = A + M$ with $A \in \mathcal{V}$, $M \in \mathcal{M}_{\mathrm{loc}}$, and $\mathsf{E}(\mathrm{Var}\, A)_\infty^p + \mathsf{E}[M]_\infty^{p/2} < \infty$.

Lemma 2.7. *The following conditions are equivalent:*
(i) $Z \in \mathcal{S}_{[0,\infty)}$;
(ii) *there exists a decomposition* $Z = A + M$ *with* $A \in \mathcal{V}_{[0,\infty)}$, $M \in \mathcal{M}_{\mathrm{loc},\,[0,\infty)}$;
(iii) *there exists an increasing sequence of stopping times* (τ_n) *such that* $\{\tau_n = \infty\} \uparrow \Omega$ *a.s. and, for each* n, *the process* $Z_t^{\tau_n^-} := Z_t I(t < \tau_n) + Z_{\tau_n -} I(t \geqslant \tau_n)$ *belongs to* $\mathcal{H}^1$;
(iv) Z *is a càdlàg* $(\mathcal{F}_t)$-*adapted process, and the collection*

$$\left\{ \int_0^\infty H_s \, \mathrm{d}Z_s : H \text{ has the form } \sum_{i=1}^n h_i I_{]\!]S_i, T_i]\!]}, \text{ where } S_1 \leqslant T_1 \leqslant \cdots \leqslant S_n \leqslant T_n \right.$$
$$\left. \text{are simple } (\mathcal{F}_t)\text{-stopping times and } h_i \in [-1,1] \right\} \tag{5}$$

is bounded in probability. (Note that $\int_0^\infty H_s \, \mathrm{d}Z_s$ here is actually a finite sum.)

Proof. (i)$\Rightarrow$(ii) This implication is obvious.
(ii)$\Rightarrow$(iii) It is sufficient to consider the stopping times $\tau_n = \inf\{t \geqslant 0 : \mathrm{Var}\, A_t \geqslant n \text{ or } [M]_t \geqslant n\}$, where $Z = A + M$ is a decomposition of Z with $A \in \mathcal{V}_{[0,\infty)}$, $M \in \mathcal{M}_{\mathrm{loc},\,[0,\infty)}$.
(iii)$\Rightarrow$(iv) Fix $\varepsilon > 0$. There exists n such that $\mathsf{P}(\tau_n = \infty) > 1 - \varepsilon$. Let $Z^{\tau_n^-} = A + M$ be a semimartingale decomposition of $Z^{\tau_n^-}$ with $\mathsf{E}(\mathrm{Var}\, A)_\infty + \mathsf{E}[M]_\infty^{1/2} < \infty$. For any process H of the form described in (5), we have

$$\int_0^\infty H_s \, \mathrm{d}Z_s^{\tau_n^-} = \int_0^\infty H_s \, \mathrm{d}A_s + \int_0^\infty H_s \, \mathrm{d}M_s.$$

Since $|H| \leqslant 1$, we have $\mathsf{E}|\int_0^\infty H_s \, \mathrm{d}A_s| \leqslant \mathsf{E}(\mathrm{Var}\, A)_\infty$. It follows from the Davis inequality (see [16; Ch. I, §9, Theorem 6]) that there exists a constant C such that, for any process H of the form described in (5), $\mathsf{E}|\int_0^\infty H_s \, \mathrm{d}M_s| \leqslant C$. Combining this with the inequalities

$$\mathsf{P}\left(\int_0^\infty H_s \, \mathrm{d}Z_s = \int_0^\infty H_s \, \mathrm{d}Z_s^{\tau_n^-} \right) \geqslant \mathsf{P}(\tau_n = \infty) > 1 - \varepsilon,$$

we get (iv).
(iv)$\Rightarrow$(i) For any bounded stopping time S, there exists a sequence of simple stopping times (S_k) such that $S_k \downarrow S$. Hence, the collection

$$\left\{ \int_0^\infty H_s \, \mathrm{d}Z_s : H \text{ has the form } \sum_{i=1}^n h_i I_{]\!]S_i, T_i]\!]}, \text{ where } S_1 \leqslant T_1 \leqslant \cdots \leqslant S_n \leqslant T_n \right.$$
$$\left. \text{are bounded } (\mathcal{F}_t)\text{-stopping times and } h_i \in [-1,1] \right\}$$

is bounded in probability.
For $a < b \in \mathbb{Q}$, $n \in \mathbb{N}$, we consider the stopping times

$$S_1 = \inf\{t \geqslant 0 : Z_t < a\} \wedge n, \qquad T_1 = \inf\{t \geqslant S_1 : Z_t > b\} \wedge n, \ldots$$
$$S_n = \inf\{t \geqslant T_{n-1} : Z_t < a\} \wedge n, \qquad T_n = \inf\{t \geqslant S_n : Z_t > b\} \wedge n.$$

Take $H^n = \sum_{i=1}^n I_{]\!]S_i,T_i]\!]}$. Then on the set $A := \{Z$ upcrosses $[a,b]$ infinitely often$\}$ we have

$$\int_0^\infty H_s^n \, dZ_s \xrightarrow[n\to\infty]{\text{a.s.}} \infty.$$

Hence, $\mathsf{P}(A) = 0$. As a and b have been chosen arbitrarily, we deduce that there exists a limit $Z_\infty := (\text{a.s.}) \lim_{t\to\infty} Z_t$.

Let us set

$$\overline{Z}_t = \begin{cases} Z_{\frac{t}{1-t}}, & t < 1, \\ Z_\infty, & t = 1. \end{cases}$$

Using the continuity of $\overline{Z}$ at $t = 1$, one easily verifies that the collection

$$\left\{ \int_0^1 H_s d\overline{Z}_s : H \text{ has the form } \sum_{i=1}^n h_i I_{]\!]S_i,T_i]\!]}, \text{ where } S_1 \leqslant T_1 \leqslant \cdots \leqslant S_n \leqslant T_n \right.$$

$$\left. \text{are simple } (\overline{\mathcal{F}}_t)\text{-stopping times and } h_i \in [-1,1] \right\}$$

is bounded in probability (here $(\overline{\mathcal{F}}_t)$ is the filtration given by (4)). By Proposition 2.6, $\overline{Z}$ is an $(\overline{\mathcal{F}}_t)$-semimartingale. This means that $Z \in \mathcal{S}_{[0,\infty)}$. $\square$

Remark. The description of $\mathcal{S}_{[0,\infty)}$ provided by (iii) is C. Stricker's definition of "semimartingales jusqu'à l'infini" (see [20]).

3 Stochastic Integrals

1. Notations and definitions. By $\mathcal{A}_{\text{loc}}^d$, $\mathcal{V}^d$, $\mathcal{M}^d$, $\mathcal{M}_{\text{loc}}^d$, $\mathcal{S}_p^d$, $\mathcal{S}^d$ we denote the corresponding spaces of d-dimensional processes.

Let $A \in \mathcal{V}^d$. There exist optional processes a^i and an increasing càdlàg $(\mathcal{F}_t)$-adapted process F such that

$$A^i = A_0^i + \int_0^{\cdot} a_s^i \, dF_s. \tag{6}$$

Consider the space

$$L_{\text{var}}(A) = \left\{ H = (H^1, \ldots, H^d) : H \text{ is predictable and,} \right.$$

$$\left. \text{for any } t \geqslant 0, \int_0^t |H_s \cdot a_s| \, dF_s < \infty \text{ a.s.} \right\},$$

where $H_s \cdot a_s := \sum_{i=1}^d H_s^i a_s^i$. Note that $L_{\text{var}}(A)$ does not depend on the choice of a^i and F that satisfy (6). For $H \in L_{\text{var}}(A)$, we set

$$\int_0^{\cdot} H_s \, dA_s := \int_0^{\cdot} H_s \cdot a_s \, dF_s.$$

This is a process with finite variation.

Let $M \in \mathcal{M}_{\text{loc}}^d$. There exist optional processes π^{ij} and an increasing càdlàg $(\mathcal{F}_t)$-adapted process F such that

$$[M^i, M^j] = \int_0^{\cdot} \pi_s^{ij} \, dF_s. \tag{7}$$

Consider the space

$$L_{\text{loc}}^1(M) = \left\{ H = (H^1, \ldots, H^d) : H \text{ is predictable} \right.$$

$$\left. \text{and} \left(\int_0^{\cdot} H_s \cdot \pi_s \cdot H_s \, dF_s \right)^{1/2} \in \mathcal{A}_{\text{loc}} \right\},$$

where $H_s \cdot \pi_s \cdot H_s := \sum_{i,j=1}^d H_s^i \pi_s^{ij} H_s^j$. Note that $L_{\text{loc}}^1(M)$ does not depend on the choice of π^{ij} and F that satisfy (7). For $H \in L_{\text{loc}}^1(M)$, one can define the stochastic integral $\int_0^{\cdot} H_s \, dM_s$ by the approximation procedure (see [19; Section 3]). This process is a local martingale.

Definition 3.1. Let $X \in \mathcal{S}^d$. A process H is X-*integrable* if there exists a decomposition $X = A + M$ with $A \in \mathcal{V}^d$, $M \in \mathcal{M}_{\text{loc}}^d$ such that $H \in L_{\text{var}}(A) \cap L_{\text{loc}}^1(M)$. In this case

$$\int_0^{\cdot} H_s \, dX_s := \int_0^{\cdot} H_s \, dA_s + \int_0^{\cdot} H_s \, dM_s.$$

The space of X-integrable processes is denoted by $L(X)$.

For the proof of the correctness of this definition and for the basic properties of stochastic integrals, see [19].

2. Predictable criterion for the integrability. Let $X \in \mathcal{S}^d$ and (B, C, ν) be the characteristics of X with respect to the truncation function $xI(|x| \leqslant 1)$ (for the definition, see [10; Ch. II, §2a]). There exist predictable processes b^i, c^{ij}, a transition kernel K from $(\Omega \times \mathbb{R}_+, \mathcal{P})$ (here $\mathcal{P}$ denotes the predictable σ-field) to $(\mathbb{R}^d, \mathcal{B}(\mathbb{R}^d))$, and an increasing predictable càdlàg process F such that

$$B^i = \int_0^{\cdot} b_s^i \, dF_s, \quad C^{ij} = \int_0^{\cdot} c_s^{ij} \, dF_s, \quad \nu(\omega, dt, dx) = K(\omega, t, dx) \, dF_t(\omega) \tag{8}$$

(see [10; Ch. II, Proposition 2.9]).

Theorem 3.2. *Let H be a d-dimensional predictable process. Set*

$$
\varphi_t(H) = \left| H_t \cdot b_t + \int_{\mathbb{R}} H_t \cdot x \big(I(|x| > 1, |H_t \cdot x| \leqslant 1) - I(|x| \leqslant 1, |H_t \cdot x| > 1) \big) K_t(\mathrm{d}x) \right|
$$
$$
+ H_t \cdot c_t \cdot H_t + \int_{\mathbb{R}} 1 \wedge (H_t \cdot x)^2 K_t(\mathrm{d}x), \quad t \geqslant 0. \tag{9}
$$

Then $H \in L(X)$ if and only if

$$
\forall t \geqslant 0, \quad \int_0^t \varphi_s(H) \, \mathrm{d}F_s < \infty \quad \text{a.s.} \tag{10}
$$

The following two statements will be used in the proof.

Proposition 3.3. *Let $X \in \mathcal{S}_p^d$ and $X = X_0 + A + M$ be the canonical decomposition of X. Let $H \in L(X)$. Then $\int_0^{\cdot} H_s \, \mathrm{d}X_s \in \mathcal{S}_p$ if and only if $H \in L_{\mathrm{var}}(A) \cap L_{\mathrm{loc}}^1(M)$. In this case*

$$
\int_0^{\cdot} H_s \, \mathrm{d}X_s = \int_0^{\cdot} H_s \, \mathrm{d}A_s + \int_0^{\cdot} H_s \, \mathrm{d}M_s
$$

is the canonical decomposition of $\int_0^{\cdot} H_s \, \mathrm{d}X_s$.

For the proof, see [9; Proposition 2].

Lemma 3.4. *Let μ be the jump measure of X and $W = W(\omega, t, x)$ be a nonnegative bounded $\mathcal{P} \times \mathcal{B}(\mathbb{R})$-measurable function. Then $(W * \mu)_\infty < \infty$ a.s. if and only if $(W * \nu)_\infty < \infty$ a.s.*

This statement is a direct consequence of the definition of the compensator (see [10; Ch. II, §1a]).

Remark. It follows from Lemma 3.4 that

$$
\forall t \geqslant 0, \quad \int_0^t \int_{\mathbb{R}} |H_s \cdot x| I(|x| > 1, |H_s \cdot x| \leqslant 1) K_s(\mathrm{d}x) \, \mathrm{d}F_s < \infty \quad \text{a.s.} \tag{11}
$$

Hence, the process $\varphi(H)$ in (10) can be replaced by a simpler process

$$
\psi_t(H) = \left| H_t \cdot b_t - \int_{\mathbb{R}} H_t \cdot x \, I(|x| \leqslant 1, |H_t \cdot x| > 1) K_t(\mathrm{d}x) \right|
$$
$$
+ H_t \cdot c_t \cdot H_t + \int_{\mathbb{R}} 1 \wedge (H_t \cdot x)^2 K_t(\mathrm{d}x), \quad t \geqslant 0.
$$

We formulate Theorem 3.2 with the process $\varphi(H)$ and not with $\psi(H)$ in order to achieve symmetry with the predictable criterion for integrability up to infinity (Theorem 4.5), where one can use only $\varphi(H)$.

Proof of Theorem 3.2. The "only if" part. Let $Y = \int_0^{\cdot} H_s \, \mathrm{d}X_s$. Consider the set $D = \{(\omega, t) : |\Delta X_t(\omega)| > 1 \text{ or } |H_t(\omega) \cdot \Delta X_t(\omega)| > 1\}$. Then D is a.s. discrete, and therefore, the processes

$$\hat{X}^i = \int_0^{\cdot} I_D \, dX_s^i, \qquad \widetilde{X}^i = X^i - \hat{X}^i,$$

$$\hat{Y} = \int_0^{\cdot} I_D \, dY_s, \qquad \widetilde{Y} = Y - \hat{Y}$$

are well defined. Obviously, $H \in L_{\mathrm{var}}(\hat{X}) \subseteq L(\hat{X})$. It follows from the equality $\Delta Y = H \cdot \Delta X$ that $\int_0^{\cdot} H_s \, d\hat{X}_s = \hat{Y}$. By linearity, $H \in L(\widetilde{X})$ and $\int_0^{\cdot} H_s \, d\widetilde{X}_s = \widetilde{Y}$.

Let μ denote the jump measure of X and X^c be the continuous martingale part of X. We have

$$X = X_0 + xI(|x| > 1) * \mu + B + xI(|x| \leqslant 1) * (\mu - \nu) + X^c$$

(see [10; Ch. II, Theorem 2.34]). Then

$$\begin{aligned} \widetilde{X} &= X_0 - xI(|x| \leqslant 1, |H \cdot x| > 1) * \mu + B + xI(|x| \leqslant 1) * (\mu - \nu) + X^c \\ &= X_0 + \widetilde{A} + \widetilde{M}, \end{aligned} \tag{12}$$

where

$$\widetilde{A} = B - xI(|x| \leqslant 1, |H \cdot x| > 1) * \nu, \tag{13}$$

$$\widetilde{M} = xI(|x| \leqslant 1, |H \cdot x| \leqslant 1) * (\mu - \nu) + X^c. \tag{14}$$

The process $\widetilde{A}$ is predictable, and therefore, the decomposition $\widetilde{X} = X_0 + \widetilde{A} + \widetilde{M}$ is the canonical decomposition of $\widetilde{X}$. By Proposition 3.3, $H \in L_{\mathrm{var}}(\widetilde{A}) \cap L^1_{\mathrm{loc}}(\widetilde{M})$.

The inclusion $H \in L_{\mathrm{var}}(\widetilde{A})$ means that

$$\forall t \geqslant 0, \quad \int_0^t \left| H_s \cdot b_s - \int_{\mathbb{R}} H_s \cdot x \, I(|x| \leqslant 1, |H_s \cdot x| > 1) K_s(dx) \right| dF_s < \infty \quad \text{a.s.} \tag{15}$$

Note that the continuous martingale part of $\widetilde{M}$ is X^c. Consequently,

$$[\widetilde{M}^i, \widetilde{M}^j] = \sum_{s \leqslant \cdot} \Delta \widetilde{M}_s^i \Delta \widetilde{M}_s^j + \int_0^{\cdot} c_s^{ij} \, dF_s \tag{16}$$

(see [10; Ch. I, Theorem 4.52]). Now, the inclusion $H \in L^1_{\mathrm{loc}}(\widetilde{M})$ implies that

$$\forall t \geqslant 0, \quad \int_0^t H_s \cdot c_s \cdot H_s \, dF_s < \infty \quad \text{a.s.} \tag{17}$$

We have

$$\forall t \geqslant 0, \quad \sum_{s \leqslant t} (H_s \cdot \Delta X_s)^2 = \sum_{s \leqslant t} \Delta Y_s^2 < \infty \quad \text{a.s.},$$

and using Lemma 3.4, we obtain

$$\forall t \geqslant 0, \quad \int_0^t \int_{\mathbb{R}} 1 \wedge (H_s \cdot x)^2 K_s(\mathrm{d}x) \, \mathrm{d}F_s < \infty \quad \text{a.s.} \tag{18}$$

Inequalities (11), (15), (17), and (18) taken together yield (10).

The "if" part. Combining condition (10) with Lemma 3.4, we get

$$\forall t \geqslant 0, \quad \sum_{s \leqslant t} (H_s \cdot \Delta X_s)^2 < \infty \quad \text{a.s.} \tag{19}$$

Hence, the set D introduced above is a.s. discrete. The processes $\hat{X}$ and $\widetilde{X}$ are well defined, and equalities (12)–(14) hold true.

Condition (10), combined with (11), implies (15), which means that $H \in L_{\mathrm{var}}(\widetilde{A})$.

It follows from (19) that

$$\forall t \geqslant 0, \quad \sum_{s \leqslant t} (H_s \cdot \Delta \widetilde{X}_s)^2 < \infty \quad \text{a.s.}$$

The inclusion $H \in L_{\mathrm{var}}(\widetilde{A})$ implies that

$$\forall t \geqslant 0, \quad \sum_{s \leqslant t} |H_s \cdot \Delta \widetilde{A}_s| < \infty \quad \text{a.s.}$$

Taking into account the equality $\Delta \widetilde{M} = \Delta \widetilde{X} - \Delta \widetilde{A}$, we get

$$\forall t \geqslant 0, \quad \sum_{s \leqslant t} (H_s \cdot \Delta \widetilde{M}_s)^2 < \infty \quad \text{a.s.}$$

Moreover, $H \cdot \Delta \widetilde{A} = {}^p (H \cdot \Delta \widetilde{X})$ (see [10; Ch. I, §2d]), which implies that $|H \cdot \Delta \widetilde{A}| \leqslant 1$, and hence, $|H \cdot \Delta \widetilde{M}| \leqslant 2$. Consequently,

$$\left(\sum_{s \leqslant \cdot} (H_s \cdot \Delta \widetilde{M}_s)^2 \right)^{1/2} \in \mathcal{A}_{\mathrm{loc}}.$$

This, combined with (16) and (17) (that is a consequence of (10)), yields the inclusion $H \in L^1_{\mathrm{loc}}(\widetilde{M})$.

As a result, $H \in L(\widetilde{X})$. Since obviously $H \in L_{\mathrm{var}}(\hat{X}) \subseteq L(\hat{X})$, we get $H \in L(X)$. $\square$

Corollary 3.5. *Let X be a one-dimensional continuous semimartingale with canonical decomposition $X = X_0 + A + M$. Then a predictable process H belongs to $L(X)$ if and only if*

$$\forall t \geqslant 0, \quad \int_0^t |H_s| \mathrm{d}(\mathrm{Var}\, A)_s + \int_0^t H_s^2 \mathrm{d}\langle M \rangle_s < \infty \quad \text{a.s.}$$

3. Application to Lévy processes. Let X be a one-dimensional $(\mathcal{F}_t)$-Lévy process, i.e. X is an $(\mathcal{F}_t)$-adapted Lévy process and, for any $s \leqslant t$, the increment $X_t - X_s$ is independent of $\mathcal{F}_s$. The notation $X \sim (b, c, \nu)_h$ means that

$$\mathsf{E}\,e^{i\lambda X_t} = \exp\left\{ t\left[i\lambda b - \frac{\lambda^2}{2}c + \int_{\mathbb{R}} \left(e^{i\lambda x} - 1 - i\lambda h(x)\right) \nu(dx)\right]\right\}.$$

For more information on Lévy processes, see [18].

The following corollary of Theorem 3.2 completes the results of O. Kallenberg [12], [13], J. Kallsen and A.N. Shiryaev [15], J. Rosinski and W. Woyczynski [17].

Corollary 3.6. *Let X be an α-stable $(\mathcal{F}_t)$-Lévy process with the Lévy measure*

$$\nu(dx) = \left(\frac{m_1 I(x < 0)}{|x|^{\alpha+1}} + \frac{m_2 I(x > 0)}{|x|^{\alpha+1}}\right) dx.$$

Let H be a predictable process.

(i) *Let $\alpha \in (0, 1)$ and $X \sim (b, 0, \nu)_0$. Then $H \in L(X)$ if and only if*

$$\forall t \geqslant 0, \ |b|\int_0^t |H_s|\,ds + (m_1 + m_2)\int_0^t |H_s|^\alpha\,ds < \infty \quad \text{a.s.}$$

(ii) *Let $\alpha = 1$ and $X \sim (b, 0, \nu)_h$, where $h(x) = xI(|x| \leqslant 1)$. Then $H \in L(X)$ if and only if*

$$\forall t \geqslant 0, \ (|b| + m_1 + m_2)\int_0^t |H_s|\,ds + |m_1 - m_2|\int_0^t |H_s|\ln|H_s|\,ds < \infty \quad \text{a.s.}$$

(iii) *Let $\alpha \in (1, 2)$ and $X \sim (b, 0, \nu)_x$. Then $H \in L(X)$ if and only if*

$$\forall t \geqslant 0, \ |b|\int_0^t |H_s|\,ds + (m_1 + m_2)\int_0^t |H_s|^\alpha\,ds < \infty \quad \text{a.s.}$$

(iv) *Let $\alpha = 2$ and $X \sim (b, c, 0)$. Then $H \in L(X)$ if and only if*

$$\forall t \geqslant 0, \ |b|\int_0^t |H_s|\,ds + c\int_0^t H_s^2\,ds < \infty \quad \text{a.s.}$$

Proof. The case $\alpha = 2$ is obvious. Let us consider the case $\alpha \in (0, 2)$. The semimartingale characteristics (B', C', ν') of X with respect to the truncation function $xI(|x| \leqslant 1)$ are given by

$$B'_t = b't, \qquad C'_t = 0, \qquad \nu'(\omega, dt, dx) = \nu(dx)\,dt.$$

The value b' is specified below. We have

$$\int_{\mathbb{R}} 1 \wedge (Hx)^2\,\nu(dx) = \frac{2(m_1 + m_2)}{\alpha(2 - \alpha)}|H|^\alpha, \quad H \in \mathbb{R}.$$

In case (i), we have

$$b' = b + \int_{\mathbb{R}} x I(|x| \leqslant 1) \, \nu(\mathrm{d}x)$$

and

$$Hb' + \int_{\mathbb{R}} Hx\big(I(|x| > 1, |Hx| \leqslant 1) - I(|x| \leqslant 1, |Hx| > 1)\big) \, \nu(\mathrm{d}x)$$
$$= Hb + \operatorname{sgn} H \, \frac{m_2 - m_1}{1 - \alpha} \, |H|^\alpha, \quad H \in \mathbb{R}.$$

In case (ii), we have $b' = b$ and

$$Hb' + \int_{\mathbb{R}} Hx\big(I(|x| > 1, |Hx| \leqslant 1) - I(|x| \leqslant 1, |Hx| > 1)\big) \, \nu(\mathrm{d}x)$$
$$= Hb + (m_1 - m_2) H \ln |H|, \quad H \in \mathbb{R}.$$

In case (iii), we have

$$b' = b - \int_{\mathbb{R}} x I(|x| > 1) \, \nu(\mathrm{d}x)$$

and

$$Hb' + \int_{\mathbb{R}} Hx\big(I(|x| > 1, |Hx| \leqslant 1) - I(|x| \leqslant 1, |Hx| > 1)\big) \, \nu(\mathrm{d}x)$$
$$= Hb + \operatorname{sgn} H \, \frac{m_2 - m_1}{1 - \alpha} \, |H|^\alpha, \quad H \in \mathbb{R}.$$

The result now follows from Theorem 3.2.   □

Corollary 3.7. *Let X be a nondegenerate strictly α-stable $(\mathcal{F}_t)$-Lévy process. Then a predictable process H belongs to $L(X)$ if and only if*

$$\forall t \geqslant 0, \quad \int_0^t |H_s|^\alpha \, \mathrm{d}s < \infty \quad \text{a.s.}$$

Corollary 3.8. *Let X be an $(\mathcal{F}_t)$-Lévy process, whose diffusion component is not equal to zero. Then a predictable process H belongs to $L(X)$ if and only if*

$$\forall t \geqslant 0, \quad \int_0^t H_s^2 \, \mathrm{d}s < \infty \quad \text{a.s.}^{\cdot}$$

Proof. This statement follows from Theorem 3.2 and the estimates

$$\left| \int_{\mathbb{R}} Hx\big(I(|x| > 1, |Hx| \leqslant 1) - I(|x| \leqslant 1, |Hx| > 1)\big) \, \nu(\mathrm{d}x) \right|$$
$$\leqslant \int_{\mathbb{R}} I(|x| > 1) \, \nu(\mathrm{d}x) + H^2 \int_{\mathbb{R}} 1 \wedge x^2 \, \nu(\mathrm{d}x), \quad H \in \mathbb{R},$$

$$\int_{\mathbb{R}} 1 \wedge (Hx)^2 \, \nu(\mathrm{d}x) \leqslant (1 \vee H^2) \int_{\mathbb{R}} 1 \wedge x^2 \, \nu(\mathrm{d}x), \quad H \in \mathbb{R}.$$

4 Stochastic Integrals up to Infinity and Improper Stochastic Integrals

1. Various definitions. Let $A \in \mathcal{V}^d$ and a^i, F satisfy (6). Consider the space

$$L_{\mathrm{var},\,[0,\infty)}(A) = \left\{ H = (H^1, \ldots, H^d) : H \text{ is predictable} \right.$$

$$\left. \text{and } \int_0^\infty |H_s \cdot a_s| \, \mathrm{d}F_s < \infty \text{ a.s.} \right\}.$$

For $H \in L_{\mathrm{var},\,[0,\infty)}(A)$, we set

$$\int_0^\infty H_s \, \mathrm{d}A_s := \int_0^\infty H_s \cdot a_s \, \mathrm{d}F_s.$$

Let $M \in \mathcal{M}_{\mathrm{loc}}^d$ and π^{ij}, F satisfy (7). Consider the space

$$L_{\mathrm{loc},\,[0,\infty)}^1(M) = \left\{ H = (H^1, \ldots, H^d) : H \text{ is predictable} \right.$$

$$\left. \text{and } \left(\int_0^\cdot H_s \cdot \pi_s \cdot H_s \, \mathrm{d}F_s \right)^{1/2} \in \mathcal{A}_{\mathrm{loc},\,[0,\infty)} \right\}.$$

For $H \in L_{\mathrm{loc},\,[0,\infty)}^1(M)$, one can define $\int_0^\infty H_s \, \mathrm{d}M_s$ by the approximation procedure similarly to the definition of $\int_0^t H_s \, \mathrm{d}M_s$.

Definition 4.1. Let $X \in \mathcal{S}^d$. We will say that a process H is X-*integrable up to infinity* if there exists a decomposition $X = A + M$ with $A \in \mathcal{V}^d$, $M \in \mathcal{M}_{\mathrm{loc}}^d$ such that $H \in L_{\mathrm{var},\,[0,\infty)}(A) \cap L_{\mathrm{loc},\,[0,\infty)}^1(M)$. In this case

$$\int_0^\infty H_s \, \mathrm{d}X_s := \int_0^\infty H_s \, \mathrm{d}A_s + \int_0^\infty H_s \, \mathrm{d}M_s.$$

The space of X-integrable up to infinity processes will be denoted by $L_{[0,\infty)}(X)$.

Remark. The above definition of $\int_0^\infty H_s \, \mathrm{d}X_s$ is correct, i.e. it does not depend on the choice of the decomposition $X = A + M$. Indeed, it follows from the definition of $\int_0^\infty H_s \, \mathrm{d}A_s$ and $\int_0^\infty H_s \, \mathrm{d}M_s$ that

$$\int_0^\infty H_s \, \mathrm{d}X_s = (\text{a.s.}) \lim_{t \to \infty} \int_0^t H_s \, \mathrm{d}X_s. \tag{20}$$

Theorem 4.2. Let $X \in \mathcal{S}^d$. Then $H \in L_{[0,\infty)}(X)$ if and only if $H \in L(X)$ and $\int_0^\cdot H_s \, \mathrm{d}X_s \in \mathcal{S}_{[0,\infty)}$.

Proof. The *"only if"* part. The inclusion $H \in L_{\mathrm{var},\,[0,\infty)}(A)$ implies that $H \in L_{\mathrm{var}}(A)$ and $\int_0^{\cdot} H_s \, dA_s \in \mathcal{V}_{[0,\infty)} \subset \mathcal{S}_{[0,\infty)}$. The inclusion $H \in L^1_{\mathrm{loc},\,[0,\infty)}(M)$ implies that $H \in L^1_{\mathrm{loc}}(M)$, $\int_0^{\cdot} H_s \, dM_s \in \mathcal{M}_{\mathrm{loc}}$, and

$$\left[\int_0^{\cdot} H_s \, dM_s \right]^{1/2} = \left(\int_0^{\cdot} H_s \cdot \pi_s \cdot H_s \, dF_s \right)^{1/2} \in \mathcal{A}_{\mathrm{loc},\,[0,\infty)}$$

(the equality here follows from [19; Lemma 4.18]). In view of Lemma 2.5, $\int_0^{\cdot} H_s \, dM_s \in \mathcal{M}_{\mathrm{loc},\,[0,\infty)} \subset \mathcal{S}_{[0,\infty)}$. As a result, $H \in L(X)$ and $\int_0^{\cdot} H_s \, dX_s \in \mathcal{S}_{[0,\infty)}$.

The *"if"* part. Proposition 2.6 and Lemma 2.7 combined together show that one can find deterministic functions $K^1, \ldots, K^d$ such that, for each i, $K^i > 0$, $K^i \in L(X^i)$ and $Y^i := \int_0^{\cdot} (K_s^i)^{-1} \, dX_s^i \in \mathcal{S}_{[0,\infty)}$. It follows from the associativity property of stochastic integrals (see [19; Theorem 4.7]) that the process $J = (K^1 H^1, \ldots, K^d H^d)$ belongs to $L(Y)$ and, for the process $Z = \int_0^{\cdot} J_s \, dY_s$, we have $Z = \int_0^{\cdot} H_s \, dX_s \in \mathcal{S}_{[0,\infty)}$. Set

$$\overline{Y}_t = \begin{cases} Y_{\frac{t}{1-t}}, & t < 1, \\ Y_{\infty}, & t = 1, \end{cases} \qquad \overline{Z}_t = \begin{cases} Z_{\frac{t}{1-t}}, & t < 1, \\ Z_{\infty}, & t = 1, \end{cases} \qquad \overline{J}_t = \begin{cases} J_{\frac{t}{1-t}}, & t < 1, \\ 0, & t = 1. \end{cases}$$

Let us prove that $\overline{J} \in L(\overline{Y})$. It will suffice to verify (see [19; Lemma 4.13]) that, for any sequences $a_n < b_n$ with $a_n \to \infty$ and any sequence $(\overline{G}^n)$ of one-dimensional $(\overline{\mathcal{F}}_t)$-predictable processes with $|\overline{G}^n| \leqslant 1$ (here $(\overline{\mathcal{F}}_t)$ is the filtration given by (4)), we have

$$\int_0^{1-1/n} \overline{G}_s^n \overline{J}_s I(a_n < |\overline{J}_s| \leqslant b_n) \, d\overline{Y}_s \xrightarrow[n \to \infty]{\mathrm{P}} 0. \tag{21}$$

We can write

$$\int_0^{1-1/n} \overline{G}_s^n \overline{J}_s I(a_n < |\overline{J}_s| \leqslant b_n) \, d\overline{Y}_s = \int_0^{n-1} G_s^n J_s I(a_n < |J_s| \leqslant b_n) \, dY_s$$

$$= \int_0^{n-1} G_s^n I(a_n < |J_s| \leqslant b_n) \, dZ_s = \int_0^{1-1/n} \overline{G}_s^n I(a_n < |\overline{J}_s| \leqslant b_n) \, d\overline{Z}_s,$$

where $G_t^n = \overline{G}_{t/1+t}^n$. Using the dominated convergence theorem for stochastic integrals (see [10; Ch. I, Theorem 4.40]), we get (21).

Thus, $\overline{J} \in L(\overline{Y})$, which means that there exists a decomposition $\overline{Y} = \overline{B} + \overline{N}$ with $\overline{B} \in \mathcal{V}^d(\overline{\mathcal{F}}_t)$, $\overline{N} \in \mathcal{M}^d_{\mathrm{loc}}(\overline{\mathcal{F}}_t)$ such that $\overline{J} \in L_{\mathrm{var}}(\overline{B}) \cap L^1_{\underline{\mathrm{loc}}}(\overline{N})$. Then $J \in L_{\mathrm{var},\,[0,\infty)}(B) \cap L^1_{\mathrm{loc},\,[0,\infty)}(M)$, where $B_t = \overline{B}_{t/1+t}$, $N_t = \overline{N}_{t/1+t}$. Consequently, $H \in L_{\mathrm{var},\,[0,\infty)}(A) \cap L^1_{\mathrm{loc},\,[0,\infty)}(M)$, where

$$A^i = \int_0^{\cdot} K_s^i \, dB_s^i, \qquad M^i = \int_0^{\cdot} K_s^i \, dN_s^i.$$

Since $X = X_0 + A + M$, the proof is completed.   □

Let us compare the notion of a stochastic integral up to infinity introduced above with the notion of an improper stochastic integral introduced below.

Definition 4.3. Let $X \in \mathcal{S}^d$. We will say that a process H is *improperly X-integrable* if $H \in L(X)$ and there exists a limit

$$(\text{a.s.}) \lim_{t \to \infty} \int_0^t H_s \, dX_s.$$

This limit is called the *improper stochastic integral* $\int_0^\infty H_s \, dX_s$. The space of improperly X-integrable processes will be denoted by $L_{\text{imp}}(X)$.

By the definition, $L_{\text{imp}}(X) \subseteq L(X)$. It follows from Theorem 4.2 and equality (20) that $L_{[0,\infty)}(X) \subseteq L_{\text{imp}}(X)$ and the two definitions of $\int_0^\infty H_s \, dX_s$ coincide for $H \in L_{[0,\infty)}(X)$. The following example shows that these two inclusions are strict.

Example 4.4. *Let $X_t = t$ and $H_t = h(t)$ be a measurable deterministic function. Then*

$$H \in L(X) \iff h \in L,$$
$$H \in L_{\text{imp}}(X) \iff h \in L_{\text{imp}},$$
$$H \in L_{[0,\infty)}(X) \iff h \in L_{[0,\infty)},$$

where the classes L, L_{imp}, and $L_{[0,\infty)}$ are defined in (1)–(3).

Proof. The first two statements follow from Theorem 3.2. The third statement is a consequence of Theorem 4.5.   □

2. Predictable criterion for integrability up to infinity. The following statement provides a description of $L_{[0,\infty)}(X)$ that is "dual" to the description of $L(X)$ provided by Theorem 3.2. We use the notation from Subsection 3.2.

Theorem 4.5. *Let H be a d-dimensional predictable process. Then $H \in L_{[0,\infty)}(X)$ if and only if*

$$\int_0^\infty \varphi_s(H) \, dF_s < \infty \quad \text{a.s.,} \tag{22}$$

where $\varphi(H)$ is given by (9).

Proposition 4.6. *Let $H \in L(X)$. Then the characteristics $(\widetilde{B}, \widetilde{C}, \widetilde{\nu})$ of $\int_0^\cdot H_s \, dX_s$ with respect to the truncation function $xI(|x| \leq 1)$ are given by*

$$\widetilde{B} = \int_0^\cdot \left(H_s \cdot b_s + \int_{\mathbb{R}} H_s \cdot x \big(I(|x| > 1, |H_s \cdot x| \leq 1) \right.$$
$$\left. - I(|x| \leq 1, |H_s \cdot x| > 1) \big) K_s(dx) \right) dF_s, \tag{23}$$

$$\widetilde{C} = \int_0^\cdot H_s \cdot c_s \cdot H_s \, dF_s, \tag{24}$$

$$\widetilde{\nu}(\omega, dt, dx) = \widetilde{K}(\omega, t, dx) \, dF_t(\omega), \tag{25}$$

where $\widetilde{K}(\omega, t, dx)$ is the image of $K(\omega, t, dx)$ under the map $\mathbb{R}^d \ni x \mapsto H_t(\omega)$.
$x \in \mathbb{R}$ and b, c, K, F satisfy (8).

For the proof, see [15; Lemma 3].

Proof of Theorem 4.5. The "only if" part. The process $Y = \int_0^\cdot H_s dX_s$ is a semimartingale up to infinity. Hence, the process

$$\widetilde{Y} = Y - \sum_{s \leqslant \cdot} \Delta Y_s I(|\Delta Y_s| > 1) \tag{26}$$

is also a semimartingale up to infinity. Since $\widetilde{Y}$ has bounded jumps, it belongs to $\mathcal{S}_{p,[0,\infty)}$, and therefore, its canonical decomposition $\widetilde{Y} = \widetilde{B} + \widetilde{N}$ satisfies $\widetilde{B} \in \mathcal{V}_{[0,\infty)}$, $\widetilde{N} \in \mathcal{M}_{\text{loc},[0,\infty)}$.

The process $\widetilde{B}$ is given by (23). The inclusion $\widetilde{B} \in \mathcal{V}_{[0,\infty)}$ means that

$$\int_0^\infty \left| H_s \cdot b_s + \int_{\mathbb{R}} H_s \cdot x \big(I(|x| > 1, |H_s \cdot x| \leqslant 1) \right.$$
$$\left. - I(|x| \leqslant 1, |H_s \cdot x| > 1) \big) K_s(dx) \right| dF_s < \infty \quad \text{a.s.} \tag{27}$$

The inclusion $\widetilde{N} \in \mathcal{M}_{\text{loc},[0,\infty)}$ implies that $[\widetilde{N}]_\infty < \infty$ a.s. Therefore, $\langle \widetilde{N}^c \rangle < \infty$ a.s. (see [10; Ch. I, Theorem 4.52]). In view of (24), this means that

$$\int_0^\infty H_s \cdot c_s \cdot H_s \, dF_s < \infty \quad \text{a.s.} \tag{28}$$

We have

$$\sum_{s \geqslant 0} (H_s \cdot \Delta X_s)^2 = \sum_{s \geqslant 0} \Delta Y_s^2 < \infty \quad \text{a.s.,}$$

and using Lemma 3.4, we obtain

$$\int_0^\infty \int_{\mathbb{R}} 1 \wedge (H_s \cdot x)^2 K_s(dx) \, dF_s < \infty \quad \text{a.s.} \tag{29}$$

Inequalities (27)–(29) taken together yield (22).

The "if" part. It follows from Theorem 3.2 that $H \in L(X)$. Set $Y = \int_0^\cdot H_s \, dX_s$ and define $\widetilde{Y}$ by (26). The process $\widetilde{B}$ in the canonical decomposition $\widetilde{Y} = \widetilde{B} + \widetilde{N}$ is given by (23).

Condition (22) implies (27), which means that $\widetilde{B} \in \mathcal{V}_{[0,\infty)}$.
Combining condition (22) with Lemma 3.4, we get

$$\sum_{s \geqslant 0} \Delta \widetilde{Y}_s^2 = \sum_{s \geqslant 0} (H_s \cdot \Delta X_s)^2 I(|H_s \cdot \Delta X_s| \leqslant 1) < \infty \quad \text{a.s.}$$

The inclusion $\widetilde{B} \in \mathcal{V}_{[0,\infty)}$ implies that

$$\sum_{s \geqslant 0} |\Delta \widetilde{B}_s| < \infty \quad \text{a.s.}$$

Taking into account the equality $\Delta \widetilde{N} = \Delta \widetilde{Y} - \Delta \widetilde{B}$, we get

$$\sum_{s \geqslant 0} \Delta \widetilde{N}_s^2 < \infty \quad \text{a.s.}$$

Since $|\Delta \widetilde{N}| \leqslant 2$ (see [10; Ch. I, Lemma 4.24]), we have

$$\left(\sum_{s \leqslant \cdot} \Delta \widetilde{N}_s^2 \right)^{1/2} \in \mathcal{A}_{\text{loc}, [0,\infty)}.$$

In view of (24),

$$\langle \widetilde{N}^c \rangle_\infty = \int_0^\infty H_s \cdot c_s \cdot H_s \, dF_s < \infty \quad \text{a.s.}$$

Thus, $[\widetilde{N}]^{1/2} \in \mathcal{A}_{\text{loc}, [0,\infty)}$. By Lemma 2.5, $\widetilde{N} \in \mathcal{M}_{\text{loc}, [0,\infty)}$.

As a result, $\widetilde{Y} \in \mathcal{S}_{[0,\infty)}$. Moreover, condition (22), together with Lemma 3.4, implies that

$$\sum_{s \geqslant 0} I(|\Delta Y_s| > 1) = \sum_{s \geqslant 0} I(|H_s \cdot \Delta X_s| > 1) < \infty \quad \text{a.s.}$$

Hence, $Y \in \mathcal{S}_{[0,\infty)}$. By Theorem 4.2, $H \in L_{[0,\infty)}(X)$.　　　　□

Corollary 4.7. *Let X be a one-dimensional continuous semimartingale with canonical decomposition $X = X_0 + A + M$. Then a predictable process H belongs to $L_{[0,\infty)}(X)$ if and only if*

$$\int_0^\infty |H_s| d(\text{Var } A)_s + \int_0^\infty H_s^2 d\langle M \rangle_s < \infty \quad \text{a.s.}$$

3. The application to Lévy processes. The following statement is "dual" to Corollary 3.6.

Corollary 4.8. *Let X be an α-stable $(\mathcal{F}_t)$-Lévy process with the Lévy measure*

$$\nu(dx) = \left(\frac{m_1 I(x < 0)}{|x|^{\alpha+1}} + \frac{m_2 I(x > 0)}{|x|^{\alpha+1}} \right) dx.$$

Let H be a predictable process.

(i) *Let $\alpha \in (0,1)$ and $X \sim (b,0,\nu)_0$. Then $H \in L_{[0,\infty)}(X)$ if and only if*

$$|b| \int_0^\infty |H_s|\,\mathrm{d}s + (m_1 + m_2) \int_0^\infty |H_s|^\alpha \,\mathrm{d}s < \infty \quad \text{a.s.}$$

(ii) *Let $\alpha = 1$ and $X \sim (b,0,\nu)_h$, where $h(x) = xI(|x| \leqslant 1)$. Then $H \in L_{[0,\infty)}(X)$ if and only if*

$$(|b| + m_1 + m_2) \int_0^\infty |H_s|\,\mathrm{d}s + |m_1 - m_2| \int_0^\infty |H_s| \ln |H_s|\,\mathrm{d}s < \infty \quad \text{a.s.}$$

(iii) *Let $\alpha \in (1,2)$ and $X \sim (b,0,\nu)_x$. Then $H \in L_{[0,\infty)}(X)$ if and only if*

$$|b| \int_0^\infty |H_s|\,\mathrm{d}s + (m_1 + m_2) \int_0^\infty |H_s|^\alpha \,\mathrm{d}s < \infty \quad \text{a.s.}$$

(iv) *Let $\alpha = 2$ and $X \sim (b,c,0)$. Then $H \in L_{[0,\infty)}(X)$ if and only if*

$$|b| \int_0^\infty |H_s|\,\mathrm{d}s + c \int_0^\infty H_s^2 \,\mathrm{d}s < \infty \quad \text{a.s.}$$

The proof is similar to the proof of Corollary 3.6.

Corollary 4.9. *Let X be a nondegenerate strictly α-stable $(\mathcal{F}_t)$-Lévy process. Then a predictable process H belongs to $L_{[0,\infty)}(X)$ if and only if*

$$\int_0^\infty |H_s|^\alpha \,\mathrm{d}s < \infty \quad \text{a.s.}$$

5 Application to Mathematical Finance

1. Fundamental Theorems of Asset Pricing. Let $\left(\Omega, \mathcal{F}, (\mathcal{F}_t)_{t \geqslant 0}, \mathsf{P}; (X_t)_{t \geqslant 0}\right)$ be a *model* of a financial market. Here X is a multidimensional $(\mathcal{F}_t)$-semimartingale. From the financial point of view, X is the discounted price process of several assets. Recall that a *strategy* is a pair (x, H), where $x \in \mathbb{R}$ and $H \in L(X)$. The discounted *capital* of this strategy is $x + \int_0^\cdot H_s \,\mathrm{d}X_s$.

The following notion was introduced by F. Delbaen and W. Schachermayer [4]. We formulate it using the notion of an improper stochastic integral introduced above (see Definition 4.3).

Definition 5.1. A sequence of strategies (x^n, H^n) realizes *free lunch with vanishing risk* if
(i) for each n, $x^n = 0$;
(ii) for each n, there exists $a_n \in \mathbb{R}$ such that $\int_0^\cdot H_s^n \,\mathrm{d}X_s \geqslant a_n$ a.s.;
(iii) for each n, $H^n \in L_{\mathrm{imp}}(X)$;
(iv) for each n, $\int_0^\infty H_s^n \,\mathrm{d}X_s \geqslant -\frac{1}{n}$ a.s.;
(v) there exists $\delta > 0$ such that, for each n, $\mathsf{P}\left(\int_0^\infty H_s^n \,\mathrm{d}X_s > \delta\right) > \delta$.
A model satisfies the *no free lunch with vanishing risk* condition if such a sequence of strategies does not exist. Notation: (*NFLVR*)

Recall that a one-dimensional process X is called a σ-*martingale* if there exists a sequence of predictable sets (D_n) such that $D_n \subseteq D_{n+1}$, $\bigcup_n D_n = \Omega \times \mathbb{R}_+$ and, for any n, the process $\int_0^\cdot I_{D_n} \, dX_s$ is a uniformly integrable martingale. For more information on σ-martingales, see [3], [6], [7], [10; Ch. III, §6e], [14], [19]. A multidimensional process X is called a σ-martingale if each its component is a σ-martingale. The space of d-dimensional σ-martingales is denoted by $\mathcal{M}_\sigma^d$.

Proposition 5.2. (First Fundamental Theorem of Asset Pricing).
A model satisfies the (NFLVR) condition if and only if there exists an equivalent σ-martingale measure, i.e. a measure $Q \sim P$ such that $X \in \mathcal{M}_\sigma^d(\mathcal{F}_t, Q)$.

This theorem was proved by F. Delbaen and W. Schachermayer [5] (compare with Yu.M. Kabanov [11]).

Definition 5.3. A model is *complete* if for any bounded $\mathcal{F}$-measurable function f, there exists a strategy (x, H) such that
(i) there exist constants a, b such that $a \leqslant \int_0^\cdot H_s \, dX_s \leqslant b$ a.s.;
(ii) $H \in L_{\mathrm{imp}}(X)$;
(iii) $f = x + \int_0^\infty H_s \, dX_s$ a.s.

Proposition 5.4. (Second Fundamental Theorem of Asset Pricing).
Suppose that a model satisfies the (NFLVR) condition. Then it satisfies the completeness condition if and only if the equivalent σ-martingale measure is unique.

This statement follows from [5; Theorem 5.14]. It can also be derived from [1] or [8; Théorème 11.2]. An explicit proof of the Second Fundamental Theorem of Asset Pricing in this form can be found in [19].

2. Stochastic integrals up to infinity in the Fundamental Theorems of Asset Pricing. If condition (iii) of Definition 5.1 and condition (ii) of Definition 5.3 are replaced by the conditions
(iii)' for each n, $H^n \in L_{[0,\infty)}(X)$,
(ii)' $H \in L_{[0,\infty)}(X)$,
respectively, then new versions of the (*NFLVR*) and of the completeness are obtained. We assert that the First and the Second Fundamental Theorems of Asset Pricing remain valid with these new versions of the (*NFLVR*) and of the completeness.

Theorem 5.5. *A model satisfies the (NFLVR) condition with the stochastic integrals up to infinity if and only if there exists an equivalent σ-martingale measure.*

Theorem 5.6. *Suppose that a model satisfies the (NFLVR) condition with the stochastic integrals up to infinity. Then it satisfies the completeness condition with the stochastic integrals up to infinity if and only if the equivalent σ-martingale measure is unique.*

Proof of Theorems 5.5, 5.6. It follows from Proposition 2.6 and Lemma 2.7 that there exist deterministic functions $K^1, \ldots, K^d$ such that, for each i, $K^i > 0$, $K^i \in L(X^i)$ and $Y^i := \int_0^{\cdot} (K_s^i)^{-1} \, dX_s^i \in \mathcal{S}_{[0,\infty)}$. Set

$$\overline{Y}_t = \begin{cases} Y_{\frac{t}{1-t}}, & t < 1, \\ Y_\infty, & t \geqslant 1, \end{cases} \qquad \overline{\mathcal{F}}_t = \begin{cases} \mathcal{F}_{\frac{t}{1-t}}, & t < 1, \\ \mathcal{F}, & t \geqslant 1. \end{cases}$$

Then each of the following conditions
(*NFLVR*) with the stochastic integrals up to infinity;
existence of an equivalent σ-martingale measure;
completeness with the stochastic integrals up to infinity;
uniqueness of an equivalent σ-martingale measure
holds or does not hold for the following models simultaneously

$$\left(\Omega, \mathcal{F}, (\mathcal{F}_t)_{t \geqslant 0}, \mathsf{P}; (X_t)_{t \geqslant 0} \right),$$
$$\left(\Omega, \mathcal{F}, (\mathcal{F}_t)_{t \geqslant 0}, \mathsf{P}; (Y_t)_{t \geqslant 0} \right),$$
$$\left(\Omega, \mathcal{F}, (\overline{\mathcal{F}}_t)_{t \geqslant 0}, \mathsf{P}; (\overline{Y}_t)_{t \geqslant 0} \right).$$

For the last of these models, the (*NFLVR*) and the completeness with the stochastic integrals up to infinity are obviously equivalent to the (*NFLVR*) and the completeness with the improper stochastic integrals. Now, the desired result follows from Propositions 5.2, 5.4.   □

Remarks. (i) Theorem 5.5 shows that the existence of an equivalent σ-martingale measure can be guaranteed by the condition weaker than (*NFLVR*).

(ii) The (*NFLVR*) condition and the completeness condition with the stochastic integrals up to infinity are more convenient than the original ones since we have a predictable description for integrability up to infinity, while there seems to be no such description for improper integrability.

References

1. *J.-P. Ansel, C. Stricker.* Couverture des actifs contingents et prix maximum.// Annales de l'Institut Henri Poincaré, **30** (1994), No. 2, p. 303–315.
2. *K. Bichteler.* Stochastic integration and L^p-theory of semimartingales.// The Annals of Probability, **9** (1981), No. 1, p. 49–89.
3. *C.S. Chou.* Caractérisation d'une classe de semimartingales.// Lecture Notes in Mathematics, **721** (1979), p. 250–252.
4. *F. Delbaen, W. Schachermayer.* A general version of the fundamental theorem of asset pricing.// Mathematische Annalen, **300** (1994), No. 3, p. 463–520.
5. *F. Delbaen, W. Schachermayer.* The fundamental theorem of asset pricing for unbounded stochastic processes.// Mathematische Annalen, **312** (1998), No. 2, p. 215–260.
6. *M. Émery.* Compensation de processus à variation finie non localement intégrables.// Lecture Notes in Mathematics, **784** (1980), p. 152–160.

7. *T. Goll, J. Kallsen.* A complete explicit solution to the log-optimal portfolio problem.// The Annals of Applied Probability **13** (2003), No. 2, p. 774–799.

8. *J. Jacod.* Calcul stochastique et problèmes de martingales.// Lecture Notes in Mathematics, **714** (1979), p. 1–539.

9. *J. Jacod.* Intégrales stochastiques par rapport à une semimartingale vectorielle et changement de filtration.// Lecture Notes in Mathematics, **784** (1980), p. 161–172.

10. *J. Jacod, A.N. Shiryaev.* Limit theorems for stochastic processes. Second edition. Springer, 2003.

11. *Yu.M. Kabanov.* On the FTAP of Kreps-Delbaen-Schachermayer.// Statistics and Control of Random Processes. The Liptser Festschrift. Proceedings of Steklov Mathematical Institute Seminar. World Scientific, 1997, p. 191–203.

12. *O. Kallenberg.* On the existence and path properties of stochastic integrals.// The Annals of Probability, **3** (1975), No. 2, p. 262–280.

13. *O. Kallenberg.* Some time change representations of stable integrals, via predictable transformations of local martingales.// Stochastic Processes and their Applications, **40** (1992), p. 199–223.

14. *J. Kallsen.* σ-localization and σ-martingales.// Rossiĭskaya Akademiya Nauk. Teoriya Veroyatnosteĭ ee Primeneniya, **48** (2003), No. 1, p. 177–188.

15. *J. Kallsen, A.N. Shiryaev.* Time change representations of stochastic integrals.// Theory of Probability and Its Applications, **46** (2001), No. 3, p. 579–585.

16. *R.S. Liptser, A.N. Shiryaev.* Theory of martingales. Kluwer Acad. Publ., Dortrecht, 1989.

17. *J. Rosinski, W. Woyczynski.* On Itô stochastic integration with respect to p-stable motion: inner clock, integrability of sample paths, double and multiple integrals.// The Annals of Probability, **14** (1986), No. 1, p. 271–286.

18. *K.-I. Sato.* Lévy processes and infinitely divisible distributions. Cambridge, 1999.

19. *A.N. Shiryaev, A.S. Cherny.* Vector stochastic integrals and the fundamental theorems of asset pricing.// Proceedings of the Steklov Mathematical Institute, **237** (2002), p. 12–56.

20. *C. Stricker.* Quelques remarques sur la topologie des semimartingales. Applications aux integrales stochastiques.// Lecture Notes in Mathematics, **850** (1981), p. 499–522.

Remarks on the true No-arbitrage Property

Yuri Kabanov[1,2] and Christophe Stricker[1]

[1] UMR 6623, Laboratoire de Mathématiques, Université de Franche-Comté,
16 Route de Gray, F-25030 Besançon Cedex, France
e-mail: kabanov@math.univ-fcomte.fr, stricker@math.univ-fcomte.fr
[2] Central Economics and Mathematics Institute, Moscow, Russia

Summary. We discuss conditions of absence of arbitrage in the classical sense (the "true" NA property) for the model given by a family of continuous value processes. In particular, we obtain a criterion for the NA property in a market model with countably many securities with continuous price processes. This result generalizes the well-known criteria due to Levental–Skorohod and Delbaen–Schachermayer.

1 Introduction

In the paper [7] Levental and Skorohod proved a criterion for the absence of arbitrage in a model of frictionless financial market with diffusion price processes. In the publication [2] Delbaen and Schachermayer suggested a necessary condition for the absence of arbitrage in a more general model where the price process is a continuous $\mathbf{R}^d$-valued semimartingale S: if the property NA holds then there is a probability measure $Q \ll P$ such that S is a local martingale with respect to Q. We analyze their proof and show that the arguments allow to conclude that there exists Q with an extra property: $Q|\mathcal{F}_0 \sim P|\mathcal{F}_0$. Now let σ runs the set of all stopping times. Since the NA property of S implies the NA property for each process $I_{]\sigma,\infty]} \cdot S$, this implies the existence of local martingale measures $^\sigma Q \ll P$ for the processes $I_{]\sigma,\infty]} \cdot S$ such that $^\sigma Q|\mathcal{F}_\sigma \sim P|\mathcal{F}_\sigma$. It turns out that this property is a necessary and sufficient condition for NA, cf. with [9].

In this note we establish a necessary condition for the absence of arbitrage in the framework where the model is given by a set of value processes and the price process even is not specified and the concept of the absolutely continuous martingale measure is replaced by that of absolutely continuous separating measure (ACSM). For the model with a continuous price process S the latter is a local martingale measure. We use intensively ideas of Delbaen and Schachermayer. In particular, we deduce the existence of ACSM from a suitable criterion for the NFLVR property. In contrast to [2], we use the fundamental theorem from [5] (a ramification of the corresponding result from [1]) and make explicit the notion of supermartingale density as a su-permartingale $Y \geqslant 0$ such that $YX + Y$ is supermartingale for every value process $X \geqslant -1$. The suggested approach allows us to avoid vector integrals and work exclusively with scalar processes and standard facts of stochastic calculus.

As usual, the difficult part is "NA $\Rightarrow$ ... ". For this we use Theorem 4 involving "technical" hypotheses. One of them, **H**, requires the existence of a supermartingale density Y and a process $\bar{X} \geqslant -1$ coinciding locally, up to the explosion time, with value processes and exploding on the set where Y hits zero. For the model generated by a scalar continuous semimartingale S the absence of immediate arbitrage (a property which is weaker than NA) implies **H** (with $\bar{X} = Y^{-1} - 1$). This can be easily verified following the same lines as in [2] (for the reader's convenience we provide a proof of Theorem 5 which is version of Theorem 3.7 from [2]).

The passage to the multidimensional case reveals an advantage to formu-late the conditions of Theorem 4 in terms of value processes. If the latter are generated by a finite or countable family of scalar continuous semimartingales $\{S^i\}$ with orthogonal martingale components, then a required supermartin-gale density can be assembled from the semimartingale densities constructed individually for each S^i. An orthogonalization procedure reduces the general case to the considered above. In this way we obtain a NA criterion for the model spanned by countably many securities. This result seems to be of inter-est for bond market models where the prices of zero coupon bonds are driven by countably many Wiener processes.

Notice that in our definition the set of value processes corresponding to the family $\{S^i\}$ is the closed linear space generated by the integrals with respect to each S^i. That is why we are not concerned by the particular structure of this space, i.e., by the question whether this is the space of vector integrals. The positive answer to this question is well-known for stock markets but for bond markets (with a continuum of securities) a suitable integration theory is still not available.

2 Preliminaries and general results

In our setting a stochastic basis $(\Omega, \mathcal{F}, \mathbf{F}, P)$ satisfying the usual conditions as well as a finite time horizon T are fixed. For the notational convenience we extend the filtration and all processes stationary after the date T.

To work comfortably within the standard framework of stochastic calculus under a measure $\tilde{P} \ll P$ we shall consider the "customized" stochastic basis $(\Omega, \tilde{\mathcal{F}}, \tilde{\mathbf{F}}, P)$ where $\tilde{\mathcal{F}}$ which is a $\tilde{P}$-completion of $\mathcal{F}$ and the filtration $\tilde{\mathbf{F}}$ is formed by the σ-algebras $\tilde{\mathcal{F}}_t$ generated by $\mathcal{F}_t$ and the $\tilde{P}$-null sets. For any $\xi \in \tilde{\mathcal{F}}_t$ there is $\xi' \in \mathcal{F}_t$ different from ξ only on a $\tilde{P}$-null set. With this remark the right-continuity of the new filtration is obvious. The processes $\xi I_{[t,\infty[}$ and $\xi' I_{[t,\infty[}$ coincide $\tilde{P}$-a.s. The monotone class argument implies that for any $\tilde{\mathbf{F}}$-predictable process H there exists a $\mathbf{F}$-predictable process H' which is $\tilde{P}$-indistinguishable from H, see details in [4].

We denote by $\mathcal{S} = \mathcal{S}(P)$ the linear space of scalar semimartingales starting from zero equipped with the Émery topology, generated, e.g., by the quasi-norm $\mathbf{D}(X) = \sup_H E|H \cdot X_T| \wedge 1$ where sup is taken over all predictable processes H with $|H| \leqslant 1$.

Let $\mathcal{X} \subseteq \mathcal{S}$ be a convex set of bounded from below semimartingales stable under the concatenation in the following sense: for any $X^1, X^2 \in \mathcal{X}$ and any bounded predictable processes H^1, H^2 with $H^1 H^2 = 0$, the sum of stochastic integrals $H^1 \cdot X^1 + H^2 \cdot X^2$, if bounded from below, belongs to $\mathcal{X}$. Obviously, $\mathcal{X}$ is a cone. For any $X \in \mathcal{X}$ and any stopping time τ the process $X^\tau = I_{[0,\tau]} \cdot X$ belongs to $\mathcal{X}$.

In the context of financial modeling the elements of $\mathcal{X}$ are interpreted as value processes; those for which $0 \leqslant X_T \neq 0$ are called arbitrage opportunities. Let $\mathcal{X}^a := \{X \in \mathcal{X} : X \geqslant -a\}$. We introduce the sets of attainable "gains" or "results" $R := \{X_T : X \in \mathcal{X}\}$ and $R^a := \{X_T : X \in \mathcal{X}^a\}$ and define also $C := (R - L_+^0) \cap L^\infty$, the set of claims hedgeable from the zero initial endowment.

The NA property of $\mathcal{X}$ means that $R \cap L_+^0 = \{0\}$ (or $C \cap L_+^\infty = \{0\}$). A stronger property, NFLVR (no free lunch with vanishing risk), means that $\bar{C} \cap L_+^\infty = \{0\}$ where $\bar{C}$ is the norm closure of C in L^∞. There is the following simple assertion relating them (Lemma 2.2 in [5] which proof is the same as of the corresponding result in [1]).

Lemma 1. *NFLVR holds iff NA holds and R^1 is bounded in L^0.*

Remark 1. Note that R^0 is a cone in R^1. If R^1 is bounded in L^0, then necessarily $R^0 = \{0\}$ and for any arbitrage opportunity X' there are $t < T$ and $\varepsilon > 0$ such that the set $\Gamma := \{X_t' \leqslant -\varepsilon\}$ is non-null. In this case the process $X := I_{\Gamma \times]t, \infty[} \cdot X'$ is an arbitrage opportunity with

$$\{X_T > 0\} = \{X_T \geqslant \varepsilon\} = \Gamma \in \mathcal{F}_t.$$

We say that $\mathcal{X}$ admits an equivalent separating measure (briefly: the ESM property holds) if there exists $\tilde{P} \sim P$ such that $\tilde{E}X_T \leqslant 0$ for all $X \in \mathcal{X}$.

Now we recall also one of the central (and difficult) results of the theory in the abstract formulation of [5], Th. 1.1 and 1.2 (cf. with that of the original paper [1] where the value processes are stochastic integrals).

Theorem 2. *Suppose that $\mathcal{X}^1$ is closed in $\mathcal{S}$. Then NFLVR holds iff ESM holds.*

We say that a supermartingale $Y \geqslant 0$ with $EY_0 = 1$ is a *supermartingale density* if $Y(X + 1)$ is a supermartingale for each $X \in \mathcal{X}^1$.

The following statement indicates that criteria for the NA property can be obtained from those for the NFLVR.

Lemma 3. *Let Y be a supermartingale density such that $Y_T > 0$ $\tilde{P}$-a.s. where $\tilde{P} \ll P$. Then the set R^1 is bounded in $L^0(\tilde{P})$.*

Proof. Let $X \in \mathcal{X}^1$. Since $EY(X + 1) \leqslant 1$, the set

$$Y_T R^1 := \{Y_T X_T : X \in \mathcal{X}^1\}$$

is bounded in $L^1(P)$, hence, it is bounded in $L^0(P)$. The absolute continuous change of measure as well as the multiplication by a finite random variable preserve the boundedness in probability. Thus, the set $R^1 = Y_T^{-1}(Y_T R^1)$ is bounded in $L^0(\tilde{P})$. $\square$

We need the following condition.

H. There exist a supermartingale density Y and a càdlàg process $\bar{X}$ with values in $[-1, \infty]$, having ∞ as an absorbing state, and such that $\bar{X}^{\theta_n} \in \mathcal{X}^1$ for every stopping time $\theta_n := \inf\{t : \bar{X}_t \geqslant n\}$ and $\{\bar{X}_T < \infty\} \subseteq \{Y_T > 0\}$ a.s.

Theorem 4. *Suppose that $\mathcal{X}^1$ is closed in $\mathcal{S}$ and the hypothesis **H** is satisfied. If NA holds then there exists an ACSM Q such that $Q|\mathcal{F}_0 \sim P|\mathcal{F}_0$.*

Proof. Clearly, $c := P(\bar{X}_T < \infty) > 0$ (otherwise $\bar{X}^{\theta_1}$ violates NA) and we can define the martingale $Z_t := c^{-1}E(I_{\{\bar{X}_T < \infty\}}|\mathcal{F}_t)$ and the probability measure $\tilde{P} := Z_T P$, the trace of P on $\{\bar{X}_T < \infty\}$.

The NA property implies that $I_{\{Z>0\}} \geqslant I_{\{\bar{X}<\infty\}}$, i.e. Z does not hit zero before the explosion of $\bar{X}$. Indeed, in the opposite case

$$B_t^N := \left\{\sup_{s \leqslant t} \bar{X}_s \leqslant N, \ Z_t = 0\right\}$$

is not a null-set for some $t < T$ and $N < \infty$. Since zero is the absorbing point for Z, $B_t^N \subseteq \{Z_T = 0\} = \{\bar{X}_T = \infty\}$ (a.s.). The process $I_{B_t^N \times]0,\infty[} \cdot \bar{X}$, bounded from below by $-N - 1$, is nontrivial only on B_t^N where it explodes. This violates the NA.

In particular, $Z_{\theta_n} > 0$, i.e. $\tilde{P}|\mathcal{F}_{\theta_n} \sim P|\mathcal{F}_{\theta_n}$. Since $\tilde{P}(\bar{X}_T < \infty) = 1$, the assumed existence of a supermartingale density ensures the boundedness of R^1 in $L^0(\tilde{P})$.

Let $\tilde{\mathcal{X}}^1$ be the closure of $\mathcal{X}^1$ in $\mathcal{S}(\tilde{P})$ and let $\tilde{\mathcal{X}} := \text{cone } \tilde{\mathcal{X}}^1$. Recall that the elements of $\mathcal{S}(\tilde{P})$, a space over the stochastic basis $(\Omega, \tilde{\mathcal{F}}, \tilde{\mathbf{F}}, \tilde{P})$, are, in fact, not processes but classes of equivalence. Notice that for any $\tilde{X} \in \tilde{\mathcal{X}}^1$ there is a process X such that $X^{\theta_n} \in \mathcal{X}^1$ and $\tilde{X}^{\theta_n} = X^{\theta_n}$ $\tilde{P}$-a.s. for every n.

One can verify that $\tilde{\mathcal{X}}$ is stable under concatenation.

From the definition of the Emery topology it follows that the set $\tilde{R}^1$ formed by the terminal values of processes from $\tilde{\mathcal{X}}^1$ is a part of the closure of R^1 in $L^0(\tilde{P})$ and, hence, $\tilde{R}^1$ is bounded in this space.

If the set $\tilde{\mathcal{X}}$ does not satisfy NA under $\tilde{P}$, we can find, according to Remark 1, a process $X = I_{\Gamma \times]t,T]} \cdot X \in \tilde{\mathcal{X}}^1$ such that the set $\Gamma \in \tilde{\mathcal{F}}_t$, $\tilde{P}(\Gamma) > 0$, and $\{X_T \geqslant \varepsilon\} = \Gamma \tilde{P}$-a.s. Choosing appropriate representatives we may assume without loss of generality that $\Gamma \in \mathcal{F}_t$ and $X^{\theta_n} \in \mathcal{X}^1$ for every n. On the stochastic interval $[0, \theta]$ the process $\bar{X}^\varepsilon := H \cdot \bar{X}$ with

$$H := (\varepsilon/2)(2 + \bar{X}_t)^{-1} I_{\Gamma \times]t, T[}$$

is well-defined. On $[0, \theta]$ the process $X + \bar{X}^\varepsilon \geqslant -1 - \varepsilon/2$; at θ it explodes to infinity in a continuous way on the set $\Gamma \cap \{\theta \leqslant T\}$ and has a finite positive limit bigger than $\varepsilon/2$ on the set $\Gamma \cap \{\theta > T\}$. Since $P(\Gamma) > 0$, an appropriate stopping yields a process in $\mathcal{X}$ which is an arbitrage opportunity. The obtained contradiction shows that $\tilde{\mathcal{X}}$ under $\tilde{P}$ satisfies NA and, by virtue of Lemma 1, also the property NFLVR.

The result follows because by Theorem 2 there exists a measure $Q \sim \tilde{P}$ separating $\tilde{K}$ and $L^0_+(\tilde{P})$.   □

3 Semimartingales with the structure property

By definition, the structure property of $X \in \mathcal{S}$ means that $X = M + h \cdot \langle M \rangle$ where $M \in \mathcal{M}^2_{loc}$ and h is a predictable process such that $|h| \cdot \langle M \rangle_T < \infty$.

The next result is a version of Theorem 3.7 from [2] and its proof is given for the reader's convenience.

Theorem 5. Let X be a continuous semimartingale with the structure property. Then there exists an integrand H such that $H \cdot X \geqslant 0$ and

$$\{H \cdot X_t > 0 \ \forall t \in \,]0, T]\} = \{h^2 \cdot \langle M \rangle_{0+} = \infty\}.$$

Proof. Without loss of generality we may assume that $\Gamma := \{h^2 \cdot \langle M \rangle_{0+} = \infty\}$ is of full measure (replacing, if necessary, P by its trace on Γ). With this assumption the main ingredient of the proof is the following assertion:

Lemma 6. Suppose that $h^2 \cdot \langle M \rangle_{0+} = \infty$ a.s. Then for any $\varepsilon > 0$, $\eta \in \,]0, 1]$ there exist $\delta > 0$ arbitrarily close to zero and a bounded integrand $H = H I_{]\delta, \varepsilon]}$ such that

(i) $H \cdot X \geqslant -1$;

(ii) $|Hh| \cdot \langle M \rangle_T + H^2 \cdot \langle M \rangle_T < 3$;

(iii) $P(H \cdot X_T \leqslant 1) \leqslant \eta$.

Proof. Let $R = 32/\eta$. Since $h^2 \cdot \langle M \rangle_{0+} = \infty$, for sufficiently small δ

$$P(h^2 I_{]\delta, \varepsilon]} I_{\{|h| \leqslant 1/\delta\}} \cdot \langle M \rangle_T \geqslant R) \geqslant 1 - \eta/2.$$

Let

$$\tau := \inf\{t \geqslant 0 : \ h^2 I_{]\delta, \varepsilon]} I_{\{|h| \leqslant 1/\delta\}} \cdot \langle M \rangle_t \geqslant R\} \wedge \varepsilon.$$

For the integrand $\tilde{H} := 2R^{-1}hI_{]\delta,\tau]}I_{\{|h|\leqslant 1/\delta\}}$ we have that $|\tilde{H}h| \cdot \langle M \rangle_T \leqslant 2$ with $P(|\tilde{H}h| \cdot \langle M \rangle_T < 2) \leqslant \eta/2$. Also,

$$\tilde{H}^2 \cdot \langle M \rangle_T \leqslant 4R^{-1} < 1$$

and, by the Chebyshev and Doob inequalities,

$$P\left(\sup_{s\leqslant T}|\tilde{H} \cdot M_s| \geqslant 1\right) \leqslant 4E(\tilde{H} \cdot M_T)^2 = 4E\tilde{H}^2 \cdot \langle M \rangle_T \leqslant 16R^{-1} \leqslant \eta/2.$$

Thus, $P(\tau_1 \leqslant T) \leqslant \eta/2$ for the stopping time $\tau_1 := \inf\{t \geqslant 0 : \tilde{H} \cdot M_t \leqslant -1\}$. The integrand $H := \tilde{H}I_{[0,\tau_1]}$ obviously meets the requirements (i) and (ii). At last, because of the inclusion $\{H \cdot X_T \leqslant 1, \ \tau_1 > T\} \subseteq \{|Hh| \cdot \langle M \rangle_T < 2\}$, we obtain that

$$P(H \cdot X_T \leqslant 1) = P(H \cdot X_T \leqslant 1, \ \tau_1 \leqslant T) + P(H \cdot X_T \leqslant 1, \ \tau_1 > T) \leqslant \eta/2 + \eta/2$$

and (iii) holds. $\qquad\qquad\qquad\qquad\qquad\qquad\qquad\qquad\qquad\qquad\qquad\qquad\square$

Using this lemma, we construct, starting, e.g., with $\varepsilon_0 = T$, a sequence of positive numbers $\varepsilon_n \downarrow 0$ and a sequence of integrands $H_n = H_nI_{]\varepsilon_{n+1},\varepsilon_n]}$ such that the conditions (i) – (iii) hold with $\eta_n = 2^{-n}$. The properties (i) and (ii) ensure that the process $G := \sum_n 3^{-n}H_n$ is integrable and $G \cdot X$ is bounded from below. By the Borel–Cantelli lemma for every ω outside a null set there is $n_0(\omega)$ such that $H_k \cdot X_{\varepsilon_k}(\omega) > 1$ for all $k > n_0(\omega)$. For $t \in]\varepsilon_{n+1}, \varepsilon_n]$ and any $n > n_0(\omega)$ we have

$$G \cdot X_t(\omega) = \sum_{k>n} 3^{-k}H_k \cdot X_{\varepsilon_k}(\omega) + 3^{-n}H_n \cdot X_t(\omega) \geqslant \sum_{k>n} 3^{-k} - 3^{-n} = \frac{1}{2}3^{-n} > 0.$$

Thus, $\sigma := \inf\{t > 0 : \ G \cdot X_t = 0\} > 0$ a.s. It follows that for the integrand $H := \sum 2^{-n}I_{[0,\sigma\wedge n^{-1}]}G$ the process $H \cdot X$ is strictly positive on $]0,T]$. $\quad\square$

4 Models based on a continuous price process

Let S be a continuous $\mathbf{R}^d$-valued semimartingale, $L(S)$ be the set of predictable processes integrable with respect to S, and $\mathcal{A}$ be the set of integrands H for which the process $H \cdot S$ is bounded from below.

We consider the model where $\mathcal{X} = \mathcal{X}(S) := \{H \cdot S : \ H \in \mathcal{A}\}$. Mémin's theorem [8] says that $\{H \cdot S, \ H \in L(S)\}$ is a closed subspace of $\mathcal{S}$. It follows immediately that $\mathcal{X}^1$ is also closed.

First, we look at the case $d = 1$. Replacing, if necessary, the generating process by a suitable integral, we may assume without loss of generality that S is a bounded continuous semimartingale starting from zero (hence, an element of $\mathcal{X}$) and even that in its canonical decomposition $S = M + A$ the martingale M and total variation of the predictable process A are both bounded.

Recall the following simple fact:

Lemma 7. *Suppose that $R^0 = \{0\}$. Then $S = M + A$ with $A = h \cdot \langle M \rangle$.*

Proof. If the claim fails, we can find, using the Lebesgue decomposition, a predictable process $H = H^2$ for which $H \cdot \langle M \rangle = 0$ and $H \cdot \mathrm{Var}\, A_T \neq 0$. Let the process G be defined as the sign of the (predictable) process $dA/d\,\mathrm{Var}\, A$. Since $(GH) \cdot S = H \cdot \mathrm{Var}\, A$, we obtain a contradiction with the assumption that $R^0 = \{0\}$. $\square$

The assumption $R^0 = \{0\}$ (no immediate arbitrage in the terminology of [2]) implies, by Theorem 5, that $h^2 \cdot \langle M \rangle_{0+}$ is finite as well as $h^2 I_{]\sigma, \infty[} \cdot \langle M \rangle_{\sigma+}$ whatever is the stopping time σ (by the same theorem applied to the process $S_{.+\sigma} - S_\sigma$ adapted to the shifted filtration $(\mathcal{F}_{t+\sigma})$). Put

$$\tau := \inf\{t \geqslant 0 : \ h^2 \cdot \langle M \rangle_t = \infty\},$$
$$\tau_n := \inf\{t \geqslant 0 : \ h^2 \cdot \langle M \rangle_t \geqslant n\}.$$

It follows that $h^2 \cdot \langle M \rangle_{\tau-} = \infty$ (a.s.) on the set $\{\tau \leqslant T\}$ (i.e. no jump to infinity). This allows us to define the process

$$Y := e^{-h \cdot M - (1/2)h^2 \cdot \langle M \rangle} I_{[0,\tau[}.$$

It follows from the law of large numbers for continuous local martingales (see Remark 2 below) that $\{Y_{\tau-} = 0\} = \{h^2 \cdot \langle M \rangle_\tau = \infty\}$ a.s., i.e. Y hits zero not by a jump. For every stopping time τ_n the stochastic exponential $Y^{\tau_n} = \mathcal{E}(-h \cdot M^{\tau_n})$ is a positive martingale and, hence, by the Fatou lemma, $Y^\tau = Y$ is a supermartingale. By the Ito formula

$$Y^{\tau_n}(H \cdot S^{\tau_n}) = Y^{\tau_n} \cdot (H \cdot M^{\tau_n}) + (H \cdot S^{\tau_n}) \cdot Y^{\tau_n}.$$

Thus, for any $X \in \mathcal{X}^1$ the process $Y^{\tau_n}(X^{\tau_n} + 1)$ is a local martingale and, again by the Fatou lemma, $Y^\tau(X^\tau + 1) = Y(X + 1)$ is a supermartingale.

At last, put $\bar{X} = Y^{-1} - 1$. Then $\{Y_T = 0\} = \{\bar{X}_T = \infty\}$ and by the Ito formula

$$\bar{X}^{\theta_n} = I_{[0,\theta_n]} Y^{-1} h \cdot S.$$

Summarizing, we come to the following:

Proposition 8. *Suppose that $R^0 = \{0\}$. Then the condition* **H** *holds.*

Remark 2. If $N \in \mathcal{M}_{loc}^c$ and $c > 0$, then

$$\left\{ \lim_{t \to \infty} (N_t - c\langle N \rangle_t) \to -\infty \right\} = \{\langle N \rangle_\infty = \infty\} \text{ a.s.},$$

see, e.g., [6], Lemma 6.5.6. The needed extension to the case where $N^{\tau_n} \in \mathcal{M}^c$ and $\tau_n \to \tau$ can be proved in the same way.

Remark 3. Though we established the above proposition only for the case $d = 1$, the extension of the arguments to the vector case when the components $S^i = M^i + A^i$ are such that $\langle M^i, M^j \rangle = 0$, $i \neq j$, is obvious: consider $Y = \mathcal{E}(-\sum_i h^i \cdot M^i)$. But without loss of generality we may always assume that S satisfies this property. It is sufficient to notice that $\mathcal{X}(S) = \mathcal{X}(\tilde{S})$ for some continuous semimartingale $\tilde{S}$ with orthogonal martingale components. This semimartingale can be constructed recursively. Namely, suppose that the orthogonality holds up to the index $n - 1$. Let $M^n = \sum_{i \leqslant n-1} H^i \cdot M^i + \tilde{M}^n$ be the Kunita–Watanabe decomposition. One can take $\tilde{S}^n = \tilde{M}^n + A^n - \sum_{i \leqslant n-1} H^i \cdot A^i$. Of course, to ensure the existence of $H^i \cdot A^i$ it may be necessary to replace first S^n by $G \cdot S^n$ with a suitable integrand G taking values in $]0, 1]$. This orthogonalization procedure works well also for a countable family $\{S^i\}_{i \in \mathbf{N}}$. Moreover, we can find bounded $\tilde{S}^i$ such that $\sum_i \tilde{S}^i$ converges in $\mathcal{S}$ to a bounded semimartingale $\tilde{S}$.

Theorem 9. *Suppose that $\mathcal{X} = \mathcal{X}(S)$ where S is a continuous $\mathbf{R}^d$-valued semimartingale. Then the NA property holds iff for any stopping time σ there exists a probability measure $^\sigma Q \ll P$ with $^\sigma Q|\mathcal{F}_\sigma \sim P|\mathcal{F}_\sigma$ such that the process $I_{]\sigma,\infty]} \cdot S \in \mathcal{M}^c_{loc}(^\sigma Q)$.*

Proof. Necessity follows from Theorem 4 and Proposition 8 applied to the process $S_{\cdot+\sigma} - S_\sigma$ adapted to the shifted filtration $(\mathcal{F}_{t+\sigma})$. As usual, the sufficiency is almost obvious. Indeed, if the claim fails, there exists a bounded process $X \in \mathcal{X}^1$ such that for the stopping time $\sigma := \inf\{t > 0 : X_t \neq 0\}$ we have $P(\sigma < T) > 0$. But then $^\sigma X := I_{]\sigma,\infty]} \cdot X$ is in $\mathcal{M}^c(^\sigma Q)$ or, equivalently, $^\sigma X Z$ is a martingale with respect to P. It starts from zero and hence is zero. The density process Z of $^\sigma Q$ with respect to P is equal to one at σ and, being right-continuous, remains strictly positive on a certain stochastic interval on which $^\sigma X$ should be zero. This contradicts to the assumption that $P(\sigma < T) > 0$. $\qquad\square$

Remark 4. Let B be a Brownian motion, $\sigma := \inf\{t \geqslant 0 : B_t = -1\}$, and $Z_t = 1 + B_{t \wedge \sigma}$. Take $S_t = B_t - B_{t \wedge \sigma} + \sqrt{(t - \sigma)^+}$. Then $S_t Z_t = 0$ and, therefore, $S \in \mathcal{M}^c_{loc}(\tilde{P})$ where $\tilde{P} := Z_T P$. Nevertheless, according to Theorem 5 there is an immediate arbitrage at σ.

In virtue of Remark 3 we obtain in the same way the following

Theorem 10. *Suppose that $\mathcal{X}$ consists of all processes bounded from below and belonging to the closed linear subspace of $\mathcal{S}$ generated by $\mathcal{X}(S^i)$, $i \in \mathbf{N}$, where S^i are continuous semimartingales. Then the NA property holds iff for any stopping time σ there exists a probability measure $^\sigma Q \ll P$ with the restriction $^\sigma Q|\mathcal{F}_\sigma \sim P|\mathcal{F}_\sigma$ such that $I_{]\sigma,\infty]} \cdot S^i \in \mathcal{M}^c_{loc}(^\sigma Q)$ for every $i \in \mathbf{N}$.*

References

1. Delbaen, F., Schachermayer, W.: A general version of the fundamental theorem of asset pricing. Math. Annalen, **300**, 463–520 (1994)
2. Delbaen, F., Schachermayer, W.: The existence of absolutely continuous local martingale measures. Ann. Applied Probab., **5** , 926–945 (1995)
3. Émery, M.: Une topologie sur l'espace de semimartingales. Séminaire de Probabilités XIII. Lect. Notes Math., **721**, 260–280 (1979)
4. Jacod, J.: Calcul Stochastique et Problèmes de Martingales. Springer, Berlin–Heidelberg–New York (1979)
5. Kabanov, Yu.M.: On the FTAP of Kreps–Delbaen–Schachermayer. Kabanov, Yu.M., Rozovskii, B.L., Shiryaev, A.N. (eds.) Statistics and Control of Random Processes. The Liptser Festschrift. Proceedings of Steklov Mathematical Institute Seminar. World Scientific, Singapour (1997)
6. Kabanov, Yu.M., Pergamenshchikov, S.: Two-Scale Stochastic Systems. Asymptotic Analysis and Control. Springer, Berlin–Heidelberg–New York (2003)
7. Levental, S., Skorohod, A.V.: A necessary and sufficient condition for absence of arbitrage with tame portfolios. Ann. Applied Probab., **5**, 906–925 (1995)
8. Mémin, J.: Espace de semimartingales et changement de probabilité. Zeitschrift für Wahrscheinlichkeitstheorie und Verw. Geb., **52**, 9–39 (1980)
9. Strasser, E.: Characterization of arbitrage-free markets. Preprint (2002)

Information-equivalence: On Filtrations Created by Independent Increments

Hans Bühler

Technische Universität zu Berlin
e-mail: buehler@math.tu-berlin.de

Summary. This article investigates filtrations created by the increments-processes of processes with independent increments: Suppose two processes create the same filtrations, when will the processes of their increments also create the same filtrations?

Our main result is that the processes of increments of two extremal continuous martingales with independent increments create the same filtrations if and only if either process admits a deterministic representation with respect to the other.

1 Introduction

Let $(\Omega, \mathcal{A}, \mathbb{P})$ be a probability space, $X = (X_t)_{0 \leqslant t \leqslant T}$ a continuous process with horizon $T \in (0, \infty]$ and $\mathbb{F}(X) := (\mathcal{F}_t(X))_{t \leqslant T}$ with $\mathcal{F}_t(X) := \sigma(X_{s \leqslant t})$ the filtration created by X (we will assume that all σ-algebras are complete and that all filtrations are right-continuous). Note that we will use abbreviations similar to $X_{a \leqslant t \leqslant b} := (X_t)_{t \in [a,b]}$ or $\sigma(X_{t \geqslant a} - X_a) := \sigma(X_t - X_a; t \geqslant a)$ throughout the article (a and b are fixed numbers). For convenience, we will also assume that all processes in this article start at zero, i.e., $X_0 \equiv 0$.

We denote by $X^a = (X_t^a)_{t \leqslant T}$ the *increment-process of X at a*, i.e.,

$$X_t^a := X_{t \vee a} - X_a = X_t - X_{t \wedge a}.$$

Clearly, if X is a martingale, X^a is a martingale as well, adapted to both $\mathbb{F}(X)$ and $\mathbb{F}(X^a)$.

Now, given two processes X and $\widetilde{X}$ which create the same filtration, we may ask how the increments of these processes are related to each other.

For example, assume that two standard Brownian motions B and $\widetilde{B}$ create the same filtration. If we take $r \in (0, T)$, since both B^r and $\widetilde{B}^r$ are independent from $\mathcal{F}_r := \mathcal{F}_r(B) = \mathcal{F}_r(\widetilde{B})$, both $\sigma(B^r) \cap \mathcal{F}_r$ and $\sigma(\widetilde{B}^r) \cap \mathcal{F}_r$ are trivial.

But, under which circumstances are $\sigma(B^r)$ and $\sigma(\widetilde{B}^r)$ equal?

Consider the following definition:

Definition 1 (Information-similarity and -equivalence). *We call two processes* $X = (X_t)_{t \leqslant T}$ *and* $\widetilde{X} = (\widetilde{X}_t)_{t \leqslant T}$

1. *information-similar, if X^0 and $\widetilde{X}^0$ create the same filtration, i.e. $\mathbb{F}(X^0) = \mathbb{F}(\widetilde{X}^0)$, and*
2. *information-equivalent, if all their processes of increments are information similar, i.e. $\mathbb{F}(X^a) = \mathbb{F}(\widetilde{X}^a)$ holds for all $a \in [0, T]$.*

Additionally, for $I \subset [0, T]$, we call X and $\widetilde{X}$

3. *information-semiequivalent (on I), if their processes of increments are information-similar for each $a \in I \cup \{0\}$, i.e., $\mathbb{F}(X^a) = \mathbb{F}(\widetilde{X}^a)$.*

In the sequel, wherever suitable, we will abbreviate *information-similarity* by *i-similarity* and so forth. Moreover, we will sometimes write *ii* for "independent increments".

Let us now illustrate the preceding definition by the following example of two Brownian motions, which are i-similar but not i-equivalent:

Assume that B is a standard Brownian motion, choose an $r \in (0, T)$ and set

$$\widetilde{B}_t := \int_0^t h_s \, \mathrm{d}B_s \qquad \text{with} \qquad h_t := 1_{t \leqslant r} + 1_{t > r} \cdot \mathrm{sgn}^-(B_r) \qquad (1)$$

where we define $\mathrm{sgn}^-(x) := 1_{x > 0} - 1_{x \leqslant 0}$.

It can be shown easily that $\widetilde{B}$ is a Brownian motion and that B and $\widetilde{B}$ create the same filtration. However, their increments at r do not:

Assume that $\sigma(B^r) = \sigma(\widetilde{B}^r)$ and set $s^- := \mathrm{sgn}^-(B_r)$. By construction, we have $\widetilde{B}_t^r = s^- \cdot B_t^r$ for all $t > r$, i.e., we can identify s^- by using the information of two random variables measurable with respect to $\sigma(B^r)$. But s^- is non-trivial and independent of $\sigma(B^r)$. Since this is impossible, neither $\sigma(B^r) \subseteq \sigma(\widetilde{B}^r)$ nor $\sigma(B^r) \supseteq \sigma(\widetilde{B}^r)$.

Remark 1. In [2], FRANK B. KNIGHT calls two continuous processes X and $\widetilde{X}$ *past-and-future equivalent* if $\sigma(X_{s \leqslant t}) = \sigma(\widetilde{X}_{s \leqslant t})$ and $\sigma(X_{s \geqslant t}) = \sigma(\widetilde{X}_{s \geqslant t})$ for all $t \in [0, T]$.

Assume X is Markov and that $\widetilde{X}_t \in \sigma(X_{s \leqslant t}) \cap \acute{\sigma}(X_{s \geqslant t})$. Then there are measurable functions g and f such that $\widetilde{X}_t = g_t(X_{s \leqslant t}) = \mathbb{E}[f_t(X_{s \geqslant t}) \mid X_{s \leqslant t}] = \mathbb{E}[f_t(X_{s \geqslant t}) \mid X_t] =: \widetilde{x}_t(X_t)$. Hence, two Markov-processes X and Y are *past-and-future equivalent*, if and only if we have a.s. $\widetilde{X}_t = \widetilde{x}_t(X_t)$ resp. $X_t = x_t(\widetilde{X}_t)$ for two suitable sequences of measurable functions x and $\widetilde{x}$ (compare lemma 1 in [2], [5], p. 113 and, for further related topics, section 17.3).

Clearly, past-and-future-equivalence and information-equivalence are different concepts; in this article, we study the latter and focus on processes with independent increments.

2 Equivalent Characterizations

Our first result shows that in case of processes with independent increments, information-equivalence is closely related to the properties of the time-reversed processes:

For a continuous ii-process $M = (M_t)_{t \leqslant T}$ with a finite time-horizon T, its *reverse process* $M'_t := M_T - M_{T-t}$ is also a continuous ii-process (note that the term *reverse process* is used differently in [2]). Then, given another ii-process $N = (N_t)_{t \leqslant T}$, we observe the following characterization:

Lemma 1. *Two continuous ii-processes M and N are i-equivalent if and only if both M and N and their time-reversed processes M' and N' are i-similar.*

Proof. We first note that M and N are i-equivalent iff for all fixed $a < b$ the relation

$$\sigma(M_b - M_{a \leqslant t \leqslant b}) = \sigma(N_b - N_{a \leqslant t \leqslant b}) \qquad (2)$$

holds. Therefore, if M and N are i-equivalent, M' and N' are trivially i-similar.

Conversely, let us assume that M and N as well as their reverse processes are i-similar. We will show that (2) holds:

First, we choose a and b with $a < b \leqslant T$ and let $\mathcal{N} := \sigma(N_{a \leqslant t \leqslant b} - N_a)$. Then, since M and N are i-similar, we have

$$\mathcal{N} \subset \sigma(N_{a \leqslant t \leqslant b} - N_a; \, N_{u \leqslant a}) = \sigma(M_{a \leqslant t \leqslant b} - M_a; \, M_{u \leqslant a}).$$

On the other hand, because also M' and N' are i-similar,

$$\mathcal{N} \subset \sigma(N_{a \leqslant t \leqslant T} - N_a) = \sigma(M'_{t \leqslant T-a}) = \sigma(M_{b \leqslant u \leqslant T} - M_b; \, M_{a \leqslant t \leqslant b} - M_a).$$

Since $\sigma(M_{t \leqslant a})$, $\sigma(M_{a \leqslant t \leqslant b} - M_a)$ and $\sigma(M_{b \leqslant t \leqslant T} - M_b)$ are independent this yields that $\mathcal{N} \subset \sigma(M_{a \leqslant t \leqslant b} - M_a)$.

If we apply the same idea to $\sigma(M_{a \leqslant t \leqslant b} - M_a)$, we find that M and N are indeed i-equivalent by (2). $\qquad \square$

Lemma 1 gives us a simple tool to check whether, say, two Brownian motions are i-equivalent. Indeed, remember our example of two Brownian motions defined in (1). There, for any $u \in (r, T)$, we find

$$\begin{aligned}
\sigma(B'_{t \leqslant T-u}) &= \sigma(B_T - B_{u \leqslant t \leqslant T}) \\
&\neq \sigma(s^- \cdot (B_T - B_{u \leqslant t \leqslant T})) \\
&= \sigma(\widetilde{B}_T - \widetilde{B}_{u \leqslant t \leqslant T}) = \sigma(\widetilde{B}'_{t \leqslant T-u}),
\end{aligned}$$

in correspondence with lemma 1.

Up to this point, M or N were general processes, not martingales. But now we want to link i-equivalence to the predictable representation property (PRP) of extremal martingales. What implication has information-equivalence for the representation of N by M?

Let $\mathbb{F}$ be a filtration, M a square-integrable martingale to $\mathbb{F}$ which starts at zero and define $L^2(M, \mathbb{F})$ as the space of $\mathbb{F}$-predictable $\mathbb{P}_M$-square-integrable integrands ($\mathbb{P}_M := \mathbb{P} \otimes d\langle M \rangle$, compare [3], p. 137). Then, M is said to have the $\mathbb{F}$-PRP (or M is $\mathbb{F}$-extremal), iff for any other $\mathcal{L}^2(\mathbb{F})$-martingale N, there exists a unique integrand $H \in L^2(M, \mathbb{F})$ such that

$$N_t = N_0 + \int_0^t H_s \, dM_s \tag{3}$$

(if we omit $\mathbb{F}$, we always refer to the filtration created by the process itself).

Remark 2. H is $\mathbb{P}_M$-a.e. not zero, if and only if N has the $\mathbb{F}$-PRP. In that case, we have $M_t = M_0 + \int_0^t H_s^{-1} \, dN_s$ with $H_s^{-1} := (1/H_s) \cdot 1_{H_s \neq 0}$.

Proof. W.l.g. assume $T < \infty$. If H is $\mathbb{P}_M$-a.e. not zero and if $dX = \xi \, dM$, holds for a martingale X, then we have $dX = \bar{\xi} \, dN$ for $\bar{\xi} := (1/H) \cdot 1_{H \neq 0} \cdot \xi$.

Conversely, if N is $\mathbb{F}(M)$-extremal, consider the square-integrable $\mathbb{F}(M)$-martingale $X_t := \int_0^t 1_{H_s=0} \, dM_s$. Then, there exists a unique K such that $X_t = \int_0^t K_s \, dN_s = \int_0^t K_s H_s \, dM_s$. Consequently, we find $\langle X, X \rangle = \int 1_{H_s=0} \cdot K_s H_s \, d\langle M \rangle \equiv 0$, i.e., X_T is zero and we get $\mathbb{P}_M[H = 0] = \mathbb{E}[X_T^2] = 0$.   □

Remark 3. A continuous $\mathcal{L}^2$-martingale M with independent increments has a deterministic bracket. In particular, M is Gaussian.

Recall that the bracket of a continuous martingale with independent increments is deterministic, thus it is a time-changed Brownian motion and therefore Gaussian.

From now on, assume that M and N are two continuous centered $\mathbb{F}(M)$-extremal $\mathcal{L}^2$-martingales and that M has independent increments (we do not assert that N has independent increments nor that N creates the same filtration as M).

Then, our main theorem states:

Theorem 1. *M and N are information-equivalent if and only if H is a.e. deterministic, i.e. there is a version of H such that*

$$\mathbb{P}_M[H \neq H^*] = 0 \qquad for \qquad H_t^* := \mathbb{E}[H_t] \tag{4}$$

(we set H_t^* to zero if $\mathbb{E}[H_t]$ does not exist).

Theorem 1 shows that for such processes the term "information-equivalence" is quite reasonable: Being i-equivalent implies that we can construct one process from the other by applying a deterministic, i.e. foreseeable rule.

Note that this result once again yields that the Brownian motions B and B' as defined in (1) are not information-equivalent. An example for a Brownian motion, which is not even information-similar (but still extremal) is TANAKA's example, $\int_0^t \mathrm{sgn}^-(B_s) \, dB_s$ (see [3], page 240).

Before we prove our theorem, we need some lemmata.

The first one is a trivial characterization of information-similarity, but nevertheless a convenient reference:

Lemma 2. *The processes M and N are information-similar if and only if H is also predictable with respect to the filtration $\mathbb{F}(N)$ created by N.*

Proof. If H is $\mathbb{F}(N)$-predictable, so is H^{-1}. Therefore $M_t = \int_0^t H_s^{-1} \, dN_s$ has to be $\mathbb{F}(N)$-adapted. The reverse assertion is also trivial. $\square$

Since M is Gaussian, its increments are also Gaussian and therefore extremal on their own filtration (proposition 4.11 in [3], p. 213):

Lemma 3. *The process M^a is extremal on $\mathbb{F}(M^a)$ for all $a \in [0, T]$.*

For our second process N we find:

Lemma 4. *The process N^a is adapted to $\mathbb{F}(M^a)$, iff it is $\mathbb{F}(M^a)$-extremal.*

Proof. First, assume that N^a is adapted to $\mathbb{F}(M^a)$. Since N^a is an $\mathbb{F}(M^a)$-martingale, we find an $\mathbb{F}(M^a)$-predictable h such that $dN^a = h \, dM^a$. The $\mathbb{P}_M$-uniqueness of $H \in L^2(M, \mathbb{F}(M))$ given $dN = H \, dM$ yields that h is a version of $H|_{(a,T]}$. Because N is extremal, it is a.s. not zero and we can write $dM^a = h^{-1} \, dN^a$.

Reversely, if N^a has the $\mathbb{F}(M^a)$-PRP, we find $dM^a = \bar{h} \, dN^a$ for some $\bar{h} \in L^2(N^a, \mathbb{F}(M^a))$. The $\mathbb{P}_N$-uniqueness of $dM = H^{-1} \, dN$ yields with the usual arguments that $dN^a = \bar{h}^{-1} \, dM^a$. $\square$

Note that we found that if N^a is $\mathbb{F}(M^a)$-extremal, we have $dN^a = \bar{h}^{-1} \, dM^a$ for $\bar{h} \in L^2(N^a, \mathbb{F}(M^a))$. Hence, independence of M^a and $\mathcal{F}_a(M)$ shows that the increment N^a is independent of $\mathcal{F}_a(N)$.

Summing up the previous results yields the following technical lemma which we will use in the proof of theorem 1:

Lemma 5. *The processes M and N are information-semiequivalent on $I = \{t_1, \ldots, t_n\}$ for with $0 < t_1 < \cdots < t_n < T$ if and only if H has an $\mathbb{F}(N)$-predictable version which can be written as*

$$H_t = \sum_{k=0}^{n} H_t^k \cdot 1_{t \in (t_k, t_{k+1}]} \tag{5}$$

with $t_0 := 0$ and $t_{n+1} := T$ and where H^k is predictable w.r.t. $\mathbb{F}(M^{t_k})$.

Let us point out that this representation means that for $t \in (t_k, t_{k+1}]$, the value H_t depends solely on the "information" provided by the increment N^{t_k} i.e. M^{t_k} up to t. We could consider the points t_k as "resets" where the information carried by H until t_k is thrown away.

Proof. Assume first that M^{t_k} and N^{t_k} create the same filtrations for all $k = 0, \ldots, n - 1$. Then, H is $\mathbb{F}(N)$-predictable by lemma 2, and N^{t_k} is adapted to $\mathbb{F}(M^{t_k})$, on which M^{t_k} is extremal by lemma 3. The uniqueness of H w.r.t. to M therefore yields that H has a version such that $H|_{(t_k, T]} =: H^k \in L^2(M^{t_k}, \mathbb{F}(M^{t_k}))$.

Conversely, if H^{t_k} is predictable w.r.t. $\mathbb{F}(M^{t_k})$, N^{t_k} is adapted to $\mathbb{F}(M^{t_k})$ and therefore extremal by lemma 4 for all k.   □

Given lemma 2 and 5, we can eventually turn to our proof of our main theorem 1. In principle, we would like to use lemma 5 and to take some kind of a limit from a finite number $t_1, \ldots, t_n$ of points towards the entire interval $[0, T]$. In practise, this is what can be done via the properties of the stochastic integral since it can be defined as a limit of "simple" processes:

Proof of theorem 1. Trivially, if the deterministic process $H^* = (H_t^*)_{t \leqslant T}$ with $H_t^* := \mathbb{E}[H_t]$ is a version of H, once again lemma 2 yields that $\mathbb{F}(M^a)$ is equal to $\mathbb{F}(N^a)$ for every $a \in [0, T]$.

Now assume conversely, that M and N are i-equivalent.

Let $\mathcal{E}_0$ be the space of "simple integrands" vanishing at zero, i.e., of all processes $\xi_t = \sum_{k=-1}^{n} f_k \cdot 1_{t \in (t_k, t_{k+1}]}$ where each f_k is an $\mathcal{F}_{t_k}(M)$-measurable, bounded random variable and with a finite sequence $0 = t_0 < t_1 < \cdots < t_{n+1} = T$

Now, H^* is an $L^2(M)$-version of H iff for any $\xi \in \mathcal{E}_0$, the scalar product of ξ and H is equal to the product of ξ and H^*.

For this purpose, let us fix a $\xi_t = \sum_k f_k \cdot 1_{t \in (t_k, t_{k+1}]} \in \mathcal{E}_0$.

Then, because of the i-equivalence of M and N, M and N are i-semiequivalent on $I = \{t_1, \ldots, t_n\}$, and lemma 5 asserts that there is a version of H which can be written as

$$H_t^\xi := \sum_{k=0}^{n} H_t^k \cdot 1_{t \in (t_k, t_{k+1}]},$$

where each H^k is $\mathbb{F}(M^{t_k})$-predictable and therefore in particular independent of $\mathcal{F}_{t_k}(M)$. This yields

$$
\begin{aligned}
(H, \xi)_{L^2(M)} &= (H^\xi, \xi)_{L^2(M)} \\
&= \mathbb{E}\left[\sum_{k=0}^{n} \int_{t_k}^{t_{k+1}} H_t^k \cdot f_k \, \mathrm{d}\langle M \rangle_t \right] \\
&= \mathbb{E}\left[\sum_{k=0}^{n} f_k \cdot \mathbb{E}\left[\int_{t_k}^{t_{k+1}} H_t^k \, \mathrm{d}\langle M^{t_k} \rangle_t \,\middle|\, \mathcal{F}_{t_k}(M) \right] \right] \\
&= \mathbb{E}\left[\sum_{k=0}^{n} f_k \cdot \mathbb{E}\left[\int_{t_k}^{t_{k+1}} H_t^k \, \mathrm{d}\langle M^{t_k} \rangle_t \right] \right]
\end{aligned}
$$

$$= \mathbb{E}\left[\sum_{k=0}^{n} f_k \cdot \mathbb{E}_{\mathbb{P}_M}\left[H^k|_{[t_k, t_{k+1}]}\right]\right]$$

$$= \mathbb{E}\left[\sum_{k=0}^{n} f_k \cdot \mathbb{E}_{\mathbb{P}_M}\left[H|_{[t_k, t_{k+1}]}\right]\right]$$

$$\overset{(*)}{=} \mathbb{E}\left[\sum_{k=0}^{n} f_k \cdot \int_{t_k}^{t_{k+1}} H_t^* \, d\langle M\rangle_t\right] = \left(H^*, \xi\right)_{L^2(M)},$$

where we have $(*)$ because M has a deterministic quadratic variation. $\quad\square$

At this stage, we want to stress the fact that i-equivalence of N and M already implies that N has independent increments: As shown in lemma 4, each N^a is extremal on $\mathbb{F}(M^a)$.

Lemma 6. *An extremal martingale X has independent increments if and only if each increment is extremal on its own filtration.*

Proof. If X^a is extremal, choose $A \in \mathcal{F}_T(X^a)$ and define the $\mathbb{F}(X^a)$-martingale $A_t^a := \mathbb{P}[A \,|\, X_{s \leqslant t}^a]$. We find a process $H^a \in L^2(X^a, \mathbb{F}(X^a))$ such that $dA^a = H^a \, dX^a$. Since X is a martingale on $\mathbb{F}(Y)$, the extension $A_t := A_{t \vee a}^a$ is also a martingale on $\mathbb{F}(X)$, and we have $\mathbb{P}[A] = A_a^a = \mathbb{E}[A_T \,|\, \mathcal{F}_a(X)] = \mathbb{P}[A \,|\, \mathcal{F}_a(X)]$, ie independence of X^a and $\mathcal{F}_a(X)$.

Reversely, if Y^a is a $\mathbb{F}(X^a)$-martingale, apply the same extension as above and see, for $t > a$, that $\mathbb{E}[Y_t^a \,|\, \mathcal{F}_a(X)] = Y_a^a$, i.e., that Y is a $\mathbb{F}(X)$-martingale. Hence we can write $dY^a = H^a \, dX^a$ and independence implies that $H^a \in L^2(X^a, \mathbb{F}(X^a))$. $\quad\square$

First of all, this lemma and its consequence that an i-equivalent N has obviously a deterministic quadratic variation, too, yields that $d\langle N\rangle = H^2 \, d\langle M\rangle$ is deterministic. Hence, the only source of randomness in H can only step from the sign of H.

A second observation is that the initial restriction to processes with independent increments does not seem too unnatural – at least regarding the property that their increments are extremal.

Note that only lemma 1 required that T is finite. In all other cases, we have proved all our claims also for an infinite time horizon. Hence, we can apply our result to a setting introduced by TSIRELSON:

2.1 Noises

In [4], BORIS TSIRELSON used the description of "noise", to investigate the possibilities to linearize Brownian motions on Polish groups. We briefly repeat his notion, simplified for our setting, to add some comments which are results of the preceding sections.

Definition 2 (Noise). *A noise is defined as a family* $\mathbb{G} = (\mathcal{G}_t^s)_{s \leqslant t; s, t \in \mathbb{R}}$ *of σ-fields and a group* $\theta = (\theta_t)_t$ *of operators on the probability space such that*

1. θ_r *sends* $\mathcal{G}_t^s$ *onto* $\mathcal{G}_{t+r}^{s+r}$ *for all* $s \leqslant t$ *and* $r \in \mathbb{R}$,
2. $\mathcal{G}_s^r$ *and* $\mathcal{G}_t^s$ *are independent for all* $r \leqslant s \leqslant t$ *and*
3. $\mathcal{G}_s^r$ *and* $\mathcal{G}_t^s$ *generate* $\mathcal{G}_t^r$ *for all* $r \leqslant s \leqslant t$.

"Noise" as defined above is the property of a filtration. However, we want to link it to adapted and generating processes:

Definition 3 (Representation of a noise). *A representation of a noise is a family* $Z = (Z_t^s)_{s \leqslant t}$ *of random variables with values in a Polish group[1] $(G, +)$ verifying*

1. θ_r *sends* Z_t^s *to* Z_{t+r}^{s+r} *for* $s \leqslant t$ *and* $r \in \mathbb{R}$ *(i.e., $Z_t^s \circ \theta_r = Z_{t+r}^{s+r}$),*
2. Z *is adapted to* $\mathcal{G}$, *i.e. Z_t^s is measurable w.r.t. $\mathcal{G}_t^s$ for $s \leqslant t$,*
3. $Z_s^r + Z_t^s = Z_t^r$ *for all* $r \leqslant s \leqslant t$ *and*
4. *for any $\delta > 0$, $\mathbb{P}[|Z_t^s| \leqslant \delta] \to 1$ for $t \downarrow s$.[2]*

We call such a representation continuous, *if*

5. *for any $\delta > 0$, $\mathbb{P}[|Z_t^s| > \delta]/(t - s) \to 0$ for $t \downarrow s$*

and we call it faithful *if*

6. $\mathcal{G}_t^0 = \sigma(Z_r^0; 0 \leqslant r \leqslant t)$ *for* $t \geqslant 0$.

The canonical example a continuous faithful representation of a noise is given by the standard Wiener space $\Omega = \mathbb{C}[\mathbb{R}]$, $\mathcal{F} = \mathbb{B}[\mathbb{R}]$ with the Wiener measure $\mathbb{P}$ and the shift-operator $\theta_t(\omega)(u) := \omega(t + u)$ for $\omega \in \Omega$.

Then, the coordinate process $X_t(\omega) := \omega(t)$ is a standard Brownian motion and obviously generates a noise with a continuous and faithful representation by virtue of our former convention $X_t^s := X_{t \vee s} - X_s$ and $\mathcal{G}_t^s := \mathcal{F}_t(X^s)$.

In fact, each $\mathbb{R}$-valued continuous faithful representation Z is in itself a sequence of continuous stationary increment processes with independent increments, i.e., a scaled Brownian motion with drift (eg. [5], page 115): Indeed, setting

$$Y_t := Z_t^0 \mathbb{1}_{t \geqslant 0} - Z_0^t \mathbb{1}_{t < 0}$$

we obtain a Brownian motion Y such that

$$Y_t^s = Y_t - Y_s = \begin{cases} -Z_0^t + Z_0^s = Z_t^s & \text{for } \mathbf{s} < t \leqslant 0, \\ Z_0^t + Z_0^s = Z_t^s & \text{for } s \leqslant 0 < t, \\ Z_t^0 - Z_s^0 = Z_t^s & \text{for } 0 < s < t. \end{cases}$$

Note that this construction also shows how to construct noises on Ω.

---

[1] On a general Polish group (i.e., a topological group with a Polish metric, cf [4]) we define a Brownian motion as a continuous centered process with independent stationary increments; on $\mathbb{R}$, this coincides with the group of scaled Brownian motions.

[2] In a general Polish space (which has a metric d), we define $|x| := d(x, 0)$ where the symbol 0 denotes unity.

Remark 4. Condition *6* in definition 3 is not symmetric since it does not yield a restriction on the way the fields $(\mathcal{G}_0^t)_{t<0}$ are generated. In order to obtain a similar "downward" condition, we shall consider the additional requirement

7. $\mathcal{G}_0^t = \sigma(Z_0^r; t \leqslant r \leqslant 0)$ for $t < 0$,

which rests on the former observation that $Z_0 := (Z_0^t)_{t;t\leqslant 0}$ is also a Brownian motion.

Given this extension, assume there is a continuous process $N = (N_t)_{t\in\mathbb{R}}$ such that $\mathcal{F}_t(N^0) = \mathcal{G}_t^0$ for $t \geqslant 0$ and $\sigma(N_0^r; t \leqslant r \leqslant 0) = \mathcal{G}_0^t$ for $t < 0$. Then,

Lemma 7. *The process N establishes a continuous faithful representation via $\tilde{Z}_t^s := N_t^s$ if and only if N^0 is information-equivalent to Z^0 and N_0 to Z_0.*

Proof. The requirements to provide a continuous faithful representation for $(\mathbb{G}, \theta)$ are obviously all met, except that N_t^s is supposed to be $\mathcal{G}_t^s$-measurable. This means that N is adapted to $\mathbb{G}$ in both directions (s resp. t).

For the case $s \geqslant 0$ fixed, this is by lemma 4 equivalent to N^0 and Z^0 being information-equivalent.

For the case $t < 0$ fixed, this is by the same reasoning equivalent to N_0 and Z_0 (as processes $(N_0^r)_{r;r\leqslant 0}$) being information-equivalent.

Since $\mathcal{G}_t^s = \mathcal{G}_0^s \vee \mathcal{G}_t^0$ for $s < 0 \leqslant t$, and $N_t^s = N_0^s + N_t^0$ this finishes the proof.   □

Note that another possible modification of condition *6* in definition 3 is to request $\mathcal{G}_t^s = \sigma(Z_r^s; s \leqslant r \leqslant t) =: \mathcal{F}_t^s(Z)$ for all $s \leqslant t$. This gives rise to an extension of the term "information-equivalence" to processes with two time variables:

Definition 4. *Two processes $X = (X_t^s)_{s\leqslant t}$ and $Y = (Y_t^s)_{s\leqslant t}$ are called information-equivalent iff they create the same filtrations, i.e., $\sigma(X_{r;s\leqslant r\leqslant t}^s) = \sigma(Y_{r;s\leqslant r\leqslant t}^s)$ for all $s \leqslant t$.*

Now, observe that $\mathcal{G}_t^s = \mathcal{F}_t^s(Z)$ for all $s \leqslant t$ is already implied when restricted to the cases where either $s = 0$ or $t = 0$ (as considered above), since $\mathcal{G}_t^s = \mathcal{G}_0^s \vee \mathcal{G}_t^0$ and $Z_t^s = Z_0^s + Z_t^0$ by definition and $\mathcal{F}_t^s(Z) = \sigma(Z_0^r + Z_u^0; s \leqslant r \leqslant 0 \leqslant u \leqslant t)$.

In other words, extending *6* by condition *7* already implies

Lemma 8. *Each continuous faithful representation N of $(\mathbb{G}, \theta)$ is information-equivalent to Z.*

Acknowledgement. Parts of this article are taken from the author's diploma thesis [1] (2001). I would like to thank ALEXANDER SCHIED and HANS FÖLLMER, for their delightful help and support and I am also very grateful to WALTER SCHACHER-MAYER for fruitful conversations and to LAURENT NGUYEN-NGOC for his continuing assistance.

References

1. Bühler, H. (2001): Zur Informationsstruktur Brownscher Bewegungen. Diploma-Thesis, Humboldt-Universität, Berlin.
2. Knight, F.B. (1995): A remark on Walsh's Brownian motions. The Journal of Fourier Analysis and Applications, Special Issue 1995, 317–324.
3. Revuz, D., Yor, M. (1999): Continuous Martingales and Brownian motion. Springer, Berlin.
4. Tsirelson, B. (1998): Unitary Brownian Motions are linearizable. From his website, http://www.math.tau.ac.il/~tsirel.
5. Yor, M. (1997): Some Aspects of Brownian motion, Part II. Birkhäuser, Basel.

Rotations and Tangent Processes
on Wiener Space

Moshe Zakai

Department of Electrical Engineering,
Technion–Israel Institute of Technology,
Haifa 32000, Israel
e-mail: zakai@ee.technion.ac.il

Dedicated to the memory of P.-A. Meyer

Summary. The paper considers (a) Representations of measure preserving transformations ("rotations") on Wiener space, and (b) The stochastic calculus of variations induced by parameterized rotations $\{T_\theta w, 0 \leqslant \theta \leqslant \varepsilon\}$: "Directional derivatives" $(\mathrm{d}F(T_\theta w)/\mathrm{d}\theta)_{\theta=0}$, "vector fields" or "tangent processes" $(\mathrm{d}T_\theta w/\mathrm{d}\theta)_{\theta=0}$ and flows of rotations.

1 Introduction

Let (W, H, μ) be an abstract Wiener space (AWS): $W = \{w\}$ is a separable Banach space, H (the Cameron–Martin space) is a separable Hilbert space densely and continuously embedded in W, $W^* \hookrightarrow H^* = H \hookrightarrow W$ and for every e in W^*, ${}_{w^*}(e, w)_w$ is $N(0, |e|_H^2)$. By the Cameron–Martin theorem, for any $h \in H$, the measure induced by $w + h$ is equivalent to the measure μ, therefore if $F(w)$ is a r.v. on the Wiener space, so is $F(w + h)$. This fact enabled the development of the stochastic calculus of variations, i.e. the Malliavin calculus which very roughly is based on the directional derivative of F in the h direction: $(\mathrm{d}F(w + \varepsilon h)/\mathrm{d}\varepsilon)_{\varepsilon=0}$. Now, let T be a measure preserving transformation on W (in short, a 'Rotation'), i.e. ${}_{w^*}(e, Tw)_w$ is also $N(0, |e|_H^2)$. Then if $F(w)$ is a r.v. so is $F(Tw)$ and if $T_\theta w\, 0 \leqslant \theta \leqslant \varepsilon$ is a smooth collection of rotations one can consider objects like $(\mathrm{d}F(T_\theta w)/\mathrm{d}\theta)_{\theta=0}$. The purpose of this paper is to survey previous work and to present new results on the following:

(i) Measure preserving transformations (rotations),
(ii) The Malliavin calculus of rotations.

The study of stochastic analysis over Riemannian manifolds showed that the Cameron–Martin space is not sufficient to represent the tangent space and as discussed in [4, 6, 5, 2, 12, 1], more general vector fields are needed. The setup of these papers was based on the model of

$$\int_0^{\cdot} \sum_{j=1}^d \sigma_{ij}(s, w) \, \mathrm{d}w_j(s), \qquad i, j \leqslant d \tag{1}$$

for "vector fields", where w is the d-dimensional $(d \geqslant 2)$ Wiener process with σ being non-anticipative and skew symmetric which induce a measure preserving transformation. This paper considers the abstract Wiener space setup presented in [8], it is restricted to flat space. A particular class of anticipative tangent processes was recently considered in [3]. A class of rotations on Wiener space introduced in [10] are different from the rotations considered here.

Section 2 summarizes results, mainly from the Malliavin Calculus, which are needed later. Rotations T are considered in section 3, where $Tw = \sum \eta_i(w)e_i$, $\eta_i(w)$ are i.i.d. $N(0, 1)$ random variables, and e_i, $i = 1, 2, \ldots$ is a complete orthonormal base in H. Rotations are introduced in Section 3. Theorem 2 ([16, 18]) presents rotations by showing that sequences of i.i.d. $N(0, 1)$ random variables can be constructed as the divergence of $R(w)e_i$ where $R(w)$ belongs to a certain class of operators. An outline of the proof is included. This is followed by new results, Proposition 6 and Theorem 3 showing that under some smoothness assumptions every sequence of i.i.d. $N(0, 1)$ random variables on the Wiener space can be represented by the construction of Theorem 2. Section 4 deals with directional derivations of the type $(\mathrm{d}F(T_\theta w)/\mathrm{d}\theta)_{\theta=0}$. A "tangent operator" is introduced and its relation to the directional derivative is indicated. Section 5 deals with "tangent process" which are Banach valued random variables, play the role of tangent vectors and induce the directional derivatives. The first part of section 6 gives a positive answer (due to Tsirelson and Glasner) to a problem raised in [8] whether the group of invertible rotations on the Wiener space are connected. The second part deals with flows of rotations, i.e. the flow induced by

$$\frac{\mathrm{d}T_t w}{\mathrm{d}t} = m(T_t w, t)$$

where $m(w, t)$ are the tangent processes introduced in section 5. The case where $m(w, t)$ is of the type of equation of (1) was considered in [1]. A more detailed proof of the result of [8] is given. The appendix deals with the following problem: In view of the results of section 3 and other results, the question arose whether the condition that $\nabla u(w)$ is quasinilpotent implies the existence of a filtration such that u is adapted. A counter example, following [14] is presented in the appendix.

Acknowledgements

We wish to express our sincere thanks to B. Tsirelson and E. Glasner for their contribution in Section 6, to A.S. Üstünel for useful discussions and to E. Mayer-Wolf for useful remarks.

2 Preliminaries

Notation

For each $e \in H$ and induced by an element of W^*, $\delta(e)$ denotes the $N(0, |e|_H^2)$ r.v. $_W \cdot (e, w)_W$. For all $e \in H$, $\delta(e)$ denotes the L^2 limit of $_W \cdot (e_n, w)_W$ as $e_n \to e$ in H.

We will not distinguish between embeddings and inclusions. For example, $e \in W^*$ will also be considered as an element of H or W; the distinction being clear from context.

For $F(w) = f(\delta e_1, \ldots, \delta e_n)$ and f smooth, ∇F is defined as $\nabla F = \sum f_i'(\delta e_1, \ldots, \delta e_n) \cdot e_i$. Let $\mathcal{X}$ be a separable Hilbert space and u an $\mathcal{X}$ valued functional. $\mathbb{D}_{p,k}(\mathcal{X})$, $p > 1$, $k \in \mathbb{N}$ will denote the Sobolev space of $\mathcal{X}$ valued functionals in $L^p(\mu, \mathcal{X})$ whose k-th order derivative $\nabla^k u$ is in $L^p(\mu, \mathcal{X} \otimes H^{\otimes k})$. $\mathbb{D}_{p,k}(R)$ will be denoted $\mathbb{D}_{p,k}$. $\mathbb{D}(\mathcal{X}) = \bigcap_{p>1} \bigcap_{k \in N} \mathbb{D}_{p,k}(\mathcal{X})$. Recall that

$$\nabla : \mathbb{D}_{p,k}(\mathcal{X}) \to \mathbb{D}_{p,k-1}(\mathcal{X} \otimes H)$$

and for δ, the adjoint of ∇ under the Wiener measure

$$\delta : \mathbb{D}_{p,k}(\mathcal{X} \otimes H) \to \mathbb{D}_{p,k-1}(\mathcal{X})$$

are continuous linear operators for any $p > 1$, $k \in \mathbb{N}$. The operator δ is the divergence or the Skorohod integral and:

(a) If $u \in \mathbb{D}_{2,1}(H)$, then

$$E[(\delta u)^2] = E[|u|_H^2] + E[\text{trace}(\nabla u)^2].$$

(b) If $F \in \mathbb{D}_{2,1}$, $u \in \mathbb{D}_{2,1}(H)$ and if $Fu \in \mathbb{D}_{2,1}(H)$, then

$$\delta(Fu) = F\delta u - (\nabla F, u)_H. \tag{2}$$

A. Exact and divergence free H-valued r.v.'s

Let $u \in \mathbb{D}_{2,1}(H)$ then (a) u is said to be "exact" if $u \in \nabla F(w)$ for some $\mathbb{D}_{2,1}$ functional $F(w)$. (b) u is said to be divergence free if $\delta u = 0$.

Set

$$U^e = U^{\text{exact}} = \{u \in \mathbb{D}_{2,1}(H) : u = \nabla F\}$$

$$U^{\text{d.f.}} = U^{\text{diverg. free}} = \{u \in \mathbb{D}_{2,1}(H) : \delta u = 0\}$$

If $u \in U^e, v \in U^{\mathrm{d.f.}}$ then $E(u,v)_H = E(\nabla f, v)_H = E(f\delta v) = 0$. Hence U^e and $U^{\mathrm{d.f.}}$ are orthogonal subspace of $\mathbb{D}_{2,1}(H)$ and $\mathbb{D}_{2,1}(H) = U^e \oplus U^{\mathrm{d.f.}}$.

Let $\mathcal{L}F$ be the Ornstein–Uhlenbeck operator: $\mathcal{L}F = \delta\nabla F$, assume that $EF = 0$ then $\mathcal{L}^{-1}$ is a bounded operator and $\mathcal{L}\mathcal{L}^{-1}F = F$. Hence, for any $F \in \mathbb{D}_{2,1}, EF = 0$, $F(w) = \delta(\nabla\mathcal{L}^{-1}F)$ and F possesses the representation $F(w) = \delta u$, $u = \nabla\mathcal{L}^{-1}F$, where $u \in \mathbb{D}_{2,1}(H)$. Note that this representation is different from the Ito-type representation of Wiener functionals.

Returning to U^e and $U^{\mathrm{d.f.}}$, for $F(w) = \delta u$, $u \in \mathbb{D}_{2,1}(H)$, $\delta(u - \nabla\mathcal{L}^{-1}\delta u) = 0$ then $\nabla\mathcal{L}^{-1}\delta u$ and $u - \nabla\mathcal{L}^{-1}\delta u$ are the projections of u on U^e and $U^{\mathrm{d.f.}}$ respectively.

We prepare, for later reference, the following lemma.

Lemma 1. *Let $u \in \mathbb{D}_{2,1}(H)$, let $\{e_i, i = 1, 2, \ldots\}$ be a CONB on H further assume that $u = \sum \delta(v_i)e_i$, $v_i \in \mathbb{D}_{2,1}(H)$. If $(v_i, e_j)_H + (v_j, e_i)_H = 0$, $i, j = 1, 2, \ldots$, then $\delta u = 0$. In particular the above result holds for $v_i = A(w)e_i$, where $A + A^T = 0$.*

Proof. For smooth F, integrating by parts we have

$$E(F \cdot \delta u) = E \sum_i \nabla_{e_i} F \delta v_i$$

$$= E \sum_i \nabla_{e_i} F \cdot \delta\left(\sum_j (v_i, e_j)e_j\right)$$

$$= E \sum_{i,j} \nabla^2_{e_i, e_j} F \cdot (v_i, e_j).$$

and $\delta u = 0$ follows since $\nabla^2_{e_i, e_j} F$ is symmetric in i and j and F is arbitrary.
□

B. Constructing a filtration on the AWS

Let (W, H, μ) be an A.W.S., we introduce a time structure i.e. a filtration and causality on it as follows: Let the projections $\{\pi_\theta,\ 0 \leqslant \theta \leqslant 1\}$ be a real continuous and strictly increasing resolution of the identity on H. Set

$$\mathcal{F}_\theta = \sigma\{\delta\pi_\theta h,\ h \in H\}. \tag{3}$$

Propositions 1–3 are from [17].

Proposition 1. *$F(w) \in \mathbb{D}_{2,1}$ is $\mathcal{F}_\theta$ measurable iff $\nabla F = \pi_\theta \nabla F$ a.s. (intuitively: if $F(w) = f(\delta h_1, \ldots, \delta h_n)$ with $h_i = \pi_\theta h_i$)*

Definition 1. *An H-valued r.v. u will be said to be $\mathcal{F}$.-adapted if for every $\theta \in [0, 1]$ and $h \in H$, $(u, \pi_\theta h)$ is $\mathcal{F}_\theta$ measurable.*

Proposition 2. $u \in \mathbb{D}_{2,1}(H)$ *is* $\mathcal{F}_\cdot$ *adapted iff*

$$\pi_\theta \nabla u \pi_\theta = \pi_\theta \nabla u$$

for all $\theta \in [0,1]$.

Definition 2. *Let* $G(w)$ *be a measurable r.v. taking values in the class of bounded transformations on* H. *Then* G *is said to be weakly adapted if for all* $h \in H$, Gh *is adapted,* G *will be said to be adapted (or causal) if* Gu *is adapted for all adapted* u.

Proposition 3. *If* $G(w)$ *satisfies* $Gu \in \mathbb{D}_{2,0}(H)$ *whenever* $u \in \mathbb{D}_{2,0}(H)$ *and* G *is weakly adapted then* G *is adapted iff* $\pi_\theta G \pi_\theta = \pi_\theta G$.

Another version of the last result is:

Proposition 4. *Under the assumptions of the previous proposition, a weakly adapted* G *is adapted iff for all* $h \in H$

$$\pi_\theta h = 0 \implies \pi_\theta G h = 0. \tag{4}$$

Proof. Let u be of the form

$$u = \sum_1^n \varphi_i (\pi_{\theta_{i+1}} - \pi_{\theta_i}) h_i \tag{5}$$

where $\theta_{i+1} > \theta_i$ and the φ_i are $\mathbb{D}_{2,0}$ r.v.'s. Then

$$Gu = \sum_{i=1}^n \varphi_i G(\pi_{\theta_{i+1}} - \pi_{\theta_i}) h_i \tag{6}$$

$h_i \in H$. Now, assume that u is adapted hence the φ_i are $\mathcal{F}_{\theta_i}$ measurable. Since G is weakly measurable then (4) implies that $\varphi_i G \pi_{\theta_{i+1}} (I - \pi_{\theta_i}) h_i$ is also adapted hence Gu is adapted and G is adapted since u of the form (5) are dense in $\mathbb{D}_2(H)$.

Conversely, again u assumed to be adapted, and since Gu is adapted, $\varphi_i G(\pi_{\theta_{i+1}} - \pi_{\theta_i}) h_i$ is adapted. Hence, since φ_i are $\mathcal{F}_{\theta_i}$ measurable, we must have

$$\pi_\theta G(\pi_{\theta_{i+1}} - \pi_{\theta_i}) h_i = 0$$

and (4) follows. $\qquad\qquad\Box$

Given a $\mathbb{D}_{2,0}$ functional $F(w)$ on (W, H, μ) and a filtration π_θ (continuous strictly increasing) then there exists a unique adapted $u \in \mathbb{D}_2(H)$ such that $u \in \mathrm{Dom}\,\delta$, and $F(w) = \delta u$, and $E(\delta u)^2 = E(|u|_H^2)$. ([19], in the classical setup this representation follows directly from the multiple Wiener integral). Hence, given $F(w) = \delta u^e$ then u^e can be lifted uniquely to $\tilde{u}$ such that $\delta \tilde{u} = \delta u^e$ and $\tilde{u}$ is adapted to a given filtration.

C. Quasinilpotent operators

An H-S operator on H is said to be quasinilpotent (q.n.p.) if any *one* of the following is satisfied (cf. [20] or [18]):

(a) trace $A^n = 0$ $\forall n \geqslant 2$.
(b) $|A^n|^{\frac{1}{n}} \longrightarrow 0$, $|.|$ is the operator norm.
(c) The spectrum of A is $\{0\}$ only.
(d) $(1 - \alpha A)^{-1} = \sum_{n=0}^{\infty} \alpha^n A^n$ $\forall \alpha$.
(e) $\det_2(I + \alpha A) = 1$ $\forall \alpha$, where $\det_2(I + A) = \prod_i (1 - \lambda_i) e^{-\lambda_i}$.

Proposition 5 ([17]). *If u is adapted, $u \in \mathbb{D}_{2,1}(H)$ then ∇u is q.n.p.*

Outline of proof. Let $\theta_{i+1} > \theta_i$, set

$$u_i = q_i(\pi_{\theta_{i+1}} - \pi_{\theta_i})h_i = q_i \tilde{h}_i.$$

Then,

$$\text{trace } \nabla u_i \nabla u_j = \text{trace}(\nabla q_i \otimes \tilde{h}_i)(\nabla q_j \otimes \tilde{h}_j)$$
$$= (\nabla q_i, \tilde{h}_j)_H (\nabla q_j, \tilde{h}_i)_H.$$

For $i = j$, $(\nabla q_i, \tilde{h}_j)_H = 0$, if $i > j$ then $(\nabla q_i, \tilde{h}_i)_H = 0$. Similarly for $i < j$.

The following question arises regarding the converse of the last result: Given u such that ∇u is q.n.p., does this imply the existence of a filtration such that u is adapted to it? The answer to this question, as shown in appendix A, is negative.

D. The Ito–Nisio theorem

Theorem 1. *Let (X_i) be a symmetric sequence of random variables (i.e. $(\pm X_1, \pm X_2, \ldots, \pm X_n)$ has the same law as $(X_1, \ldots, X_n)$ for any n) with values in a separable Banach space B. Denote by μ_n the distribution of the partial sum $S_n = \sum_{i=1}^{n} X_i$. The following are equivalent:*

(i)   *the sequence (S_n) converges almost surely in the Banach norm;*
(ii) *(S_n) converges in probability;*
(iii) *(μ_n) converges weakly;*
(iv) *there exists a B-valued r.v. γ such that $(S_n, f) \xrightarrow{P} (\gamma, f)$ for all f in B';*
(v)   *there exists a probability measure μ in $\mathcal{P}(B)$ sucht that $\mu_n \circ f^{-1} \to \mu \circ f^{-1}$ weakly for every f in B'.*

Cf. [11] for (i), (ii), (iii) and (v); (iv) follows from (i) and implies (v).

Let (W, H, μ) be AWS, then for any complete orthonormal basis (CONB), $\{e_i\}$ and any i.i.d. $N(0, 1)$ random variables η_i then

$$y_n = \sum_{i=1}^{n} \eta_i e_i$$

converges a.s. in the Banach norm (in particular $|w - \sum_{i=1}^{n} \delta e_i e_i|_W \xrightarrow{\text{a.s.}} 0$). Hence denoting $y = \lim y_n$, then $y = Tw$ is a measure preserving transformation and $\eta_i = {}_W(Tw, e_i)_{W^*}$.

3 Rotations on Wiener space

Let $\eta_i = $ i.i.d. $N(0,1)$ r.v.'s and $\{e_n\}$ a CONB induced by W^* then by the Ito–Nisio theorem

$$Tw = \sum \eta_i e_i \tag{7}$$

is a measure preserving transformation on W, we will refer to it as a rotation.

Theorem 2 ([16, 18]). *Let $w \mapsto R(w)$ be a strongly measurable random variable on W with values in the space of bounded linear operators on H. Assume that R is almost surely an isometry on H, i.e. $|R(w)h|_H = |h|_H$ a.s. for all $h \in H$. Further assume that for some $p > 1$ and for all $h \in H$, $Rh \in \mathbb{D}_{p,2}(H)$, and $\nabla Rh \in \mathbb{D}_{p,1}(H \otimes H)$ is a quasi-nilpotent operator on H. If moreover, either*
(a) $(I + i\nabla Rh)^{-1} \cdot Rh$ is in $L^q(\mu, H)$, $q > 1$, for any $h \in H$ (here q may depend on $h \in H$) or,
(b) $Rh \in \mathbb{D}(H)$ for a dense set in H.
    Then

$$E\big[\exp i\delta(Rh)\big] = \exp\Big(-\frac{1}{2}|h|_H^2\Big).$$

Namely, if $(e_n, n \in \mathbb{N})$ is a complete, orthonormal basis in H then $(\delta(Re_n), n \in \mathbb{N})$ are independent $N(0,1)$-random variables and consequently $\sum_i \delta(Re_i)e_i$ defines a measure preserving transformation of W.

The map R satisfying the conditions for this theorem with $p = 2$ and under (a) with $q = 2$ will be said to *satisfy the rotation conditions*.

Outline of proof. Let $u : B \to H$ be "an H-C^1 map" and $T = w + u$, assume that T is a.s. invertible then [18].

$$E\big(F(Tw) \cdot |\Lambda(w)|\big) = EF(w)$$

where

$$\Lambda(w) = \det{}_2(I_H + \nabla u)\exp\Big(-\delta u - \frac{1}{2}|u|_H^2\Big).$$

In particular, $F(w) = 1$, $u = \nabla Rh$, then since $\det_2(I_H + \nabla u) = 1$ and $|Rh|_H^2 = |h|_H^2$, hence

$$E\exp\Big(-\delta(Rh) - \frac{1}{2}|h|_H^2\Big) = 1$$

or

$$E \exp\bigl(-\delta(Rh)\bigr) = \exp\Bigl(\frac{1}{2}\,|h|_H^2\Bigr)$$

consequently $\delta(Re_i), i = 1, 2, \ldots$ are i.i.d., $N(0,1)$ and $\sum \delta(Re_i)e_i$ is a rotation by the Ito–Nisio theorem.

The conditions on R in the rotation theorem are obviously not necessary since if for all i, $u_i \in U^{\mathrm{d.f.}}$ and defining $\rho : H \to H$ by

$$\rho e_i = u_i, \qquad i = 1, 2, \ldots$$

then $\delta \rho h = 0$ and if R induces a rotation so does $R + \rho$. We have, however, the following two results, which yield a converse to the rotation theorem.

Proposition 6. *Let $R(w)$ be an a.s. bounded operator on H. Assume that $R(w)$ is weakly adapted with respect to a filtration induced by a continuous increasing $\pi_\cdot$. Further assume for all $h \in H$, $R(w)h$ is in the domain of δ and the probability law of $\delta(Rh)$ is $N(0, |h|_H^2)$ then:*

1. *If $h_1, h_2 \in H$ and $(h_1, h_2)_H = 0$ then $\delta(Rh_1)$ and $\delta(Rh_2)$ are independent.*
2. *$R(w)$ is a.s. an isometry on H.*
3. *$\sum_i \delta(Re_i)e_i$ is a rotation and if e_i, $i = 1, 2, \ldots$ and h_i, $i = 1, 2, \ldots$ are CONB's on H then, a.s.*

$$\sum_i \delta(Rh_i)h_i = \sum_i \delta(Re_i)e_i.$$

Proof.

1. $\begin{aligned}[t] E\bigl(\exp i\alpha\,\delta(Rh_1)\exp i\beta\,\delta(Rh_2)\bigr) &= E \exp i\delta\bigl(R(\alpha h_1 + \beta h_2)\bigr) \\ &= \exp\Bigl(-\frac{\alpha^2}{2}|h_1|_H^2 - \frac{\beta^2}{2}|h_2|_H^2\Bigr) \\ &= E \exp\bigl(i\alpha\,\delta Rh_1\bigr)\, E \exp\bigl(i\beta\,\delta(Rh_2)\bigr). \end{aligned}$

2. By part 1, $y_\theta = \delta(R\pi_\theta h)$ is a Gaussian process of independent increments. Hence it is Gaussian martingale and its quadratic variation satisfies

$$\langle y, y\rangle_\theta = E y_\theta^2.$$

and by our assumption $E y_\theta^2 = |\pi_\theta h|_H^2$. But

$$\langle y, y\rangle_\theta = (R\pi_\theta h, R\pi_\theta h)_H$$

and $R^T R = I$ follows.

3. Follows from the Ito–Nisio theorem.   □

Theorem 3. *Let $Tw = \sum \eta_i e_i$ be a rotation. Let $\mathcal{F}.$ be a filtration induced by a continuous increasing resolution of the identity. Then there exists a unique $\mathcal{F}.$ weakly adapted $R(w) : H \to H$ which is an isometry and*

$$Tw = \sum_i \delta(Re_i)e_i.$$

If, moreover, $\eta_i \in \mathbb{D}_{2,2}$ then ∇Rh is q.n.p.

Proof. By our assumptions, every η_i can be uniquely represented as $\eta_i = \delta u_i$ where the u_i are adapted, in the domain of δ, and $u_i \in \mathbb{D}_2(H)$. Define R by

$$R(w)e_i = u_i$$

then $R(w)$ is weakly adapted, and satisfies the assumptions of the previous result. Hence R is an isometry and $Tw = \sum \delta(Re_i)e_i$. If moreover the $\eta_i \in \mathbb{D}_{2,2}$ then $\nabla Rh \in \mathbb{D}_{2,1}(H)$ and is q.n.p. since it is adapted. $\qquad\square$

Remark 1. For a given rotation $Tw = \sum \delta(Re_i)e_i$, $R(w)$ is "highly non unique"; instead of representing η_i as the divergence of adapted processes, we can define $\eta_i = \delta v_i$ with $v_i \in U^e$ to yield a unique $R^e(w)$ such that $R^e h$ is exact for all $h \in H$ and $Tw = \sum \delta(R^e e_i)e_i$. In other words, given any R satisfying the assumptions of the theorem we can construct an R^e such that $\delta(R^e h) = \delta(Rh)$ and $R^e h \in U^e$ for all $h \in H$. Thus R^e will not necessarily be an isometry. Also, we can "lift" R^e to another $\tilde{R}$ which is weakly adapted with respect to another filtration.

4 Tangent operators

Let $T_t w = \sum_i \delta(R_t e_i)e_i$ where R_t, $t \in [0,\delta]$ is unitary, satisfies the rotation condition, and let $(e_i, i = 1, 2, \dots)$ be a CONB on H induced by W^*. Assume that $(R_{t+\varepsilon} - R_t)/\varepsilon$ converges a.s. in the operator norm to $B_t(w)$ as $\varepsilon \to 0$, then $(R_{t+\varepsilon}^T - R_t^T)/\varepsilon$ converges to B_t^T and $0 = \frac{d(R_t^T R_t)}{dt} = \frac{d(R_t R_t^T)}{dt}$. Hence

$$B_t^T R_t + R_t^T B_t = 0 \qquad \text{and} \qquad B_t R_t^T + R_t B_t^T = 0. \tag{8}$$

Setting $R_0 = I$ and $A = (dR_t/dt)$ at $t = 0$ yields

$$A + A^T = 0. \tag{9}$$

Let $f(x_1, \dots, x_n)$ be a smooth function on $\mathbb{R}_n$. Set $F(w) = f(\delta e_1, \dots, \delta e_n)$. Since $_W(T_t w, e_i)_{W^*} = \delta(R_t e_i)$ and $\delta e_i \circ T_t w = {}_W(T_t w, e_i)_{W^*}$, hence with $A = (dR_t/dt)_{t=0}$:

$$\frac{dF(T_t w)}{dt}\Big|_{t=0} = \sum_i \left(f_i'(\delta(R_t e_1), \dots, \delta(R_t e_n)) \delta\left(\frac{dR_t}{dt} e_i\right)\right)_{t=0}$$

$$= \sum_i f_i'(\delta e_1, \dots, \delta e_n) \delta(A e_i)$$

$$= \sum_i \delta(f_i'(\,.\,) A e_i) - \sum_{ij} f_{ij}''(\,.\,)(e_j, A e_i)$$

The second term vanished since f_{ij}'' is symmetric and A is skew symmetric. Hence

$$\frac{dF(T_t w)}{dt}\Big|_{t=0} = \delta(A \nabla F) \tag{10}$$

Motivated by (10) we define

Definition 3 ([8]). *Let $w \to Q(w)$ be a weakly measurable mapping taking values in the space of bounded operators on H. Assume that for all $F \in \mathbb{D}_{2,1}$, $Q \nabla F \in \mathbb{D}_{2,1}(H)$. For every Q satisfying these conditions and $u \in \mathbb{D}_{2,1}(H)$ we define*

$$\mathcal{L}_{Q,u} F = \delta(Q \nabla F) + \nabla_u F$$

and denote it as the Tangent Operator *induced by (Q, u). Also $\mathcal{L}_{Q,0} F =: \mathcal{L}_Q F$.*

The following summarizes some properties of the tangent operator (cf. [8] for proofs).

1) $\mathcal{L}_{Q,u}$ is closeable in H (i.e. if $F_n \to 0$ in H a.s. and $\mathcal{L} F_n$ exist, then $\mathcal{L} F_n \to 0$).
2) The adjoint of $\mathcal{L}_{Q,u}$ satisfies $\mathcal{L}_{Q,u}^* F = \mathcal{L}_Q F + \delta(F u)$.
3) If $Q = A$ where $A^T + A = 0$, then

$$\mathcal{L}_A F_1 F_2 = F_1 \mathcal{L}_A F_2 + F_2 \mathcal{L}_A F_1$$

namely, $\mathcal{L}_A$ is a derivation (i.e. behaves as a first order operator).
4) Let $g : R \to R$ be twice differentiable, set $g'(x) = \frac{dg(x)}{dx}$. Assume that both $F(w)$ and $g(F(w))$ are in $\mathbb{D}_{2,1}$ and both $A \nabla F$ and $A \nabla g(F)$ are in $\mathbb{D}_{2,1}(H)$. Then, for $A + A^T = 0$ we have

$$\mathcal{L}_A g(F) = \delta(g'(F) A \nabla F)$$
$$= g'(F) \mathcal{L}_A F + g''(F)(\nabla F, A \nabla F)$$
$$= g'(F) \mathcal{L}_A F.$$

5) Cf. [8] for results for $\mathcal{L}_A(\delta u)$ and $[\mathcal{L}_A, \mathcal{L}_B]$.

Let R be unitary and satisfy the rotation condition. Let $R_{t,k}(w)$ denote $R(w + t \cdot k)$, $t \in [0, 1]$, $k \in H$. Then $R_{t,k}$ is also a.s. unitary and $\nabla R_{t,k} h$ is also a.s. quasinilpotent. Assume that $R_{t,k}$ satisfies condition a or b of the rotation theorem then $R_{t,k}$ also induces a rotation, let $T_{t,k}$ denote this rotation. Setting

$$X_k^R F(w) = \left. \frac{\mathrm{d}F(T_{t,k}w)}{\mathrm{d}t} \right|_{t=0}$$

$$= \mathcal{L}_{\dot{R}_k} F$$

where

$$R_k = \left. \frac{\mathrm{d}R_{t,k}}{\mathrm{d}t} \right|_{t=0}.$$

As shown in [16] and [18]

$$\nabla(F \circ T) = R(\nabla F \circ T) + X^R F \tag{11}$$

i.e.

$$\nabla_h(F \circ T) = \big(R(\nabla F \circ T), h\big)_H + X_h^R F$$

and when $u : W \to H$ is a cylindrical map

$$(\delta u) \circ T = \delta\big(R(u \circ T)\big) + \sum_i \big(X^R(u, e_i), Re_i\big)$$

$$= \delta\big(R(u \circ T)\big) + \mathrm{trace}(R^{-1} X^R u). \tag{12}$$

If $g(x), x \in \mathbb{R}$ is smooth then

$$0 = \frac{\mathrm{d}}{\mathrm{d}t} Eg\big(F(T_{t,k}w)\big)$$

hence

$$0 = E\Big\{ g'\Big(F\big(T_{t,k}(w)\big)\Big) \cdot X_{t,k}^R F \Big\}.$$

5 Tangent processes

Let R_t satisfy the rotation condition, set $\mathcal{L}_{R_t} w = T_t w = \sum_i \delta(R_t e_i) e_i$. In order to represent the "directional derivative" $\mathcal{L}_A F$ as the action of a "tangent vector" on F, the "vector field" $(\mathrm{d}T_t w / \mathrm{d}t)_{t=0}$ is needed. Formally, for $A = \big(\frac{\mathrm{d}R_t}{\mathrm{d}t}\big)_{t=0}$,

$$\left. \frac{\mathrm{d}T_t w}{\mathrm{d}t} \right|_{t=0} = \sum_i \delta(Ae_i) e_i$$

which motivates the following definition:

Definition 4. *Let $Q(w)$ be a weakly measurable H operator valued transformation on H. Assume that $Q(w)h \in \mathbb{D}_{2,1}(H)$ for all $h \in H$. Let $e_i, i = 1, 2, \ldots$ be a CONB induced by elements of W^*, if $\sum_i \delta(Qe_i) e_i$ converges weakly in the Banach space as $n \to \infty$; namely, if there exists a W-valued random variable Y such that $\sum_i \delta(Qe_i)(e_i, \tilde{\alpha})_H$ converges in probability ($\tilde{\alpha}$ is the image of α in H under the canonical injection from W^* to H) to $_W\langle Y, \alpha \rangle_{W^*}$ for all $\alpha \in W^*$. The limit Y will be denoted by $Y = \mathcal{L}_Q w$ and will be called the tangent process induced by Q and $\{e_i\}$).*

Remark 2. The definition given here is somewhat different from that in [8] as Y may depend on $\{e_i\}$.

Remark 3. It is often necessary to consider the case where the series $\sum \delta(Qe_i)e_i$ satisfy a stronger convergence condition; several cases assuring a.s. convergence are

a. The case where the $\delta(Qe_i)$ satisfy the conditions of the extended Ito–Nisio theorem.
b. The case where Q is a bounded non random operator ([9, Theorem 1.14]).
c. Let $\varphi \in \mathbb{D}_{2,1}$ and $Q^*\nabla\varphi \in \mathbb{D}_{2,0}(H)$, then $\mathcal{L}_{\varphi Q}w$ exists in the sense of a.s. convergence in W if and only if $\varphi\mathcal{L}_Q w$ exists in the corresponding sense and then

$$\mathcal{L}_{\varphi Q}w = \varphi\mathcal{L}_Q w - Q^T\nabla\varphi.$$

The proof of c. follows directly from

$$\sum_{i=1}^{N} \delta(\varphi Qe_i)e_i = \sum_{i=1}^{N} \varphi\delta(Q_i e_i)e_i - \sum_{i=1}^{N} (Q^T\nabla\varphi, e_i) \cdot e_i.$$

The relation between the tangent process $\mathcal{L}_Q w$ and the tangent operator is reflected in the following lemma.

Lemma 2. *Assume that Q satisfies the requirements of Definition 4 and $\sum \delta(Qe_i)e_i$ converges a.s. Further assume that $u \in \mathbb{D}_{2,1}(H)$, $Qu \in \mathbb{D}_{2,1}(H)$ and u is the image in H of $\underset{\sim}{u}(w)$ which is W^* valued and $\nabla u Q$ is of trace class on H. Set $(\mathrm{trace})_e K = \overline{\lim}_{n\to\infty} \sum_1^n (e_i, Ke_i)_H$. Then*
(a) $(\mathrm{trace})_e(\nabla u Q)$ exists and

$$w\big(\mathcal{L}_Q w, \underset{\sim}{u}(w)\big)_{W^*} = \delta(Qu) + (\mathrm{trace})_e(\nabla u q) \tag{13}$$

(b) If we also assume that $u = \nabla F$ and Q is skew-symmetric then

$$w(\mathcal{L}_Q w, \underset{\sim}{\nabla F})_{W^*} = \mathcal{L}_Q F \tag{14}$$

(and then $\mathcal{L}_Q w$ acts as a vector field on F with $\mathcal{L}_Q F$ being the directional derivative along the tangent process).

Proof. Setting $(u, e_i) = v_i$ and $u_n = \sum_1^n v_i e_i$, where $\{e_i\}$ is a CONB induced by W^*.

$$
\begin{aligned}
w\big(\mathcal{L}_\theta w, \underset{\sim}{u}(w)\big)_{W^*} &= \sum_i^n \delta(Qe_i)v_i \\
&= \sum_i^n \delta(Qv_i e_i) + \sum_i^n (\nabla v_i, Qe_i) \\
&= \mathcal{L}_Q u_n + \sum_i^n \sum_j^n \nabla_{e_j} v_i(e_j, Qe_i)
\end{aligned}
$$

The left hand side is a continuous functional on W^* and the last equation converges on both sides to (13) which proves (a). (b) follows since $\nabla_{e_j} u_i = \nabla^2_{e_i,e_j} F$ is symmetric. □

Note that by Lemma 1, if $\mathcal{L}_Q w$ exist in $\mathbb{D}_{2,1}(H)$, and if $Q + Q^T = 0$ then:

$$\delta(\mathcal{L}_Q w) = 0.$$

The tangent processes that were considered in [1]–[5] were of the form of the right hand side of (1) in the introduction with $\{\sigma_{ij}\}$ skew symmetric and nonanticipative. The relation to the $\mathcal{L}_Q w$ formulation will now be pointed out. Consider the case of the d-dimensional Brownian motion, $h = \int_0^\cdot h'_s \, ds$, $h' \in L^2([0,1], \mathbb{R}^d)$, then we have

Proposition 7.
(A) Let $\mathbf{q}$ *denote the matrix* $\{q_{ij}(\theta, u), i \leqslant i, j \leqslant d, \theta, u \in [0,1]\}$ *and set*

$$(Qh)_i = \sum_{j=0}^n \int_0^\cdot \int_0^1 q_{ij}(\theta, u) h'_j(u) \, du \, d\theta, \qquad i = 1, \ldots, d.$$

Assume that the $q_{i,j}(\theta, u)$ *are* $\mathcal{F}_\theta$ *adapted for all* $u \in [0,1]$ *and* $E|Qh|^2_H \leqslant K \cdot |h|^2_H$.
Then $\mathcal{L}_Q w$ *exists and as a* W*-valued r.v.*

$$\mathcal{L}_Q w = \int_0^\cdot \left(\int_0^1 \mathbf{q}(\theta, u) \, du \right) dw_\theta \tag{15}$$

(B) If $\mathbf{b} = \{b_{i,j}(s), 1 \leqslant i, j \leqslant d, s \in [0,1]\}$ *where* $b_{i,j}(s)$ *are* $\mathcal{F}_s$ *adapted and*

$$E \sum_{i,j} b^2_{i,j}(s) < \infty.$$

Setting

$$(Bh)_i = \int_0^\cdot b_{ij}(s) h'_j(s) \, ds$$

then

$$\mathcal{L}_B w = \int_0^\cdot \mathbf{b}(s) \, dw_s. \tag{16}$$

Proof. Let $(e_i, i \geqslant 1)$ be an orthonormal basis of H. Then

$$Qe_i = \int_0^\cdot \left(\int_0^1 \mathbf{q}(\tau, u) \dot{e}_i(u) \, du \right) d\tau,$$

We have

$$\sum_{i=1}^{n} \delta(Qe_i)\langle e_i, \alpha \rangle = \sum_{i=1}^{n} \int_0^1 \left(\int_0^1 \mathbf{q}(\tau, u)\dot{e}_i(u)\, du \right) dw_\tau \int_0^1 \dot{e}_i(s)\dot{\alpha}(s)\, ds$$

$$= \int_0^1 \int_0^1 \mathbf{q}(\tau, u) \left(\sum_{i=1}^{n} \int_0^1 \dot{e}_i(s)\dot{\alpha}(s)\, ds \dot{e}_i(u) \right) du\, dw_\tau$$

$$= \int_0^1 \int_0^1 \mathbf{q}(\tau, u)\beta_n(u)\, du\, dw_\tau, \tag{17}$$

where

$$\beta_n(u) = \sum_{i=1}^{n} \int_0^1 \dot{e}_i(s)\dot{\alpha}(s)\, ds\, \dot{e}_i(u).$$

It is clear that β_n converges to $\dot{\alpha}$ in $L^2[0,1]$. Since A is of Hilbert–Schmidt, we see that

$$\int_0^1 \mathbf{q}(\tau, u)\beta_n(u)\, du$$

converges to $\int_0^1 \mathbf{a}(\tau, u)\dot{\alpha}(u)\, du$ in $L^2[0,1]$. Consequently, (17)) converges to

$$\int_0^1 \int_0^1 \mathbf{q}(\tau, u)\dot{\alpha}(u)\, du\, dw_\tau$$

in L^2 hence in probability and (17) follows.

In order to prove (16) we have to show that (15) holds for the case where

$$\mathbf{q}(\theta, u) = \mathbf{b}(\theta)\delta(\theta - \mathbf{u})$$

where δ is the Dirac delta function. Setting

$$\delta_\varepsilon(u) = \frac{1}{\varepsilon} \quad \text{for } u \in [0, \varepsilon]$$

$$= 0 \quad \text{otherwise}$$

then, (16) follows since by (15)

$$\mathcal{L}_{Q_n} w = \int_0^\cdot \frac{1}{\varepsilon} \left(\int_{\theta - \varepsilon}^\theta b(u)\, du \right) dw_\theta$$

and

$$E \int_0^1 \left(b - \frac{1}{\varepsilon} \int_{\theta - \varepsilon}^\theta b(u)\, du \right)^2 d\theta \xrightarrow[\varepsilon \to 0]{} 0.$$

$\square$

6 Groups of rotations

A. Theorem 4 ([15], [7]). *Let T be an invertible measure preserving transformation on the Wiener space, then there exists a family $T_\theta, \theta \in [0,1]$ of measure preserving transformations such that $T_0 w = w$, $T_1 = T$ and for every $\theta \in [0,1]$, $E|T_\eta w - T_\theta w|_W \to 0$ as $\eta \to \theta$.*

The proof was first shown to us by Tsirelson [15], the proof given here is a shorter proof due to Glasner [7].

Proof. Since the Wiener measure on $C[0,1]$ and the Lebesgue measure on $[0,1]$ are isomorphic, it suffices to prove the result for the Lebesgue measure. Set $T_a(X) = aT(X/a)$ in $(0,a)$ and $T_a(X) = X$ in $(a,1)$ which is measure preserving, $T_1 = T$ and T_0 is the identity. Now,

$$E\left|T_a(X) - T_{a+\varepsilon}(X)\right| \leqslant E\left\{\left|aT\left(\frac{X}{a}\right) - (a+\varepsilon)T\left(\frac{X}{a+\varepsilon}\right)\right| \cdot 1_{X \leqslant a}\right\} + a$$

$$\leqslant E\left\{\left|aT\left(\frac{X}{a}\right) - aT\left(\frac{X}{a+\varepsilon}\right)\right| \cdot 1_{X \leqslant a}\right\} + 2\varepsilon.$$

Applying Lusin's theorem to approximate $T(X)$ by a continuous $\tau_\theta(X)$, yields

$$\left|E\left(T_a(X) - T_{a+\varepsilon}(X)\right)\right| \leqslant a\left|E\left(\tau_\theta\left(\frac{X}{a}\right) - \tau_\theta\left(\frac{X}{a=\varepsilon}\right)\right)\right| + 2\theta + 2\varepsilon$$

and continuity in ε follows by dominated convergence since θ is arbitrary.   □

B. Flows

We want to show that for $A_t + A_t^T = 0$ and additional conditions the equation

$$\frac{dT_t w}{dt} = \left(\mathcal{L}_{A_t(w)}w\right) \circ T_t w, \qquad T_0 w = w \tag{18}$$

defines a flow of rotations. The case where W is the d-dimension Wiener space and A is adapted:

$$\left(\mathcal{L}_A w\right)_i = \int_0^{\cdot} \sum_j^d a_{ij}(s,w) \, dw_j(s), \qquad i = 1, 2, \ldots, d$$

was considered by Cipriano and Cruzeiro [1]. The general result presented in the next theorem is from [8] and is followed by a more detailed proof.

Theorem 5 ([8]). *Assume that for all $t \geqslant 0$, $A = A_t(w)$: is a skew symmetric strongly measurable mapping and for any $h \in H$, $A_t h \in \mathbb{D}_{p,1}(H)$, for some $p > 1$ and a.a. $t \in [0,T]$ where $T > 0$ is fixed. Further assume that:*

1. The series

$$B_t = \sum_{i=1}^{\infty} \delta(A_t e_i) e_i$$

converges in $L^p(d\mu \times dt, W)$ (as a W-valued random variable), where $(e_i, i \geqslant 1)$ is a fixed orthonormal basis of H.

2. Let p_n denote the orthogonal projection onto the span of $\{e_1, \ldots, e_n\}$ and V_n denote the sigma-algebra generated by $\{\delta e_1, \ldots, \delta e_n\}$. Assume that the sequence of vector fields $(B^n, n \geqslant 1)$ defined by

$$B_t^n = \sum_{i=1}^n \delta E[p_n A_t e_i \,|\, V_n] e_i \tag{19}$$

or, as will be shown later to be the same as

$$B_t^n = \sum_{i=1}^n E[\delta(p_n A_t e_i) \,|\, V_n] e_i \tag{20}$$

converges to B in $L^p(\mathrm{d}\mu \times \mathrm{d}t, W)$.

3. Assume that for a given $\varepsilon > 0$, we have

$$\int_0^T E\{\exp \varepsilon \|\|\nabla B_t\|\|\} \mathrm{d}t = \Gamma_{0,T} < \infty, \tag{21}$$

the norm above is defined as

$$\|\|\nabla B_t\|\| = \sup\{ \,|\nabla_h(w \langle B_t, \alpha \rangle_{W^\star})| \,:\, h \in B_1, \, \alpha \in W_1^\star\}$$

where $B_1 = \{h \in H : |h|_W = 1\}$ and $W_1^\star$ is the unit ball of $W^\star$. Further assume that (21) also holds for B_t replaced by B_t^n (this holds, e.g., when $\{e_i\}$ is a Schauder basis of W).

Then the equation

$$\phi_{s,t}(w) = w + \int_s^t B_r(\phi_{s,r}) \, \mathrm{d}r, \; s < t, \tag{22}$$

defines a flow of measure preserving diffeomorphisms of W whose almost sure inverse is denoted by $(\psi_{s,t}, 0 \leqslant s \leqslant t \leqslant T)$ and satisfies

$$\mu\{w : \phi_{s,t} \circ \psi_{s,t}(w) = \psi_{s,t} \circ \phi_{s,t}(w) = w\} = 1.$$

Moreover the inverse flow is the unique solution of the equation

$$\psi_{s,t}(w) = w - \int_s^t B_r(\psi_{r,t}) \, \mathrm{d}r \tag{23}$$

and $\phi_{s,t}$ (hence $\psi_{s,t}$) leaves the Wiener measure invariant, i.e. $\phi_{s,t}^\star \mu = \mu$ for any $s < t \in [0, T]$.

Proof. We start with showing the equality (19) and (20). Set $\alpha^m = p_m A p_m$ and $a_{ij}^m = (e_j, \alpha^m e_i)$, then for $i, j \leqslant m$

$$E\{\delta(p_m Ae_i) \mid V_m\} = E\{\delta(\alpha^m e_i) \mid V_m\}$$

$$= \sum_{j=1}^{m} E\{\delta(a_{ij}^m e_j) \mid V_m\}$$

$$= \sum_{j=1}^{m} \delta e_j E\{a_{i,j} \mid V_m\} - \sum_j E\{\nabla_{e_j} a_{ij} \mid V_m\}$$

$$= \sum_{j=1}^{m} \delta e_j E\{a_{ij} \mid V_m\} - \sum_j \nabla_{e_j} E\{a_{ij} \mid V_m\}$$

$$= \sum_{j=1}^{m} \delta \big(E\{a_{ij} \mid V_m\} \cdot e_j \big)$$

$$= \delta E(\alpha^m e_i \mid V_m)$$

and (20) follows. Also, since A_t is skew symmetric, so are the matrices a_{ij}^m and $E(a_{ij}^m \mid V_m)$ hence by Lemma 1, $\delta B_t^m = 0$. Consequently, (cf., e.g., [18, Theorem 5.3.1]) the claimed results of Theorem 4 hold for B_t replaced by B_t^m. Therefore, denoting by $\phi_{s,t}^n$, $s \leqslant t \in [0,T]$, the flow associated to the cylindrical vector field B^n and by $\psi_{s,t}^n$, $s \leqslant t \in [0,T]$, its inverse, then in particular we have

$$\frac{\mathrm{d}\phi_{s,t}^{n\star}\mu}{\mathrm{d}\mu} = \exp \int_s^t (\delta B_r^n) \circ (\psi_{r,t}^n) \, \mathrm{d}r = 1$$

and

$$\frac{\mathrm{d}\psi_{s,t}^{n\star}\mu}{\mathrm{d}\mu} = \exp - \int_s^t (\delta B_r^n) \circ (\phi_{s,r}^n) \, \mathrm{d}r = 1. \qquad \Box$$

Let $e_1, e_2, \ldots$ be a fixed CONB of H induced by elements of W^*. Let M denote the following class of cylindrical operator Q on H. Let $q_{ij} = (e_j, Qe_i)$ then, for some m

(a) $q_{i,j} = 0$ for $i > m$ or $j > m$
(b) $q_{i,j} = -q_{ji}$
(c) $q_{ij} = f_{ij}(\delta e_1, \ldots, \delta e_m)$ and f_{ij} possesses bounded first derivatives.

Set $B_r(w) = \sum_1^m \delta(Q_r(w)e_i)e_i$, $Q_r \in M$. The following version of [18, Theorem 5.2.1] is needed to complete the proof of the theorem.

Proposition 8. Let $Q_r^a, Q_r^b \in M$, $r \in [0,T]$ and assume that for some $\varepsilon > 0$

$$E \int_0^T \left(\exp \varepsilon \|\|\nabla B_r^a\|\| + \exp \varepsilon \|\|\nabla B_r^b\|\| \right) \mathrm{d}r \leqslant \Gamma_{0,T} < \infty. \qquad (24)$$

Let $\varphi_{s,t}^a, \varphi_{s,t}^b$ denote the flows induced by Q^a and Q^b. Then for $s < t$, $(t-s)$ sufficiently small

$$E \sup_{u \in [s,t]} |\varphi_{s,u}^a - \varphi_{s,u}^b|_W \leqslant E \left(\int_s^t |B_r^a - B_r^b|_W^p \, dr \right)^{1/p} \Gamma_{s,t}^{\frac{1}{q}} \left(\frac{1}{t-s} \right)^{1/q}$$

where $\Gamma_{s,t}$ is defined as $\Gamma_{0,T}$ (equation (21)) with $0, T$ replaced by s, t, and $(t-s)q \leqslant \varepsilon$.

Proof of the proposition. Set $D_r = B_r^a - B_r^b$ and let $\varphi_{s,t}^\lambda$, $\lambda \in [0,1]$ be the solution to

$$\varphi_{s,t}^\lambda(w) = w + \int_s^t \left(\lambda B_r^a + (1-\lambda) B_r^b \right) \circ \varphi_{s,r}^\lambda \, dr.$$

Then $\varphi_{s,t}^\lambda$ is also a rotation. Set $Z_{s,t}^\lambda = \frac{d\varphi_{s,t}^\lambda}{d\lambda}$, then

$$\varphi_{s,t}^a - \varphi_{s,t}^b = \int_0^1 Z_{s,t}^\lambda \, d\lambda$$

and

$$Z_{s,t}^\lambda = \int_s^t D_r \circ \varphi_{s,r}^\lambda \, d\lambda + \int_s^t \left[(\nabla (B_r^b + \lambda D_r)) \right] \circ \varphi_{s,r}^\lambda Z_{s,r}^\lambda \, dr.$$

By Gronwall's lemma

$$|Z_{s,t}^\lambda|_W \leqslant \left(\int_s^t |D_r \circ \varphi_{s,r}^\lambda|_W \, dr \right) \exp \int_s^t \|\nabla B_r^b + \lambda \nabla D_r\| \circ \varphi_{s,r}^\lambda \, dr.$$

Therefore, since φ^λ is measure preserving:

$$E \sup_{r \in [s,t]} |\varphi_{s,r}^a - \varphi_{s,r}^b|_W$$

$$\leqslant E \int_0^1 |Z_s^\lambda|_W \, d\lambda$$

$$\leqslant E \left\{ \int_0^1 \left(\int_s^t |D_r(w)|_W \, dr \right) \exp \int_s^t \|\nabla B_r^b + \lambda \nabla D_r\| \, dr \, d\lambda \right\}$$

$$\leqslant E \left\{ \int_s^t |D_r|_W \, dr \exp \int_s^t \left(\|\nabla B_r^b\| + \|\nabla B_r^a\| \right) dr \right\}$$

$$\leqslant \left(E \int_s^t |D_r|_W^p \, dr \right)^{1/p} \left(E \exp q \int_s^t \left(\|\nabla B_r^a\| + \|\nabla B_r^b\| \right) dr \right)^{1/q}$$

$$\leqslant \left(E \int_s^t |D_r|_W^p \, dr \right)^{1/p} \left(E \frac{1}{t-s} \int_s^t \exp q(t-s) \left(\|\nabla B_r^a\| + \|B_r^b\| \right) dr \right)^{1/q}$$

which proves the proposition. $\qquad \Box$

Returning to the proof of the theorem, setting

$$Q_r^a = E(p_m A_r p_m \,|\, V_m)$$

and similarly, with m replaced by n, for Q_r^b yields

$$E \sup_{u \in [s,t]} \left| \varphi_{s,u}^m - \varphi_{s,u}^n \right|_W \leqslant \left(\Gamma_{s,t} \frac{1}{t-s} \right)^{1/q} E \left(\int_s^t \left| B_r^m - B_r^n \right|_W^p \, dr \right)^{1/p}. \quad (25)$$

This result implies the convergence of $(\phi_{s,u}^n, u \in [s,t])$ in $L^1(\mu, W)$ uniformly with respect to u for the intervals $[s,t]$ and the limit $(\phi_{s,u}, u \in [s,t])$ is the unique solution of the equation (23)). Equation (25)) implies the uniqueness of the equation since if $(\phi_{s,u}')$ is another solution, then its finite dimensional approximations must coincide with $(\phi_{s,u}^n)$. Now, using Lemmas 5.3.1, 5.3.2 and 5.3.5 of [18], it can be shown that the constructions of $(\phi_{s,u})$ on the different small intervals can be patched together to give the entire flow. For the inverse flow the same reasoning applies also. □

Appendix

As discussed in section 2, a necessary condition for $u \in \mathbb{D}_{2,1}(H)$ to be adapted to a given filtration is that ∇u be a.s. quasinilpotent. Now, given a $u \in \mathbb{D}_{2,1}(H)$ such that ∇u is quasinilpotent, the question arises whether this assumes the existence of a filtration for which u is adapted. The answer is negative as seen from the following example:

Let

$$u(w) = \sum_{i=1}^{\infty} 2^{-i\alpha} \delta(e_{i+1}) e_i$$

where e_i is a CONB in H. Hence, for $h \in H$

$$(\nabla u)h = \sum_{i=1}^{\infty} 2^{-i\alpha} (e_{i+1}, h) e_i$$

$$(\nabla u)^r h = \sum_{i=1}^{\infty} \beta_{i,r} (e_{i+r}, h) e_i$$

where $\beta_{i,r} = 2^{-i\alpha}$. $\beta_{i+1,r-1} = 2^{-\alpha r(2i+r-1)/2}$. Therefore ∇u is quasinilpotent and by lemma 4.1 of Ringrose [14], for any $h \in H$, it holds that for r large enough

$$\left| \left((\nabla u)^r h, e_1 \right)_H \right| > \frac{1}{2} \left| (\nabla u)^r h \right|_H. \quad (A.1)$$

Lemma A. *Let E. be a continuous strictly monotone resolution of the identity. Given $\delta > 0$ and some e in H with $|e|_H = 1$, then there exists an N such that for all h for which $(I - E_{1-\frac{1}{N}})h = h$, we have $|(h,e)_H| \leqslant \delta |h|_H$.*

Proof. Given $e \in H$ and $\delta > 0$, then for any given h and N large enough

$$\left|\left(h, (1 - E_{1-\frac{1}{N}})e\right)_H\right| \leqslant |h|_H \cdot \left|(1 - E_{1-\frac{1}{N}})e\right|_H$$
$$\leqslant |h|_H \cdot \delta.$$

Hence

$$\left|\left((1 - E_{1-\frac{1}{N}})h, e\right)_H\right| = \left|\left(h, (1 - E_{1-\frac{1}{N}})e\right)\right|_H$$
$$\leqslant |h|_H \cdot \delta.$$

If h satisfies $(1 - E_{1-\frac{1}{N}})h = h$ then

$$|(he)_H| \leqslant \delta \cdot |h|_H.$$ □

Assume now that u is adapted to $\mathcal{F}.$ induced by $E.$, then

$$E_\lambda \cdot \nabla u = E_\lambda \cdot \nabla u \cdot E_\lambda.$$

Hence

$$(\nabla u)(1 - E_\lambda) = (1 - E_\lambda)\,\nabla u(1 - E_\lambda).$$

Consequently if h satisfies $h = (1 - E_\lambda)h$ then it also holds that

$$(\nabla u)^r h = (1 - E_\lambda)(\nabla u)^r h.$$

Hence for $\lambda = 1 - \frac{1}{N}$ and $\delta = \frac{1}{3}$

$$\left|(\nabla u)^r h, e_1)_H\right| \leqslant \frac{1}{3}\left|(\nabla u)^r h\right|_H$$

which contradicts (A.1). Consequently u cannot be adapted to any continuous filtration.

References

1. Cipriano, F., Cruzeiro, A.B.: Flows associated to tangent processes on Wiener space. J. Funct. Anal., **166**, 310–331 (1999).
2. Cruzeiro, A.B., Malliavin, P.: Renormalized differential geometry on path space, structural equation, curvature. J. Funct. Anal., **139**, 119–181 (1996).
3. Cruzeiro, A.B., P. Malliavin, P.: A class of anticipative tangent processes on the Wiener space. C.R. Acad. Sci. Paris, **333(1)**, 353–358 (2001).
4. Driver, B.K.: A Cameron-Martin type quasi-invariance theorem for Brownian motion on a compact manifold. J. Funct. Anal., **110**, 272–376 (1992).
5. Driver, B.K.: Towards calculus and geometry on path spaces. Symp. Proc. Pure Math., **57**, 405–422 (1995).
6. Fang, S., P. Malliavin, P.: Stochastic analysis on the path space of a Riemannian manifold. I. Markovian stochastic calculus. J. Funct. Anal., **118**, 249–274 (1993).
7. Glasner, E.: Personal Communication.

8. Hu, Y., Üstünel, A.S., Zakai M.: Tangent processes on Wiener space. J. Funct. Anal., **192**, 234–270 (2002).
9. Kusuoka, S.: Analysis on Wiener spaces. I. Nonlinear maps. J. Funct. Anal., **98**, 122–168 (1991).
10. Kusuoka, S.: Nonlinear transformations containing rotation and Gaussian measure. Preprint.
11. Ledoux, M., Talagrand, M.: Probability in Banach Spaces. Springer (1991).
12. Malliavin, P.: Stochastic Analysis. Springer-Verlag, Berlin/New York (1997).
13. Riesz, F., Nagy, B.Sz.: Functional Analysis. Ungar Publishing Co., New York (1955).
14. Ringrose, R.: Operators of Volterra type. J. London Math. Society, **33**, 418–424 (1958).
15. Tsirelson, B.: Personal communication.
16. Üstünel, A.S., Zakai, M.: Random rotations of the Wiener path. Probab. Theory Related Fields. **103**, 409–430 (1995).
17. Üstünel, A.S., Zakai, M.: The construction of filtrations on abstract Wiener space. J. Funct. Anal., **143** (1997).
18. Üstünel, A.S., Zakai, M.: Transformation of Measure on Wiener Space. Springer-Verlag, New York/Berlin (1999).
19. Üstünel, A.S., Zakai, M.: Embedding the abstract Wiener space in a probability space. J. Func. Anal., **171**, 124–138 (2000).
20. Zakai, M., Zeitouni, O.: When does the Ramer formula look like the Girsanov formula? Annals of Probability, **20**, 1436–1440 (1992).

L^p Multiplier Theorem
for the Hodge–Kodaira Operator

Ichiro Shigekawa[*]

Department of Mathematics,
Graduate School of Science, Kyoto University
Kyoto 606-8502, JAPAN
e-mail: ichiro@math.kyoto-u.ac.jp

Summary. We discuss the L^p multiplier theorem for a semigroup acting on vector valued functions. A typical example is the Hodge–Kodaira operator on a Riemannian manifold. We give a probabilistic proof. Our main tools are the semigroup domination and the Littlewood–Paley inequality.

1 Introduction

We discuss the L^p multiplier theorem. In L^2 setting, it is well known that $\varphi(-L)$ is bounded if and only if φ is bounded where L is a non-positive self-adjoint operator. In L^p setting, the criterion above is no more true in general.

E. M. Stein [9] gave a sufficient condition when L is a generator of a symmetric Markov process. It reads as follows: define a function φ on $[0, \infty)$ by

$$\varphi(\lambda) = \lambda \int_0^\infty e^{-2t\lambda} m(t)\, dt. \tag{1.1}$$

Here we assume that m is a bounded function. A typical example is $\varphi(\lambda) = \lambda^{i\alpha}$ ($\alpha \in \mathbb{R}$). Then Stein proved that $\varphi(-L)$ is a bounded operator in L^p for $1 < p < \infty$. He also proved that the operator norm of $\varphi(-L)$ depends only on p and the bound of m.

In the meanwhile we consider the Hodge–Kodaira operator on a compact Riemannian manifold M. It is of the form $\mathbf{L} = -(dd^* + d^*d)$ where d is the exterior differentiation. A typical feature is that $\mathbf{L}$ acts on vector valued functions, to be precise, differential forms on M. In this case, we can get the following theorem:

---

[*] This research was partially supported by the Japanese Ministry of Education, Culture, Sports, Science and Technology, Grant-in-Aid for Scientific Research (B), No. 11440045, 2001.

Theorem 1.1. *For sufficiently large κ, $\varphi(\kappa - \mathbf{L})$ is a bounded operator in L^p. Further the operator norm is estimated in terms of m and p only.*

To show this theorem, we use the following facts.

1. The semigroup domination.
2. The Littlewood–Paley inequality.

As for the first, we can show that

$$\left| e^{t(\mathbf{L}-\kappa)}\theta \right| \leqslant e^{tL}|\theta|. \tag{1.2}$$

Here L is the Laplace–Beltrami operator on M and the inequality holds pointwisely. This inequality can be shown by means of Ouhabaz criterion ([3]). To use the criterion, the following inequality is essential.

$$L|\theta|^2 - 2(\mathbf{L}\theta, \theta) + \kappa|\theta|^2 \geqslant 0.$$

As for the second, we need the Littlewood–Paley function. This is somehow different from the usual one. We may call it the Littlewood–Paley function of parabolic type. It is defined as follows:

$$\mathcal{P}\theta(x) = \left\{ \int_0^\infty |\nabla \mathbf{T}_t \theta(x)|^2 \, \mathrm{d}t \right\}^{1/2}.$$

Here $\mathbf{T}_t$ denotes the semigroup $e^{t(\mathbf{L}-\kappa)}$. We can show the following inequality: there exists a positive constant C independent of θ such that

$$\|\mathcal{P}\theta\|_p \leqslant C\|\theta\|_p$$

where $\|\,.\,\|_p$ stands for the L^p-norm. This inequality is called the Littlewood–Paley inequality.

Combining these two inequality we can show that

$$\left| \left(\varphi(\kappa - \mathbf{L})\theta, \eta \right) \right| \leqslant C_1 \|\mathcal{P}\theta\|_p \|\mathcal{P}\eta\|_q \leqslant C_2 \|\theta\|_p \|\eta\|_q.$$

Here q is the conjugate exponent of p. Now the desired result follows easily.

The organization of the paper is as follows. We discuss this problem in the general framework of a symmetric diffusion process. We give this formulation in §2. We introduce the square field operator not only in the scalar valued case but also in the vector valued case. We give conditions to ensure the semigroup domination which plays an important role in the paper. In §3, we discuss the Littlewood–Paley inequality. We use the Littlewood–Paley function of parabolic type. After these preparartions, we give a proof of the multiplier theorem. In §4, we give an example—obtained in Proposition 4.3—of the Hodge–Kodaira operator. The crucial issue is the intertwining property of these operators.

2 Symmetric Markov processes and the semigroup domination

In the introduction, we stated the theorem for the Hodge–Kodaira operator but it can be discussed under more general setting. We give it in the framework of symmetric Markov diffusion process.

Let (M, μ) be a measure space and suppose that we are given a conservative diffusion process (X_t, P_x) on M. Here P_x denotes a measure on $C([0, \infty) \to M)$ that stands for the law of the diffusion process starting at $x \in M$. We assume that (X_t) is symmetric with respect to μ and hence the semigroup $\{T_t\}$ defined by

$$T_t f(x) = E_x[f(X_t)], \tag{2.1}$$

is a strongly continuous symmetric semigroup in $L^2(m)$. Here E_x stands for the expectation with respect to P_x. We denote the associated Dirichlet form by $\mathcal{E}$ and the generator by L. We assume further that there exists a continuous bilinear map $\Gamma \colon \mathrm{Dom}(\mathcal{E}) \times \mathrm{Dom}(\mathcal{E}) \to L^1(m)$ such that

$$2 \int_M \Gamma(f, g) h \, d\mu = \mathcal{E}(fg, h) - \mathcal{E}(f, gh) - \mathcal{E}(g, fh),$$
$$\text{for } f, g, h \in \mathrm{Dom}(\mathcal{E}) \cap L^\infty. \tag{2.2}$$

Γ is called the square field operator ("opérateur carré du champ" in French literature) and we impose on Γ the following derivation property:

$$\Gamma(fg, h) = f\Gamma(g, h) + g\Gamma(f, h), \qquad \text{for } f, g, h \in \mathrm{Dom}(\mathcal{E}) \cap L^\infty. \tag{2.3}$$

We are dealing with a semigroup acting on vector valued functions (to be precise, sections of a vector bundle) and so we are given another semigroup $\{\mathbf{T}_t\}$. The semigroup acts on L^2-sections of a vector bundle E. Here E is equipped with a metric $(\,.\,,\,.\,)_E$ and L^2-sections are measurable sections θ with

$$\|\theta\|_2^2 = \int_M |\theta(x)|_E^2 \, \mu(dx) < \infty.$$

The norm $|\,.\,|_E$ is defined by $|\theta|_E = \sqrt{(\theta, \theta)_E}$. We denote the set of all L^2-sections by $L^2\Gamma(E)$. The typical example of E is a exterior bundle of T^*M over a Riemannian manifold M and in this case $L^2\Gamma(E)$ is the set of all square integrable differential forms. $\mathbf{L}$ denotes the generator of $\{\mathbf{T}_t\}$ and $\mathcal{E}$ denotes the associated bilinear form. We assume that $\mathbf{L}$ is decomposed as

$$\mathbf{L} = \hat{L} - \kappa - R. \tag{2.4}$$

Here R is a symmetric section of $\mathrm{Hom}(E; E)$ and κ is a positive constant. Later κ will be taken to be large enough. $\hat{L}$ is self-adjoint and non-negative definite. It generates a contraction semigroup which we denote by $\{\hat{T}_t\}$. $\hat{L}$

and L satisfy the following relation: there exists a square field operator $\hat{\Gamma}\colon \mathrm{Dom}(\hat{\mathcal{E}}) \times \mathrm{Dom}(\hat{\mathcal{E}}) \to L^1(\mu)$ such that

$$2\int_M \hat{\Gamma}(\theta,\eta)h\,\mathrm{d}\mu = \hat{\mathcal{E}}\big((\theta,\eta)_E,h\big) - \hat{\mathcal{E}}(\theta,h\eta) - \hat{\mathcal{E}}(h\theta,\eta),$$

$$\text{for } \theta,\ \eta \in \mathrm{Dom}(\hat{\mathcal{E}}) \cap L^\infty,\ h \in \mathrm{Dom}(\mathcal{E}) \cap L^\infty. \quad (2.5)$$

We assume that $\hat{\Gamma}$ enjoys the positivity $\hat{\Gamma}(\theta,\theta) \geqslant 0$ and

$$2h\mathbf{\Gamma}(\theta,\eta) = -\Gamma\big(h,(\theta,\eta)_E\big) + \mathbf{\Gamma}(\theta,h\eta) + \mathbf{\Gamma}(h\theta,\eta) \quad (2.6)$$

for $\theta,\ \eta \in \mathrm{Dom}(\hat{\mathcal{E}}) \cap L^\infty,\ h \in \mathrm{Dom}(\mathcal{E}) \cap L^\infty$. These properties lead to the semigroup domination (see e.g., [5]):

$$\big|\hat{T}_t\theta\big|_E \leqslant T_t|\theta|_E. \quad (2.7)$$

Since R is bounded, we may assume that $\kappa + R$ is non-negative definite at any point of M by taking κ large enough. We assume further that there exists a positive constant $\delta > 0$ such that

$$\kappa(\theta,\theta)_E + (R\theta,\theta)_E \geqslant \delta(\theta,\theta)_E. \quad (2.8)$$

Then the semigroup domination for $\{\mathbf{T}_t\}$ also holds as follows:

$$|\mathbf{T}_t\theta|_E \leqslant \mathrm{e}^{-\delta t}T_t|\theta|_E. \quad (2.9)$$

We give a correspondence to the Hodge–Kodaira operator when M is a Riemannian manifold. $L = \Delta$ (i.e., the Laplace–Beltrami operator), $E = \bigwedge^q T^*M$ (the exterior product of the cotangent bundle) and $L^2\Gamma(E)$ is the set of all square integrable q-forms. $\hat{L} = -\nabla^*\nabla$ is the covariant Laplacian (Bochner Laplacian), $\mathbf{L} = -(dd^* + d^*d) - \kappa = \hat{L} - \kappa - R_{(q)}$. The explicit form of $R_{(q)}$ is given by the Weitzenböck formula and can be written in terms of the curvature tensor. We do not give the explicit form because we do not need it. We only need the boundedness of $R_{(q)}$. $\hat{\Gamma}$ is given by

$$\hat{\Gamma}(\theta,\eta) = \frac{1}{2}\{\Delta(\theta,\eta)_E + (\nabla^*\nabla\theta,\eta)_E + (\theta,\nabla^*\nabla\eta)_E\} = (\nabla\theta,\nabla\eta)_{E\otimes T^*M}.$$

The positivity of $\hat{\Gamma}$ clearly holds and (2.6) follows from the derivation property of ∇.

We now return to the general framework. We assume that $\hat{\Gamma}$ is expressed as

$$\hat{\Gamma}(\theta,\eta) = (D\theta, D\eta) \quad (2.10)$$

for some operator D. For instance, the covariant Laplacian satisfies this condition. In this case, D is the covariant derivation ∇. Later we need this condition when the exponent p is greater than 2.

D is an operator from $L^2\Gamma(E)$ to $L^2\Gamma(\widetilde{E})$, $\widetilde{E}$ being another vector bundle over M. The domain of D is not necessarily the whole space $L^2\Gamma(E)$ but we do assume that D is a closed operator. Our last assumption is the following intertwining property: there exists a self-adjoint operator $\mathbf{\Lambda}$ satisfying

$$D\mathbf{L} = \mathbf{\Lambda}D + K \tag{2.11}$$

where K is a bounded section of $\mathrm{Hom}(E; E')$. For $\mathbf{\Lambda}$, we assume the same conditions as $\mathbf{L}$. In particular, we need the semigroup domination for $\mathbf{S}_t = \mathrm{e}^{t\mathbf{\Lambda}}$:

$$|\mathbf{S}_t\xi|_{\widetilde{E}} \leqslant \mathrm{e}^{-\delta t}T_t|\xi|_{\widetilde{E}}, \qquad \xi \in L^2\Gamma(\widetilde{E}). \tag{2.12}$$

Due to the boundedness of K, this is possible by taking κ large enough. Moreover the intertwining property (2.11) implies

$$D\mathbf{T}_t\theta = \mathbf{S}_tD\theta + \int_0^t \mathbf{S}_{t-s}K\mathbf{T}_s\theta\,\mathrm{d}s, \qquad \forall\theta \in \mathrm{Dom}(D), \tag{2.13}$$

(see [8]).

3 Littlewood–Paley inequality

We introduce the Littlewood–Paley function of parabolic type. They are given as follows:

$$\mathcal{P}\theta(x) = \left\{ \int_0^\infty \hat{\Gamma}(\mathbf{T}_t\theta, \mathbf{T}_t\theta)(x)\,\mathrm{d}t \right\}^{1/2}, \tag{3.1}$$

$$\mathcal{H}\theta(x) = \left\{ \int_0^\infty T_t\hat{\Gamma}(\mathbf{T}_t\theta, \mathbf{T}_t\theta)(x)\,\mathrm{d}t \right\}^{1/2}. \tag{3.2}$$

We fix a time N and set

$$u(x,t) = T_{N-t}\theta(x), \qquad 0 \leqslant t \leqslant N.$$

Then we have

$$\begin{aligned}
(\partial_t &+ L)|u(x,t)|_E^2 \\
&= (\partial_t + L)(\mathbf{T}_{N-t}\theta, \mathbf{T}_{N-t}\theta) \\
&= -2(\mathbf{L}\mathbf{T}_{N-t}\theta, \mathbf{T}_{N-t}\theta) + 2(\hat{L}\mathbf{T}_{N-t}\theta, \mathbf{T}_{N-t}\theta) + 2\hat{\Gamma}(\mathbf{T}_{N-t}\theta, \mathbf{T}_{N-t}\theta) \\
&= -2\big((\hat{L} - \kappa - R)\mathbf{T}_{N-t}\theta, \mathbf{T}_{N-t}\theta\big) \\
&\quad + 2(\hat{L}\mathbf{T}_{N-t}\theta, \mathbf{T}_{N-t}\theta) + 2\hat{\Gamma}(\mathbf{T}_{N-t}\theta, \mathbf{T}_{N-t}\theta) \\
&= 2\big((\kappa + R)\mathbf{T}_{N-t}\theta, \mathbf{T}_{N-t}\theta\big) + 2\hat{\Gamma}(\mathbf{T}_{N-t}\theta, \mathbf{T}_{N-t}\theta)
\end{aligned}$$

For notational simplicity, we use the following convention. We write $\|A\theta\|_p \lesssim \|\theta\|_p$ if there exists a constant C such that $\|A\theta\|_p \leqslant C\|\theta\|_p$. C is independent of θ but may depend on p and A. We use this convention without mention. Now we have the following.

Proposition 3.1. *For $1 < p \leqslant 2$, it holds that*

$$\|\mathcal{P}\theta\|_p \lesssim \|\theta\|_p. \tag{3.3}$$

Proof. Define a martingale (M_t) by

$$
M_t = |u(X_t, t)|_E^2 - |u(X_0, 0)|_E^2 - \int_0^t (\partial_s + L)|u(X_s, s)|_E^2 \, ds
$$

$$
= |u(X_t, t)|_E^2 - |u(X_0, 0)|_E^2
$$
$$
- 2 \int_0^t \big\{ \big((\kappa + R) \mathbf{T}_{N-s} \theta(X_s), \mathbf{T}_{N-s} \theta(X_s) \big)
$$
$$
+ \hat{\Gamma} \big(\mathbf{T}_{N-s} \theta(X_s), \mathbf{T}_{N-s} \theta(X_s) \big) \big\} \, ds.
$$

Then the quadratic variation of (M_t) is written as

$$
\langle M, M \rangle_t = 2 \int_0^t \Gamma \big(|u(\,\cdot\,, s)|_E^2, |u(\,\cdot\,, s)|_E^2 \big)(X_s) \, ds
$$
$$
= 8 \int_0^t |u(X_s, s)|_E^2 \Gamma \big(|u(\,\cdot\,, s)|_E, |u(\,\cdot\,, s)|_E \big)(X_s) \, ds.
$$

In particular, $Z_t = |u(X_t, t)|_E^2$ is a non-negative submartingale:

$$Z_t = |u(X_0, 0)|_E^2 + M_t + B_t \tag{3.4}$$

where an increasing process B_t is given by

$$
B_t = 2 \int_0^t \big\{ \big((\kappa + R) \mathbf{T}_{N-s} \theta(X_s), \mathbf{T}_{N-s} \theta(X_s) \big)
$$
$$
+ \hat{\Gamma} \big(\mathbf{T}_{N-s} \theta(X_s), \mathbf{T}_{N-s} \theta(X_s) \big) \big\} \, ds. \tag{3.5}
$$

Take any $\varepsilon > 0$ and apply the Itô formula to $(|u|_E^2 + \varepsilon)^{p/2}$, we have

$$
d\big(|u|_E^2 + \varepsilon \big)^{p/2} = \frac{p}{2} \big(|u|_E^2 + \varepsilon \big)^{p/2-1} d\big(|u|_E^2 + \varepsilon \big)
$$
$$
+ \frac{1}{2} \frac{p}{2} \Big(\frac{p}{2} - 1 \Big) \big(|u|_E^2 + \varepsilon \big)^{p/2-2} d\langle M, M \rangle_t
$$
$$
= \frac{p}{2} \big(|u|_E^2 + \varepsilon \big)^{p/2-1} dM_t
$$
$$
+ \Big[\frac{p}{2} \big(|u|_E^2 + \varepsilon \big)^{p/2-1} 2\big\{ \big((\kappa + R)u, u \big) + \hat{\Gamma}(u, u) \big\}
$$
$$
+ p(p-2) \big(|u|_E^2 + \varepsilon \big)^{p/2-2} |u|_E^2 \Gamma(|u|_E, |u|_E) \Big] dt.
$$

Here, in the above identity, $u(X_t, t)$ is simply denoted by u. Therefore

$$
\big(|u(X_t, t)|_E^2 + \varepsilon \big)^{p/2} = \big(|u(X_0, 0)|_E^2 + \varepsilon \big)^{p/2} + \int_0^t \frac{p}{2} \big(|u|_E^2 + \varepsilon \big)^{p/2-1} dM_s + A_t.
$$

Here A_t is defined by

$$
A_t = \int_0^t \Big[p\big(|u|_E^2 + \varepsilon\big)^{p/2-1} \big\{ ((\kappa + R)u, u) + \hat{\Gamma}(u, u) \big\}
$$
$$
+ p(p - 2)\big(|u|_E^2 + \varepsilon\big)^{p/2-2} |u|_E^2 \Gamma(|u|_E, |u|_E) \Big]\, dt.
$$

(A_t) is an increasing process. To see this, recalling the inequality

$$
\Gamma(|u|_E, |u|_E) \leqslant \hat{\Gamma}(u, u),
$$

we have

$$
\begin{aligned}
dA_t &\geqslant p\big(|u|_E^2 + \varepsilon\big)^{p/2-1} \big\{ ((\kappa + R)u, u) + \hat{\Gamma}(u, u) \big\} \\
&\quad + p(p - 2)\big(|u|_E^2 + \varepsilon\big)^{p/2-2} |u|^2 \hat{\Gamma}(u, u) \\
&\geqslant (p + p(p - 2))\big(|u|_E^2 + \varepsilon\big)^{p/2-1} \hat{\Gamma}(u, u) \\
&\quad + p\big(|u|_E^2 + \varepsilon\big)^{p/2-1} ((\kappa + R)u, u) \\
&\geqslant p(p - 1)\big(|u|_E^2 + \varepsilon\big)^{p/2-1} \hat{\Gamma}(u, u)
\end{aligned}
$$

which implies that A_t is increasing. By taking expectation of $(|u(X_N, N)|_E^2 + \varepsilon)^{p/2}$, we obtain

$$
\begin{aligned}
p(p - 1)E\left[\int_0^N \big(|u|^2 + \varepsilon\big)^{p/2-1} \hat{\Gamma}(u, u)\, dt \right] &\leqslant E\left[\big(|u(X_N, N)|_E^2 + \varepsilon\big)^{p/2} \right] \\
&\leqslant E\left[\big(|\theta(X_N)|_E^2 + \varepsilon\big)^{p/2} \right] \\
&\leqslant \left\| \big(|\theta|_E^2 + \varepsilon\big)^{1/2} \right\|_p^p .
\end{aligned}
$$

We proceed to the estimation of the left hand side. By the semigroup domination

$$
|\mathbf{T}_{N-t}\theta(x)| \leqslant \mathbf{T}_{N-t}|\theta|(x) \leqslant \sup_{s \geqslant 0} T_s|\theta|(x) =: \theta^*(x)
$$

The maximal ergodic theorem implies $\|\theta^*\|_p \lesssim \|\theta\|_p$. Now, noting $p/2 - 1 \leqslant 0$,

$$
\begin{aligned}
E\left[\int_0^N \big(|u|^2 + \varepsilon\big)^{p/2-1} \hat{\Gamma}(u, u)\, dt \right] &= \left\| \int_0^N \big(|u|^2 + \varepsilon\big)^{p/2-1} \hat{\Gamma}(\mathbf{T}_{N-t}\theta, \mathbf{T}_{N-t}\theta)\, dt \right\|_1 \\
&\gtrsim \left\| \big((\theta^*)^2 + \varepsilon\big)^{p/2-1} \int_0^N \hat{\Gamma}(\mathbf{T}_t\theta, \mathbf{T}_t\theta)\, dt \right\|_1 .
\end{aligned}
$$

Letting $N \to \infty$,

$$
\left\| \big(|\theta|^2 + \varepsilon\big)^{1/2} \right\|_p^p \gtrsim \left\| \big((\theta^*)^2 + \varepsilon\big)^{p/2-1} \int_0^\infty \hat{\Gamma}(\mathbf{T}_t\theta, \mathbf{T}_t\theta)\, dt \right\|_1
$$

$$\gtrsim \left\| ((\theta^*)^2 + \varepsilon)^{p/2-1} \mathcal{P}\theta^2 \right\|_1 = \left\| ((\theta^*)^2 + \varepsilon)^{(p-2)/4} \mathcal{P}\theta \right\|_2^2.$$

Therefore

$$\|\mathcal{P}\theta\|_p = \left\| ((\theta^*)^2 + \varepsilon)^{(2-p)/4} ((\theta^*)^2 + \varepsilon)^{(p-2)/4} \mathcal{P}\theta \right\|_p$$

$$\leqslant \left\| ((\theta^*)^2 + \varepsilon)^{(2-p)/4} \right\|_{2p/(2-p)} \left\| ((\theta^*)^2 + \varepsilon)^{(p-2)/4} \mathcal{P}\theta \right\|_2$$

$$\left(\text{since } \frac{1}{p} = \frac{2-p}{2p} + \frac{1}{2} \right)$$

$$\lesssim \left\| ((\theta^*)^2 + \varepsilon)^{1/2} \right\|_p^{(2-p)/2} \left\| (|\theta|^2 + \varepsilon)^{1/2} \right\|_p^{p/2}.$$

Finally, letting $\varepsilon \to 0$, we obtain

$$\|\mathcal{P}\theta\|_p \lesssim \|\theta^*\|_p^{(2-p)/2} \|\theta\|_p^{p/2} \lesssim \|\theta\|_p^{(2-p)/2} \|\theta\|_p^{p/2} = \|\theta\|_p.$$

The proof is complete. □

Next we show the case $p \geqslant 2$. First we need the following easy lemma.

Lemma 3.1. *Let j be a non-negative fonction on $M \times [0, N]$. Then it holds that*

$$E_\mu \left[\int_0^N j(X_t, t) \, dt \,\Big|\, X_N = x \right] = \int_0^N T_t \big(j(\,.\,, N - t) \big)(x) \, dt. \tag{3.6}$$

Here E_μ stands for the integration with respect to $P_\mu = \int_M P_x \, \mu(dx)$.

Proof. It is enough to show that

$$E_\mu \left[\left\{ \int_0^N j(X_t, t) \, dt \right\} f(X_N) \right] = \int_E \left\{ \int_0^N T_{N-t} j(x, t) \, dt \right\} f(x) \, \mu(dx) \tag{3.7}$$

for any non-negative function f. To see this,

$$E_\mu \left[\left\{ \int_0^N j(X_t, t) \, dt \right\} f(X_N) \right] = \int_0^N E_\mu[j(X_t, t) f(X_N)] \, dt$$

$$= \int_0^N E_\mu[j(X_t, t) E_\mu[f(X_N)|\mathcal{F}_t]] \, dt$$

$$= \int_0^N E_\mu[j(X_t, t) T_{N-t} f(X_t)] \, dt$$

(by the Markov property)

$$= \int_0^N dt \int_M j(x, t) T_{N-t} f(x) \, \mu(dx)$$

$$= \int_0^N \mathrm{d}t \int_M T_{N-t}(j(\,.\,,t))(x) f(x)\,\mu(\mathrm{d}x)$$

(by symmetry)

$$= \int_M \left\{ \int_0^N T_{N-t}(j(\,.\,,t))(x)\,\mathrm{d}t \right\} f(x)\,\mu(\mathrm{d}x)$$

which shows (3.7). $\square$

Proposition 3.2. *For* $p \geqslant 2$, *we have*

$$\|\mathcal{H}\theta\|_p \lesssim \|\theta\|_p. \tag{3.8}$$

Proof. We consider a submartingale $Z_t = |u(X_t,t)|^2$. As was seen in (3.4), Z_t is decomposed as

$$Z_t = |u(X_0,0)|_E^2 + M_t + B_t.$$

Then the following inequality is well-known (see, [2]): for $q \geqslant 1$,

$$E[B_N^q] \lesssim E[Z_N^q]. \tag{3.9}$$

Using Lemma 3.1, we have

$$\int_M \mu(\mathrm{d}x) \left\{ \int_0^N T_t \hat{\Gamma}(\mathbf{T}_t\theta, \mathbf{T}_t\theta)(x)\,\mathrm{d}t \right\}^{p/2}$$

$$= \int_M \mu(\mathrm{d}x) E_\mu \left[\int_0^N \hat{\Gamma}(\mathbf{T}_{N-t}\theta, \mathbf{T}_{N-t}\theta)(X_t)\,\mathrm{d}t \ \bigg|\ X_N = x \right]^{p/2}$$

$$\leqslant \int_M \mu(\mathrm{d}x) E_\mu \left[\left\{ \int_0^N \hat{\Gamma}(\mathbf{T}_{N-t}\theta, \mathbf{T}_{N-t}\theta)(X_t)\,\mathrm{d}t \right\}^{p/2} \bigg|\ X_N = x \right]$$

(by the Jensen inequality)

$$= E_\mu \left[\left\{ \int_0^N \hat{\Gamma}(\mathbf{T}_{N-t}\theta, \mathbf{T}_{N-t}\theta)(X_t)\,\mathrm{d}t \right\}^{p/2} \right]$$

$$\leqslant E_\mu \left[\left\{ \int_0^N \left\{ ((\kappa + R)\mathbf{T}_{N-t}\theta(X_s), \mathbf{T}_{N-t}\theta(X_s)) \right. \right. \right.$$

$$\left. \left. \left. + \hat{\Gamma}(\mathbf{T}_{N-t}\theta(X_s), \mathbf{T}_{N-t}\theta(X_s)) \right\}\,\mathrm{d}t \right\}^{p/2} \right]$$

$$\lesssim E[B_N^{p/2}] \quad \text{(thanks to (3.5))}$$

$$\lesssim E[Z_N^{p/2}] \quad \text{(thanks to (3.9))}$$

$$= E[|\theta(X_N)|^p]$$

$$= \|\theta\|_p^p.$$

Now $\mathcal{H}\theta$ can be estimated as follows:

$$\|\mathcal{H}\theta\|_p^p = \left\|\left\{\int_0^\infty T_t\hat{\Gamma}(\mathbf{T}_t\theta, \mathbf{T}_t\theta)(x)\,\mathrm{d}t\right\}^{p/2}\right\|_1$$

$$= \lim_{N\to\infty}\int_M \mu(\mathrm{d}x)\left\{\int_0^N T_t\hat{\Gamma}(\mathbf{T}_t\theta, \mathbf{T}_t\theta)(x)\,\mathrm{d}t\right\}^{p/2}$$

$$\lesssim \|\theta\|_p^p.$$

This completes the proof. $\qquad\qquad\qquad\qquad\qquad\qquad\qquad\qquad\qquad\qquad\square$

Let us proceed to the estimation of $\mathcal{P}\theta$.

Proposition 3.3. *For $p \geqslant 2$, we have*

$$\mathcal{P}\theta(x) \leqslant \sqrt{2}\,\mathcal{H}\theta(x) + \frac{\|K\|_\infty}{4\delta^{3/2}}\,\theta^*(x) \qquad\qquad (3.10)$$

Proof. We have

$$\mathcal{P}\theta(x) = \left\{\int_0^\infty \hat{\Gamma}(\mathbf{T}_t\theta, \mathbf{T}_t\theta)(x)\,\mathrm{d}t\right\}^{1/2}$$

$$= \left\{\int_0^\infty |D\mathbf{T}_t\theta(x)|_{\tilde{E}}^2\,\mathrm{d}t\right\}^{1/2} \qquad \text{(thanks to (2.10))}$$

$$= \left\{2\int_0^\infty |D\mathbf{T}_{2t}\theta(x)|_{\tilde{E}}^2\,\mathrm{d}t\right\}^{1/2}$$

$$= \left\{2\int_0^\infty |\mathbf{T}_t D\mathbf{T}_t\theta(x)|_{\tilde{E}}^2\,\mathrm{d}t\right\}^{1/2}$$

$$= \left\{2\int_0^\infty \left|\mathbf{S}_t D\mathbf{T}_t\theta(x) + \int_0^t \mathbf{S}_{t-s}K\mathbf{T}_s\mathbf{T}_t\theta(x)\,\mathrm{d}s\right|_{\tilde{E}}^2\,\mathrm{d}t\right\}^{1/2} \qquad \text{(by (2.13))}$$

$$\leqslant \sqrt{2}\left\{\int_0^\infty |\mathbf{S}_t D\mathbf{T}_t\theta(x)|_{\tilde{E}}^2\,\mathrm{d}t\right\}^{1/2}$$

$$\quad + \sqrt{2}\left\{\int_0^\infty \left\{\int_0^t |\mathbf{S}_{t-s}K\mathbf{T}_{s+t}\theta(x)|_{\tilde{E}}\,\mathrm{d}s\right\}^2\,\mathrm{d}t\right\}^{1/2}$$

$$\leqslant \sqrt{2}\left\{\int_0^\infty T_t|D\mathbf{T}_t\theta(x)|_{\tilde{E}}^2\,\mathrm{d}t\right\}^{1/2}$$

$$\quad + \sqrt{2}\left\{\int_0^\infty \left\{\int_0^t e^{-\delta(t-s)}T_{t-s}|K\mathbf{T}_{s+t}\theta(x)|_{\tilde{E}}\,\mathrm{d}s\right\}^2\right\}^{1/2} \qquad \text{(by (2.8))}$$

$$= \sqrt{2}\,\mathcal{H}\theta(x)$$

$$\quad + \sqrt{2}\left\{\int_0^\infty \left\{\int_0^t \|K\|_\infty e^{-\delta(t-s)}T_{t-s}e^{-\delta(s+t)}T_{s+t}|\theta|_E(x)\,\mathrm{d}s\right\}^2\,\mathrm{d}t\right\}^{1/2}$$

$$= \sqrt{2}\,\mathcal{H}\theta(x) + \sqrt{2}\left\{\int_0^\infty \|K\|_\infty^2 e^{-4\delta t}\big(T_{2t}|\theta|_E(x)\big)^2 t^2\,\mathrm{d}t\right\}^{1/2}$$

$$= \sqrt{2}\,\mathcal{H}\theta(x) + \sqrt{2}\,\|K\|_\infty \theta^*(x)\left\{\int_0^\infty t^2 e^{-4\delta t}\,dt\right\}^{1/2}$$

$$= \sqrt{2}\,\mathcal{H}\theta(x) + \frac{1}{4\delta^{3/2}}\,\|K\|_\infty \theta^*(x)$$

which is the desired result. □

Combining these two propositions and the maximal ergodic inequality, we easily obtain the following.

Proposition 3.4. *For $p \geqslant 2$, we have*

$$\|\mathcal{P}\theta\|_p \lesssim \|\theta\|_p. \tag{3.11}$$

Before proving the theorem, we give an expression of $\varphi(-\mathbf{L})$. Recall that

$$\varphi(\lambda) = \lambda \int_0^\infty e^{-2t\lambda} m(t)\,dt.$$

There exists the following correspondence:

$$-\mathbf{L} \longleftrightarrow \lambda$$
$$e^{t\mathbf{L}} \longleftrightarrow e^{-t\lambda}.$$

Therefore $\varphi(-\mathbf{L})$ is expressed as

$$\varphi(-\mathbf{L}) = -\mathbf{L}\int_0^\infty \mathbf{T}_{2t}\,m(t)\,dt.$$

Proof of Theorem 1.1. Using the expression above, we have

$$(\varphi(-\mathbf{L})\theta, \eta) = \left(-\mathbf{L}\int_0^\infty \mathbf{T}_{2t}\,m(t)\,dt\,\theta, \eta\right)$$

$$= \int_0^\infty \int_M (-\mathbf{L}\mathbf{T}_{2t}\theta, \eta)_E\,\mu(dx)m(t)\,dt$$

$$= \int_0^\infty m(t)\,dt \int_M \{(-\hat{L}\mathbf{T}_t\theta, \mathbf{T}_t\eta)_E + ((\kappa + R)\mathbf{T}_t\theta, \mathbf{T}_t\eta)_E\}\,\mu(dx)$$

$$= \int_0^\infty m(t)\,dt \int_M \hat{\Gamma}(\mathbf{T}_t\theta, \mathbf{T}_t\eta)_E$$

$$+ \int_0^\infty m(t)\,dt\big((\kappa + R)\mathbf{T}_t\theta, \mathbf{T}_t\eta\big)_E\,\mu(dx).$$

We estimate two terms on the right hand side respectively.

For the first term,

$$\left|\int_0^\infty m(t)\,dt \int_M \hat{\Gamma}(\mathbf{T}_t\theta, \mathbf{T}_t\eta)\,\mu(dx)\right|$$

$$\leqslant \|m\|_\infty \int_0^\infty \mathrm{d}t \int_M \hat{\Gamma}(\mathbf{T}_t\theta, \mathbf{T}_t\theta)^{1/2} \hat{\Gamma}(\mathbf{T}_t\eta, \mathbf{T}_t\eta)^{1/2} \mu(\mathrm{d}x)$$

(thanks to the Schwarz inequality for $\hat{\Gamma}$)

$$\leqslant \|m\|_\infty \int_M \left\{ \int_0^\infty \hat{\Gamma}(\mathbf{T}_t\theta, \mathbf{T}_t\theta) \, \mathrm{d}t \right\}^{1/2} \left\{ \int_0^\infty \hat{\Gamma}(\mathbf{T}_t\eta, \mathbf{T}_t\eta) \, \mathrm{d}t \right\}^{1/2} \mu(\mathrm{d}x)$$

$$= \|m\|_\infty \int_M \mathcal{P}\theta(x) \mathcal{P}\eta(x) \, \mu(\mathrm{d}x)$$

$$\leqslant \|m\|_\infty \|\mathcal{P}\theta\|_p \|\mathcal{P}\eta\|_q$$

$$\lesssim \|m\|_\infty \|\theta\|_p \|\eta\|_q.$$

For the second term,

$$\left| \int_0^\infty m(t) \, \mathrm{d}t \big((\kappa + R)\mathbf{T}_t\theta, \mathbf{T}_t\eta \big)_E \, \mu(\mathrm{d}x) \right|$$

$$\leqslant \|m\|_\infty \int_0^\infty \mathrm{d}t \int_M \|\kappa + R\|_\infty \, |\mathbf{T}_t\theta|_E \, |\mathbf{T}_t\eta|_E \, \mu(\mathrm{d}x)$$

$$\leqslant \|m\|_\infty \int_0^\infty \mathrm{d}t \int_M \|\kappa + R\|_\infty \, \mathrm{e}^{-2\delta t} T_t|\theta|_E \, T_t|\eta|_E \, \mu(\mathrm{d}x)$$

$$\leqslant \|m\|_\infty \|\kappa + R\|_\infty \frac{1}{2\delta} \|\theta\|_p \|\eta\|_q.$$

Thus we have shown that

$$|(\varphi(-\mathbf{L})\theta, \eta)| \lesssim \|\theta\|_p \|\eta\|_q$$

which implies that $\varphi(-\mathbf{L})$ is bounded in L^p.   □

4 Hodge–Kodaira operator

In this section we consider the the Hodge–Kodaira operator $-(dd^* + d^*d)$ acting on differential forms. What remains to show is the defective intertwining property. We have to seek for operators Λ and K that satisfy

$$-\nabla(dd^* + d^*d)\theta = \Lambda\nabla\theta + K\theta.$$

Even if θ is a differential form, $\nabla\theta$ is no longer a differential form. So we discuss the issue in the framework of tensor fields. Let M be a Riemannian manifold and ∇ be the Levi-Civita connection. The Riemannian curvature tensor is defined by

$$R(X, Y)Z = \nabla_X\nabla_Y Z - \nabla_Y\nabla_X Z - \nabla_{[X,Y]}Z$$

where X, Y, $Z \in \Gamma(TM)$. Here Γ denotes the set of all smooth sections of a vector bundle. In this case, $\Gamma(TM)$ is the set of vector fields. Let $T_nM =$

$\underbrace{T^*M \otimes \cdots \otimes T^*M}_{n}$ be a tensor bundle of type $(0,n)$. The exterior bundle is denoted by $\bigwedge^p T^*M = \underbrace{T^*M \wedge \cdots \wedge T^*M}_{n}$. We define an operator Δ^{HK} on $\Gamma(T_nM)$ as follows. $u_1 \otimes \cdots \otimes u_n$, $u_i \in \Gamma(T^*M)$ is a typical form of an element of $\Gamma(T_nM)$. Any element of $\Gamma(T_nM)$ can be written as a linear combination of them. We are given a Riemmanian metric g and there exists a natural isomophism $\sharp \colon T^*M \to TM$ e.g.,

$$\langle \omega, X \rangle = g(\omega^\sharp, X), \qquad \omega \in T^*M, \, X \in TM.$$

In the sequel, we omit g and denote the inner product $g(X,Y)$ by (X,Y). The inner product in T^*M is also denoted by (ω, η). The natural pairing between T^*M and TM is denoted by $\langle \omega, X \rangle$. We take a local orthonormal basis $\{e_1, \ldots, e_n\}$ and let $\{\omega^1, \ldots, \omega^n\}$ be its dual basis. We introduce linear opetators $S_{p,q}^{(n)}$ $1 \leqslant p,q \leqslant n$ on $\Gamma(T_n)$ as follows; for $p \neq q$,

$$S_{p,q}^{(n)}(u_1 \otimes \cdots \otimes u_n)$$
$$= \big(R(u_p^\sharp, e_k)u_q^\sharp, e_l\big)u_1 \otimes \cdots \otimes \overset{p}{\omega^k} \otimes \cdots \otimes \overset{q}{\omega^l} \otimes \cdots \otimes u_n. \quad (4.1)$$

Here we used the Einstein rule: we omit the summation sign for repeated indices. For example, in the equation above $\sum_{k,l=1}^n$ is omitted. For $p = q$, we define

$$S_{p,p}^{(n)}(u_1 \otimes \cdots \otimes u_n) = \big(\mathrm{Ric}\, u_p^\sharp, e_k\big)u_1 \otimes \cdots \otimes \overset{p}{\omega^k} \otimes \cdots \otimes u_n$$
$$= \big(R(u_p^\sharp, e_i)e_i, e_k\big)u_1 \otimes \cdots \otimes \overset{p}{\omega^k} \otimes \cdots \otimes u_n. \quad (4.2)$$

Ric denotes the Ricci tensor.

We now define the operator Δ^{HK} by

$$\Delta^{\mathrm{HK}}v = -\nabla^*\nabla v - \sum_{p,q=1}^n S_{p,q}^{(n)}v. \quad (4.3)$$

Here the superscript HK stands for Hodge–Kodaira. This notation is justified by the following proposition.

Proposition 4.1. *For $\theta \in \Gamma(\bigwedge^p T^*M)$, it holds that*

$$\Delta^{\mathrm{HK}}\theta = -(dd^* + d^*d)\theta. \quad (4.4)$$

Proof. We first note the following identity: for $u_1, \ldots, u_n \in \Gamma(T^*M)$,

$$u_1 \wedge u_2 \wedge \cdots \wedge u_n := \sum_\sigma \mathrm{sgn}\,\sigma\, u_{\sigma(n)} \otimes \cdots \otimes u_{\sigma(n)}$$

$$= \sum_\alpha (-1)^{\alpha-1} u_\alpha \otimes (u_1 \wedge \overset{\alpha}{\overset{\vee}{\cdots}} \wedge u_n).$$

Here σ runs over the set of all permutations of order n, sgn σ is the sign of σ and $\overset{\alpha}{\vee}$ means that u_α is deleted. Similarly we have

$$u_1 \wedge u_2 \wedge \cdots \wedge u_n = \sum_{\alpha<\beta} (-1)^{\alpha+\beta-1} (u_\alpha \otimes u_\beta - u_\beta \otimes u_\alpha)(u_1 \wedge \overset{\alpha\ \beta}{\overset{\vee\ \vee}{\cdots}} \wedge u_n).$$

Next let us compute $\sum_{p,q} S_{p,q}^{(n)}$. First, for $\sum_{p \neq q} S_{p,q}^{(n)}$

$$\sum_{p \neq q} S_{p,q}^{(n)}(u_1 \wedge u_2 \wedge \cdots \wedge u_n)$$

$$= \sum_{p \neq q} S_{p,q}^{(n)} \sum_\sigma \operatorname{sgn} \sigma\, u_{\sigma(1)} \otimes \cdots \otimes u_{\sigma(n)}$$

$$= \sum_{p \neq q} \sum_\sigma \operatorname{sgn} \sigma\, (R(u_{\sigma(p)}^\sharp, e_k) u_{\sigma(q)}^\sharp, e_l) u_{\sigma(1)} \otimes \cdots \otimes \overset{p}{\overset{\vee}{\omega^k}} \otimes \cdots \otimes \overset{q}{\overset{\vee}{\omega^l}} \otimes \cdots \otimes u_{\sigma(n)}.$$

Here p-th $u_{\sigma(p)}$ is replaced by ω^k and q-th $u_{\sigma(q)}$ is replaced by ω^l. By exchanging the order of summation, we have

$$\sum_{p \neq q} S_{p,q}^{(n)}(u_1 \wedge u_2 \wedge \cdots \wedge u_n)$$

$$= \sum_{\alpha \neq \beta} \sum_\sigma \operatorname{sgn} \sigma\, (R(u_\alpha^\sharp, e_k) u_\beta^\sharp, e_l) u_{\sigma(1)} \otimes \cdots \otimes \overset{\sigma^{-1}(\alpha)}{\overset{\vee}{\omega^k}} \otimes \cdots \otimes \overset{\sigma^{-1}(\beta)}{\overset{\vee}{\omega^l}} \otimes \cdots \otimes u_{\sigma(n)}$$

$$= \sum_{\alpha \neq \beta} (R(u_\alpha^\sharp, e_k) u_\beta^\sharp, e_l) \sum_\sigma \operatorname{sgn} \sigma\, u_{\sigma(1)} \otimes \cdots \otimes \overset{\sigma^{-1}(\alpha)}{\overset{\vee}{\omega^k}} \otimes \cdots \otimes \overset{\sigma^{-1}(\beta)}{\overset{\vee}{\omega^l}} \otimes \cdots \otimes u_{\sigma(n)}$$

$$= \sum_{\alpha \neq \beta} (R(u_\alpha^\sharp, e_k) u_\beta^\sharp, e_l) u_1 \wedge \cdots \wedge \overset{\alpha}{\overset{\vee}{\omega^k}} \wedge \cdots \wedge \overset{\beta}{\overset{\vee}{\omega^l}} \wedge \cdots \wedge u_n$$

$$= \sum_{\alpha < \beta} (R(u_\alpha^\sharp, e_k) u_\beta^\sharp, e_l) \omega^k \wedge \omega^l \wedge u_1 \wedge \overset{\alpha\ \beta}{\overset{\vee\ \vee}{\cdots}} \wedge u_n$$

$$\quad + \sum_{\alpha > \beta} (R(u_\alpha^\sharp, e_k) u_\beta^\sharp, e_l) \omega^l \wedge \omega^k \wedge u_1 \wedge \overset{\alpha\ \beta}{\overset{\vee\ \vee}{\cdots}} \wedge u_n$$

$$= \sum_{\alpha < \beta} (R(u_\alpha^\sharp, e_k) u_\beta^\sharp, e_l) \omega^k \wedge \omega^l \wedge u_1 \wedge \overset{\alpha\ \beta}{\overset{\vee\ \vee}{\cdots}} \wedge u_n$$

$$\quad + \sum_{\alpha > \beta} (R(u_\beta^\sharp, e_l) u_\alpha^\sharp, e_k) \omega^l \wedge \omega^k \wedge u_1 \wedge \overset{\alpha\ \beta}{\overset{\vee\ \vee}{\cdots}} \wedge u_n$$

$$= 2 \sum_{\alpha < \beta} (R(u_\alpha^\sharp, e_k) u_\beta^\sharp, e_l) \omega^k \wedge \omega^l \wedge u_1 \wedge \overset{\alpha\ \beta}{\overset{\vee\ \vee}{\cdots}} \wedge u_n.$$

Similarly we have

$$\sum_p S_{p,p}^{(n)}(u_1 \wedge u_2 \wedge \cdots \wedge u_n) = \sum_p S_{p,p}^{(n)} \sum_\sigma \operatorname{sgn}\sigma\, u_{\sigma(1)} \otimes \cdots \otimes u_{\sigma(n)}$$

$$= \sum_p \sum_\sigma \operatorname{sgn}\sigma\, (\operatorname{Ric} u_{\sigma(p)}^\sharp, e_k) u_{\sigma(1)} \otimes \cdots \otimes \overset{p}{\omega^k} \otimes \cdots \otimes u_{\sigma(n)}.$$

Here the p-th $u_{\sigma(p)}$ is replaced by ω^k. Exchanging the order of summation, we have

$$\sum_p S_{p,p}^{(n)}(u_1 \wedge u_2 \wedge \cdots \wedge u_n)$$

$$= \sum_\alpha \sum_\sigma \operatorname{sgn}\sigma\, (\operatorname{Ric} u_\alpha^\sharp, e_k) u_{\sigma(1)} \otimes \cdots \otimes \overset{\sigma^{-1}(\alpha)}{\omega^k} \otimes \cdots \otimes u_{\sigma(n)}$$

$$= \sum_\alpha (\operatorname{Ric} u_\alpha^\sharp, e_k) u_1 \wedge \cdots \wedge \overset{\alpha}{\omega^k} \wedge \cdots \wedge u_n$$

$$= \sum_\alpha (-1)^{\alpha-1}(\operatorname{Ric} u_\alpha^\sharp, e_k) \omega^k \wedge u_1 \wedge \overset{\alpha}{\cdots} \wedge u_n$$

Using this identity, we can calculate $-(dd^* + d^*d)$. Before that we have to recall the Weitzenböck formula:

$$- (dd^* + d^*d) = -\nabla^*\nabla$$
$$+ \big(R(e_l, e_j)e_k, e_i\big)\omega^l \wedge \omega^k \wedge i(e_j)i(e_i) - (\operatorname{Ric} e_k, e_i)\omega^k \wedge i(e_i).$$

Here $i(\,.\,)$ denotes the interior product, i.e., $i(X)\theta = \theta(X, .\,,\ldots, .\,)$. Now we have

$$-(dd^* + d^*d)(u_1 \wedge \cdots \wedge u_n)$$
$$= -\nabla^*\nabla(u_1 \wedge \cdots \wedge u_n) + \big(R(e_l, e_j)e_k, e_i\big)\omega^l \wedge \omega^k \wedge i(e_j)i(e_i)$$
$$\times \sum_{\alpha<\beta}(-1)^{\alpha+\beta-1}(u_\alpha \otimes u_\beta - u_\beta \otimes u_\alpha)(u_1 \wedge \overset{\alpha\;\beta}{\cdots} \wedge u_n)$$
$$- (\operatorname{Ric} e_k, e_i)\omega^k \wedge i(e_i)\sum_\alpha (-1)^{\alpha-1}u_\alpha \otimes (u_1 \wedge \overset{\alpha}{\cdots} \wedge u_n)$$
$$= -\nabla^*\nabla(u_1 \wedge \cdots \wedge u_n)$$
$$+ \sum_{\alpha<\beta}(-1)^{\alpha+\beta-1}\{\langle u_\alpha, e_i\rangle\langle u_\beta, e_j\rangle - \langle u_\beta, e_i\rangle\langle u_\alpha, e_j\rangle\}$$
$$\times \big(R(e_l, e_j)e_k, e_i\big)\omega^l \wedge \omega^k \wedge u_1 \wedge \overset{\alpha\;\beta}{\cdots} \wedge u_n$$
$$- \sum_\alpha (-1)^{\alpha-1}(\operatorname{Ric} e_k, e_i)\langle u_\alpha, e_i\rangle i(e_i)\omega^k \wedge u_1 \wedge \overset{\alpha}{\cdots} \wedge u_n$$
$$= -\nabla^*\nabla(u_1 \wedge \cdots \wedge u_n) + \sum_{\alpha<\beta}(-1)^{\alpha+\beta-1}$$

$$\times \left\{\left(R(e_l, u_\beta^\sharp)e_k, u_\alpha^\sharp\right) - \left(R(e_l, u_\alpha^\sharp)e_k, u_\beta^\sharp\right)\right\}\omega^l \wedge \omega^k \wedge u_1 \wedge \overset{\alpha}{\overset{\vee}{\cdots}} \overset{\beta}{\overset{\vee}{\cdots}} \wedge u_n$$

$$- \sum_\alpha (-1)^{\alpha-1}(\mathrm{Ric}\, e_k, u_\alpha)\omega^k \wedge u_1 \wedge \overset{\alpha}{\overset{\vee}{\cdots}} \wedge u_n$$

$$= -\nabla^*\nabla(u_1 \wedge \cdots \wedge u_n)$$

$$+ 2\sum_{\alpha<\beta} (-1)^{\alpha+\beta-1}\left(R(e_l, u_\beta^\sharp)e_k, u_\alpha^\sharp\right)\omega^l \wedge \omega^k \wedge u_1 \wedge \overset{\alpha}{\overset{\vee}{\cdots}} \overset{\beta}{\overset{\vee}{\cdots}} \wedge u_n$$

$$- \sum_\alpha (-1)^{\alpha-1}(\mathrm{Ric}\, e_k, u_\alpha)\omega^k \wedge u_1 \wedge \overset{\alpha}{\overset{\vee}{\cdots}} \wedge u_n$$

$$= -\nabla^*\nabla(u_1 \wedge \cdots \wedge u_n)$$

$$- 2\sum_{\alpha<\beta} (-1)^{\alpha+\beta-1}\left(R(u_\alpha^\sharp, e_k)u_\beta^\sharp, e_l\right)\omega^k \wedge \omega^l \wedge u_1 \wedge \overset{\alpha}{\overset{\vee}{\cdots}} \overset{\beta}{\overset{\vee}{\cdots}} \wedge u_n$$

$$- \sum_\alpha (-1)^{\alpha-1}(\mathrm{Ric}\, e_k, u_\alpha)\omega^k \wedge u_1 \wedge \overset{\alpha}{\overset{\vee}{\cdots}} \wedge u_n$$

$$= -\nabla^*\nabla(u_1 \wedge \cdots \wedge u_n) - \sum_{p,q} S_{p,q}^{(n)}(u_1 \wedge \cdots \wedge u_n)$$

$$= \Delta^{\mathrm{HK}}(u_1 \wedge \cdots \wedge u_n)$$

which is the required identity.   □

We are interested in the intertwining property for the Hodge–Kodaira operator $-(dd^* + d^*d)$. By the above proposition, it is enough to calculate Δ^{HK}. We first show the intertwining property for $\nabla^*\nabla$.

Proposition 4.2. *It holds that*

$$-\nabla(\nabla^*\nabla)u - (\nabla^*\nabla)\nabla u = \sum_{j=2}^{n+1}\{S_{1,j}^{(n+1)}\nabla u + S_{j,1}^{(n+1)}\nabla u\}$$
$$+ S_{1,1}^{(n+1)}\nabla u + \omega^k \otimes \nabla_i R^{(n)}(e_i, e_k)u \qquad (4.5)$$

Proof. Pick a point $x \in M$ and fix it. We take a normal coordinate at x. Then there exists a local frame $\{e_1, e_2, \ldots, e_n\}$ of TM so that $\nabla_{e_i} e_j(x) = 0$. To avoid complexity, we simply denote ∇_i in place of ∇_{e_i}. Let $\{\omega^1, \omega^2, \ldots, \omega^n\}$ be the dual frame. Due to our choice of a local frame, at the point x it holds that $\nabla_{i,j}^2 = \nabla_i\nabla_j$, $[e_i, e_j] = 0$ and $\nabla_i\omega^k = 0$. Moreover we have the following identity at x:

$$-[e_i, \nabla_j e_k] = \nabla_i\nabla_j e_k, \qquad (4.6)$$
$$\nabla_i\nabla_{\nabla_j e_k} = \nabla_{\nabla_i\nabla_j e_k}, \qquad (4.7)$$
$$\langle \nabla_i\nabla_i e_k, \omega^l \rangle = -\langle e_k, \nabla_i\nabla_i\omega^l \rangle \qquad (4.8)$$

Here (4.7) is the identity for $T_n M$.

To see (4.6) we note that the torsion is free and so we have

$$[e_i, \nabla_j e_k] = \nabla_i \nabla_j e_k - \nabla_{\nabla_j e_k} e_i = \nabla_i \nabla_j e_k.$$

As for (4.7), we use the definition of the curvature $R^{(n)}$.

$$\begin{aligned}
\nabla_i \nabla_{\nabla_j e_k} &= R^{(n)}(e_i, \nabla_j e_k) + \nabla_{\nabla_j e_k} \nabla_i + \nabla_{[e_i, \nabla_j e_k]} \\
&= \nabla_{\nabla_i \nabla_j e_k}. \qquad \text{(thanks to (4.6))}
\end{aligned}$$

(4.8) can be shown as

$$\begin{aligned}
0 &= \nabla_i \nabla_i \langle e_k, \omega^l \rangle \\
&= \langle \nabla_i \nabla_i e_k, \omega^l \rangle + 2 \langle \nabla_i e_k, \nabla_i \omega^l \rangle + \langle e_k, \nabla_i \nabla_i \omega^l \rangle \\
&= \langle \nabla_i \nabla_i e_k, \omega^l \rangle + \langle e_k, \nabla_i \nabla_i \omega^l \rangle.
\end{aligned}$$

We use these identities freely. From now on all equations are evaluated at the point x. Now

$$\begin{aligned}
-(\nabla^* & \nabla) \nabla u + \nabla(\nabla^* \nabla) u \\
&= \nabla_i \nabla_i (\omega^k \otimes \nabla_k u) - \nabla_{\nabla_i e_i}(\omega^k \otimes \nabla_k u) - \nabla(\nabla_i \nabla_i u - \nabla_{\nabla_i e_i} u) \\
&= \nabla_i \nabla_i \omega^k \otimes \nabla_k u + 2 \nabla_i \omega^k \otimes \nabla_i \nabla_k u + \omega^k \otimes \nabla_i \nabla_i \nabla_k u \\
&\quad - \omega^k \otimes \nabla_k \nabla_i \nabla_i u + \omega^k \otimes \nabla_k \nabla_{\nabla_i e_i} u \\
&= \nabla_i \nabla_i \omega^k \otimes \nabla_k u + \omega^k \otimes \nabla_i \nabla_i \nabla_k u - \omega^k \otimes \nabla_k \nabla_i \nabla_i u + \omega^k \otimes \nabla_k \nabla_{\nabla_i e_i} u \\
&= \nabla_i \nabla_i \omega^k \otimes \nabla_k u + \omega^k \otimes \nabla_i \{ R^{(n)}(e_i, e_k) u + \nabla_k \nabla_i u + \nabla_{[e_i, e_k]} u \} \\
&\quad - \omega^k \otimes \{ R^{(n)}(e_k, e_i) \nabla_i u + \nabla_i \nabla_k \nabla_i u + \nabla_{[e_k, e_i]} \nabla_i u \} + \omega^k \otimes \nabla_k \nabla_{\nabla_i e_i} u \\
&= \nabla_i \nabla_i \omega^k \otimes \nabla_k u + \omega^k \otimes \{ \nabla_i R^{(n)}(e_i, e_k) u + R^{(n)}(\nabla_i e_i, e_k) u \\
&\quad + R^{(n)}(e_i, \nabla_i e_k) u + R^{(n)}(e_i, e_k) \nabla_i u + \nabla_i \nabla_{[e_i, e_k]} u \} \\
&\quad - \omega^k \otimes R^{(n)}(e_k, e_i) \nabla_i u + \omega^k \otimes \nabla_k \nabla_{\nabla_i e_i} u \\
&= \nabla_i \nabla_i \omega^k \otimes \nabla_k u + \omega^k \otimes \nabla_i R^{(n)}(e_i, e_k) u + 2 \omega^k \otimes R^{(n)}(e_i, e_k) \nabla_i u \\
&\quad + \omega^k \otimes \{ \nabla_i \nabla_{[e_i, e_k]} u + \nabla_k \nabla_{\nabla_i e_i} u \}.
\end{aligned}$$

On the other hand, using (4.6), (4.7) and (4.8), we have

$$\begin{aligned}
\omega^k \otimes & \{ \nabla_i \nabla_{[e_i, e_k]} u + \nabla_k \nabla_{\nabla_i e_i} u \} \\
&= \omega^k \otimes (\nabla_i \nabla_{\nabla_i e_k} - \nabla_i \nabla_{\nabla_k e_i} + \nabla_k \nabla_{\nabla_i e_i}) u \\
&= \omega^k \otimes (\nabla_{\nabla_i \nabla_i e_k} + \nabla_{R(e_k, e_i) e_i} + \nabla_{\nabla_{[e_k, e_i]} e_i}) u \\
&= \omega^k \otimes (\nabla_{\nabla_i \nabla_i e_k} + \nabla_{R(e_k, e_i) e_i}) u \\
&= \omega^k \otimes (\langle \nabla_i \nabla_i e_k, \omega^l \rangle \nabla_l u + \langle R(e_k, e_i) e_i, \omega^l \rangle \nabla_l u) \\
&= \omega^k \otimes \langle \nabla_i \nabla_i e_k, \omega^l \rangle \nabla_l u + \omega^k \otimes \langle \text{Ric}\, e_k, \omega^l \rangle \nabla_l u
\end{aligned}$$

$$= -\omega^k \otimes \langle e_k, \nabla_i \nabla_i \omega^l \rangle \nabla_l u + \omega^k \otimes \langle \mathrm{Ric}\, e_k, \omega^l \rangle \nabla_l u \qquad \text{(thanks to (4.8))}$$
$$= -\nabla_i \nabla_i \omega^l \otimes \nabla_l u + \omega^k \otimes \langle \mathrm{Ric}\, e_k, \omega^l \rangle \nabla_l u.$$

Combining all of them, we have

$$-(\nabla^* \nabla)\nabla u + \nabla(\nabla^* \nabla)u$$
$$= \omega^k \otimes \nabla_i R^{(n)}(e_i, e_k)u + 2\omega^k \otimes R^{(n)}(e_i, e_k)\nabla_i u + \omega^k \otimes \langle \mathrm{Ric}\, e_k, \omega^l \rangle \nabla_l u$$
$$= \omega^k \otimes \nabla_i R^{(n)}(e_i, e_k)u + \sum_{j=2}^{n+1} (S_{1,j}^{(n+1)} \nabla u + S_{j,1}^{(n+1)} \nabla u) + S_{1,1}^{(n+1)} \nabla u.$$

This completes the proof. $\square$

We are now ready to prove the intertwining property for Δ^{HK}.

Proposition 4.3. *Take any local orthonormal frame $\{e_1, e_2, \ldots, e_d\}$ and its dual frame $\{\omega^1, \omega^2, \ldots, \omega^d\}$. Then it holds that*

$$\nabla \Delta_n^{\mathrm{HK}} u = \Delta_{n+1}^{\mathrm{HK}} \nabla u - \sum_{p,q=1}^{n} \omega^k \otimes (\nabla_k S_{p,q}^{(n)})u - \omega^k \otimes (\nabla_i R^{(n)}(e_i, e_k))u. \quad (4.9)$$

Proof. We recall (4.3). Then

$$\nabla \Delta_n^{\mathrm{HK}} u - \Delta_{n+1}^{\mathrm{HK}} \nabla u$$
$$= -\nabla \left(\nabla^* \nabla + \sum_{p,q=1}^{n} S_{p,q}^{(n)} \right) + \left(\nabla^* \nabla + \sum_{p,q=1}^{n+1} S_{p,q}^{(n+1)} \right) \nabla u$$
$$= -\sum_{j=2}^{n+1} (S_{1,j}^{(n+1)} \nabla u + S_{j,1}^{(n+1)} \nabla u) - S_{1,1}^{(n+1)} \nabla u - \omega^k \otimes \nabla_i R^{(n)}(e_i, e_k)u$$
$$\quad - \sum_{p,q=1}^{n} \omega^k \otimes (\nabla_k S_{p,q}^{(n)})u - \sum_{p,q=1}^{n} \omega^k \otimes S_{p,q}^{(n)} \nabla_k u + \sum_{p,q=1}^{n+1} S_{p,q}^{(n+1)} \nabla u$$
$$= -\sum_{j=2}^{n+1} (S_{1,j}^{(n+1)} \nabla u + S_{j,1}^{(n+1)} \nabla u) - S_{1,1}^{(n+1)} \nabla u - \omega^k \otimes \nabla_i R^{(n)}(e_i, e_k)u$$
$$\quad - \sum_{p,q=1}^{n} \omega^k \otimes (\nabla_k S_{p,q}^{(n)})u - \sum_{p,q \geqslant 2}^{n+1} S_{p,q}^{(n+1)} \nabla u + \sum_{p,q=1}^{n+1} S_{p,q}^{(n+1)} \nabla u$$
$$= -\sum_{p,q=1}^{n} \omega^k \otimes (\nabla_k S_{p,q}^{(n)})u - \omega^k \otimes \nabla_i R^{(n)}(e_i, e_k)u$$

which completes the proof. $\square$

The above intertwining property for Δ^{HK} is defective, i.e., it satisfies the identity of the type (2.11). The defective term is removed if we replace ∇ with the exterior derivative d. To define the exterior derivative, we need to introduce the alternating operation A as follows. For a tensor u of type $(0, n)$, we define $A^{(n)}$ by

$$A^{(n)}u(X_1, \ldots, X_n) = \sum_\sigma \mathrm{sgn}\, \sigma\, u(X_{\sigma(1)}, \ldots, X_{\sigma(n)}).$$

The exterior derivative is defined by

$$d = A^{(n+1)}\nabla u.$$

This definition is consistent with the usual definition for differential forms. Now we have the following *intertwining property*.

Proposition 4.4. *For $u \in \Gamma(T_n(M))$, it holds that*

$$d\Delta_n^{\mathrm{HK}}u = \Delta_{n+1}^{\mathrm{HK}}du. \tag{4.10}$$

Proof. By Proposition 4.3, we have

$$d\Delta_n^{\mathrm{HK}}u = \Delta_{n+1}^{\mathrm{HK}}du$$
$$- \sum_{p,q=1}^n A^{(n+1)}\left(\omega^k \otimes (\nabla_k S_{p,q}^{(n)})u\right) - A^{(n+1)}\left(\omega^k \otimes (\nabla_i R^{(n)}(e_i, e_k))u\right).$$

We have to show that the additional terms vanish. Before proving this, we recall the Bianchi identity for the Riemannian curvature:

$$-\mathfrak{S}R(X, Y)Z = 0, \tag{4.11}$$
$$\mathfrak{S}\nabla_X R(Y, Z) = 0. \tag{4.12}$$

Here $\mathfrak{S}$ stands for the cyclic sum, e.g.,

$$\mathfrak{S}R(X, Y)Z = R(X, Y)Z + R(Y, Z)X + R(Z, X)Y.$$

(4.11) is called the first Bianchi identity and (4.12) is called the second Bianchi identity.

We may assume that $u = u_1 \otimes \cdots \otimes u_n$. For $p = q$, we have

$$\sum_{p=1}^n A^{(n+1)}\left(\omega^k \otimes (\nabla_k S_{p,p}^{(n)})u\right) - A^{(n+1)}\left(\omega^k \otimes (\nabla_i R^{(n)}(e_i, e_k))u\right)$$
$$= \sum_{p=1}^n A^{(n+1)}\big(\omega^k \otimes u_1 \otimes \cdots \otimes \nabla_k \mathrm{Ric}\, u_p^\sharp \otimes \cdots \otimes u_n$$
$$+ \omega^k \otimes u_1 \otimes \cdots \otimes \nabla_i R(e_i, e_k)u_p^\sharp \otimes \cdots \otimes u_n\big)$$

$$= \sum_{p=1}^{n} (-1)^p \big\{ \nabla_k \operatorname{Ric} u_p^\sharp \wedge \omega^k \wedge u_1 \wedge \overset{p}{\check\cdots} \wedge u_n$$
$$+ \nabla_i R(e_i, e_k) u_p^\sharp \wedge \omega^k \wedge u_1 \wedge \overset{p}{\check\cdots} \wedge u_n \big\}.$$

We need to compute $\nabla_k \operatorname{Ric} u_p^\sharp \wedge \omega^k + \nabla_i R(e_i, e_k) u_p^\sharp \wedge \omega^k$. To do this,

$$\nabla_k \operatorname{Ric} u_p^\sharp \wedge \omega^k + \nabla_i R(e_i, e_k) u_p^\sharp \wedge \omega^k$$
$$= \nabla_k R(u_p^\sharp, e_i) e_i \wedge \omega^k + \nabla_i R(e_i, e_k) u_p^\sharp \wedge \omega^k$$
$$= \big(\nabla_k R(u_p^\sharp, e_i) e_i, e_l \big) \omega^l \wedge \omega^k + \big(\nabla_i R(e_i, e_k) u_p^\sharp, e_l \big) \omega^l \wedge \omega^k$$
$$= \big\{ -\big(\nabla_{u_p^\sharp} R(e_i, e_k) e_i, e_l \big) - \big(\nabla_i R(e_k, u_p^\sharp) e_i, e_l \big)$$
$$+ \big(\nabla_i R(e_i, e_k) u_p^\sharp, e_l \big) \big\} \omega^l \wedge \omega^k \qquad \text{(by the 2nd Bianchi identity)}$$
$$= -\big(\nabla_{u_p^\sharp} R(e_i, e_k) e_i, e_l \big) \omega^l \wedge \omega^k$$
$$+ \big\{ \big(\nabla_i R(e_i, e_l) u_p^\sharp, e_k \big) + \big(\nabla_i R(e_i, e_k) u_p^\sharp, e_l \big) \big\} \omega^l \wedge \omega^k$$
$$= 0.$$

Here, in the last line, we used that the coefficients are symmetric with respect to k and l.

For $p \neq q$, we may assume $p < q$.

$$A^{(n+1)} \big(\omega^k \otimes \big(\nabla_k S_{p,q}^{(n)} \big) u \big)$$
$$= A^{(n+1)} \big(\omega^k \otimes \big(\nabla_k R(u_p^\sharp, e_l) u_q^\sharp, e_m \big) u_1 \otimes \cdots \otimes \overset{p}{w^l} \otimes \cdots \otimes \overset{q}{\omega^m} \otimes \cdots u_n \big)$$
$$= \big(\nabla_k R(u_p^\sharp, e_l) u_q^\sharp, e_m \big) \omega^k \wedge u_1 \wedge \cdots \wedge \overset{p}{\omega^l} \wedge \cdots \wedge \overset{q}{\omega^m} \wedge \cdots \wedge u_n$$
$$= \big(\nabla_k R(u_p^\sharp, e_l) u_q^\sharp, e_m \big) \omega^k \wedge \omega^l \wedge \omega^m \wedge u_1 \wedge \overset{p}{\check{}} \overset{q}{\check{}} \cdots \wedge u_n.$$

To calculate $\big(\nabla_k R(u_p^\sharp, e_l) u_q^\sharp, e_m \big) \omega^k \wedge \omega^l \wedge \omega^m$, we have

$$\big(\nabla_k R(u_p^\sharp, e_l) u_q^\sharp, e_m \big) \omega^k \wedge \omega^l \wedge \omega^m$$
$$= \big\{ \big(\nabla_{u_p^\sharp} R(e_l, e_k) u_q^\sharp, e_m \big) - \big(\nabla_l R(e_k, u_p^\sharp) u_q^\sharp, e_m \big) \big\} \omega^k \wedge \omega^l \wedge \omega^m$$
$$\qquad \text{(by the 2nd Bianchi identity)}$$
$$= \big(\nabla_{u_p^\sharp} R(e_l, e_k) e_m, u_q^\sharp \big) \omega^k \wedge \omega^l \wedge \omega^m - \big(\nabla_l R(e_k, u_p^\sharp) u_q^\sharp, e_m \big) \omega^k \wedge \omega^l \wedge \omega^m$$
$$= \big(\nabla_l R(u_p^\sharp, e_k) u_q^\sharp, e_m \big) \omega^k \wedge \omega^l \wedge \omega^m \qquad \text{(by the first Bianchi identity)}$$
$$= \big(\nabla_k R(u_p^\sharp, e_l) u_q^\sharp, e_m \big) \omega^l \wedge \omega^k \wedge \omega^m \qquad \text{(by relabeling)}$$
$$= -\big(\nabla_k R(u_p^\sharp, e_l) u_q^\sharp, e_m \big) \omega^k \wedge \omega^l \wedge \omega^m.$$

The last term is just the same as the original one with the opposite sign. Thus we have

$$\big(\nabla_k R(u_p^\sharp, e_l) u_q^\sharp, e_m \big) \omega^k \wedge \omega^l \wedge \omega^m = 0$$

as desired. $\qquad\qquad\qquad\qquad\qquad\qquad\qquad\qquad\qquad\qquad\qquad\qquad\quad\square$

References

1. P. A. Meyer, Retour sur la théorie de Littlewood–Paley, *Séminaire de Prob.* XV, Lecture Notes in Math., vol. 850, 151–166, Springer-Verlag, Berlin-Heidelberg-New York, 1981.
2. E. Lenglart, D. Lépingle and M. Pratelli, Présentation unifiée de certaines inégalités de théorie des martingales, *Séminaire de Prob.* XIV, Lecture Notes in Math., vol. 784, 26–48. Springer-Verlag, Berlin-Heidelberg-New York, 1980.
3. E. Ouhabaz, Invariance of closed convex sets and domination criteria for semigroups *Potential Analysis*, **5** (1996), 611–625.
4. I. Shigekawa, L^p contraction semigroups for vector valued functions, *J. Funct. Anal.*, **147** (1997), 69–108.
5. I. Shigekawa, Semigroup domination on a Riemannian manifold with boundary, *Acta Applicandae Math.*, **63** (2000), 385–410.
6. I. Shigekawa, The domain of a generator and the intertwining property, in *Stochastics in Finite and Infinite Dimensions*, ed. by T. Hida et al, 401–410, Birkhäuser, Boston, 2001.
7. I. Shigekawa, Littlewood–Paley inequality for a diffusion satisfying the logarithmic Sobolev inequality and for the Brownian motion on a Riemannian manifold with boundary, preprint.
8. I. Shigekawa, Defective intertwining property and generator domain, preprint
9. E. M. Stein, "*Topics in harmonic analysis, related to Littlewood–Paley theory,*" Annals of Math. Studies, 63, Princeton Univ. Press, 1974.
10. N. T. Varopoulos, Aspects of probabilistic Littlewood–Paley theory, *J. Funct. Anal.*, **38** (1980), 25–60.

Gaussian Limits for Vector-valued Multiple Stochastic Integrals

Giovanni Peccati and Ciprian A. Tudor

Laboratoire de Statistique Théorique et Appliquée,
Université de Paris VI, 175, rue du Chevaleret, 75013 Paris, France
e-mail: giovanni.peccati@libero.it
and
Laboratoire de Probabilités et Modèles Aléatoires,
Universités de Paris VI & VII, 175, rue du Chevaleret, 75013 Paris, France
email: tudor@ccr.jussieu.fr

Summary. We establish necessary and sufficient conditions for a sequence of d-dimensional vectors of multiple stochastic integrals $\mathbf{F}_d^k = (F_1^k, \ldots, F_d^k)$, $k \geqslant 1$, to converge in distribution to a d-dimensional Gaussian vector $\mathbf{N}_d = (N_1, \ldots, N_d)$. In particular, we show that if the covariance structure of $\mathbf{F}_d^k$ converges to that of $\mathbf{N}_d$, then componentwise convergence implies joint convergence. These results extend to the multidimensional case the main theorem of [10].

Key words: Multiple stochastic integrals, Limit theorems, Weak convergence, Brownian motion.

AMS Subject classification: 60F05, 60H05.

1 Introduction

For $d \geqslant 2$, fix d natural numbers $1 \leqslant n_1 \leqslant \cdots \leqslant n_d$ and, for every $k \geqslant 1$, let $\mathbf{F}_d^k = (F_1^k, \ldots, F_d^k)$ be a vector of d random variables such that, for each $j = 1, \ldots, d$, F_j^k belongs to the n_jth Wiener chaos associated to a real valued Gaussian process. The aim of this paper is to prove necessary and sufficient conditions to have that the sequence $\mathbf{F}_d^k$ converges in distribution to a given d-dimensional Gaussian vector, when k tends to infinity. In particular, our main result states that, if for every $1 \leqslant i, j \leqslant d$, $\lim_{k \to +\infty} \mathbb{E}[F_i^k F_j^k] = \delta_{ij}$, where δ_{ij} is the Kronecker symbol, then the following two conditions are equivalent: (i) $\mathbf{F}_d^k$ converges in distribution to a standard centered Gaussian vector $\mathbf{N}_d(0, \mathbf{I}_d)$ ($\mathbf{I}_d$ is the $d \times d$ identity matrix), (ii) for every $j = 1, \ldots, d$, F_j^k converges in distribution to a standard Gaussian random variable. Now suppose that, for every $k \geqslant 1$ and every $j = 1, \ldots, d$, the random variable F_j^k is the multiple Wiener–Itô stochastic integral of a square integrable kernel $f_j^{(k)}$, for instance on $[0,1]^{n_j}$. We recall that, according to the main result of [10], condition (ii) above is equivalent to either one of the following: (iii) $\lim_{k\to+\infty} \mathbb{E}[(F_j^k)^4] = 3$ for every j, (iv) for every j and every $p = 1,\ldots,n_j - 1$ the contraction $f_j^{(k)} \otimes_p f_j^{(k)}$ converges to zero in $L^2([0,1]^{2(n_j-p)})$. Some other necessary and sufficient conditions for (ii) to hold are stated in the subsequent sections, and an extension is provided to deal with the case of a Gaussian vector $\mathbf{N}_d$ with a more general covariance structure.

Besides [10], our results should be compared with other central limit theorems (CLT) for non linear functionals of Gaussian processes. The reader is referred to [2], [6], [7], [8], [15] and the references therein for several results in this direction. As in [10], the main tool in the proof of our results is a well known time-change formula for continuous local martingales, due to Dambis, Dubins and Schwarz (see e.g. [13, Chapter 5]). In particular, this technique enables to obtain our CLTs, by estimating and controlling expressions that are related uniquely to the fourth moments of the components of each vector $\mathbf{F}_d^k$.

The paper is organized as follows. In Section 2 we introduce some notation and discuss preliminary results; in Section 3 our main theorem is stated and proved; finally, in Section 4 we present some applications, to the weak convergence of *chaotic martingales* (that is, martingales admitting a multiple Wiener integral representation), and to the convergence in law of random variables with a finite chaotic decomposition.

2 Notation and preliminary results

Let H be a separable Hilbert space. For every $n \geqslant 1$, we define $H^{\otimes n}$ to be the nth tensor product of H and write $H^{\odot n}$ for the nth symmetric tensor product of H, endowed with the modified norm $\sqrt{n!}\,\|.\|_{H^{\otimes n}}$. We denote by $X = \{X(h) : h \in H\}$ an *isonormal process* on H, that is, X is a centered H-indexed Gaussian family, defined on some probability space $(\Omega, \mathcal{F}, \mathbb{P})$ and such that

$$\mathbb{E}[X(h)X(k)] = \langle h, k \rangle_H, \qquad \text{for every } h,\, k \in H.$$

For $n \geqslant 1$, let $\mathcal{H}_n$ be the nth Wiener chaos associated to X (see for instance [9, Chapter 1]): we denote by I_n^X the isometry between $\mathcal{H}_n$ and $H^{\odot n}$. For simplicity, in this paper we consider uniquely spaces of the form $H = L^2(T, \mathcal{A}, \mu)$, where $(T, \mathcal{A})$ is a measurable space and μ is a σ-finite and atomless measure. In this case, I_n^X can be identified with the multiple Wiener–Itô integral with respect to the process X, as defined e.g. in [9, Chapter 1]. We also note that, by some standard Hilbert space argument, our results can be immediately extended to a general H. The reader is referred to [10, Section 3.3] for a discussion of this fact.

Let $H = L^2(T, \mathcal{A}, \mu)$; for any $n,\, m \geqslant 1$, every $f \in H^{\odot n}$, $g \in H^{\odot m}$, and $p = 1,\ldots,n \wedge m$, the pth *contraction* between f and g, noted $f \otimes_p g$, is defined to be the element of $H^{\otimes m+n-2p}$ given by

$$f \otimes_p g(t_1, \ldots, t_{n+m-2p}) = \int_{T^p} f(t_1, \ldots, t_{n-p}, s_1, \ldots, s_p) \times$$
$$\times g(t_{n-p+1}, \ldots, t_{m+n-2p}, s_1, \ldots, s_p) \, d\mu(s_1) \ldots d\mu(s_p);$$

by convention, $f \otimes_0 g = f \otimes g$ denotes the tensor product of f and g. Given $\phi \in H^{\otimes n}$, we write $(\phi)_s$ for its canonical symmetrization. In the special case $T = [0,1]$, $\mathcal{A} = \mathcal{B}([0,1])$ and $\mu = \lambda$, where λ is Lebesgue measure, some specific notation is needed. For any $0 < t \leqslant 1$, Δ_t^n stands for the symplex contained in $[0,t]^n$, i.e. $\Delta_t^n := \{(t_1, \ldots, t_n) : 0 < t_n < \cdots < t_1 < t\}$. Given a function f on $[0,1]^n$ and $t \in [0,1]$, f_t denotes the application on $[0,1]^{n-1}$ given by

$$(s_1, \ldots, s_{n-1}) \longmapsto f(t, s_1, \ldots, s_{n-1}).$$

For any $n, m \geqslant 1$, for any pair of functions f, g such that $f \in L^2([0,1]^n, \mathcal{B}([0,1]^n), d\lambda^{\otimes n}) := L^2([0,1]^n)$ and $g \in L^2([0,1]^m)$, and for every $1 < t \leqslant 1$ and $p = 1, \ldots, n \wedge m$, we write $f \otimes_p^t g$ for the pth contraction of f and g on $[0,t]$, defined as

$$f \otimes_p^t g(t_1, \ldots, t_{n+m-2p}) = \int_{[0,t]^p} f(t_1, \ldots, t_{n-p}, s_1, \ldots, s_p) \times$$
$$\times g(t_{n-p+1}, \ldots, t_{m+n-2p}, s_1, \ldots, s_p) \, d\lambda(s_1) \ldots d\lambda(s_p);$$

as before, $f \otimes_0^t g = f \otimes g$. Eventually, we recall that if $H = L^2([0,1], \mathcal{B}([0,1]), d\lambda)$, then X coincides with the Gaussian space generated by the standard Brownian motion

$$t \longmapsto W_t := X(\mathbf{1}_{[0,t]}), \qquad t \in [0,1]$$

and this implies in particular that, for every $n \geqslant 2$, the multiple Wiener–Itô integral $I_n^X(f)$, $f \in L^2([0,1]^n)$, can be rewritten in terms of an iterated stochastic integral with respect to W, that is: $I_n^X(f) = I_n^1((f)_s) = n! J_n^1((f)_s)$, where

$$J_n^t((f)_s) = \int_0^t \cdots \int_0^{u_{n-1}} (f(u_1, \ldots, u_n))_s \, dW_{u_n} \ldots dW_{u_1}$$
$$I_n^t((f)_s) = n! J_n^t((f)_s), \qquad t \in [0,1].$$

3 d-dimensional CLT

The following facts will be used to prove our main results. Let $H = L^2(T, \mathcal{A}, \mu)$, $f \in H^{\odot n}$ and $g \in H^{\odot m}$. Then,

F1: (see [1, p. 211] or [9, Proposition 1.1.3])

$$I_n^X(f)I_m^X(g) = \sum_{p=0}^{n \wedge m} p! \binom{n}{p} \binom{m}{p} I_{n+m-2p}^X(f \otimes_p g); \qquad (1)$$

F2: (see [16, Proposition 1])

$$(n+m)! \, \|(f \otimes_0 g)_s\|^2_{H^{\otimes n+m}} = m! \, n! \, \|f\|^2_{H^{\otimes n}} \|g\|^2_{H^{\otimes m}}$$
$$+ \sum_{q=1}^{n \wedge m} \binom{n}{q}\binom{m}{q} n! \, m! \, \|f \otimes_q g\|^2_{H^{\otimes n+m-2q}} \quad (2)$$

F3: (see [10])

$$\mathbb{E}[I_n^X(f)^4] = 3(n!)^2 \|f\|^4_{H^{\otimes n}} + \sum_{p=1}^{n-1} \frac{(n!)^4}{(p! \, (n-p)!)^2} \left[\|f \otimes_p f\|^2_{H^{\otimes 2(n-p)}} \right.$$
$$\left. + \binom{2n-2p}{n-p} \|(f \otimes_p f)_s\|^2_{H^{\otimes 2(n-p)}} \right]. \quad (3)$$

Let V_d be the set of all $(i_1, i_2, i_3, i_4) \in (1, \ldots, d)^4$, such that one of the following conditions is satisfied: (a) $i_1 \neq i_2 = i_3 = i_4$, (b) $i_1 \neq i_2 = i_3 \neq i_4$ and $i_4 \neq i_1$, (c) the elements of $(i_1, \ldots, i_4)$ are all distinct. Our main result is the following.

Theorem 1. *Let $d \geqslant 2$, and consider a collection $1 \leqslant n_1 \leqslant \cdots \leqslant n_d < +\infty$ of natural numbers, as well as a collection of kernels*

$$\left\{ (f_1^{(k)}, \ldots, f_d^{(k)}) : k \geqslant 1 \right\}$$

such that $f_j^{(k)} \in H^{\odot n_j}$ for every $k \geqslant 1$ and every $j = 1, \ldots, d$, and

$$\lim_{k \to \infty} j! \, \|f_j^{(k)}\|^2_{H^{\otimes n_j}} = 1, \qquad \forall j = 1, \ldots, d,$$
$$\lim_{k \to \infty} \mathbb{E}\left[I_{n_i}^X(f_i^{(k)}) I_{n_l}^X(f_l^{(k)}) \right] = 0, \qquad \forall 1 \leqslant i < l \leqslant d. \quad (4)$$

Then, the following conditions are equivalent:

(i) *for every $j = 1, \ldots, d$*

$$\lim_{k \to \infty} \|f_j^{(k)} \otimes_p f_j^{(k)}\|_{H^{\otimes 2(n_j - p)}} = 0$$

for every $p = 1, \ldots, n_j - 1$;

(ii) $\lim_{k \to \infty} \mathbb{E}\left[\left(\sum_{i=1,\ldots,d} I_{n_i}^X(f_i^{(k)}) \right)^4 \right] = 3d^2$, *and*

$$\lim_{k \to \infty} \mathbb{E}\left[\prod_{l=1}^4 I_{n_{i_l}}^X(f_{i_l}^{(k)}) \right] = 0$$

for every $(i_1, i_2, i_3, i_4) \in V_d$;

(iii) *as k goes to infinity, the vector $\left(I_{n_1}^X\left(f_1^{(k)}\right),\ldots,I_{n_d}^X\left(f_d^{(k)}\right)\right)$ converges in distribution to a d-dimensional standard Gaussian vector $N_d(0,\mathbf{I}_d)$;*

(iv) *for every $j = 1,\ldots,d$, $I_{n_j}^X\left(f_j^{(k)}\right)$ converges in distribution to a standard Gaussian random variable;*

(v) *for every $j = 1,\ldots,d$,*

$$\lim_{k\to\infty}\mathbb{E}\left[I_{n_j}^X\left(f_j^{(k)}\right)^4\right] = 3.$$

Proof. We show the implications

$$\text{(iii)}\Longrightarrow\text{(ii)}\Longrightarrow\text{(i)}\Longrightarrow\text{(iii)}\qquad\text{and}\qquad\text{(iv)}\Longleftrightarrow\text{(v)}\Longleftrightarrow\text{(i)}$$

(iii) $\Longrightarrow$ (ii). First notice that, for every $k \geqslant 1$, the multiple integrals $I_{n_1}^X\left(f_1^{(k)}\right),\ldots,I_{n_d}^X\left(f_d^{(k)}\right)$ are contained in the sum of the first n_d chaoses associated to the Gaussian measure X. As a consequence, condition (4) implies (see e.g. [3, Chapter V]) that for every $M \geqslant 2$ and for every $j = 1,\ldots,d$

$$\sup_{k\geqslant 1}\mathbb{E}\left[\left|I_{n_j}^X\left(f_j^{(k)}\right)\right|^M\right] < +\infty$$

and the conclusion is obtained by standard arguments.

(ii) $\Longrightarrow$ (i). The key of the proof is the following simple equality

$$\mathbb{E}\left[\left(\sum_{i=1}^d I_{n_i}^X\left(f_i^{(k)}\right)\right)^4\right] = \sum_{i=1}^d\mathbb{E}\left[I_{n_i}^X\left(f_i^{(k)}\right)^4\right]$$
$$+ 6\sum_{1\leqslant i<j\leqslant d}\mathbb{E}\left[I_{n_j}^X\left(f_j^{(k)}\right)^2 I_{n_i}^X\left(f_i^{(k)}\right)^2\right] + \sum_{(i_1,\ldots,i_4)\in V_d}\mathbb{E}\left[\prod_{l=1}^4 I_{n_{i_l}}^X\left(f_{i_l}^{(k)}\right)\right].$$

By the multiplication formula (1), for every $1 \leqslant i < j \leqslant d$,

$$I_{n_i}^X\left(f_i^{(k)}\right)I_{n_j}^X\left(f_j^{(k)}\right) = \sum_{q=0}^{n_i} q!\binom{n_i}{q}\binom{n_j}{q}I_{n_i+n_j-2q}^X\left(f_i^{(k)}\otimes_q f_j^{(k)}\right)$$

and therefore

$$\mathbb{E}\left[I_{n_i}^X\left(f_i^{(k)}\right)^2 I_{n_j}^X\left(f_j^{(k)}\right)^2\right]$$
$$= \sum_{q=0}^{n_i}\left[q!\binom{n_i}{q}\binom{n_j}{q}\right]^2 (n_i+n_j-2q)!\left\|\left(f_i^{(k)}\otimes_q f_j^{(k)}\right)_s\right\|^2_{H^{\otimes n_i+n_j-2q}}.$$

Now, relations (2) and (3) imply that

$$\mathbb{E}\left[\left(\sum_{i=1}^d I_{n_i}^X\left(f_i^{(k)}\right)\right)^4\right] = T_1(k) + T_2(k) + T_3(k)$$

where

$$T_1(k) = \sum_{i=1}^{d} \left\{ 3(n_i!)^2 \left\| f_i^{(k)} \right\|_{H^{\otimes n_i}}^4 + \sum_{p=1}^{n_i-1} \frac{(n_i!)^4}{(p!\,(n_i-p)!)^2} \left[\left\| f_i^{(k)} \otimes_p f_i^{(k)} \right\|_{H^{\otimes 2(n_i-p)}}^2 \right. \right. $$
$$\left. \left. + \binom{2n_i-2p}{n_i-p} \left\| (f_i^{(k)} \otimes_p f_i^{(k)})_s \right\|_{H^{\otimes 2(n_i-p)}}^2 \right] \right\}$$

$$T_2(k) = 6 \sum_{1 \leqslant i < j \leqslant d} \left\{ n_i!\, n_j! \left\| f_i^{(k)} \right\|_{H^{\otimes n_i}}^2 \left\| f_j^{(k)} \right\|_{H^{\otimes n_j}}^2 \right.$$
$$+ \sum_{q=1}^{n_i} \left[\left(q! \binom{n_i}{q} \binom{n_j}{q} \right)^2 (n_i + n_j - 2q)! \left\| (f_i^{(k)} \otimes_q f_j^{(k)})_s \right\|_{H^{\otimes n_i + n_j - 2q}}^2 \right.$$
$$\left. \left. + \binom{n_i}{q} \binom{n_j}{q} n_i!\, n_j! \left\| f_i^{(k)} \otimes_q f_j^{(k)} \right\|_{H^{\otimes n_j + n_i - 2q}}^2 \right] \right\},$$

and

$$T_3(k) = \sum_{(i_1,\ldots,i_4) \in V_d} \mathbb{E}\left[\prod_{l=1}^{4} I_{n_{i_l}}^X \left(f_{i_l}^{(k)} \right) \right].$$

But

$$3 \sum_{i=1}^{d} (n_i!)^2 \left\| f_i^{(k)} \right\|_{H^{\otimes n_i}}^4 + 6 \sum_{1 \leqslant i < j \leqslant d} n_i!\, n_j! \left\| f_i^{(k)} \right\|_{H^{\otimes n_i}}^2 \left\| f_j^{(k)} \right\|_{H^{\otimes n_j}}^2$$
$$= 3 \left[\sum_{i=1}^{d} n_i! \left\| f_i^{(k)} \right\|_{H^{\otimes n_i}}^2 \right]^2$$

and the desired conclusion is immediately obtained, since condition (4) ensures that the right side of the above expression converges to $3d^2$ when k goes to infinity.

(i) $\Longrightarrow$ (iii). We will consider the case

$$H = L^2\left([0,1], \mathcal{B}([0,1]), \mathrm{d}x\right), \tag{5}$$

where $\mathrm{d}x$ stands for Lebesgue measure, and use the notation introduced at the end of Section 2. We stress again that the extension to a general, separable Hilbert space H can be done by following the line of reasoning presented in [10, Section 3.3.] and it is not detailed here. Now suppose (i) and (5) hold. The result is completely proved, once the asymptotic relation

$$\sum_{i=1}^{d} \lambda_i I_{n_i}^X \left(f_i^{(k)} \right) = \sum_{i=1}^{d} \lambda_i n_i!\, J_{n_i}^1 \left(f_i^{(k)} \right) \xrightarrow[k \uparrow +\infty]{\text{Law}} \left\| \lambda_d \right\|_{\mathbb{R}^d} \times N(0,1)$$

is verified for every vector $\lambda_d = (\lambda_1, \ldots, \lambda_d) \in \mathbb{R}^d$. Thanks to the Dambis–Dubins–Schwarz Theorem (see [13, Chapter V]), we know that for every k, there exists a standard Brownian motion $W^{(k)}$ (which depends also on λ_d) such that

$$
\sum_{i=1}^{d} \lambda_i n_i! \, J_{n_i}^1\big(f_i^{(k)}\big) = W^{(k)}\left[\int_0^1 \left(\sum_{i=1}^{d} \lambda_i n_i! \, J_{n_i-1}^t\big(f_{i,t}^{(k)}\big)\right)^2 dt\right]
$$

$$
= W^{(k)}\left[\sum_{i=1}^{d} \lambda_i^2 \int_0^1 \left(n_i! \, J_{n_i-1}^t\big(f_{i,t}^{(k)}\big)\right)^2 dt\right.
$$

$$
\left. + 2 \sum_{1 \leqslant i < j \leqslant d} \lambda_i \lambda_j n_i! \, n_j! \int_0^1 \left[J_{n_i-1}^t\big(f_{i,t}^{(k)}\big) J_{n_j-1}^t\big(f_{j,t}^{(k)}\big)\right] dt\right].
$$

Now, since (4) implies

$$
\mathbb{E}\left[\left(n_i! \, J_{n_i}^1\big(f_i^{(k)}\big)\right)^2\right] \xrightarrow[k \uparrow +\infty]{} 1
$$

for every i, condition (i) yields—thanks to Proposition 3 in [10]—that

$$
\sum_{i=1}^{d} \lambda_i^2 \int_0^1 \left(n_i! \, J_{n_i-1}^t\big(f_{i,t}^{(k)}\big)\right)^2 dt \xrightarrow[k \uparrow +\infty]{L^2} \|\lambda_d\|_{\mathbb{R}^d}^2.
$$

To conclude, we shall verify that (i) implies also that for every $i < j$

$$
\int_0^1 \left[J_{n_i-1}^1\big(f_{i,t}^{(k)}\big) J_{n_j-1}^1\big(f_{j,t}^{(k)}\big)\right] dt = \int_0^1 \frac{\left[I_{n_i-1}^t\big(f_{i,t}^{(k)}\big) I_{n_j-1}^t\big(f_{j,t}^{(k)}\big)\right]}{(n_i-1)! \, (n_j-1)!} \, dt \xrightarrow[k \uparrow +\infty]{L^2} 0.
$$

To see this, use once again the multiplication formula (1) to write

$$
\int_0^1 dt \left[I_{n_i-1}^t\big(f_{i,t}^{(k)}\big) I_{n_j-1}^t\big(f_{j,t}^{(k)}\big)\right]
$$

$$
= \sum_{q=0}^{n_i-1} (n_i + n_j - 2(q+1))! \, q! \binom{n_i-1}{q}\binom{n_j-1}{q} \times
$$

$$
\times \int_{\Delta_1^{n_i+n_j-2(q+1)}} \left[\int_{s_1}^1 dt \big(f_{i,t}^{(k)} \otimes_q^t f_{j,t}^{(k)}\big)_s \big(s_1, \ldots, s_{n_i+n_j-2(q+1)}\big)\right]
$$

$$
dW_{s_1} \ldots dW_{s_{n_i+n_j-2(q+1)}},
$$

when $n_i < n_j$, or, when $n_i = n_j$

$$\int_0^1 dt \left[I_{n_i-1}^t \big(f_{i,t}^{(k)} \big) I_{n_j-1}^t \big(f_{j,t}^{(k)} \big) \right]$$

$$= \int_0^1 dt \, \mathbb{E} \left[I_{n_i-1}^t \big(f_{i,t}^{(k)} \big) I_{n_i-1}^t \big(f_{j,t}^{(k)} \big) \right] + \sum_{q=0}^{n_i-2} (2n_i - 2(q+1))! \, q! \binom{n_i-1}{q}^2 \times$$

$$\times \int_{\Delta_1^{n_i+n_j-2(q+1)}} \left[\int_{s_1}^1 dt \big(f_{i,t}^{(k)} \otimes_q^t f_{j,t}^{(k)} \big)_s \big(s_1, \ldots, s_{n_i+n_j-2(q+1)} \big) \right]$$

$$dW_{s_1} \ldots dW_{s_{n_i+n_j-2(q+1)}}.$$

In what follows, for every $m \geqslant 2$, we write $\mathbf{t}_m$ to indicate a vector $(t_1, \ldots, t_m) \in \mathbb{R}^m$, whereas $d\mathbf{t}_m$ stands for Lebesgue measure on $\mathbb{R}^m$; we shall also use the symbol $\widehat{\mathbf{t}}_m = \max_i(t_i)$. Now fix $q < n_i - 1 \leqslant n_j - 1$, and observe that, by writing $p = q + 1$,

$$\int_{\Delta_1^{n_i+n_j-2(q+1)}} ds_1 \ldots ds_{n_i+n_j-2(q+1)}$$

$$\times \left[\int_{s_1}^1 dt \big(f_{i,t}^{(k)} \otimes_q^t f_{j,t}^{(k)} \big)_s \big(s_1, \ldots, s_{n_i+n_j-2(q+1)} \big) \right]^2$$

$$\leqslant \int_{[0,1]^{n_i-p}} ds_{n_i-p} \int_{[0,1]^{n_j-p}} d\boldsymbol{\tau}_{n_j-p}$$

$$\times \left[\int_{\widehat{\mathbf{s}}_{n_i-p} \vee \widehat{\boldsymbol{\tau}}_{n_j-p}}^1 dt \int_{[0,t]^{p-1}} d\mathbf{u}_{p-1} f_j^{(k)}(t, \boldsymbol{\tau}_{n_j-p}, \mathbf{u}_{p-1}) f_i^{(k)}(t, \mathbf{s}_{n_i-p}, \mathbf{u}_{p-1}) \right]^2$$

$$= C(k)$$

and moreover

$$C(k)^2 = \left\{ \int_0^1 dt \int_{[0,1]^{p-1}} d\mathbf{u}_{p-1} \int_0^1 dt' \int_{[0,1]^{p-1}} d\mathbf{v}_{p-1} \, \mathbf{1}_{(\widehat{\mathbf{u}}_{p-1} \leqslant t, \widehat{\mathbf{v}}_{p-1} \leqslant t')} \right.$$

$$\times \left[\int_{[0,t \wedge t']^{n_i-p}} d\mathbf{s}_{n_i-p} f_i^{(k)}(t, \mathbf{s}_{n_i-p}, \mathbf{u}_{p-1}) f_i^{(k)}(t', \mathbf{s}_{n_i-p}, \mathbf{v}_{p-1}) \right]$$

$$\left. \times \left[\int_{[0,t \wedge t']^{n_j-p}} d\boldsymbol{\tau}_{n_j-p} f_j^{(k)}(t, \boldsymbol{\tau}_{n_j-p}, \mathbf{u}_{p-1}) f_j^{(k)}(t', \boldsymbol{\tau}_{n_j-p}, \mathbf{v}_{p-1}) \right] \right\}^2$$

$$\leqslant C_i(k) \times C_j(k)$$

where, for $\gamma = i, j$

$$C_\gamma(k) = \int_0^1 dt \int_{[0,1]^{p-1}} d\mathbf{u}_{p-1} \int_0^1 dt' \int_{[0,1]^{p-1}} d\mathbf{v}_{p-1}$$

$$\times \left[\int_{[0,t \wedge t']^{n_\gamma-p}} d\mathbf{s}_{n_\gamma-p} f_\gamma^{(k)}(t, \mathbf{s}_{n_\gamma-p}, \mathbf{u}_{p-1}) f_\gamma^{(k)}(t', \mathbf{s}_{n_\gamma-p}, \mathbf{v}_{p-1}) \right]^2$$

and the calculations contained in [10] imply immediately that both $C_j(k)$ and $C_i(k)$ converge to zero whenever (i) is verified. On the other hand, when

$$q = n_i - 1 < n_j - 1$$

$$\int_{\Delta_1^{n_j-n_i}} \left[\int_{s_1}^1 dt \, \left(f_{i,t}^{(k)} \otimes_{n_i-1}^t f_{j,t}^{(k)} \right)_s (s_1, \ldots, s_{n_j-n_i}) \right]^2 ds_1 \ldots ds_{n_j-n_i}$$

$$\leqslant \int_{[0,1]^{n_j-n_i}} d\boldsymbol{\tau}_{n_j-n_i}$$

$$\times \left[\int_{\widehat{\boldsymbol{\tau}}_{n_j-n_i}}^1 dt \int_{[0,t]^{n_i-1}} d\mathbf{u}_{n_i-1} \, f_j^{(k)}(t, \boldsymbol{\tau}_{n_j-n_i}, \mathbf{u}_{n_i-1}) f_i^{(k)}(t, \mathbf{u}_{n_i-1}) \right]^2$$

$$= D(k)$$

and also

$$D(k)^2 \leqslant D_1(k) \times D_2(k)$$

where

$$D_1(k) = \int_0^1 dt \int_{[0,1]^{n_i-1}} d\mathbf{u}_{n_i-1} \int_0^1 dt' \int_{[0,1]^{n_i-1}} d\mathbf{v}_{n_i-1}$$

$$\times \left[\int_{[0,t\wedge t']^{n_j-n_i}} d\boldsymbol{\tau}_{n_j-n_i} \, f_j^{(k)}(t, \boldsymbol{\tau}_{n_j-n_i}, \mathbf{u}_{n_i-1}) \, f_j^{(k)}(t', \boldsymbol{\tau}_{n_j-n_i}, \mathbf{v}_{n_i-1}) \right]^2$$

and

$$D_2(k) = \int_0^1 dt \int_{[0,1]^{n_i-1}} d\mathbf{u}_{n_i-1} \int_0^1 dt' \int_{[0,1]^{n_i-1}} d\mathbf{v}_{n_i-1}$$

$$\times \left(f_i^{(k)}(t, \mathbf{u}_{n_i-1}) f_i^{(k)}(t', \mathbf{v}_{n_i-1}) \right)^2 = \left\| f_i^{(k)} \right\|_{H^{\otimes n_i}}^4$$

so that the conclusion is immediately achieved, due to (4). Finally, recall that for $n_i = n_j$

$$\int_0^1 dt \, \mathbb{E}\left[I_{n_i-1}^t \left(f_{i,t}^{(k)} \right) I_{n_i-1}^t \left(f_{j,t}^{(k)} \right) \right]$$

$$= (n_i - 1)! \int_0^1 dt \int_{[0,t]^{n_i-1}} d\mathbf{u}_{n_i-1} \, f_j^{(k)}(t, \mathbf{u}_{n_i-1}) f_i^{(k)}(t, \mathbf{u}_{n_i-1})$$

$$= \left((n_i - 1)! \right)^2 \int_{\Delta_1^{n_i}} dt \, d\mathbf{u}_{n_i-1} \, f_j^{(k)}(t, \mathbf{u}_{n_i-1}) f_i^{(k)}(t, \mathbf{u}_{n_i-1})$$

$$= \left[\frac{(n_i - 1)!}{n_i!} \right]^2 \mathbb{E}\left[I_{n_i}^X(f_i^{(k)}) I_{n_i}^X(f_j^{(k)}) \right] \xrightarrow[k\uparrow+\infty]{} 0$$

again by assumption (4). The proof of the implication is concluded.

(iv) $\Longleftrightarrow$ (v) $\Longleftrightarrow$ (i). This is a consequence of Theorem 1 in [10]. $\quad\square$

In what follows, $\mathbf{C}_d = \{ C_{ij} : 1 \leqslant i, j \leqslant d \}$ indicates a $d \times d$ positive definite symmetric matrix. In the case of multiple Wiener integrals of the same order, a useful extension of Theorem 1 is the following

Proposition 1. *Let $d \geqslant 2$, and fix $n \geqslant 2$ as well as a collection of kernels*

$$\left\{ \left(f_1^{(k)}, \ldots, f_d^{(k)} \right) : k \geqslant 1 \right\}$$

such that $f_j^{(k)} \in H^{\odot n}$ for every $k \geqslant 1$ and every $j = 1, \ldots, d$, and

$$\lim_{k \to \infty} j! \left\| f_j^{(k)} \right\|_{H^{\otimes n}}^2 = C_{jj}, \qquad \forall j = 1, \ldots, d,$$
$$\lim_{k \to \infty} \mathbb{E} \left[I_n^X \left(f_i^{(k)} \right) I_n^X \left(f_j^{(k)} \right) \right] = C_{ij}, \qquad \forall 1 \leqslant i < j \leqslant d. \tag{6}$$

Then, the following conditions are equivalent:

(i) *as k goes to infinity, the vector $\left(I_n^X \left(f_1^{(k)} \right), \ldots, I_n^X \left(f_d^{(k)} \right) \right)$ converges in distribution to a d-dimensional Gaussian vector $N_d(0, \mathbf{C}_d) = (N_1, \ldots, N_d)$ with covariance matrix $\mathbf{C}_d$;*

(ii)

$$\lim_{k \to \infty} \mathbb{E} \left[\left(\sum_{i=1,\ldots,d} I_n^X \left(f_i^{(k)} \right) \right)^4 \right] = 3 \left(\sum_{i=1}^d C_{ii} + 2 \sum_{1 \leqslant i < j \leqslant d} C_{ij} \right)^2 = \mathbb{E} \left[\left(\sum_{i=1}^d N_i \right)^4 \right],$$

and

$$\lim_{k \to \infty} \mathbb{E} \left[\prod_{l=1}^4 I_n^X \left(f_{i_l}^{(k)} \right) \right] = \mathbb{E} \left[\prod_{l=1}^4 N_{i_l} \right]$$

for every $(i_1, i_2, i_3, i_4) \in V_d$;

(iii) *for every $j = 1, \ldots, d$, $I_n^X \left(f_j^{(k)} \right)$ converges in distribution to N_j, that is, to a centered Gaussian random variable with variance C_{jj};*

(iv) *for every $j = 1, \ldots, d$,*

$$\lim_{k \to \infty} \mathbb{E} \left[I_n^X \left(f_j^{(k)} \right)^4 \right] = 3 C_{jj}^2;$$

(v) *for every $j = 1, \ldots, d$*

$$\lim_{k \to \infty} \left\| f_i^{(k)} \otimes_p f_i^{(k)} \right\|_{H^{\otimes 2(n-p)}} = 0,$$

for every $p = 1, \ldots, n-1$.

Sketch of the proof. The main idea is contained in the proof of Theorem 1. We shall discuss only implications (ii) $\Longrightarrow$ (v) and (v) $\Longrightarrow$ (i). In particular, one can show that (ii) implies (v) by adapting the same arguments as in the proof of Theorem 1 to show that

$$\mathbb{E} \left[\left(\sum_{i=1}^d I_n^X \left(f_i^{(k)} \right) \right)^4 \right] = V_1(k) + V_2(k) + V_3(k)$$

where

$$V_1(k) = \sum_{i=1}^{d} \left\{ 3(n!)^2 \|f_i^{(k)}\|_{H^{\otimes n}}^4 + \sum_{p=1}^{n-1} \frac{(n!)^4}{(p!(n-p)!)^2} \left[\|f_i^{(k)} \otimes_p f_i^{(k)}\|_{H^{\otimes 2(n-p)}}^2 \right. \right.$$
$$\left. \left. + \binom{2n-2p}{n-p} \|(f_i^{(k)} \otimes_p f_i^{(k)})_s\|_{H^{\otimes 2(n-p)}}^2 \right] \right\}$$

$$V_2(k) = 6 \sum_{1 \leqslant i < j \leqslant d} \left\{ (n!)^2 \|f_i^{(k)}\|_{H^{\otimes n}}^2 \|f_j^{(k)}\|_{H^{\otimes n}}^2 \right.$$
$$+ \sum_{q=1}^{n-1} \left[\left(q! \binom{n}{q}^2 \right)^2 (2n-2q)! \left\| (f_i^{(k)} \otimes_q f_j^{(k)})_s \right\|_{H^{\otimes 2n-2q}}^2 \right.$$
$$\left. \left. + \binom{n}{q}^2 (n!)^2 \|f_i^{(k)} \otimes_q f_j^{(k)}\|_{H^{\otimes 2n-2q}}^2 \right] \right\}$$
$$+ 12(n!)^2 \sum_{1 \leqslant i < j \leqslant d} \langle f_i^{(k)}, f_j^{(k)} \rangle_{H^{\otimes n}}^2$$

and

$$V_3(k) = \sum_{(i_1,\ldots,i_4) \in V_d} \mathbb{E}\left[\prod_{l=1}^{4} I_n^X(f_{i_l}^{(k)}) \right].$$

But (6) yields

$$3(n!)^2 \sum_{i=1}^{d} \|f_i^{(k)}\|_{H^{\otimes n}}^4 + 6 \sum_{1 \leqslant i < j \leqslant d} \left[(n!)^2 \|f_i^{(k)}\|_{H^{\otimes n}}^2 \|f_j^{(k)}\|_{H^{\otimes n}}^2 \right.$$
$$\left. + 2(n!)^2 \langle f_i^{(k)}, f_j^{(k)} \rangle_{H^{\otimes n}}^2 \right] \xrightarrow[k \uparrow +\infty]{} 3 \sum_{i=1}^{d} C_{ii}^2 + 6 \sum_{1 \leqslant i < j \leqslant d} [C_{ii}C_{jj} + 2C_{ij}^2]$$

and the conclusion is obtained, since

$$\mathbb{E}\left[\left(\sum_{i=1}^{d} N_i \right)^4 \right] = 3 \sum_{i=1}^{d} C_{ii}^2 + 6 \sum_{1 \leqslant i < j \leqslant d} [C_{ii}C_{jj} + 2C_{ij}^2] + \sum_{(i_1,\ldots,i_4) \in V_d} \mathbb{E}\left[\prod_{l=1}^{4} N_{i_l} \right].$$

Now keep the notations of the last part of the proof of Theorem 1. The implication (v) $\Rightarrow$ (i) follows from the calculations therein contained, implying, thanks to (6), that the quantity

$$\int_0^1 \left(\sum_{i=1}^{d} \lambda_i n! J_{n-1}^t(f_{i,t}^{(k)}) \right)^2 dt$$

converges in L^2 to $\sum_{i=1,\ldots,d} \lambda_i^2 C_{ii} + 2 \sum_{1 \leqslant i < j \leqslant d} \lambda_i \lambda_j C_{ij}$, and therefore the desired conclusion. The remaining details can be easily provided by the reader.

$\square$

4 Applications

In this section, we will present some consequences of our results. We mention that our list of applications is by no means exhaustive; for instance, the weak convergence results for quadratic functionals of (fractional) Brownian motion given in [10], [11] and [12] can be immediately extended to the multidimensional case. An example is given in the following generalization of the results contained in [12].

Proposition 2. *Let W be a standard Brownian motion on $[0,1]$ and, for every $d \geqslant 2$, define the process*

$$t \longmapsto W_t^{\otimes d} := \int_0^t \cdots \int_0^{s_{d-1}} dW_{s_d} \dots dW_{s_1}, \qquad t \in [0,1].$$

Then: (a) for every $d \geqslant 1$ the vector

$$\frac{1}{\sqrt{\log(1/\varepsilon)}} \left(\int_\varepsilon^1 \frac{da}{a^2} W_a^{\otimes 2}, \int_\varepsilon^1 \frac{da}{a^3} W_a^{\otimes 4}, \dots, \int_\varepsilon^1 \frac{da}{a^{d+1}} W_a^{\otimes 2d} \right)$$

converges in distribution, as $\varepsilon \to 0$, to

$$\left(N_1(0,1), 2\sqrt{3!}\, N_2(0,1), \dots, d\sqrt{(2d-1)!}\, N_d(0,1) \right)$$

where the $N_j(0,1)$, $j = 1, \dots, d$, are standard, independent Gaussian random variables; (b) by defining, for every $d \geqslant 1$ and for every $j = 0, \dots, d$, the positive constant

$$c(d,j) = \frac{(2d)!}{2^{d-j}(d-j)!},$$

for every $d \geqslant 1$ the vector

$$\frac{1}{\sqrt{\log(1/\varepsilon)}} \left(\int_\varepsilon^1 \frac{da}{a^2} W_a^2 - c(1,0) \log \frac{1}{\varepsilon}, \int_\varepsilon^1 \frac{da}{a^3} W_a^4 - c(2,0) \log \frac{1}{\varepsilon}, \dots \right.$$

$$\left. \dots, \int_\varepsilon^1 \frac{da}{a^{d+1}} W_a^{2d} - c(d,0) \log \frac{1}{\varepsilon} \right)$$

converges in distribution to a Gaussian vector $(G_1, \dots, G_d)$ with the following covariance structure:

$$\mathbb{E}[G_{k'} G_k] = \sum_{j=1}^{k'} c(k,j) c(k',j) j^2 (2j-1)!$$

for every $1 \leqslant k' \leqslant k \leqslant d$.

Proof. From Proposition 4.1 in [12], we obtain immediately that for every $j = 1, \ldots, d$,

$$\frac{1}{\sqrt{\log(1/\varepsilon)}} \int_\varepsilon^1 \frac{da}{a^{j+1}} \, W_a^{\otimes 2j} \xrightarrow{(d)} j\sqrt{(2j-1)!} \, N_j(0,1),$$

and the asymptotic independence follows from Theorem 1, since for every $i \neq j$

$$\mathbb{E}\left[\int_\varepsilon^1 \frac{da}{a^{j+1}} W_a^{\otimes 2j} \int_\varepsilon^1 \frac{db}{b^{i+1}} W_b^{\otimes 2i} \right] = \int_\varepsilon^1 \frac{da}{a^{j+1}} \int_\varepsilon^1 \frac{db}{b^{i+1}} \, \mathbb{E}\left[W_a^{\otimes 2j} W_b^{\otimes 2i} \right] = 0.$$

To prove point (b), use for instance Stroock's formula (see [14]) to obtain that for every $k = 1, \ldots, d$

$$\int_\varepsilon^1 \frac{da}{a^{k+1}} \, W_a^{2k} = \sum_{j=1}^k c(k,j) \int_\varepsilon^1 \frac{da}{a^{j+1}} \, W_a^{\otimes 2j} + c(k,0) \log\frac{1}{\varepsilon},$$

so that the result derives immediately from point (a). □

In what follows, we prove a new asymptotic version of *Knight's theorem*—of the kind discussed e.g. in [13, Chapter XIII]—and a necessary and sufficient condition for a class of random variables living in a finite sum of chaoses—and satisfying some asymptotic property—to have a Gaussian weak limit. Further applications will be explored in a subsequent paper.

More specifically, we are interested in an asymptotic Knight's theorem for *chaotic martingales*, which, in our terminology, are martingales having a multiple Wiener integral representation (we stress that there is no relation with *normal martingales with the chaotic representation property*, as discussed e.g. in [1, Chapter XXI]). To this end, take $d \geqslant 2$ integers

$$1 \leqslant n_1 \leqslant n_2 \leqslant \cdots \leqslant n_d,$$

and, for $j = 1, \ldots, d$ and $k \geqslant 1$ take a class

$$\{\phi_{j,k}^t : t \in [0,1]\}$$

of elements of $H^{\odot n_j}$, such that there exists a filtration $\{\mathcal{F}_t : t \in [0,1]\}$, satisfying the usual conditions and such that, for every k and for every j, the process

$$t \longmapsto M_{j,k}(t) = I_{n_j}^X(\phi_{j,k}^t), \qquad t \in [0,1],$$

is a $\mathcal{F}_t$-continuous martingale on $[0,1]$, vanishing at zero. We note $\langle M_{j,k}, M_{j,k} \rangle$ and $\langle M_{j,k}, M_{i,k} \rangle$, $1 \leqslant i,j \leqslant d$, the corresponding quadratic variation and covariation processes, whereas $\beta_{j,k}$ is the Dambis–Dubins–Schwarz Brownian motion associated to $M_{j,k}$. Then, we have the following

Proposition 3 (Asymptotic Knight's theorem for chaotic martingales). *Under the above assumptions and notation, suppose that for every $j = 1, \ldots, d$,*

$$\langle M_{j,k}, M_{j,k} \rangle \xrightarrow[k \to +\infty]{(d)} T_j, \tag{7}$$

where $t \mapsto T_j(t)$ is a deterministic, continuous and non-decreasing process. If in addition

$$\lim_{k \to +\infty} \mathbb{E}[\langle M_{i,k}, M_{j,k} \rangle_t] = 0 \tag{8}$$

for every $i \neq j$ and for every t, then $\{M_{j,k} : 1 \leqslant j \leqslant d\}$ converges in distribution to

$$\{B_j \circ T_j : 1 \leqslant j \leqslant d\},$$

where $\{B_j : 1 \leqslant j \leqslant d\}$ is a d-dimensional standard Brownian motion.

Proof. Since

$$M_{j,k}(t) = \beta_{j,k}(\langle M_{j,k}, M_{j,k} \rangle_t), \qquad t \in [0,1],$$

and $\langle M_{j,k}, M_{j,k} \rangle$ weakly converges to T_j, we immediately obtain that $M_{j,k}$ converges in distribution to the Gaussian process $B_j \circ T_j$. Thanks to Theorem 1, it is now sufficient to prove that, for every $i \neq j$ and for every $s, t \in [0,1]$, the quantity $\mathbb{E}[M_{j,k}(s)M_{i,k}(t)]$ converges to zero. But

$$\mathbb{E}[M_{j,k}(s)M_{i,k}(t)] = \mathbb{E}[\langle M_{i,k}, M_{j,k} \rangle_{t \wedge s}]$$

and assumption (8) yields the result. □

Remark. An analogue of Proposition 4 for general martingales verifying (7) can be found in [13, Exercise XIII.1.16], but in this case (8) has to be replaced by

$$\langle M_{j,k}, M_{i,k} \rangle \xrightarrow[k \to +\infty]{(d)} 0$$

for every $i \neq j$. Since chaotic martingales have a very explicit covariance structure (due to the isometric properties of multiple integrals), condition (8) is usually quite easy to verify. We also recall that—according e.g. to [13, Theorem XIII.2.3]—if condition (7) is dropped, to prove the asymptotic independence of the Brownian motions $\{\beta_{j,k} : 1 \leqslant j \leqslant d\}$ one has to check the condition

$$\lim_{k \to +\infty} \langle M_{i,k}, M_{j,k} \rangle_{\tau_j^k(t)} = \lim_{k \to +\infty} \langle M_{i,k}, M_{j,k} \rangle_{\tau_i^k(t)} = 0$$

in probability for every $i \neq j$ and for every t, where τ_j^k and τ_i^k are the stochastic time-changes associated respectively to $\langle M_{j,k}, M_{j,k} \rangle$ and $\langle M_{i,k}, M_{i,k} \rangle$.

We conclude the paper by stating a result on the weak convergence of random variables belonging to a finite sum of Wiener chaoses to a standard normal random variable (the proof is a direct consequence of the arguments contained in the proof of Theorem 1).

Proposition 4. *Let* $1 \leqslant n_1 < \cdots < n_d$, $d \geqslant 2$, *and let* $f_j^{(k)} \in H^{\odot n_j}$, *for every* $k \geqslant 1$ *and* $1 \leqslant j \leqslant d$. *Assume that*

$$n_j! \lim_{k \uparrow +\infty} \left\| f_j^{(k)} \right\|_{H^{\otimes n_j}}^2 = 1, \qquad j = 1, \ldots, d, \tag{9}$$

and

$$\lim_{k \uparrow +\infty} \sum_{(i_1, \ldots, i_4) \in V_d} \mathbb{E} \left[\prod_{l=1}^4 I_{n_{i_l}}^X \left(f_{i_l}^{(k)} \right) \right] \geqslant 0. \tag{10}$$

Define moreover $S_d^{(k)} = \sum_{j=1,\ldots,d} I_{n_j}^X \left(f_j^{(k)} \right)$. *Then, the following conditions are equivalent:*

(i) *the sequence* $d^{-1/2} S_d^{(k)}$ *converges in distribution to a standard Gaussian random variable, as* k *tends to infinity;*

(ii) *for every* $j = 1, \ldots, d$,

$$\lim_{k \uparrow +\infty} \left\| f_j^{(k)} \otimes_p f_j^{(k)} \right\|_{H^{\otimes 2(n_j - p)}}^2 = 0, \qquad p = 1, \ldots, n_j - 1;$$

(iii) *for every* $j = 1, \ldots, d$, $I_{n_j}^X \left(f_j^{(k)} \right)$ *converges in law to a standard Gaussian random variable, as* k *goes to infinity.*

An interesting consequence of the above result is the following

Corollary 1. *Let* $1 \leqslant n_1 < \cdots < n_d$, $d \geqslant 2$, $f_j^{(k)} \in H^{\odot n_j}$, $k \geqslant 1$ *and* $1 \leqslant j \leqslant d$. *Assume moreover that* (9) *is verified and that, for every* k, *the random variables* $I_{n_j}^X \left(f_j^{(k)} \right)$, $j = 1, \ldots, d$, *are pairwise independent. Then, the sequence* $d^{-1/2} S_d^{(k)}$, $k \geqslant 1$, *defined as before, converges in law to a standard Gaussian random variable* $N(0,1)$ *if, and only if, for every* j, $I_{n_j}^X \left(f_j^{(k)} \right)$ *converges in law to* $N(0,1)$.

Proof. We know from [16] (see also [4]) that, in the case of multiple stochastic integrals, pairwise independence implies mutual independence, so that condition (10) is clearly verified. □

Remarks. (i) If we add the assumption that, for every j, the sequence $I_{n_j}^X \left(f_j^{(k)} \right)$, $k \geqslant 1$, admits a weak limit, say μ_j, then the conclusion of Corollary 6 can be directly deduced from [5, p. 248]. As a matter of fact, in such a reference the following implication is proved: if the d probability measures μ_j, $j = 1, \ldots, d$, are such that (a) $\int x \, d\mu_j(x) = 0$ for every j, and (b) $\mu_1 \star \cdots \star \mu_d$, where $\star$ indicates convolution, is Gaussian, then each μ_j is necessarily Gaussian.

(ii) Condition (10) is also satisfied when $d = 2$ and $n_1 + n_2$ is odd.

References

1. Dellacherie, C., Maisonneuve, B. and Meyer, P.A. (1992), *Probabilités et Potentiel*, Chapitres XVII à XXIV, Hermann, Paris.
2. Giraitis, L. and Surgailis, D. (1985), "CLT and Other Limit Theorems for Functionals of Gaussian Processes", *Z. Wahr. verw. Gebiete* **70** (2), 191–212.
3. Janson, S. (1997), *Gaussian Hilbert spaces*, Cambridge University Press, Cambridge.
4. Kallenberg, O. (1991), "On an independence criterion for multiple Wiener integrals", *Ann. Probab.* **19** (2), 483–485.
5. Lukacs, E. (1983), *Developments in characteristic functions*, MacMillan Co., New York.
6. Major, P. (1981), *Multiple Wiener–Itô Integrals*, Lecture Notes in Mathematics **849**, Springer-Verlag, New York.
7. Maruyama, G. (1982), "Applications of the multiplication of the Ito–Wiener expansions to limit theorems", *Proc. Japan Acad.* **58**, 388–390.
8. Maruyama, G. (1985), "Wiener functionals and probability limit theorems, I: the central limit theorem", *Osaka Journal of Mathematics* **22**, 697–732.
9. Nualart, D. (1995), *The Malliavin Calculus and Related Topics*, Springer, Berlin–Heidelberg–New York.
10. Nualart, D. and Peccati, G. (2004), "Central limit theorems for sequences of multiple stochastic integrals", to appear in *The Annals of Probability*.
11. Peccati, G. and Yor, M. (2004a), "Four limit theorems for quadratic functionals of Brownian motion and Brownian bridge", to appear in the volume: *Asymptotic Methods in Stochastics*, American Mathematical Society, Communication Series.
12. Peccati, G. and Yor, M. (2004b), "Hardy's inequality in $L^2([0,1])$ and principal values of Brownian local times", to appear in the volume: *Asymptotic Methods in Stochastics*, American Mathematical Society, Communication Series.
13. Revuz, D. and Yor, M. (1999), *Continuous Martingales and Brownian Motion*, Springer, Berlin–Heidelberg–New York.
14. D. W. Stroock (1987), "Homogeneous chaos revisited", in: *Séminaire de Probabilités XXI*, Springer, Berlin, LNM **1247**, 1–8.
15. Surgailis, D. (2003), "CLTs for Polynomials of Linear Sequences: Diagram Formula with illustrations", in: *Theory and Applications of Long Range Dependence*, Birkhäuser, Boston.
16. Üstünel, A. S. and Zakai, M. (1989), "Independence and conditioning on Wiener space", *The Annals of Probability* **17** (4), 1441–1453.

Derivatives of Self-intersection Local Times

Jay Rosen*

Department of Mathematics
College of Staten Island, CUNY
Staten Island, NY 10314
e-mail: jrosen3@earthlink.net

Summary. We show that the renormalized self-intersection local time $\gamma_t(x)$ for both the Brownian motion and symmetric stable process in R^1 is differentiable in the spatial variable and that $\gamma_t'(0)$ can be characterized as the continuous process of zero quadratic variation in the decomposition of a natural Dirichlet process. This Dirichlet process is the potential of a random Schwartz distribution. Analogous results for fractional derivatives of self-intersection local times in R^1 and R^2 are also discussed.

1 Introduction

In their study of the intrinsic Brownian local time sheet and stochastic area integrals for Brownian motion, [14, 15, 16], Rogers and Walsh were led to analyze the functional

$$A(t, B_t) = \int_0^t 1_{[0,\infty)}(B_t - B_s)\, \mathrm{d}s \tag{1}$$

where B_t is a 1-dimensional Brownian motion. They showed that $A(t, B_t)$ is not a semimartingale, and in fact showed that

$$A(t, B_t) - \int_0^t L_s^{B_s}\, \mathrm{d}B_s \tag{2}$$

has finite non-zero $4/3$-variation. Here L_s^x is the local time at x, which is formally $L_s^x = \int_0^s \delta(B_r - x)\, \mathrm{d}r$, where $\delta(x)$ is Dirac's 'δ-function'. A formal application of Ito's lemma, using $\frac{\mathrm{d}}{\mathrm{d}x} 1_{[0,\infty)}(x) = \delta(x)$ and $\frac{\mathrm{d}^2}{\mathrm{d}x^2} 1_{[0,\infty)}(x) = \delta'(x)$, yields

$$A(t, B_t) - \int_0^t L_s^{B_s}\, \mathrm{d}B_s = t + \frac{1}{2} \int_0^t \int_0^s \delta'(B_s - B_r)\, \mathrm{d}r\, \mathrm{d}s \tag{3}$$

---

* This research was supported, in part, by grants from the National Science Foundation and PSC-CUNY.

which motivates the subject matter of this paper. We study the process which is formally defined as

$$\gamma_t' = -\int_0^t \int_0^s \delta'(X_s - X_r)\, dr\, ds \tag{4}$$

where δ' is the derivative of the delta-function, and X_t is Brownian motion or, more generally, a symmetric stable process in R^1. The process γ_t' is related to the self-intersection local time process which is formally defined as

$$\alpha_t = \int_0^t \int_0^s \delta(X_s - X_r)\, dr\, ds. \tag{5}$$

If we set

$$\alpha_t(y) = \int_0^t \int_0^s \delta(X_s - X_r - y)\, dr\, ds \tag{6}$$

we obtain a 'near intersection' local time, and formally differentiating in y suggests that (4) is the derivative $\frac{d}{dy}\alpha_t(y)\big|_{y=0}$.

The process α_t has not been studied much in one dimension since it can be expressed in terms of the local time L_t^y of the process X_t: $\alpha_t = \frac{1}{2}\int (L_t^y)^2\, dy$ and $\alpha_t(y) = \iint_0^t L_s^{x-y}\, d_s L_s^x\, dx$. The fact that L_t^y is not differentiable in the spatial variable y indicates that the existence of (4) and its identification as a derivative requires some care.

In two dimensions, even for Brownian motion, α_t does not exist and must be 'renormalized' by subtracting off a counterterm. This was first done by Varadhan [22], and has been the subject of a large literature, see Dynkin [4], Le Gall [12], Bass and Khoshnevisan [1], Rosen [20, 21]. The resulting renormalized self-intersection local time turns out to be the right tool for the solution of certain "classical" problems such as the asymptotic expansion of the area of the Wiener and stable sausage in the plane and fluctuations of the range of stable random walks. (See Le Gall [11, 10], Le Gall–Rosen [13] and Rosen [19]).

The process γ_t' in R^1, in a certain sense, is even more singular than self-intersection local time in R^2, but as we shall see, due to the symmetry properties of δ', there is no need for a counterterm. We begin with a precise definition of γ_t', show that it exists, is the spatial derivative of the renormalized self-intersection local time $\gamma_t(x) = \alpha_t(x) - E\big(\alpha_t(x)\big)$, and has zero quadratic variation. We then show how it can be characterized as the continuous process of zero quadratic variation in the decomposition of a natural Dirichlet process. This Dirichlet process is the potential of a random Schwartz distribution.

Let

$$\alpha_{t,\epsilon}(y) \stackrel{\text{def}}{=} \int_0^t \int_0^s f_\epsilon(X_s - X_r - y)\, dr\, ds \tag{7}$$

and

$$\alpha_{t,\epsilon}'(y) \stackrel{\text{def}}{=} -\int_0^t \int_0^s f_\epsilon'(X_s - X_r - y)\, dr\, ds \tag{8}$$

where f_ϵ is an approximate δ-function at zero, i.e. $f_\epsilon(x) = f(x/\epsilon)/\epsilon$ with f a positive, C^1, even function of x supported in the unit interval with $\int f \, dx = 1$. We then define

$$\alpha_t(y) = \lim_{\epsilon \to 0} \alpha_{t,\epsilon}(y), \qquad \alpha'_t(y) = \lim_{\epsilon \to 0} \alpha'_{t,\epsilon}(y) \tag{9}$$

$$\gamma_t(y) = \lim_{\epsilon \to 0} \Big(\alpha_{t,\epsilon}(y) - E\big(\alpha_{t,\epsilon}(y)\big) \Big) \tag{10}$$

and

$$\gamma'_t(y) = \lim_{\epsilon \to 0} \Big(\alpha'_{t,\epsilon}(y) - E\big(\alpha'_{t,\epsilon}(y)\big) \Big) \tag{11}$$

whenever the limit exists. We set $\gamma'_t = \gamma'_t(0)$. Let

$$h_\beta(x) = \begin{cases} c(\beta) \, \mathrm{sgn}(x) |x|^{\beta-2} & \text{if } x \neq 0 \\ 0 & \text{if } x = 0 \end{cases} \tag{12}$$

where $c(\beta) = -\pi^{-1} \Gamma(2-\beta) \cos((1-\beta)\pi/2)$ if $1 < \beta < 2$ and $c(2) = -1$. Note that $h_\beta(x)$ is not continuous at $x = 0$.

Theorem 1. *Let X_t denote the symmetric stable process of order $\beta > 3/2$ in R^1. Then $\alpha_{t,\epsilon}(y)$ and $\alpha'_{t,\epsilon}(y)$ converge a.s. and in all L^p spaces as $\epsilon \to 0$ for any $(t,y) \in R_+ \times R^1$.*

The following hold almost surely:

1. For any continuous function $g(y)$ we have

$$\int_0^t \int_0^s g(X_s - X_r) \, dr \, ds = \int g(y) \alpha_t(y) \, dy. \tag{13}$$

and if $g \in C^1$

$$\int_0^t \int_0^s g'(X_s - X_r) \, dr \, ds = -\int g(y) \alpha'_t(y) \, dy. \tag{14}$$

2. $\alpha'_t(y) - h_\beta(y) t$ is continuous in y.
3. $\{\gamma_t(y), (t,y) \in R_+ \times R^1\}$ and $\{\gamma'_t(y), (t,y) \in R_+ \times R^1\}$ are continuous and $\gamma_t(y)$ is differentiable in y with $\gamma'_t(y) = \frac{d}{dy} \gamma_t(y)$.
4. $\alpha'_t(0) = \gamma'_t(0)$.

We see from (13) and (14) that $\alpha'_t(y)$ is the distributional derivative of $\alpha_t(y)$. However, we see from 2. that $\alpha'_t(y)$ is not continuous at $y = 0$ so that these equations do not allow us to characterize $\alpha'_t(0)$.

For any function g_t and any sequence $\tau = \{\tau_n\}$ of partitions $\tau_n = \{0 = t_0 < t_{n,1} < \cdots < t_{n,n} = T\}$ of $[0,T]$, with mesh size $|\tau_n| = \max_i |t_{n,i} - t_{n,i-1}|$ going to 0, we set

$$V_p(g; \tau) = \lim_{n \to \infty} \sum_{i=1}^n |g_{t_{n,i}} - g_{t_{n,i-1}}|^p \tag{15}$$

whenever it exists.

In [16], Rogers and Walsh show that for Brownian motion $V_{4/3}(\gamma';\tau)$ is a finite non-zero constant, independent of τ. Let β' denote the usual conjugate exponent to β, i.e. $\frac{1}{\beta} + \frac{1}{\beta'} = 1$.

Theorem 2. *Let X be a symmetric stable process of order $\beta > 3/2$ in R^1. Then $V_p(\gamma';\tau) = 0$ for any τ and any $p > \frac{2\beta'}{3}$.*

Note that for Brownian motion this shows that $V_p(\gamma';\tau) = 0$ for any $p > 4/3$. We conjecture that for the symmetric stable process of order $\beta > 3/2$ in R^1 we have that $V_{\frac{2\beta'}{3}}(\gamma';\tau)$ is a finite non-zero constant, independent of τ.

We now obtain an intrinsic characterization for γ'_t, which doesn't involve limits. The key idea is that γ'_t has zero quadratic variation. In the case of renormalized intersection local time γ_t for Brownian motion in the plane this was observed by Bertoin, [2] and extended by us in [21].

Recall that a continuous adapted process Z_t is said to have zero quadratic variation, if for each $T > 0$ and any sequence of partitions $\tau_n = \{0 = t_0 < t_1 < \cdots < t_n = T\}$ of $[0,T]$, with mesh size $|\tau_n| = \max_i |t_i - t_{i-1}|$ going to 0

$$\lim_{n\to\infty} E\left(\sum_{t_i \in \tau_n} (Z_{t_i} - Z_{t_{i-1}})^2 \right) = 0. \tag{16}$$

Föllmer [7] has coined the term "Dirichlet process" to refer to any process which can be written as the sum of a martingale and a process of zero quadratic variation. It is important to note that such a decomposition is unique. The class of Dirichlet processes is much wider than the class of semimartingales.

We use Y_t to denote our stable process X_t killed at an independent exponential time λ. In the following theorem γ'_t will be defined for the process Y_t in place of X_t.

Let us begin with a special case of the Doob–Meyer decomposition for semimartingales. Let L^μ_t denotes the continuous additive functional of X_t with Revuz measure μ. Using the additivity of L^μ_t and the Markov property we have

$$E^x(L^\mu_\lambda \,|\, \mathcal{F}_t) = L^\mu_{t\wedge\lambda} + U^1\mu(Y_t) \tag{17}$$

where $\mathcal{F}_t = \sigma(Y_s, \, s \leqslant t)$. Equivalently, $U^1\mu(Y_t) = M_t - L^\mu_{t\wedge\lambda}$ where $M_t = E^x(L^\mu_\lambda \,|\, \mathcal{F}_t)$ is a martingale. This is the Doob–Meyer decomposition for the potential $U^1\mu(Y_t)$. We will show that γ'_t arises in a similar decomposition for the potential of a random Schwartz distribution. This new potential will no longer be a semimartingale but a Dirichlet process, and γ'_t will correspond to the process of zero quadratic variation in the decomposition of this Dirichlet process.

Let F_t be the random additive distribution-valued process defined by

$$F_t(g) = -\lim_{\epsilon\to 0} \int_0^t \int g(y + Y_r) f'_\epsilon(y) \, dy \, dr \tag{18}$$

whenever the limit exists. It follows by integration by parts that for all $g \in C_0^\infty(R^1)$

$$F_t(g) = \int_0^t g'(Y_r) \, dr. \tag{19}$$

With this definition it is natural to set

$$U^1 F_t(x) = -\lim_{\epsilon \to 0} \int_0^t \int u^1(x - (y + Y_r)) f_\epsilon'(y) \, dy \, dr \tag{20}$$

whenever the limit exists.

Note that formally

$$U^1 F_t(x) = -\int_0^t \frac{du^1}{dx}(x - Y_s) \, ds. \tag{21}$$

We have the following analogue of the Doob–Meyer decomposition.

Theorem 3. *Let Y be a symmetric stable process of order $\beta > 3/2$ in R^1, killed at an independent exponential time λ. Then γ_t' is continuous a.s. with zero quadratic variation and*

$$U^1 F_t(Y_t) = M_t - \gamma_t' \tag{22}$$

where M_t is the martingale $E^x(\gamma_\infty' \mid \mathcal{F}_t)$.

In view of Theorem 3 we can characterize the renormalized intersection local time γ_t' as the continuous process of zero quadratic variation in the decomposition of the random potential $U^1 F_t(Y_t)$ which is formally $-\int_0^t \frac{du^1}{dx}(Y_t - Y_s) \, ds$.

1.1 Fractional derivatives

All our results can be extended to fractional derivatives. There are in fact several natural candidates for the fractional derivative of order $0 < \rho < 1$ in R^d: $g^{(\rho)}(x) = (2\pi)^{-d} \int e^{ipx} w(p) \hat{g}(p) \, dp$ with $w(p)$ positively homogeneous of index ρ, i.e. $w(\lambda p) = |\lambda|^\rho w(p)$ for all $\lambda > 0$. Our results can be extended for any such $w(p)$, but for simplicity we work with symmetric fractional derivatives: $w(-p) = -w(p)$ which allows us to avoid introducing counterterms. In one dimension this determines $w(p)$ up to a constant factor: $w(p) = \mathrm{sgn}(p)|p|^\rho$. Then we study

$$\gamma_t^{(\rho)} = -\int_0^t \int_0^s \delta^{(\rho)}(X_s - X_r) \, dr \, ds. \tag{23}$$

More precisely, let

$$\alpha_{t,\epsilon}^{(\rho)}(y) \stackrel{\text{def}}{=} -\int_0^t \int_0^s f_\epsilon^{(\rho)}(X_s - X_r - y) \, dr \, ds, \tag{24}$$

$$\alpha_t^{(\rho)}(y) = \lim_{\epsilon \to 0} \alpha_{t,\epsilon}^{(\rho)}(y) \tag{25}$$

and

$$\gamma_t^{(\rho)}(y) = \lim_{\epsilon \to 0} \left(\alpha_{t,\epsilon}^{(\rho)}(y) - E(\alpha_{t,\epsilon}^{(\rho)}(y)) \right) \tag{26}$$

whenever the limit exists. We set $\gamma_t^{(\rho)} = \gamma_t^{(\rho)}(0)$.

Theorem 4. *Let X_t denote the symmetric stable process of order $\beta > 1 \vee (\rho + 1/2)$ in R^1. Then $\alpha_{t,\epsilon}^{(\rho)}(y)$ converges a.s. and in all L^p spaces as $\epsilon \to 0$ for any $(t,y) \in R_+ \times R^1$.*
  The following hold almost surely:

1. *For any C^ρ function $g(x)$ we have*

$$\int_0^t \int_0^s g^{(\rho)}(X_s - X_r) \, dr \, ds = -\int g(y) \alpha_t^{(\rho)}(y) \, dy. \tag{27}$$

2. *$\alpha_t^{(\rho)}(y) - h_{\beta+1-\rho}(y)t$ is continuous in y.*
3. *$\{\gamma_t^{(\rho)}(y), \, (t,y) \in R_+ \times R^1\}$ is a.s. continuous and $\gamma_t^{(\rho)}(y)$ is the derivative of order ρ in y of $\gamma_t(y)$.*
4. *$\alpha_t^{(\rho)}(0) = \gamma_t^{(\rho)}(0)$.*
5. *$V_p(\gamma^{(\rho)}; \tau) = 0$ for any τ and any $p > \frac{2\beta}{3\beta-1-2\rho}$.*

Once again, we use Y_t to denote our R^1 valued Lévy process X_t killed at an independent exponential time λ and in the following theorem $\gamma_t^{(\rho)}$ will be defined for the process Y_t in place of X_t. Let now Φ_t be the random additive distribution-valued process defined by

$$\Phi_t(g) = -\lim_{\epsilon \to 0} \int_0^t \int g(y + Y_r) f_\epsilon^{(\rho)}(y) \, dy \, dr \tag{28}$$

whenever the limit exists. It is easy to check that for all $g \in C_0^\infty(R^1)$

$$\Phi_t(g) = \int_0^t g^{(\rho)}(Y_r) \, dr. \tag{29}$$

With this definition it is natural to set

$$U^1 \Phi_t(x) = -\lim_{\epsilon \to 0} \int_0^t \int u^1 \left(x - (y + Y_r) \right) f_\epsilon^{(\rho)}(y) \, dy \, dr \tag{30}$$

whenever the limit exists.
  Note that formally

$$U^1 \Phi_t(x) = -\int_0^t (u^1)^{(\rho)}(x - Y_r) \, dr. \tag{31}$$

We have the following analogue of the Doob–Meyer decomposition.

Theorem 5. *Let Y be a symmetric stable process of order $\beta > \rho + 1/2$ in R^1, killed at an independent exponential time λ. Then $\gamma_t^{(\rho)}$ is continuous a.s. with zero quadratic variation and*

$$U^1 \Phi_t(Y_t) = M_t - \gamma_t^{(\rho)} \tag{32}$$

where M_t is the martingale $E^x(\gamma_\lambda^{(\rho)} \mid \mathcal{F}_t)$.

We leave to the interested reader the task of formulating analogous results for fractional derivatives of renormalized self-intersection local times in the plane.

This paper is organized as follows. In section 2 we prove Theorem 1 and in section 3 prove Theorem 2. Section 4 contains the short proof of Theorem 3. The proofs of Theorems 4 and 5 are similar and left to the reader.

Acknowledgement. I would like to thank J. Walsh for some very helpful conversations.

2 Existence of $\alpha_t'(x)$

Proof of Theorem 1. For any $x \in R^1$ and bounded Borel set $B \subseteq R_+^2$ let

$$\alpha_\epsilon'(x, B) = -\int_B \int f_\epsilon'(X_s - X_r - x) \, dr \, ds. \tag{33}$$

$\alpha_\epsilon'(x, B)$ is clearly continuous in all parameters as long as $\epsilon > 0$. We use $|B|$ to denote the Lebesgue measure of $B \subseteq R_+^2$. For any random variable Y we set $\{Y\}_0 = Y - E(Y)$.

The following Lemma will be proven at the end of this section.

Lemma 1. *Let X be the symmetric stable process of index $\beta > 3/2$ in R^1. Then for some $\zeta > 0$*

$$\left| E\left(\{\alpha_\epsilon'(x, B) - \alpha_{\epsilon'}'(x', B)\}_0^j\right) \right| \leqslant c_0(\zeta, j) \, |(\epsilon, x) - (\epsilon', x')|^{j\zeta} \tag{34}$$

and

$$\left| E\left(\{\alpha_\epsilon'(x, B)\}_0^j\right) \right| \leqslant c_0(\zeta, j) \, |B|^{j\zeta} \tag{35}$$

for all $j \in Z_+$, $\epsilon, \epsilon' \in (0, 1]$, $x, x' \in R^1$ and all Borel sets $B \subseteq A_1^1 =: [0, 1/2] \times [1/2, 1]$.

Let

$$A_k^n = [(2k - 2)2^{-n}, (2k - 1)2^{-n}] \times [(2k - 1)2^{-n}, (2k)2^{-n}]. \tag{36}$$

Using the scaling $X_{\lambda t} \overset{\mathcal{L}}{=} \lambda^{1/\beta} X_t$ and $f_{\lambda\epsilon}'(x) = \frac{1}{\lambda^2} f_\epsilon'(x/\lambda)$ we have

$$\alpha'_\epsilon(x, B) \overset{\mathcal{L}}{=} 2^{-n(2-2/\beta)} \alpha'_{2^n/\beta_\epsilon}(2^{n/\beta} x, 2^n B) \tag{37}$$

so that from (34) and (35) we have that for all Borel sets $B \subseteq A_k^{n+1}$

$$\left| E\big(\{\alpha'_\epsilon(x, B) - \alpha'_{\epsilon'}(x', B)\}_0^j\big) \right| \leqslant c_0(\zeta, j) 2^{-nj(2-2/\beta-\zeta/\beta)} |(\epsilon, x) - (\epsilon', x')|^{\zeta j} \tag{38}$$

and

$$\left| E\big(\{\alpha'_\epsilon(x, B)\}_0^j\big) \right| \leqslant c_0(\zeta, j) 2^{-nj(2-2/\beta-2\zeta)} |B|^{\zeta j}. \tag{39}$$

For any B let

$$\gamma'_\epsilon(x, B) = \{\alpha'_\epsilon(x, B)\}_0. \tag{40}$$

Following Le Gall [12] we write

$$\gamma'_\epsilon(x, B) = \sum_{n,k} \gamma'_\epsilon(x, B \cap A_k^n) \tag{41}$$

for any $B \subseteq \{0 \leqslant r \leqslant s \leqslant 1\}$. Using (38) and (39) together with independence we then have, (see Prop. 3.5.2 of [8]),

$$\left| E\left(\left\{\sum_{k=1}^{2^n} \gamma'_\epsilon(x, B \cap A_k^{n+1}) - \gamma'_{\epsilon'}(x', B \cap A_k^{n+1})\right\}^j\right) \right| \tag{42}$$

$$\leqslant c_0(\zeta, j) 2^{nj/2} 2^{-nj(2-2/\beta-\zeta/\beta)} |(\epsilon, x) - (\epsilon', x')|^{\zeta j},$$

and

$$\left| E\left(\left\{\sum_{k=1}^{2^n} \gamma'_\epsilon(x, B \cap A_k^{n+1})\right\}^j\right) \right| \leqslant c_0(\zeta, j) 2^{nj/2} 2^{-nj(2-2/\beta-2\zeta)} |B|^{\zeta j} \tag{43}$$

so that

$$\|\gamma'_\epsilon(x, B) - \gamma'_{\epsilon'}(x', B)\|_j \leqslant c |(\epsilon, x) - (\epsilon', x')|^\zeta \tag{44}$$

and

$$\|\gamma'_\epsilon(x, B)\|_j \leqslant c |B|^\zeta \tag{45}$$

if we choose $\zeta > 0$ so that $1/2 - 2 + 2/\beta + \zeta(2 + 1/\beta) < 0$. This is possible for $\beta > 4/3$ (and we are assuming that $\beta > 3/2$).

Let $B_t = \{0 \leqslant r \leqslant s \leqslant t\}$ and set $\gamma'_{\epsilon,t}(x) =: \gamma'_\epsilon(x, B_t)$. If $t, t' \leqslant M < \infty$ then $|B_t - B_{t'}| \leqslant M|t - t'|$ so that by (45) for some $c < \infty$

$$\|\gamma'_{\epsilon,t}(x) - \gamma'_{\epsilon,t'}(x)\|_j \leqslant c |t - t'|^\zeta \tag{46}$$

and combined with (44) this shows that for some $\zeta > 0$

$$\|\gamma'_{\epsilon,t}(x) - \gamma'_{\epsilon',t'}(x')\|_j \leqslant c |(\epsilon, x, t) - (\epsilon', x', t')|^\zeta. \tag{47}$$

Kolmogorov's lemma then shows that locally

$$|\gamma'_{\epsilon,t}(x) - \gamma'_{\epsilon',t'}(x')| \leqslant c_\omega \, |(\epsilon, x, t) - (\epsilon', x', t')|^{\zeta'}, \qquad \epsilon, \epsilon' > 0 \qquad (48)$$

for some $\zeta' > 0$, which assures us of a locally uniform and hence continuous limit

$$\gamma'_t(x) = \lim_{\epsilon \to 0} \gamma'_{\epsilon,t}(x). \qquad (49)$$

The next Lemma is proven in section 5.

Lemma 2. *For the symmetric stable process of index $\beta > 1$ we can find a continuous function $h(x)$ with $h(0) = 0$ such that for each x*

$$\lim_{\epsilon \to 0} E\big(\alpha'_{\epsilon,t}(x)\big) - h(x) = \begin{cases} c(\beta) \operatorname{sgn}(x)|x|^{\beta-2} t & \text{if } x \neq 0 \\ 0 & \text{if } x = 0 \end{cases} \qquad (50)$$

where $c(\beta) = -2\Gamma(2 - \beta)\cos((1 - \beta)\pi/2)$ if $1 < \beta < 2$ and $c(2) = -1$. Furthermore, (50) converges locally uniformly in x away from 0 and locally in L^1.

Using this Lemma and the locally uniform convergence (49) we see that

$$\alpha'_t(x) = \lim_{\epsilon \to 0} \alpha'_{\epsilon,t}(x) \qquad (51)$$

exists for all x, t, and $\alpha'_t(0) = \gamma'_t(0)$. Furthermore the convergence is locally uniform in x away from 0 and locally in L^1.

A similar and simpler analysis shows that locally

$$|\gamma_{\epsilon,t}(x) - \gamma_{\epsilon',t'}(x')| \leqslant c_\omega \, |(\epsilon, x, t) - (\epsilon', x', t')|^{\zeta'}, \qquad \epsilon, \epsilon' > 0 \qquad (52)$$

for some $\zeta' > 0$, which assures us of a locally uniform and hence continuous limit

$$\gamma_t(x) = \lim_{\epsilon \to 0} \gamma_{\epsilon,t}(x). \qquad (53)$$

Using the bound $p_s(y) \leqslant c/s^{1/\beta}$ we can check that $\int_0^t \int_0^s p_{s-r}(y)\,dr\,ds$ is bounded and continuous in y and that uniformly in x

$$\lim_{\epsilon \to 0} E\big(\alpha_{\epsilon,t}(x)\big) = \lim_{\epsilon \to 0} \int_0^t \int_0^s \left(\int f_\epsilon(y - x) p_{s-r}(y)\,dy \right) dr\,ds \qquad (54)$$

$$= \lim_{\epsilon \to 0} \int f_\epsilon(y - x) \left(\int_0^t \int_0^s p_{s-r}(y)\,dr\,ds \right) dy$$

$$= \int_0^t \int_0^s p_{s-r}(x)\,dr\,ds$$

Together with the locally uniform convergence of (53) we see that

$$\alpha_t(x) = \lim_{\epsilon \to 0} \alpha_{\epsilon,t}(x). \qquad (55)$$

locally uniformly.

Now let $g(x)$ be a C^1 function with compact support. The locally L^1 convergence in (51) and the fact that $\{X_s \,;\, 0 \leqslant s \leqslant t\}$ is bounded a.s. shows that

$$
\int g(x)\alpha_t'(x)\,\mathrm{d}x = \lim_{\epsilon \to 0} \int g(x)\alpha_{\epsilon,t}'(x)\,\mathrm{d}x \tag{56}
$$

$$
= -\lim_{\epsilon \to 0} \int g(x)\left(\int_0^t \int_0^s f_\epsilon'(X_s - X_r - x)\,\mathrm{d}r\,\mathrm{d}s\right)\mathrm{d}x
$$

$$
= -\lim_{\epsilon \to 0} \int_0^t \int_0^s \left(\int g'(x)f_\epsilon(X_s - X_r - x)\,\mathrm{d}x\right)\mathrm{d}r\,\mathrm{d}s
$$

$$
= -\lim_{\epsilon \to 0} \int_0^t \int_0^s f_\epsilon * g'(X_s - X_r)\,\mathrm{d}r\,\mathrm{d}s
$$

$$
= -\int_0^t \int_0^s g'(X_s - X_r)\,\mathrm{d}r\,\mathrm{d}s.
$$

Since the path $\{X_s \,;\, 0 \leqslant s \leqslant t\}$ is bounded a.s. we have that $\alpha_t'(x)$ has compact support a.s. so that (56) holds for all C^1 functions $g(x)$.

Similarly, using the locally uniform convergence (55) we see that for any continuous function $h(x)$ we have

$$
\int h(x)\alpha_t(x)\,\mathrm{d}x = \int_0^t \int_0^s h(X_s - X_r)\,\mathrm{d}r\,\mathrm{d}s. \tag{57}
$$

Therefore

$$
\int g(x)\alpha_t'(x)\,\mathrm{d}x = -\int g'(x)\alpha_t(x)\,\mathrm{d}x \tag{58}
$$

holds for all C^1 functions $g(x)$.

It is clear that $\frac{\mathrm{d}}{\mathrm{d}x}\gamma_{\epsilon,t}(x) = \gamma_{\epsilon,t}'(x)$ for any $\epsilon > 0$ and hence

$$
\gamma_{\epsilon,t}(x) = \gamma_{\epsilon,t}(y) + \int_y^x \gamma_{\epsilon,t}'(z)\,\mathrm{d}z. \tag{59}
$$

The locally uniform convergence shown above then implies that

$$
\gamma_t(x) = \gamma_t(y) + \int_y^x \gamma_t'(z)\,\mathrm{d}z \tag{60}
$$

and therefore $\frac{\mathrm{d}}{\mathrm{d}x}\gamma_t(x) = \gamma_t'(x)$. This completes the proof of Theorem 1.

Proof of Lemma 1. We begin by showing how to find a bound

$$
\left| E\left(\left(\alpha_\epsilon'(x, B)\right)^j\right)\right| \leqslant c^j \tag{61}
$$

uniform in $\epsilon \in (0, 1]$, $x \in R^1$ and $B \subseteq A_1^1$.

We use

$$f'_\epsilon(x) = \frac{i}{2\pi} \int e^{ipx} p \widehat{f}(\epsilon p) \, dp \tag{62}$$

and independence to write that

$$E\left((\alpha'_\epsilon(x, B))^j\right) \tag{63}$$

$$= \frac{1}{(2\pi i)^j} \iint_{B^j} \exp\left(-ix \sum_{k=1}^{j} p_k\right) E\left(\exp\left(\sum_{k=1}^{j} i p_k (X_{s_k} - X_{r_k})\right)\right)$$

$$\prod_{k=1}^{j} p_k \widehat{f}(\epsilon p_k) \, ds_k \, dr_k \, dp_k$$

$$= \frac{1}{(2\pi i)^j} \iint_{B^j} \exp\left(-ix \sum_{k=1}^{j} p_k\right) E\left(\exp\left(\sum_{k=1}^{j} i p_k (X_{1/2} - X_{r_k})\right)\right)$$

$$E\left(\exp\left(\sum_{k=1}^{j} i p_k (X_{s_k} - X_{1/2})\right)\right) \prod_{k=1}^{j} p_k \widehat{f}(\epsilon p_k) \, ds_k \, dr_k \, dp_k.$$

We write

$$\sum_{k=1}^{j} p_k (X_{1/2} - X_{r_k}) = \sum_{k=1}^{j} v_k (X_{t_{k+1}} - X_{t_k}) \tag{64}$$

where the $t_1, \ldots, t_j$ are the r_i's relabeled so that $t_1 \leqslant t_2 \leqslant \cdots \leqslant t_j \leqslant t_{j+1} \stackrel{\text{def}}{=} 1/2$ and $v_i = \sum_{l:r_l \leqslant t_i} p_l$ so that the v_i's span R^j. Similarly we rewrite

$$\sum_{k=1}^{j} p_k (X_{s_k} - X_{1/2}) = \sum_{k=1}^{j} v'_k (X_{t'_k} - X_{t'_{k-1}}) \tag{65}$$

with $t'_0 \stackrel{\text{def}}{=} 1/2$. Then using (63) and independence we have

$$E\left((\alpha'_\epsilon(x, B))^j\right) \tag{66}$$

$$= \frac{1}{(2\pi i)^j} \iint_{B^j} \exp\left(-ix \sum_{k=1}^{j} p_k\right) \exp\left(-\sum_{k=1}^{j} |v_k|^\beta (t'_{k+1} - t'_k)\right)$$

$$\exp\left(-\sum_{k=1}^{j} |v'_k|^\beta (t_k - t_{k-1})\right) \prod_{k=1}^{j} p_k \widehat{f}(\epsilon p_k) \, ds_k \, dr_k \, dp_k.$$

Using this and the simple bound

$$\int_0^1 e^{-t|v|^\beta} \, dt \leqslant \frac{c}{1 + |v|^\beta} \tag{67}$$

we have the uniform bound

$$\left| E\left((\alpha'_\epsilon(x, B))^j\right)\right| \leqslant c^j \int \prod \frac{1}{1 + |v_k|^\beta} \prod \frac{1}{1 + |v'_k|^\beta} \prod_{k=1}^j |p_k| \, dp_k \quad (68)$$

$$\leqslant c^j \left\| \prod \frac{|p_k|^{1/2}}{1 + |v_k|^\beta} \right\|_2 \left\| \prod \frac{|p_k|^{1/2}}{1 + |v'_k|^\beta} \right\|_2.$$

Since $v_i = \sum_{l:r_l \leqslant t_i} p_l$, we see that each p_k can be represented as the difference $p_k = v_i - v_{i-1}$ for some i, and each v_i appears in the representation of at most two p_k's. Thus

$$\left\| \prod \frac{|p_k|^{1/2}}{1 + |v_k|^\beta} \right\|_2^2 = \int \prod \frac{|p_k|}{1 + |v_k|^{2\beta}} \, dp_k \quad (69)$$

$$\leqslant c^j \int \prod \frac{1 + |v_k| + |v_k|^2}{1 + |v_k|^{2\beta}} \, dv_k$$

which is bounded if $\beta > 3/2$.

We can now establish (34). To handle the variation in x we replace (66) by

$$E\left((\alpha'_\epsilon(x, B) - \alpha'_\epsilon(x', B))^j\right) \quad (70)$$

$$= \frac{1}{(2\pi i)^j} \iint_{B^j} \left(\prod_{k=1}^j \{\exp(-i\, p_k x) - \exp(-i\, p_k x')\} \right)$$

$$\exp\left(-\sum_{k=1}^j |v_k|^\beta (t'_{k+1} - t'_k) \right) \exp\left(-\sum_{k=1}^j |v'_k|^\beta (t_k - t_{k-1}) \right)$$

$$\prod_{k=1}^j p_k \widehat{f}(\epsilon p_k) \, ds_k \, dr_k \, dp_k$$

and use the bound

$$|\exp(-i\, p_k x) - \exp(-i\, p_k x')| \leqslant C |p_k|^\zeta |x - x'|^\zeta \quad (71)$$

for any $0 \leqslant \zeta \leqslant 1$.

Similarly to handle the variation in ϵ we replace (66) by

$$E\left((\alpha'_\epsilon(x, B) - \alpha'_{\epsilon'}(x, B))^j\right) \quad (72)$$

$$= \frac{1}{(2\pi i)^j} \iint_{B^j} \exp\left(-i\, x \sum_{k=1}^j p_k \right) \exp\left(-\sum_{k=1}^j |v_k|^\beta (t'_{k+1} - t'_k) \right)$$

$$\exp\left(-\sum_{k=1}^j |v'_k|^\beta (t_k - t_{k-1}) \right) \prod_{k=1}^j p_k \left(\widehat{f}(\epsilon p_k) - \widehat{f}(\epsilon' p_k) \right) ds_k \, dr_k \, dp_k$$

and use the bound

$$|\widehat{f}(\epsilon p_k) - \widehat{f}(\epsilon' p_k)| \leqslant C|p_k|^\varsigma |\epsilon - \epsilon'|^\varsigma \tag{73}$$

for any $0 \leqslant \varsigma \leqslant 1$. Since (70)-(72) hold also when $j = 1$, we obtain (34). To prove (35) we first apply Holder's inequality to (66):

$$E\Big((\alpha'_\epsilon(x, B))^j\Big) \tag{74}$$

$$= \frac{1}{(2\pi i)^j} \iint_{B^j} \exp\Big(-ix\sum_{k=1}^{j} p_k\Big) \exp\Big(-\sum_{k=1}^{j} |v_k|^\beta (t'_{k+1} - t'_k)\Big)$$

$$\exp\Big(-\sum_{k=1}^{j} |v'_k|^\beta (t_k - t_{k-1})\Big) \prod_{k=1}^{j} p_k \widehat{f}(\epsilon p_k) \, ds_k \, dr_k \, dp_k$$

$$\leqslant c^j |B|^{j/a} \int \Big(\int_{B^j} \exp\Big(-a'\sum_{k=1}^{j} |v_k|^\beta (t'_{k+1} - t'_k)\Big)$$

$$\exp\Big(-a'\sum_{k=1}^{j} |v'_k|^\beta (t_k - t_{k-1})\Big) \prod_{k=1}^{j} ds_k \, dr_k\Big)^{1/a'} \prod_{k=1}^{j} |p_k| \, dp_k$$

for any $1/a + 1/a' = 1$. The last integral can be bounded as before if a' is chosen close to 1. As in the proof of (34), this completes the proof of (35) and therefore of Lemma 1.

3 p-variation of γ'_t

Proof of Theorem 2. Since we know that $\alpha'_{t,\epsilon} \to \gamma'_t$ in L^2, we have

$$E\big((\gamma'_t - \gamma'_{t'})^2\big) \tag{75}$$

$$= \lim_{\epsilon \to 0} E\big((\alpha'_{t,\epsilon} - \alpha'_{t',\epsilon})^2\big)$$

$$= \lim_{\epsilon \to 0} E\bigg(\Big(\int_{t'}^{t} \int_0^s f'_\epsilon(X_s - X_r) \, dr \, ds\Big)^2\bigg)$$

$$= \lim_{\epsilon \to 0} \iint_{\left\{\substack{0 \leqslant s_1 \leqslant s_2 \leqslant s_3 \leqslant s_4 \\ t' \leqslant s_3 \leqslant t \,;\, t' \leqslant s_4 \leqslant t}\right\}} E\Big(\exp(-ip(X_{s_2} - X_{s_1}) - i(p+q)(X_{s_3} - X_{s_2}))$$

$$\exp(-iq(X_{s_4} - X_{s_3}))\Big) \prod_{k=1}^{4} ds_k \, pq\widehat{f}(\epsilon p)\widehat{f}(\epsilon q) \, dp \, dq$$

$$+ \lim_{\epsilon \to 0} \iint_{\left\{\substack{0 \leqslant s_1 \leqslant s_2 \leqslant s_3 \leqslant s_4 \\ t' \leqslant s_3 \leqslant t \,;\, t' \leqslant s_4 \leqslant t}\right\}} E\Big(\exp(-ip(X_{s_2} - X_{s_1}) - i(p+q)(X_{s_3} - X_{s_2}))$$

$$\exp(-ip(X_{s_4} - X_{s_3}))\Big) \prod_{k=1}^{4} ds_k \, pq\widehat{f}(\epsilon p)\widehat{f}(\epsilon q) \, dp \, dq$$

$$= \lim_{\epsilon \to 0} \iint_{\left\{\substack{0 \leqslant s_1 \leqslant s_2 \leqslant s_3 \leqslant s_4 \\ t' \leqslant s_3 \leqslant t \,;\, t' \leqslant s_4 \leqslant t}\right\}} e^{-(s_2 - s_1)|p|^\beta - (s_3 - s_2)|p+q|^\beta - (s_4 - s_3)|q|^\beta}$$

$$\prod_{k=1}^{4} \mathrm{d}s_k \, pq\widehat{f}(\epsilon p)\widehat{f}(\epsilon q) \, \mathrm{d}p \, \mathrm{d}q$$

$$+ \lim_{\epsilon \to 0} \iint_{\left\{ \substack{0 \leqslant s_1 \leqslant s_2 \leqslant s_3 \leqslant s_4 \\ t' \leqslant s_3 \leqslant t \,;\, t' \leqslant s_4 \leqslant t} \right\}} \mathrm{e}^{-(s_2-s_1)|p|^\beta - (s_3-s_2)|p+q|^\beta - (s_4-s_3)|p|^\beta}$$

$$\prod_{k=1}^{4} \mathrm{d}s_k \, pq\widehat{f}(\epsilon p)\widehat{f}(\epsilon q) \, \mathrm{d}p \, \mathrm{d}q$$

Consider the first summand on the right hand side of (75). It is bounded by

$$\iint_{\left\{ \substack{t' \leqslant r_1+r_2+r_3 \leqslant t \\ t' \leqslant r_1+r_2+r_3+r_4 \leqslant t} \right\}} \mathrm{e}^{-r_2|p|^\beta - r_3|p+q|^\beta - r_4|q|^\beta} \prod_{k=1}^{4} \mathrm{d}r_k |p||q| \, \mathrm{d}p \, \mathrm{d}q \qquad (76)$$

$$\leqslant C(t-t')^{1+a} \int \frac{1}{1+|p|^\beta} \frac{1}{1+|p+q|^\beta} \frac{1}{1+|q|^{(1-a)\beta}} |p||q| \, \mathrm{d}p \, \mathrm{d}q$$

where we first integrated with respect to $\mathrm{d}r_4$ using Hölder's inequality with $0 \leqslant a \leqslant 1$

$$\int_{t'-(r_1+r_2+r_3)}^{t-(r_1+r_2+r_3)} \mathrm{e}^{-r_4|q|^\beta} \, \mathrm{d}r_4 \leqslant (t-t')^a \left(\int_0^t \mathrm{e}^{-r_4|q|^\beta/(1-a)} \, \mathrm{d}r_4 \right)^{(1-a)}, \qquad (77)$$

then with respect to $\mathrm{d}r_1$ using $\int_{t'-(r_2+r_3)}^{t-(r_2+r_3)} \mathrm{d}r_1 \leqslant t - t'$, and finally the $\mathrm{d}r_2$, $\mathrm{d}r_3$ integrals are bounded using (67). It is easily seen that (76) is bounded as long as $\beta - 1 + (1-a)\beta - 1 > 1$, i.e. $a < 2 - 3/\beta$.

Now consider the second summand on the right hand side of (75). An attempt to use a bound similar to (76) where we bound pq by $|pq|$ would be fatal. Rather, we first observe that $\int \mathrm{e}^{-(s_3-s_2)|q|^\beta} q \, \mathrm{d}q = 0$ and use this to rewrite the second summand on the right hand side of (75) as

$$\lim_{\epsilon \to 0} \iint_{\left\{ \substack{0 \leqslant s_1 \leqslant s_2 \leqslant s_3 \leqslant s_4 \\ t' \leqslant s_3 \leqslant t \,;\, t' \leqslant s_4 \leqslant t} \right\}} \mathrm{e}^{-(s_2-s_1)|p|^\beta - (s_4-s_3)|p|^\beta} \qquad (78)$$

$$\left\{ \mathrm{e}^{-(s_3-s_2)|p+q|^\beta} - \mathrm{e}^{-(s_3-s_2)|p|^\beta} \mathrm{e}^{-(s_3-s_2)|q|^\beta} \right\} \prod_{k=1}^{4} \mathrm{d}s_k \, pq\widehat{f}(\epsilon p)\widehat{f}(\epsilon q) \, \mathrm{d}p \, \mathrm{d}q.$$

Now we bound this as in (76) by

$$\iint_{\left\{ \substack{t' \leqslant r_1+r_2+r_3 \leqslant t \\ t' \leqslant r_1+r_2+r_3+r_4 \leqslant t} \right\}} \mathrm{e}^{-r_2|p|^\beta - r_4|p|^\beta} \qquad (79)$$

$$\left| \mathrm{e}^{-r_3|p+q|^\beta} - \mathrm{e}^{-r_3|p|^\beta} \mathrm{e}^{-r_3|q|^\beta} \right| \prod_{k=1}^{4} \mathrm{d}r_k |p||q| \, \mathrm{d}p \, \mathrm{d}q$$

$$\leqslant C(t-t')^{1+a} \int \frac{1}{1+|p|^{(2-a)\beta}} \left| \frac{1}{1+|p+q|^\beta} - \frac{1}{1+|p|^\beta + |q|^\beta} \right| |p||q| \, \mathrm{d}p \, \mathrm{d}q.$$

Here we proceeded as in (76) except that for the dr_3 integral we used

$$\int e^{-r}\left|e^{-r|p+q|^\beta} - e^{-r(|p|^\beta+|q|^\beta)}\right| dr \leqslant \left|\frac{1}{1+|p+q|^\beta} - \frac{1}{1+|p|^\beta+|q|^\beta}\right| \quad (80)$$

by arguing separately depending on whether or not $|p|^\beta + |q|^\beta > |p+q|^\beta$. We claim that once again the integral on the right hand side of (79) is finite whenever $\beta - 1 + (1-a)\beta - 1 > 1$, i.e. $a < 2 - 3/\beta$. This is clear in the region where $|q| \leqslant 2|p|$ since we can use the bound $\frac{1}{1+|p|^{(2-a)\beta}} \leqslant \frac{1}{1+|p|^\beta}\frac{c}{1+|q|^{(1-a)\beta}}$. If, however, $|q| > 2|p|$, we can use the bound

$$\left|\frac{1}{1+|p+q|^\beta} - \frac{1}{1+|p|^\beta+|q|^\beta}\right| = \frac{\left||p+q|^\beta - |q|^\beta - |p|^\beta\right|}{(1+|p+q|^\beta)(1+|p|^\beta+|q|^\beta)} \quad (81)$$

$$\leqslant \frac{c|p||q|^{\beta-1}+|p|^\beta}{(1+|p+q|^\beta)(1+|p|^\beta+|q|^\beta)}$$

$$\leqslant \frac{c|p||q|^{\beta-1}}{(1+|q|^\beta)^2}.$$

This allows us to bound the resulting integral from the right hand side of (79) by

$$c\int \frac{|p|^2}{1+|p|^{(2-a)\beta}} \frac{1}{(1+|q|^\beta)} \, dp \, dq \quad (82)$$

which leads to the same result as before.

Finally, for $h \leqslant 2$ we have

$$E\left((\gamma'_t - \gamma'_{t'})^h\right) \leqslant \left\{E\left((\gamma'_t - \gamma'_{t'})^2\right)\right\}^{h/2} \leqslant C(t - t')^{(1+a)h/2}. \quad (83)$$

We will have h-variation 0 when $(1 + a)h/2 > 1$ for some $a < 2 - 3/\beta$, i.e. $(3 - 3/\beta)h/2 > 1$. Thus we will have h-variation 0 when $h > (2/3)\beta'$. This completes the proof of Theorem 2.

4 The Doob–Meyer decomposition

Proof of Theorem 3. It is easy to check that the proofs of Theorems 1 and 2 go through with X replaced by Y and in that case

$$\gamma'_\infty = \lim_{\epsilon\to 0} \gamma'_{\epsilon,\infty} \quad \text{and} \quad \gamma'_t = \lim_{\epsilon\to 0} \gamma'_{\epsilon,t} \quad (84)$$

converge a.s. and in all L^p spaces. Let

$$U^1 F_{\epsilon,t}(y) = -\int_0^t \int u^1\left(x - (y - Y_r)\right) f'_\epsilon(x) \, dx \, dr. \quad (85)$$

The proof of Theorem 3 then follows from (84) and the next Lemma.

Lemma 3. *Let* $\{Y_s \, ; \, s \in R_+^1\}$ *be the exponentially killed symmetric stable process of index* $\beta > 3/2$ *in* R^1. *Then for any* $\epsilon > 0$

$$U^1 F_{\epsilon,t}(Y_t) = E^x(\gamma'_{\epsilon,\infty} \,|\, \mathcal{F}_t) - \gamma'_{\epsilon,t} \tag{86}$$

Proof of Lemma 3. We have

$$\gamma'_{\epsilon,t} = -\left\{ \int_0^t \int_0^s f'_\epsilon(Y_s - Y_r) \, dr \, ds \right\} \tag{87}$$

so that

$$\gamma'_{\epsilon,\infty} = -\left\{ \int_0^\infty \int_0^s f'_\epsilon(Y_s - Y_r) \, dr \, ds \right\} \tag{88}$$

$$= \gamma'_{\epsilon,t} - \left\{ \int_t^\infty \int_t^s f'_\epsilon(Y_s - Y_r) \, dr \, ds \right\}$$

$$- \left\{ \int_t^\infty \int_0^t f'_\epsilon(Y_s - Y_r) \, dr \, ds \right\}.$$

Now, using the fact that $\{Y_s \, ; \, s \in R_+^1\}$ has independent increments

$$E\left(\left\{ \int_t^\infty \int_t^s f'_\epsilon(Y_s - Y_r) \, dr \, ds \right\} \,\bigg|\, \mathcal{F}_t \right) \tag{89}$$

$$= E\left(\int_t^\infty \int_t^s f'_\epsilon(Y_s - Y_r) \, dr \, ds \,\bigg|\, \mathcal{F}_t \right) - E\left(\int_t^\infty \int_t^s f'_\epsilon(Y_s - Y_r) \, dr \, ds \right) = 0$$

and

$$E\left(\left\{ \int_t^\infty \int_0^t f'_\epsilon(Y_s - Y_r) \, dr \, ds \right\} \,\bigg|\, \mathcal{F}_t \right) \tag{90}$$

$$= E\left(\int_t^\infty \int_0^t f'_\epsilon(Y_s - Y_r) \, dr \, ds \,\bigg|\, \mathcal{F}_t \right) - E\left(\int_t^\infty \int_0^t f'_\epsilon(Y_s - Y_r) \, dr \, ds \right).$$

We have

$$E\left(\int_t^\infty \int_0^t f'_\epsilon(Y_s - Y_r) \, dr \, ds \,\bigg|\, \mathcal{F}_t \right) \tag{91}$$

$$= E\left(\int_t^\infty \int_0^t f'_\epsilon((Y_s - Y_t) + (Y_t - Y_r)) \, dr \, ds \,\bigg|\, \mathcal{F}_t \right)$$

$$= \bar{E}\left(\int_0^\infty \int_0^t f'_\epsilon(\bar{Y}_s + (Y_t - Y_r)) \, dr \, ds \right)$$

where $\{\bar{Y}_s \, ; \, s \in R_+^1\}$ is an independent copy of $\{Y_s \, ; \, s \in R_+^1\}$ and $\bar{E}$ denotes expectation with respect to $\{\bar{Y}_s \, ; \, s \in R_+^1\}$. Hence

$$E\left(\int_t^\infty \int_0^t f'_\epsilon(Y_s - Y_r)\,dr\,ds \;\Big|\; \mathcal{F}_t\right) = \int_0^t \int f'_\epsilon(x + Y_t - Y_r)u^1(x)\,dx\,dr \quad (92)$$

$$= \int_0^t \int u^1\big(x - (Y_t - Y_r)\big)f'_\epsilon(x)\,dx\,dr.$$

The same sort of argument shows that

$$E\left(\int_t^\infty \int_0^t f'_\epsilon(Y_s - Y_r)\,dr\,ds\right) = \int_0^t \int f'_\epsilon(x + y)u^1(x)p_r(y)\,dx\,dy\,dr \quad (93)$$

which is zero by symmetry. This concludes the proof of Lemma 3.

5 Proof of Lemma 2

Proof of Lemma 2. We have

$$E\big(\alpha'_{\epsilon,t}(x)\big) = -\int_0^t \int_0^s \left(\int f'_\epsilon(y - x)p_{s-r}(y)\,dy\right)dr\,ds \quad (94)$$

We first consider the case $1 < \beta < 2$. Using the fact that f'_ϵ is an odd fuction, Plancherel and then Fubini

$$-\int_0^t \int_0^s \left(\int f'_\epsilon(y - x)p_{s-r}(y)\,dy\right)dr\,ds \quad (95)$$

$$= i(2\pi)^{-1}\int_0^t \int_0^s \left(\int e^{ipx}p\widehat{f}(\epsilon p)e^{-(s-r)|p|^\beta}\,dp\right)dr\,ds$$

$$= i(2\pi)^{-1}\int e^{ipx}p\widehat{f}(\epsilon p)\left(\int_0^t \int_0^s e^{-(s-r)|p|^\beta}\,dr\,ds\right)dp$$

and

$$\int_0^t \int_0^s e^{-(s-r)|p|^\beta}\,dr\,ds \quad (96)$$

$$= \int_0^t \int_r^t e^{-(s-r)|p|^\beta}\,ds\,dr = \int_0^t \left(\int_0^{t-r} e^{-s|p|^\beta}\,ds\right)dr$$

$$= \int_0^t \int_0^r e^{-s|p|^\beta}\,ds\,dr = t\int_0^\infty e^{-s|p|^\beta}\,ds - \int_0^t \int_r^\infty e^{-s|p|^\beta}\,ds\,dr$$

$$= t|p|^{-\beta} - \int_0^t e^{-r|p|^\beta}|p|^{-\beta}\,dr = t|p|^{-\beta} - (1 - e^{-t|p|^\beta})|p|^{-2\beta}.$$

Hence

$$E\big(\alpha'_{\epsilon,t}(x)\big) = i(2\pi)^{-1}t\int e^{ipx}\,\mathrm{sgn}(p)|p|^{1-\beta}\widehat{f}(\epsilon p)\,dp \quad (97)$$

$$-i(2\pi)^{-1}\int e^{ipx}\,\mathrm{sgn}(p)(1 - e^{-t|p|^\beta})|p|^{1-2\beta}\widehat{f}(\epsilon p)\,dp.$$

It is easily checked that $(1 - e^{-t|p|^\beta})|p|^{1-2\beta} \in L^1(R^1)$ so that the last term converges uniformly as $\epsilon \to 0$ to a continuous limit. On the other hand, it follows from [9, formula (13), page 173] that

$$i(2\pi)^{-1} \int e^{ipx} \operatorname{sgn}(p)|p|^{1-\beta} \widehat{f}(\epsilon p) \, dp \tag{98}$$

$$= -\pi^{-1}\Gamma(2-\beta)\cos\big((1-\beta)\pi/2\big) \int \operatorname{sgn}(y)|y|^{\beta-2} f_\epsilon(y-x) \, dy.$$

Since $\operatorname{sgn}(y)|y|^{\beta-2}$ is locally in L^1 and continuous away from 0, this last term converges locally uniformly away from 0 and locally in L^1 to $c(\beta)\operatorname{sgn}(x)|x|^{\beta-2}$. On the other hand, when $x = 0$, (94) is 0 by symmetry. This completes the proof of Lemma 2 for $\beta \neq 2$.

For Brownian motion, we proceed differently. We first write

$$-\int_0^t \int_0^s \left(\int f_\epsilon'(y-x) p_{s-r}(y) \, dy \right) dr \, ds \tag{99}$$

$$= \int_0^t \int_0^s \left(\int f_\epsilon(y-x) p_{s-r}'(y) \, dy \right) dr \, ds.$$

We have $p_s'(y) = \frac{-1}{(2\pi)^{1/2}s^{3/2}} y e^{-y^2/2s}$ so that, for $y \neq 0$, $\{|p_s'(y)| ; s \geq 0\}$ is the density of $T_{|y|}$, the first hitting time of $|y|$ for Brownian motion. Hence

$$\int_0^\infty |p_s'(y)| \, ds = 1, \qquad y \neq 0. \tag{100}$$

This justifies the use of Fubini

$$\int_0^t \int_0^s \left(\int f_\epsilon(y-x) p_{s-r}'(y) \, dy \right) dr \, ds \tag{101}$$

$$= \int f_\epsilon(y-x) \left(\int_0^t \int_0^s p_{s-r}'(y) \, dr \, ds \right) dy$$

and then, just as in (96)

$$\int_0^t \int_0^s p_{s-r}'(y) \, dr \, ds = t \int_0^\infty p_s'(y) \, ds - \int_0^t \int_r^\infty p_s'(y) \, ds \, dr, \qquad y \neq 0. \tag{102}$$

Just as in (100) we see that the first integral is $-\operatorname{sgn}(y)$ for $y \neq 0$. Now $|p_s'(y)| \leq \frac{1}{(2\pi)^{1/2}s^{3/2}} |y|$, so that by the dominated convergence theorem the second integral is continuous in y. As before, when $y = 0$, the left hand side of (99) is 0 by symmetry. Lemma 2 now follows.

References

1. R. Bass and D. Khoshnevisan, *Intersection local times and Tanaka formulas*, Ann. Inst. H. Poincaré Prob. Stat. **29** (1993), 419–452.
2. J. Bertoin, *Applications des processus de Dirichlet aux temps locaux et temps locaux d'intersection d'un mouvement Brownien*, Prob. Theory Related Fields **80** (1989), 433–460.
3. R. Blumenthal and R. Getoor, *Some theorems on stable processes*, Trans. Amer. Math. Soc. **95** (1960), 261–273.
4. E. B. Dynkin, *Self-intersection gauge for random walks and for Brownian motion*, Ann. Probab. **16** (1988), 1–57.
5. P. Fitzsimmons and T. Salisbury, *Capacity and energy for multiparameter Markov processes*, Ann. Inst. H. Poincaré Prob. Stat. **25** (1989), 325–350.
6. P. Fitzsimmons and R. Getoor, *Limit theorems and variation properties for fractional derivatives of the local time of a stable process*, Ann. Inst. H. Poincaré Prob. Stat. **28** (1992), 311–333.
7. H. Föllmer, *Dirichlet processes*, Proceedings Durham 1980, Lecture Notes Math, vol. 851, Springer-Verlag, Berlin, 1981, 476–478.
8. A. Garsia, *Topics in Almost Everywhere Convergence*, Markham Publishing Company, Chicago (1970).
9. I. M. Gelfand and G. E. Shilov, *Generalized Functions*, Vol. 1, Academic Press, N. Y. (1964).
10. J.-F. Le Gall, *Propriétés d'intersection des marches aléatoires, I*, Comm. Math. Phys. **104** (1986), 471–507.
11. _____, *Fluctuation results for the Wiener sausage*, Ann. Probab. **16** (1988), 991–1018.
12. _____, *Some properties of planar Brownian motion*, École d'Été de Probabilités de St. Flour XX, 1990, Lecture Notes in Mathematics, vol. 1527, Springer-Verlag, Berlin, 1992.
13. J.-F. Le Gall and J. Rosen, *The range of stable random walks*, Ann. Probab. **19** (1991), 650–705.
14. L. C. G. Rogers and J. B. Walsh, *Local time and stochastic area integrals*, Ann. Probab. **19** (1991), 457–482.
15. L. C. G. Rogers and J. B. Walsh, *The intrinsic local time sheet of Brownian motion*, Prob. Theory Related Fields **80** (1991), 433–460.
16. L. C. G. Rogers and J. B. Walsh, *$A(t, B_t)$ is not a semimartingale*, Seminar on Stochastic Processes, Birkhäuser, Boston. (1991), 457–482.
17. J. Rosen, *Joint continuity and a Doob-Meyer type decomposition for renormalized intersection local times*, Ann. Inst. H. Poincaré Prob. Stat. **35** (1999), 143–176.
18. _____, *Continuity and singularity of the intersection local time of stable processes in $\mathbf{R}^2$*, Ann. Probab. **16** (1988), 75–79.
19. _____, *The asymptotics of stable sausages in the plane*, Ann. Probab. **20** (1992), 29–60.
20. _____, *Joint continuity of renormalized intersection local times*, Ann. Inst. H. Poincaré Prob. Stat. **32** (1996), 671–700.
21. _____, *Dirichlet processes and an intrinsic characterization of renormalized intersection local times*, Ann. Inst. H. Poincaré Prob. Stat. **37** (2001), 403–420.
22. S. R. S. Varadhan, *Appendix to Euclidian quantum field theory by K. Symanzyk*, Local Quantum Theory (R. Jost, ed.), Academic Press (1969).

On Squared Fractional Brownian Motions

Nathalie Eisenbaum and Ciprian A. Tudor

Laboratoire de Probabilités et Modèles Aléatoires,
UMR 7599, CNRS - Université de Paris VI
4, Place Jussieu, 75252, Paris Cedex 05, France
e-mail: nae@ccr.jussieu.fr, tudor@ccr.jussieu.fr

Summary. We have proved recently that fractional Brownian motions with Hurst parameter H in $(0, 1/2]$ satisfy a remarkable property: their squares are infinitely divisible. In the Brownian motion case (the case $H = 1/2$), this property is completely understood thanks to stochastic calculus arguments. We try here to take advantage of the stochastic calculus recently developed with respect to fractional Brownian motion, to construct analogous explanations of this property in the case $H \neq 1/2$.

1 Introduction

A process $(X_t, t \geqslant 0)$ is infinitely divisible if for any $n \in \mathbb{N}^\star$ there exist n i.i.d. processes $X^{(1)}, \ldots, X^{(n)}$ such that

$$X \stackrel{\text{(law)}}{=} X^{(1)} + \cdots + X^{(n)}.$$

The question of the infinite divisibility of squared fractional Brownian motions has been recently solved (see [4]). The answer depends on the value of the Hurst parameter H. More precisely, let $(\gamma_t, t \geqslant 0)$ be a fractional Brownian motion with parameter H in $(0, 1)$;

- if $H > 1/2$, γ^2 is not infinitely divisible;
- if $H \leqslant 1/2$, γ^2 is infinitely divisible.

The proof is based on a criterion established by Griffiths [6] and consists in computing the sign of each cofactor of the covariance matrix of $(\gamma_{t_1}, \gamma_{t_2}, \ldots, \gamma_{t_n})$ for every $(t_1, t_2, \ldots, t_n)$ in $\mathbb{R}_n^+$. More recently, the difference between the cases $H \leqslant 1/2$ and $H > 1/2$ has been clearly explained in [5]: in the first case the covariance is the Green function of a Markov process and in the second case it is not. But we think that the property of infinite divisibility should also be understandable from the point of view of stochastic calculus. More precisely, for any value of H in $(0, 1)$ we can set $Y_{1,0} = \gamma^2$ and, for any $x \geqslant 0$, $Y_{1,x} = (\gamma + \sqrt{x})^2$. More generally for any $d \in \mathbb{N}^*$ we set

$$Y_{d,0} = \sum_{i=1}^{d} (\gamma^{(i)})^2$$

where $\gamma^{(1)}, \ldots, \gamma^{(d)}$ are d independent fractional Brownian motions with Hurst parameter H.

We can similarly define for any $d \in \mathbb{N}^*$ and any $x \geqslant 0$

$$Y_{d,x} = \sum_{i=1}^{d} (\gamma^{(i)} + b_i)^2$$

for any $(b_1, b_2, \ldots b_d)$ in $\mathbb{R}^d$ such that $\sum_{i=1}^{d} b_i^2 = x$.

When $H \leqslant 1/2$, thanks to the infinite divisibility of γ^2, the existence of $Y_{d,x}$ for non integers d and $x \geqslant 0$ satisfying the following property has been shown in [4]

$$Y_{d,x} + Y_{d',x'} \overset{\text{(law)}}{=} Y_{d+d',x+x'} \tag{A}$$

for any (d, d', x, x') in $\mathbb{R}_+^4$ and for $Y_{d,x}$ and $Y_{d',x'}$ chosen independently.

The property (A) is called the additivity property. In the case $H = 1/2$, the family $(Y_{d,x})_{d,x \geqslant 0}$ was already known and called the family of the squared Bessel processes. Squared Bessel processes can also be defined as solutions of stochastic differential equations driven by a real valued Brownian motion. Indeed for a fixed couple (d, x), $Y_{d,x}$ is the unique solution of

$$X(t) = x + 2 \int_0^t \sqrt{X(s)} \, dB_s + dt \tag{1}$$

where B is real valued Brownian motion.

The additivity property of $(Y_{d,x})_{d,x \geqslant 0}$ and the infinite divisibility of each $Y_{d,x}$ can then easily be derived from that definition (see for example Revuz and Yor [10]).

Since stochastic calculus with respect to fractional Brownian motion has been extensively developed these past years, it is then natural to ask, when $H \leqslant 1/2$, whether similarly to the case $H = 1/2$, $Y_{d,x}$ could be obtained as the solution of a stochastic differential equation driven by a fractional Brownian motion with parameter H.

To start solving that question, we present here the most natural attempt. Namely we ask whether $Y_{d,x}$ is a solution of the equation

$$X(t) = x + 2 \int_0^t \sqrt{X(s)} \, dZ_s^H + dt^{2H}$$

where Z^H is a fractional Brownian motion with parameter H.

Note that the same question can be asked in the case $H > 1/2$ for d in $\mathbb{N}^*$.

We show that in any case the answer is negative. Nevertheless we think that the family $(Y_{d,x})_{d,x \geqslant 0}$ should correspond to a family of stochastic differential equations indexed by (d, x).

Besides, from our proof one can deduce that for every $d \in \mathbb{N}^*$, $d \geqslant 2$, the process $\int_0^t \sum_{i=1}^d \frac{\gamma_s^{(i)}}{\sqrt{Y_{d,0}(s)}} \, d\gamma_s^{(i)}$ is not a fractional Brownian motion. This latter question has been raised up by Guerra and Nualart in a recent preprint [7] where the process $Y_{d,0}$ is called the squared d-dimensional fractional Bessel process.

In Section 2, we recall some preliminaries on stochastic integration with respect to fractional Brownian motions. In Section 3, our result is properly stated and established.

We mention that after the completion of this work we received the paper [8] where an analogous result is obtained with a totally different proof based on chaos expansion.

2 Malliavin Calculus with respect to the fractional Brownian motion

In this section, we give only the elements of the Malliavin calculus that are essential to the proof given in Section 3. For a complete exposition one should consult the paper of Alòs et al. [1] and the book of Nualart [9]. Let T be the time interval $[0,1]$ and let $(\gamma_t^H)_{t \in T}$ the one-dimensional fractional Brownian motion with Hurst parameter $H \in (0,1)$. This means by definition that γ^H is a centered Gaussian process with covariance

$$r(t,s) = E(\gamma_s^H \gamma_t^H) = \frac{1}{2}(t^{2H} + s^{2H} - |t-s|^{2H}).$$

This process has the Wiener integral representation (see e.g. [3]) $\gamma_t^H = \int_0^t K^H(t,s) \, dW_s$, where $W = \{W_t : t \in T\}$ is a Wiener process, and $K^H(t,s)$ is a deterministic kernel (we will not need its expression). In the sequel we will work with a fixed H so we will omit to write it.

Let K^* be the operator on $L^2(T)$ defined as follows

if $H > \frac{1}{2}$, $(K^*\varphi)(s) = \int_s^1 \varphi(r) \frac{\partial K}{\partial r}(r,s) \, dr$, $\forall s \in T$

if $H < \frac{1}{2}$, $(K^*\varphi)(s) = K(1,s)\varphi(s) + \int_s^1 (\varphi(r) - \varphi(s)) \frac{\partial K}{\partial r}(r,s) \, dr$, $\forall s \in T$.

Let $\mathcal{H}$ be the canonical Hilbert space of the fractional Brownian motion. This space is actually the closure of the vector space of indicator functions $\{\mathbf{1}_{[0,t]}, t \in T\}$ with respect to the scalar product

$$\langle \mathbf{1}_{[0,t]}, \mathbf{1}_{[0,s]} \rangle_{\mathcal{H}} = r(t,s).$$

Let D^γ and δ^γ denote respectively the derivative operator and the anticipating (Skorohod) integral with respect to γ that can be defined using the space $\mathcal{H}$

(see [1] for more details). Since $\gamma \in \text{Dom}(\delta^\gamma)$ if and only if $H > 1/4$ (see [2]), an extended divergence integral was needed to include the case of any parameter $H \in (0,1)$. We refer to [2] for the construction of the extended integral. Let $\mathcal{H}'$ be the Hilbert space $\mathcal{H}' = K^{*,a}K^*(L^2(0,1))$, where $K^{*,a}$ denotes the adjoint operator of K^*. We can construct a Malliavin derivative using $\mathcal{H}'$ instead of $\mathcal{H}$. A process $u \in L^2(T \times \Omega)$ belongs to the extended domain of the divergence integral $(u \in \text{Dom}^*(\delta^\gamma))$ if there exists a random variable $\delta'^{,\gamma}(u)$ satisfying

$$\int_0^1 E(u_s K^{*,a}K^* D_s F) \, ds = E\big(\delta'^{,\gamma}(u)F\big). \tag{2}$$

for every "smooth" random variable F. This way, $\text{Dom}(\delta^\gamma) \subset \text{Dom}^*(\delta^\gamma)$ and $\delta'^{,\gamma}$ restricted to $\text{Dom}(\delta^\gamma)$ coincides with δ^γ. It has been proved in [2] that for any $H \in (0,1)$ and $f \in C^2(\mathbb{R})$ satisfying a growth condition, $f'(\gamma) \in \text{Dom}^*(\delta^\gamma)$ and the following Itô's formula holds

$$f(\gamma_t) = f(0) + \int_0^t f'(\gamma_s) \, d\gamma_s + H \int_0^t f''(\gamma_s) s^{2H-1} \, ds. \tag{3}$$

3 On the integral representation for the fractional Bessel processes

In Section 1, the process $Y_{n,0}$, that we will simply denote by Y_n, has been introduced for any $n \in \mathbb{N}^*$ and any $H \in (0,1)$ as follows

$$Y_n = \sum_{i=1}^n \left(\gamma^{(i)}\right)^2$$

where $\gamma^{(1)}, \ldots, \gamma^{(n)}$ are n independent fractional Brownian motions with Hurst parameter H.

Thanks to the Itô formula (3) we have for every i

$$\left(\gamma_t^{(i)}\right)^2 = 2 \int_0^t \gamma_s^{(i)} \, d\gamma_s^{(i)} + t^{2H}$$

which leads to

$$Y_n(t) = 2 \int_0^t \sum_{i=1}^n \gamma_s^{(i)} \, d\gamma_s^{(i)} + n t^{2H}.$$

Moreover, for $n \geqslant 2$, in [7] the authors established the following representation for the fractional Bessel processes (see also [8])

$$\sqrt{Y_n(t)} = Z_t^H + H(n-1) \int_0^t \frac{s^{2H-1}}{\sqrt{Y_n(s)}} \, ds \tag{4}$$

where the process Z^H is defined by

$$Z_t^H = \int_0^t \sum_{i=1}^n \frac{\gamma_s^{(i)}}{\sqrt{Y_n(s)}} \, d\gamma_s^{(i)}. \tag{5}$$

In view of the representation of the squared Bessel processes given by (1) in the introduction and of the relation (4), it is hence natural to ask whether Y_n satisfies

$$Y_n(t) = 2 \int_0^t \sqrt{Y_n(s)} \, dZ_s^H + nt^{2H} \tag{6}$$

where the process Z^H is given by (5). If Z^H was a semimartingale then (6) would be necessarily satisfied. Since Z^H is not a semimartingale, it makes no sense to check (6) without assuming the nature of Z^H. Therefore, we ask the following question for $n \in \mathbb{N}^*$ and $H \in (0,1)$:

Question Q. Is the following statement true: *The process Z^H given by (5) is a fractional Brownian motion and the equation (6) is satisfied?*

We prove that the above statement is true if and only if $H = 1/2$.

Proposition 1. *For any value of (n, H) in $\mathbb{N}^* \times (0,1) \setminus \{\frac{1}{2}\}$, the answer to question Q is negative.*

The case $n = 1$ is the most delicate to treat because of the appearance of the local time. Actually for the other cases we have the following result:

Proposition 2. *For every $n \geqslant 2$ and for every $H \in (0,1) \setminus \{\frac{1}{2}\}$ the process Z^H given by (5) is not a fractional Brownian motion.*

Proof of Proposition 2. Assuming that Z^H is a fractional Brownian motion we can use Itô's formula (3) to compute $\left(\sqrt{Y_n(t)}\right)^2$ thanks to (4). Then Y_n has to satisfy (6) and that leads to a contradiction by applying Proposition 1. $\square$

Proof of Proposition 1. There exist n independent real valued Brownian motions $W^{(1)}, W^{(2)}, \ldots, W^{(n)}$ such that for every i, $\gamma^{(i)}$ admits the following representation $\gamma_t^{(i)} = \int_0^t K(t,s) \, dW_s^{(i)}$.

We will need the following basic properties of $Y_n(t)$:

- The process Y_n is $2H$-self-similar. This follows immediately from the H-self-similarity of the fBm.
- Since γ_t/t^H is a standard normal random variable, $Y_n(t)/t^{2H} = \left(\gamma_t^{(1)}/t^H\right)^2 + \cdots + \left(\gamma_t^{(n)}/t^H\right)^2$ has a chi-square distribution with n-degrees of freedom which probability density function is $p_n(x) = \frac{(1/2)^{n/2}}{\Gamma(n/2)} x^{n/2-1} e^{-x/2}$, $x > 0$. As a consequence, if $m + n > 0$

$$E\left(Y_n(t)\right)^{m/2} = t^{Hm} \frac{(1/2)^{n/2}}{\Gamma(n/2)} \int_0^\infty x^{\frac{m+n}{2}-1} e^{-x/2} \, dx$$

$$= t^{Hm} \frac{2^{m/2}}{\Gamma(n/2)} \Gamma\left(\frac{m+n}{2}\right). \tag{7}$$

We will use the identity:

$$K^{*,a}K^*\mathbf{1}_{[0,t]}(s) = \frac{d}{ds}r(t,s) = H\big((t-s)^{2H-1} + s^{2H-1}\big) \tag{8}$$

which follows from (2) since

$$r(t,s) = E(\gamma_t\gamma_s) = E\left(\gamma_t\int_0^1 \mathbf{1}_{[0,s]}(\alpha)\,d\gamma_\alpha\right) = \int_0^s K^{*,a}K^*\mathbf{1}_{[0,t]}(\alpha)\,d\alpha.$$

Suppose first $n \geqslant 2$ and assume that the answer to question Q is positive. Multiplying both sides of (4) by Z_t^H we hence have

$$Z_t^H\big(Y_n(t) - nt^{2H}\big) = 2Z_t^H\int_0^t \sqrt{Y_n(s)}\,dZ_s^H.$$

On one hand, by (2),

$$E\left(2Z_t^H\int_0^t \sqrt{Y_n(s)}\,dZ_s^H\right) = 2\int_0^t E\big(\sqrt{Y_n(s)}\big)K^{*,a}K^*\mathbf{1}_{[0,t]}(s)\,ds$$

$$= 2H\,E\big(\sqrt{Y_n(1)}\big)\int_0^t r^H\big((t-r)^{2H-1} + r^{2H-1}\big)\,dr. \tag{9}$$

On the other hand, using (2) and the chain rule for the derivative operator (see [9], Prop. 1.2.3, p. 30),

$$E\big(Z_t^H\big(Y_n(t) - nt^{2H}\big)\big) = E\left(Y_n(t)\sum_{i=1}^n\int_0^t \frac{\gamma_s^{(i)}}{\sqrt{Y_n(s)}}\,d\gamma_s^{(i)}\right)$$

$$= \sum_{i=1}^n\int_0^t E\left(\frac{\gamma_s^{(i)}}{\sqrt{Y_n(s)}}K^{*,a}K^*D_s^{\gamma,i}Y_n(t)\right)ds$$

where $D^{\gamma,i}$ denotes the derivation with respect to $\gamma^{(i)}$. Since $D_s^{\gamma,i}Y_n(t) = 2\gamma_t^{(i)}\mathbf{1}_{[0,t]}(s)$, we obtain

$$E\big(Z_t^H\big(Y_n(t) - nt^{2H}\big)\big) = 2\sum_{i=1}^n\int_0^t E\left(\frac{\gamma_t^{(i)}\gamma_s^{(i)}}{\sqrt{Y_n(s)}}\right)K^{*,a}K^*\mathbf{1}_{[0,t]}(s)\,ds \tag{10}$$

Using again (2) and the chain rule for the derivative operator, we have

$$E\left(\sum_{i=1}^n \frac{\gamma_t^{(i)}\gamma_s^{(i)}}{\sqrt{Y_n(s)}}\right) = \sum_{i=1}^n E\int_0^t K^{*,a}K^*D_\alpha^{\gamma,i}\left(\frac{\gamma_s^{(i)}}{\sqrt{Y_n(s)}}\right)d\alpha$$

$$= r(t,s)\sum_{i=1}^n\left(\frac{1}{\sqrt{Y_n(s)}} - \frac{(\gamma_s^i)^2}{Y_n(s)^{3/2}}\right)$$

$$= (n-1)r(t,s)E\left(\frac{1}{\sqrt{Y_n(s)}}\right)$$

$$= r(t,s)\frac{E\big(\sqrt{Y_n(s)}\big)}{r^{2H}} = E\big(\sqrt{Y_n(1)}\big)\frac{r(t,s)}{r^H}.$$

Consequently, thanks to (10),

$$E\big(Z_t^H\big(Y_n(t) - nt^{2H}\big)\big)$$
$$= 2H\, E\big(\sqrt{Y_n(1)}\big) \int_0^t \frac{r(t,s)}{s^H}\big((t-s)^{2H-1} + s^{2H-1}\big)\, \mathrm{d}s. \quad (11)$$

By (9) and (11) we finally obtain

$$\int_0^t \left(\frac{r(t,s)}{s^H} - s^H\right)\big((t-s)^{2H-1} + s^{2H-1}\big)\, \mathrm{d}s = 0$$

or, by doing the change of variable $z = s/t$,

$$t^{3H} \int_0^1 \frac{1 - (1-z)^{2H} - z^{2H}}{z^H}\big((1-z)^{2H-1} + z^{2H-1}\big)\, \mathrm{d}z = 0.$$

Note that the function $f(z) = 1 - (1-z)^{2H} - z^{2H}$ keeps a constant sign on $[0,1]$ (it is positive if $H > 1/2$ and negative if $H < 1/2$). The integral is zero if the integrand is zero almost-everywhere and this is false unless $H = 1/2$.

Suppose now $n = 1$ and assume that the the answer to question Q is positive. In this case $Z_t^H = \int_0^t \mathrm{sign}(\gamma_s)\, \mathrm{d}\gamma_s$. The equation (6) yields, by multiplying both sides by Z_t^H, to the relation

$$\int_0^t E|\gamma_s| K^{*,a} K^* \mathbf{1}_{[0,t]}(s)\, \mathrm{d}s = \int_0^t E\big(\mathrm{sign}(\gamma_s)|\gamma_t|\big) K^{*,a} K^* \mathbf{1}_{[0,t]}(s)\, \mathrm{d}s.$$

Using the self-similarity of $|\gamma_s|$ and the joint distribution of (γ_t, γ_s) we obtain $E|\gamma_s| = s^H E|\gamma_1|$ and $E\big(\mathrm{sign}(\gamma_s)|\gamma_t|\big) = \frac{r(t,s)}{s^H} E|\gamma_1|$. A repetition of the previous arguments finishes the proof. $\qquad\square$

Remark. • Note that in [8] the authors showed using a different method that the process Z^H given by (5) is not a fractional Brownian motion for *every* $n \geqslant 1$ and $H \in (0,1)$.
• Although the process (5) is not a fractional Brownian motion, it satisfies some properties that the fractional Brownian motion enjoys: it is H-self-similar, has long range dependence (if $H > 2/3$)·and has the same $1/H$ variation as the fractional Brownian motion with Hurst parameter H (see [8] and [7]). Nevertheless the question of the nature of the law of Z^H is still open. In particular, we still don't know whether it is Gaussian or not.

References

1. E. Alos, O. Mazet, D. Nualart (2001): *Stochastic calculus with respect to Gaussian processes.* Ann. Probab., **29**, 766–801.

2. P. Cheridito, D. Nualart (2002): *Stochastic integral of divergence type with respect to fractional Brownian motion with Hurst parameter $H \in (0, \frac{1}{2})$.* Preprint no. 314, IMUB.
3. L. Decreusefond, A.S. Ustunel (1998): *Stochastic analysis of the fractional Brownian motion.* Potential Analysis, **10** , 117–214.
4. N. Eisenbaum (2003): *On the infinite divisibility of squared Gaussian processes.* Probab. Theory and Relat. Fields., **125**, 381–392.
5. N. Eisenbaum, H. Kaspi (2003): *A characterization of the infinitely divisible squared Gaussian processes.* Preprint.
6. R.C. Griffiths (1984): *Characterization of infinitely divisible multivariate gamma distributions.* Jour. Multivar. Anal. **15**, 12–20.
7. J.M.E. Guerra, D. Nualart (2003): *The $\frac{1}{H}$-variation of the divergence integral with respect to the fBm for $H > \frac{1}{2}$ and fractional Bessel processes.* Preprint No. 339, IMUB.
8. Y.Z. Hu, D. Nualart (2003): *Some processes associated with fractional Bessel processes.* Preprint no. 348, IMUB.
9. D. Nualart (1995): *Malliavin calculus and related topics.* Springer.
10. D. Revuz, M. Yor (1994): *Continuous martingales and Brownian motion.* Springer.

Regularity and Identification of Generalized Multifractional Gaussian Processes

Antoine Ayache[1], Albert Benassi[2], Serge Cohen[3], and Jacques Lévy Véhel[4]

[1] Université Paul Sabatier
  UFR MIG, Laboratoire de Statistique et de Probabilités
  118, Route de Narbonne, 31062 Toulouse, France
  e-mail: ayache@math.ups-tlse.fr
[2] Université Blaise Pascal (Clermont-Ferrand II),
  LaMP, CNRS UPRESA 6016, 63177 Aubière Cedex, France
  e-mail: A.Benassi@opgc.univ-bpclermont.fr
[3] Université Paul Sabatier
  UFR MIG, Laboratoire de Statistique et de Probabilités
  118, Route de Narbonne, 31062 Toulouse, France
  e-mail: Serge.Cohen@math.ups-tlse.fr
[4] INRIA-Groupe Fractales,
  B.P. 105 Domaine de Voluceau
  78153 Le Chesnay Cedex, France
  e-mail: Jacques.Levy_Vehel@inria.fr

Summary. In this article a class of multifractional processes is introduced, called Generalized Multifractional Gaussian Process (GMGP). For such multifractional models, the Hurst exponent of the celebrated Fractional Brownian Motion is replaced by a function, called the multifractional function, which may be irregular. The main aim of this paper is to show how to identify irregular multifractional functions in the setting of GMGP. Examples of discontinuous multifractional functions are also given.

Key words: Gaussian processes, identification, multifractional function.

AMS classification (2000): 60G15, 62G05.

1 Introduction

Fractional Brownian Motion, that was introduced in [16] and studied in [19], is the continuous mean-zero Gaussian process $\{B_H(t)\}_{t \in \mathbb{R}}$, depending on a parameter $H \in (0, 1)$, called the Hurst index, with covariance kernel

$$\mathbb{E}\big(B_H(s)B_H(t)\big) = c\big(|s|^{2H} + |t|^{2H} - |s - t|^{2H}\big), \tag{1}$$

where s, t are arbitrary reals and $c > 0$ is a constant. This process has important applications in modeling [6]. One of its main interests is that its pointwise Hölder regularity can be prescribed via its Hurst parameter. Indeed, the Hölder exponent of FBM, at any point, is equal to H, almost surely. Recall that, for a stochastic process X, the Hölder exponent at a point t_0, is defined as

$$\alpha_X(t_0) = \sup\left\{\alpha, \limsup_{h \to 0} \frac{|X(t_0 + h) - X(t_0)|}{|h|^\alpha} = 0\right\}. \tag{2}$$

However, the Hölder exponent of FBM remains the same all along its trajectory and this can be restrictive in some situations. For this reason, various models have been proposed to replace the Hurst index $H \in (0, 1)$ by a function $h(t)$. These so-called multifractional models[5] are useful in various domains of applications, see for instance [1, 15, 18, 11, 17]. The starting point of these generalizations is very often the harmonizable representation of the FBM:

$$B_H(t) = \int_{\mathbb{R}} \frac{e^{-it \cdot \xi} - 1}{|\xi|^{\frac{1}{2} + H}} W(\mathrm{d}\xi) \tag{3}$$

where $W(\mathrm{d}\xi)$ is a Wiener measure such that X is a real valued process (cf. [12] or page 1138 in [1] for a discussion on the Wiener measure). For example we may recall the Multifractional Brownian Motion (MBM), whose harmonizable representation is given by:

$$B_h(t) = \int_{\mathbb{R}} \frac{e^{-it \cdot \xi} - 1}{|\xi|^{\frac{1}{2} + h(t)}} W(\mathrm{d}\xi) \tag{4}$$

where h is a Hölder function ranging in $[a, b] \subset (0, 1)$. The function h is called the multifractional function. This process has been introduced independently in [22] (where the denomination MBM was introduced) and in [10]. Similarly to FBM, MBM is continuous and its Hölder exponent at any point t_0 is equal to $h(t_0)$, almost surely. However, one of the main problems with these models is that their Hölder exponents cannot be irregular, because h must be Hölderian. Recall that when the function h is discontinuous then the trajectories of the MBM are themselves, with probability 1, discontinuous (see Proposition 1 of [1]).

The continuity of h is a real drawback for some applications, for instance image classification and segmentation: The Hölder regularity is expected to vary abruptly on textured zones and around edges. On the other hand, the identification of h has been performed under the assumption that h is continuously differentiable [8].

---

[5] One should not confuse *multifractional* and *multifractal*. A multifractal process has a multifractal spectrum that takes non trivial values on a set of positive measure. A multifractional process is a process with non constant pointwise Hölder exponent. Multifractality implies multifractionality, while the converse is false.

Recently, two continuous Gaussian processes with irregular Hölder exponents have been introduced. These processes are the Generalized Multifractional Brownian Motion (in short GMBM) [5, 4] and the Step Fractional Brownian Motion (in short SFBM) [7]. The GMBM is an extension of the MBM while the SFBM is an extension of the FBM. In this article we will study a Gaussian process that is slightly more general than the GMBM, so let us first recall the definition of the GMBM. Let $\lambda > 1$ be an arbitrary fixed real, and $(h) = (h_n)_{n \in \mathbb{N}}$ be a sequence of Hölder functions with values in $[a, b] \subset (0, 1)$. The GMBM with parameters λ and (h) is the continuous Gaussian process $\{Y_{(h)}(t)\}_{t \in \mathbb{R}}$ defined for every real t as

$$Y_{(h)}(t) = \int_{0 < \xi < 1} \frac{e^{-it \cdot \xi} - 1}{\xi^{\frac{1}{2} + h_0(t)}} W(d\xi) + \sum_{n=1}^{+\infty} \int_{\lambda^{n-1} \leqslant \xi < \lambda^n} \frac{e^{-it \cdot \xi} - 1}{\xi^{\frac{1}{2} + h_n(t)}} W(d\xi).$$

One of the main properties of the GMBM is that its Hölder exponent at any point t_0 is almost surely equal to $\liminf_{n \to \infty} h_n(t_0)$, under the assumption that $\|h_n\| = o(2^{na})$, where $\|h_n\|$ denotes the Hölder norm of h_n (see [2, 5, 4]). Therefore, the Hölder exponent of the GMBM is of the most general form: any liminf of positive continuous functions (see [14, 3, 2]). In contrast, the one of SFBM is a step function (see subsection 1.3.1 in [7]). As a consequence, the Hölder exponent of the GMBM may be discontinuous everywhere while that of the SFBM can only be discontinuous on a finite set. Likewise, the Hölder exponent of the GMBM may be constant except on a finite set (see Proposition 1 of [5]), while this is impossible with the SFBM. On the other hand, the parameters of the SFBM have been identified in [7], and we present in this article an identifiable model which has many desirable properties of the GMBM.

Let us define a Generalized Multifractional Gaussian Process (in short GMGP) by its harmonizable representation:

$$X(t) = \int_0^{+\infty} \frac{e^{-it \cdot \xi} - 1}{\xi^{\frac{1}{2} + H(t,\xi)}} W(d\xi). \tag{5}$$

where $W(d\xi)$ is a Wiener measure such that X is real valued process and $H : \mathbb{R} \times (0, +\infty) \to [a, b] \subset (0, 1)$ is called the Hurst function. Clearly the MBM is a GMGP for which $H(t, \xi)$ does not depend on the frequency variable ξ and is always equal to the multifractional function $h(t)$. Actually the GMGP model is very close to the GMBM: On the one hand, every GMBM with parameters λ and $(h_n)_{n \in \mathbb{N}}$ is a GMGP with Hurst function $H(t, \xi) = \mathbf{1}_{[0,1)}(\xi)h_0(t) + \sum_{n=1}^{+\infty} \mathbf{1}_{[\lambda^{n-1}, \lambda^n)}(\xi)h_n(t)$. On the other hand, if $H : \mathbb{R} \times (0, +\infty) \to [a, b] \subset (0, 1)$ is continuous in almost all ξ, then it is not hard to see that the high frequencies part of the GMBM of parameter λ and $(H(., \lambda^n))_{n \in \mathbb{N}}$ converges, in distribution to that of the GMGP with Hurst function $H(., .)$, when $\lambda \to 1$.

As for the SFBM, it is also, in some sense, similar to the GMGP, although, contrarily to the SFBM, no wavelet expansion is needed to introduce the GMGP. We refer to [13] for a comparison between SFBM and GMBM with parameter $\lambda = 2$.

The sequel of this paper is organized as follows. In section 2, we first obtain the pointwise Hölder exponent of the GMGP. We then define our estimator and we give the precise assumptions ensuring its almost sure convergence. Section 3 is devoted to three examples where the Hurst functions $H(t, \xi)$ converge to multifractional functions of particular interest. The proof of the results are performed in section 4.

2 Statement of the results

Following the correspondence between the GMBM and the GMGP, the multifractional function of a GMGP is defined to be:

$$h(t) = \lim_{\xi \to +\infty} H(t, \xi), \tag{6}$$

when the limit exists. This definition is coherent with the fact that the local properties of the sample paths are given by the high frequencies of the harmonizable representation. In this paper we will show that, similarly to the GMBM, at any point t_0, the Hölder exponent of the GMGP is almost surely equal to $h(t_0)$. To identify the multifractional function that governs the regularity of the sample paths, we will then use the generalized quadratic variation

$$V_N = \sum_{p=0}^{N-2} \left(X\left(\frac{p+2}{N}\right) - 2X\left(\frac{p+1}{N}\right) + X\left(\frac{p}{N}\right) \right)^2. \tag{7}$$

Actually first the infimum of the multifractional function $\inf_{t \in (0,1)} h(t) = h_*$ will be identified. Before stating the main results and making precise the technical assumptions we need to explain why h_* is naturally identified in this framework. As a rule of thumb one may assume for multifractional models that

$$\mathbb{E} V_N \approx \sum_{p=0}^{N-2} \frac{1}{N^{2h(p/N)}} \tag{8}$$

and then it is classical to have

$$\lim_{N \to +\infty} \frac{\ln(\mathbb{E} V_N)}{\ln(N)} = 1 - 2h_*. \tag{9}$$

In this article we will prove a Law of the Large Numbers for V_N and the preceding heuristic suggests to take as an estimator:

$$\widehat{h}_N = 1/2 \left(1 - \frac{\ln(V_N)}{\ln(N)} \right). \tag{10}$$

Let us now state more precisely the technical assumptions needed to determine the Hölder regularity of GMGP and to show that $\widehat{h}_N$ is a consistent estimator of h_*. First of all we state a weak assumption on the multifractional function h that will entail (9):

A 1. $\exists C > 0, \quad \forall N \geqslant 1$

$$\frac{1}{C} N^{1-2h_*} \leqslant \sum_{p=0}^{N-2} \frac{1}{N^{2h(p/N)}} \leqslant CN^{1-2h_*}. \tag{11}$$

Note that C is a generic constant that may change in this article and that the last inequality of (11) is always satisfied. The lower bound in (11) means that there is a sufficiently large number of points p/N such that $h(p/N)$ is close to h_*.

The asymptotic behavior (11) is clearly true under mild assumptions of smoothness for $t \to h(t)$ and we will see in the examples that it is fulfilled for non trivial functions $h(t)$.

One needs then to strengthen the existence of the limit (6) to ensure that (8) is satisfied. The following technical assumption states that the speed of convergence of $H(t, \xi)$ to $h(t)$ is compatible with (8).

A 2. $\forall 0 \leqslant t \leqslant 1, \quad \forall \xi \geqslant 0, \quad H(t, \xi) \geqslant h(t).$
$\exists \alpha < 1$ *such that*

$$\sum_{p=0}^{N-2} \int_{N^\alpha}^{+\infty} \frac{|h(p/N) - H(p/N, \xi)|}{\xi^{2h(p/N)+1}} \ln(\xi)\, d\xi = o(N^{1-2h_*}). \tag{12}$$

Let us make some comments to explain (12). In our setting, it will yield that $\mathbb{E}\, V_N$ is sufficiently close to N^{1-2h_*} to have (9). Now why is (12) a reasonable assumption? Let us suppose that $|H(p/N) - H(p/N, \xi)| \leqslant C_N$. Then:

$$\left| \sum_{p=0}^{N-2} \int_{N^\alpha}^{+\infty} \frac{|h(p/N) - H(p/N, \xi)|}{\xi^{2h(p/N)+1}} \ln(\xi)\, d\xi \right| \leqslant \sum_{p=0}^{N-2} \frac{C_N \ln N}{N^{2\alpha h(p/N)}}.$$

If we assume that $\lim\limits_{N \to +\infty} C_N \ln N N^{2\alpha b} = 0$, then (12)' is satisfied because of **A 1**. Hence **A 2** can be interpreted as a way of expressing that $H(t, \xi)$ is converging uniformly fast enough to $h(t)$ when $\xi \to +\infty$.

The last assumption is concerned with the behaviour of the functions $H(\cdot, \xi)$ when ξ is finite. It expresses the fact that these functions must be smooth enough so that the Hölder regularity of the GMGP is indeed governed by the multifractional function.

A 3. • *For* $0 < \xi \leqslant 4\pi/3$,

$$H(t, \xi) = b.$$

- $\exists g$ such that $0 < g < \min(a/2, 1/4)$ (where $H(t, \xi) \in [a, b]$) and $\exists \beta$ such that $b < \beta \leqslant 1$ such that for $4\pi/3 < \xi$

$$\left| \frac{\partial^k H}{\partial \xi^k}(t, \xi) - \frac{\partial^k H}{\partial \xi^k}(t', \xi) \right| \leqslant C |\xi|^{g-k} |t - t'|^\beta \tag{13}$$

for $t, t' \in [0, 1]$ and $k = 0, 1, 2$. Moreover

$$\left| \frac{\partial^k H}{\partial \xi^k}(t, \xi) \right| \leqslant C \xi^{g-k} \tag{14}$$

for $t \in [0, 1]$ and $k = 0, 1, 2$.

Of course, in these assumptions, $4\pi/3$ is arbitrary and can be replaced by any non-negative constant. The inequality $0 < g < \min(a/2, 1/4)$ has a deeper meaning: it expresses the fact that the Hölder constant of $H(t, \xi)$ cannot grow too fast when $\xi \to +\infty$. Finally, the inequality $b < \beta$ is similar to the condition $r > \sup(a(x))$ in the Definition 1.4, p. 41, of [10] or to the condition $\sup_t H(t) < \beta$ in Definition 3.1, p. 11, of [4]; we refer to [13] section 4.2 or to [1] section 2.2 for a more complete discussion.

At this point, we can state the three main results of this paper.

Theorem 1. *Under assumption A 3, the Hölder exponent of the GMGP at each point t_0 satisfies almost surely,*

$$\alpha_X(t_0) = h(t_0). \tag{15}$$

Theorem 2. *Under assumptions A 1, A 2, A 3,*

$$\lim_{N \to +\infty} \widehat{h}_N = h_* \qquad \text{almost surely.} \tag{16}$$

Actually, the identification of the infimum h_* of the multifractal function h is the result needed to obtain the identification of h itself. The classical technique to have full identification is the localization of the generalized quadratic variations. Basically if one wants to estimate $h(t)$ for a given $t \in (0, 1)$, one has first to localize V_N on an (ε, N)-neighborhood of t defined by

$$\mathcal{V}_{\varepsilon, N}(t) = \left\{ p \in \mathbb{Z}, \left| \frac{p}{N} - t \right| \leqslant \varepsilon \right\}.$$

Then the localized generalized variation is defined by

$$\mathbf{V}_{\varepsilon, N}(t) = \sum_{p \in \mathcal{V}_{\varepsilon, N}(t)} \left(X\left(\frac{p+1}{N}\right) - 2X\left(\frac{p}{N}\right) + X\left(\frac{p-1}{N}\right) \right)^2. \tag{17}$$

We define the localized estimator

$$\widehat{h}_{\varepsilon, N} = 1/2 \left(1 - \frac{\ln(\mathbf{V}_{\varepsilon, N}(t))}{\ln(N)} \right), \tag{18}$$

and we get

$$\lim_{N\to+\infty} \widehat{h}_{\varepsilon,N} = \inf_{\{|s-t|<\varepsilon\}} h(s) \qquad \text{almost surely}$$

as in Remark 1 in [8].

The next step to identify h at point t is to let $\varepsilon \to 0^+$. For instance, taking $\varepsilon = N^{-\gamma}$ for $\gamma > 0$, one obtains the following result as a corollary of Theorem 2:

Corollary 1. *Fix* $0 < \gamma < 1/2$. *Under assumptions* **A 1**, **A 2** *and* **A 3**,

$$\lim_{N\to+\infty} 1/2\left(1 - \gamma - \frac{\ln(V_{N^{-\gamma},N}(t))}{\ln(N)}\right) = \liminf_{s\to t} h(s) \qquad a.s. \qquad (19)$$

The statement of this corollary is an example of localization of Theorem 2. Remark that if h is lower semi-continuous and satisfies **A 1**, **A 2**, and **A 3**, then it is identifiable.

3 Examples

In this section we present three examples of application of Theorem 2, that illustrate the way this theorem can be used. In each of these examples, one assumes given a precise multifractional function. Then, one constructs a GMGP by exhibiting a Hurst function $H(t,\xi)$ such that the assumptions **A 1**, **A 2** and **A 3** are fulfilled. Techniques similar to the ones used in this section were put to work in the proofs of Propositions 1 and 2 of [5].

3.1 Hölder continuous multifractional functions

Remark first that if the multifractional function h is β-Hölder continuous with $\beta > b$ and satisfies **A 1**, one may choose a Hurst function $H(t,\xi)$ that interpolates a constant for $0 \leqslant \xi \leqslant 4\pi/3$ and the multifractional function $h(t)$ when $\xi \geqslant 8\pi/3$. One can take for instance

$$H(t,\xi) = b \qquad \text{if} \qquad 0 \leqslant \xi \leqslant \frac{4\pi}{3}$$

and

$$H(t,\xi) = \left(2 - \frac{3\xi}{4\pi}\right)b + \left(\frac{3\xi}{4\pi} - 1\right)h(t) \qquad \text{if} \qquad \frac{4\pi}{3} \leqslant \xi \leqslant \frac{8\pi}{3},$$

and $H(t,\xi) = h(t)$ if $\xi \geqslant 8\pi/3$. In this case the GMGP exhibits the same properties as a Multifractional Brownian Motion and the Theorem 2 is the analogue of the Remark 1 in [8] where the identification of a multifractional function is performed by localizing the generalized quadratic variations as in [8]. Please note that an alternate method to corollary 1 has recently been proposed in [21] to identify the multifractional function of a MBM in the Hölderian case.

3.2 Multifractional functions with one jump

In this example we consider a multifractional function that has only one discontinuity and that is piecewise constant. This very simple example can be easily extended to the case of piecewise constant functions with a finite number of jumps. The GMGP associated to this kind of multifractional function are comparable to SFBM for the applications. Let us suppose that

$$h(t) = b\,\mathbf{1}_{[0,t_0]} + a\,\mathbf{1}_{(t_0,1]} \tag{20}$$

where $t_0 \in (0,1)$ and $0 < a < b < 1$. **A 1** is clearly satisfied with $h_* = a$. Then one can construct a Hurst function that admits such a multifractional function through the following procedure:

- For $0 < \xi \leqslant 4\pi/3$,
$$H(t,\xi) = b \qquad \forall t \in [0,1].$$

- for $4\pi/3 < \xi$ let us choose $0 < f < \min(a/6, 1/12)$

$$H(t,\xi) = \begin{cases} b & \text{if } t \leqslant t_0 \\ a + (b-a)\sin^4\left(\dfrac{\pi\xi^f}{2}\left(t - t_0 - \dfrac{1}{\xi^f}\right)\right) & \text{if } t \in \left[t_0, t_0 + \dfrac{1}{\xi^f}\right) \\ a & \text{if } t \geqslant t_0 + \dfrac{1}{\xi^f}. \end{cases}$$

An elementary computation shows that **A 3** is fulfilled with $g = 3f$ and $\beta = 1$. To check **A 2**, let us remark that, if $\sharp$ denotes the cardinality of a set,

$$\left|\sum_{p=0}^{N-2} \int_{N^\alpha}^{+\infty} \frac{|h(p/N) - H(p/N,\xi)|}{\xi^{2h(p/N)+1}} \ln(\xi)\,d\xi\right|$$

$$\leqslant (b-a)\,\sharp\left\{p \text{ such that } t_0 \leqslant \frac{p}{N} \leqslant t_0 + \frac{1}{N^{\alpha f}}\right\} \times \int_{N^\alpha}^{+\infty} \frac{\ln(\xi)}{\xi^{2a+1}}\,d\xi.$$

Hence

$$\left|\sum_{p=0}^{N-2} \int_{N^\alpha}^{+\infty} \frac{|h(p/N) - H(p/N,\xi)|}{\xi^{2h(p/N)+1}} \ln(\xi)\,d\xi\right| \leqslant \frac{CN\ln N}{N^{\alpha f}}\,\frac{1}{N^{2h_*\alpha}}.$$

Then α is chosen such that $\alpha(f + 2h_*) > 2h_*$ to have (12).

3.3 Multifractional functions with accumulation of jumps

Let us address the more interesting case of piecewise constant multifractional functions which have an infinite number of jumps. In this example let us suppose the multifractional function is defined by

$$h(t) = \begin{cases} 1/2 & \text{if } t \leqslant 1/2 \\ \dfrac{1}{2} + \dfrac{1}{k+1} & \text{if } \dfrac{1}{2} + \dfrac{1}{k+1} < t \leqslant \dfrac{1}{2} + \dfrac{1}{k} \\ \dfrac{1}{2} + \dfrac{1}{3} & \text{if } 1 \leqslant t \end{cases} \tag{21}$$

where $k \geqslant 2$. This multifractional function fulfills **A 1** with $h_* = 1/2$, and $t = 1/2$ is the time where the jumps accumulate. Let us now exhibit a Hurst function corresponding to (21) by using the construction of the previous example which is valid when the multifractional function has only a finite number of jumps. Let us define the function (21) where only k_0 jumps are considered:

$$h_{k_0}(t) = \begin{cases} 1/2 & \text{if } t \leqslant 1/2 \\ \dfrac{1}{2} + \dfrac{1}{k+1} & \text{if } \dfrac{1}{2} + \dfrac{1}{k+1} < t \leqslant \dfrac{1}{2} + \dfrac{1}{k}, \text{ for } k \leqslant k_0 \\ \dfrac{1}{2} + \dfrac{1}{k_0+1} & \text{if } \dfrac{1}{2} + \dfrac{1}{k_0} \leqslant t. \end{cases} \tag{22}$$

Then one can construct, with the same technique as in the example of section 3.2, a Hurst function corresponding to h_{k_0} which is denoted by $H_{k_0}(t, \xi)$ and is defined for $(k_0(k_0+1))^{1/f} \leqslant \xi$. If we fix $f < \min(a/6, 1/12)$ in the previous example, then this last condition simply means that $[\frac{1}{2} + \frac{1}{k}, \frac{1}{2} + \frac{1}{k} + \frac{1}{\xi^f}]$ is included in the intervals where h_{k_0} is constant. Hence this condition allows us to manage the jumps of h_{k_0} separately. Let us now introduce a Hurst function corresponding to (21) by taking more and more jumps into account when $\xi \to +\infty$. Define

$$K(\xi) = \sup\{k \in \mathbb{N}^* \text{ such that } (k(k+1))^{1/f} \leqslant \xi\} \tag{23}$$

and define the Hurst function with the H_{k_0}'s

$$H(t, \xi) = H_{K(\xi)}(t, \xi). \tag{24}$$

It is not difficult to check that this Hurst function satisfies **A 3** since

$$\frac{\partial^m H}{\partial \xi^m}(t, \xi) = \frac{\partial^m H_{k_0}}{\partial \xi^m}(t, \xi) \,,$$

when $(k_0(k_0+1))^{1/f} \leqslant \xi \leqslant ((k_0+1)(k_0+2))^{1/f}$. To check **A 2** let us write

$$\int_{N^\alpha}^{+\infty} \sum_{p=0}^{N-2} \frac{|h(p/N) - H(p/N, \xi)|}{\xi^{2h(p/N)+1}} \ln(\xi) \, d\xi = I + J$$

where

$$I = \int_{N^\alpha}^{+\infty} \sum_{p=0}^{[\frac{N}{2} + \frac{N}{K(\xi)+1}]} \frac{\frac{1}{K(\xi)+1}}{\xi^{2h(p/N)+1}} \ln(\xi) \, d\xi,$$

and where $[x]$ stands for the integral part of x. $|h(p/N) - H(p/N, \xi)|$ has been replaced by its constant value for the considered p's. The second part is

$$J = \int_{N^\alpha}^{+\infty} \sum_{p=[\frac{N}{2}+\frac{N}{K(\xi)+1}]+1}^{N-2} \frac{|h_{K(\xi)}(p/N) - H_{K(\xi)}(p/N, \xi)|}{\xi^{2h(p/N)+1}} \ln(\xi) \, d\xi.$$

Let us upper-bound I by remarking that if $N^\alpha \leqslant \xi$ then

$$\frac{1}{K(\xi)+1} \leqslant CN^{\frac{-\alpha f}{2}}.$$

Hence

$$|I| \leqslant N \times CN^{\frac{-\alpha f}{2}} \int_{N^\alpha}^{+\infty} \frac{\ln(\xi)}{\xi^2} \, d\xi$$

and α is chosen such that

$$|I| \leqslant C \frac{N \ln(N)}{N^{\alpha(1+f/2)}} = o(1)$$

which is the required estimate since $h_* = 1/2$.

To upper-bound J let us remark that for a fixed ξ, $h_{K(\xi)}$ has $K(\xi)$ jumps at time $\frac{1}{2} + \frac{1}{k+1}$ of magnitude $\frac{1}{k} - \frac{1}{k+1}$. Hence

$$\sum_{p=[\frac{N}{2}+\frac{N}{K(\xi)+1}]+1}^{N-2} |h_{K(\xi)}(p/N) - H_K(\xi)(p/N, \xi)| \leqslant CN^{1-\alpha f} \sum_{k=1}^{K(\xi)} \left(\frac{1}{k} - \frac{1}{k+1} \right)$$

$$\leqslant CN^{1-\alpha f}$$

where the factor $CN^{1-\alpha f}$ comes from $\sharp\{p$ such that $t_0 \leqslant p/N \leqslant t_0 + \frac{1}{N^{\alpha f}}\}$ for each jump t_0 as in the previous example. Then

$$|J| \leqslant CN^{1-\alpha f} \frac{\ln(N)}{N^\alpha}$$

which yields (12) for a convenient α.

4 Proof of the results

In this section, Theorems 1, 2 and Corollary 1 are proved. First let us describe the main steps of the proofs. In the definition (5) of the GMGP we remark that the integrand

$$\frac{e^{-it\cdot\xi} - 1}{|\xi|^{\frac{1}{2}+H(t,\xi)}}$$

depends of the variable t twice. To study the generalized variation V_N a new process is considered

$$Y(s,t) = \int_0^{+\infty} \frac{e^{-is\cdot\xi} - 1}{|\xi|^{\frac{1}{2}+H(t,\xi)}} W(d\xi) \tag{25}$$

where this dependence is split into two variables (s,t). Then a variation W_N associated to this process Y is introduced

$$W_N = \sum_{p=0}^{N-2} \left(Y\left(\frac{p+2}{N}, \frac{p}{N}\right) - 2Y\left(\frac{p+1}{N}, \frac{p}{N}\right) + Y\left(\frac{p}{N}, \frac{p}{N}\right) \right)^2, \tag{26}$$

W_N is simpler to study than V_N because its variation does not involve the variable t that appears in the Hurst function. Moreover one can see easily that

$$V_N = \sum_{p=0}^{N-2} \left(Y\left(\frac{p+2}{N}, \frac{p+2}{N}\right) - 2Y\left(\frac{p+1}{N}, \frac{p+1}{N}\right) + Y\left(\frac{p}{N}, \frac{p}{N}\right) \right)^2. \tag{27}$$

A result very similar to Proposition 2.1 in [2] (namely Proposition 1 below) shows that the process Y is Hölder continuous in the second variable uniformly with respect to the first variable. This fact yields the regularity result of Theorem 1 and that V_N can be replaced by W_N. This is the first step in the proof of Theorem 2. Then it is shown that

$$\lim_{N\to+\infty} \frac{\ln(\mathbb{E}\, W_N)}{\ln(N)} = 1 - 2h_*$$

in the second step. The third step consists in a study of the asymptotics of the variance of W_N that allows to replace $\mathbb{E}\, W_N$ by W_N in the previous limit. Finally Corollary 1 is proved.

4.1 Study of Y

In this part we prove the Hölder continuity of Y. Although the main lines of the proof of the following Proposition are similar to those of Proposition 2.1 in [2], we will give this proof for completeness.

Proposition 1. *Under assumption* **A 3**, *there exists an almost surely positive random variable $C(\omega)$ such that:*

$$\sup_{s\in[0,1]} |Y(s,t) - Y(s,t')| \leqslant C(\omega)|t - t'|^\beta \tag{28}$$

for $t,\ t' \in [0,1]$ where β is defined in (13).

Proof. To get (28) we will apply a series expansion of Y in the orthonormal basis of L^2 given by the Lemarié–Meyer wavelets (cf. [20]). This technique is classical to study the smoothness of multifractal process. As in [2] for the GMF (see Relation (2.2)) or in [10] for the MBM we get

$$Y(s,t) = \sum_{j=0}^{+\infty} \sum_{k=-\infty}^{+\infty} s_{j,k}(s,t) \eta_{j,k} \tag{29}$$

where $\eta_{j,k}$ are identically distributed standard Gaussian variables and where the coefficients are:

$$s_{j,k}(s,t) = 2^{-j/2} \int_0^{+\infty} \frac{e^{-is\cdot\xi} - 1}{\xi^{\frac{1}{2}+H(t,\xi)}} \overline{\psi_{j,k}}(\xi) \, d\xi. \tag{30}$$

Please note that equation (29) is only true up to an additive smooth process. In (30) $(\psi_{j,k})_{j,k\in\Lambda^+}$ is the Lemarié–Meyer basis indexed by $\Lambda^+ = \{j,k,\ j \geqslant 0,$ and $k \in \mathbb{Z}\}$ as explained in [20] or in page 29 of [10]. Then the Hölder continuity of Y derives from estimates on the coefficients $s_{j,k}$ (cf. [2]).

$\exists \varepsilon,\ \forall k \in \mathbb{Z}$

$$|s_{j,k}(s,t)| \leqslant C2^{-j\varepsilon} \left\{ \frac{1}{(1+|2^j s - k|^2)} + \frac{1}{1+k^2} \right\}, \tag{31}$$

for $j \geqslant 1$

$$|s_{j,k}(s,t) - s_{j,k}(s,t')| \leqslant Cj2^{-j\varepsilon}|t-t'|^\beta \left\{ \frac{1}{(1+|2^j s - k|^2)} + \frac{1}{1+k^2} \right\}. \tag{32}$$

Note that the estimates (31) and (32) are quite similar to those in Lemma 2.5 in [2]. This Lemma is, to a certain extent, inspired from Proposition 4.1, page 79 in [10]. The estimate (31) and classical almost sure bounds for the supremum of Gaussian variables yield that the series in (29) converges almost surely. Then (32) yields the Hölder continuity of Y. Remark that

$$Y(s,t) - Y(s,t') = \sum_{j=1}^{+\infty} \sum_{k=-\infty}^{+\infty} \left(s_{j,k}(s,t) - s_{j,k}(s,t') \right) \eta_{j,k}$$

where the terms for $j = 0$ are dropped since the Hurst function is constant on $\{\xi \leqslant 4\pi/3\}$. The main ingredient to get these estimates are integrations by parts in the integral of (30). In the framework of GMGP, to carry these integrations by parts, the only new point compared to [2] is the dependence of the Hurst function $H(t,\xi)$ on a continuous parameter ξ instead of a discrete parameter n. In the following argument we only stress how the assumption **A 3** allows us to control the additional terms caused by this dependence.

Let us study the Hölder continuity of the coefficients $s_{j,k}$'s. One can write

$$s_{j,k}(s,t) - s_{j,k}(s,t') = 2^{-j/2} \int_0^{+\infty} (e^{-is\xi} - 1)e^{i2^{-j}k\xi}$$
$$\times \left(\xi^{-(\frac{1}{2}+H(t,\xi))} - \xi^{-(\frac{1}{2}+H(t',\xi))} \right) \overline{\widehat{\psi}}(2^{-j}\xi) \, d\xi$$

where we have only used the translation-dilatation structure of the Lemarié–Meyer basis

$$\psi_{j,k}(s) = 2^{j/2}\psi(2^j s - k),$$

for $j \geqslant 1$. The factor $e^{-is\xi} - 1$ in the previous integral can be split since the support of $\widehat{\psi}$ is contained in $[2\pi/3, 8\pi/3]$ and the first part leads to

$$2^{-j/2}\int_0^{+\infty} e^{i(k-2^j s)2^{-j}\xi}\big(\xi^{-(\frac{1}{2}+H(t,\xi))} - \xi^{-(\frac{1}{2}+H(t',\xi))}\big)\overline{\widehat{\psi}}(2^{-j}\xi)\,d\xi.$$

The second part is of the same form but $e^{i(k-2^j s)2^{-j}\xi}$ is replaced by 1. Hence this second part yields the second terms in the brackets in the right hand side of (31) and of (32). Let us define

$$g(t,t',\xi) = \big(\xi^{-(\frac{1}{2}+H(t,\xi))} - \xi^{-(\frac{1}{2}+H(t',\xi))}\big)\overline{\widehat{\psi}}(2^{-j}\xi).$$

After performing two integrations by parts on the last integral one gets for $s \neq 2^j k$

$$2^{-j/2}\int_0^{+\infty} \frac{e^{i(k-2^j s)2^{-j}\xi}}{i^2(k-2^j s)^2 2^{-2j}}\,\frac{\partial^2 g(t,t',\xi)}{\partial \xi^2}\,d\xi.$$

Then the Leibniz rule is applied to the preceding partial derivative

$$\frac{\partial^2 g(t,t',\xi)}{\partial \xi^2} = \sum_{p=0}^{2}\binom{2}{p}\frac{\partial^p}{\partial \xi^p}\big(\xi^{-(\frac{1}{2}+H(t,\xi))} - \xi^{-(\frac{1}{2}+H(t',\xi))}\big)$$

$$\times 2^{-(2-p)j}\frac{\partial^{2-p}\overline{\widehat{\psi}}}{\partial \xi^{2-p}}(2^{-j}\xi). \quad (33)$$

Because of the support of $\overline{\widehat{\psi}}$ the partial derivatives of $\overline{\widehat{\psi}}$ can be bounded independently of j, p, hence we have only to bound integrals that comes from the factors $(\xi^{-(\frac{1}{2}+H(t,\xi))} - \xi^{-(\frac{1}{2}+H(t',\xi))})$. For instance for $p = 0$

$$I_j = \int_{2\pi 2^j/3}^{8\pi 2^j/3} \big(\xi^{-(\frac{1}{2}+H(t,\xi))} - \xi^{-(\frac{1}{2}+H(t',\xi))}\big)\,d\xi$$

$$\leqslant \int_{2\pi 2^j/3}^{8\pi 2^j/3} |H(t,\xi) - H(t',\xi)|\frac{\ln(\xi)}{\xi^{a+1/2}}\,d\xi. \quad (34)$$

Then applying (13) for $k = 0$ yields

$$I_j \leqslant Cj2^{-j(a+\frac{1}{2}-g-1)}|t - t'|^\beta \quad (35)$$

which is exactly the estimate we need for (32) if we take $0 < \varepsilon < a - g$.

Let us now consider the contribution of the term $p = 1$ in the sum (33). First

$$\frac{\partial}{\partial \xi}\xi^{-(\frac{1}{2}+H(t,\xi))} = -\frac{1}{2}\xi^{-(\frac{3}{2}+H(t,\xi))} - H(t,\xi)\xi^{-(\frac{3}{2}+H(t,\xi))}$$

$$- \ln(\xi)\frac{\partial H}{\partial \xi}(t,\xi)\xi^{-(\frac{1}{2}+H(t,\xi))}. \quad (36)$$

And the first term of (36) leads to

$$J_{j,0} = \frac{-1}{2} \int_{2\pi 2^j/3}^{8\pi 2^j/3} \left(\xi^{-(\frac{3}{2}+H(t,\xi))} \right) - \xi^{-(\frac{3}{2}+H(t',\xi))} \right) d\xi \leqslant Cj2^{j(-\varepsilon-1/2)} |t-t'|^{\beta}.$$

This bound gives the same contribution as I_j to (32). To conclude the case $p = 1$ we have to bound

$$J_{j,1} = \int_{2\pi 2^j/3}^{8\pi 2^j/3} \left(H(t',\xi) \xi^{-(\frac{3}{2}+H(t',\xi))} - H(t,\xi) \xi^{-(\frac{3}{2}+H(t,\xi))} \right) \ln(\xi) \, d\xi$$

and

$$J_{j,2} = \int_{2\pi 2^j/3}^{8\pi 2^j/3} \left(\frac{\partial H}{\partial \xi}(t',\xi) \xi^{-(\frac{1}{2}+H(t',\xi))} - \frac{\partial H}{\partial \xi}(t,\xi) \xi^{-(\frac{1}{2}+H(t,\xi))} \right) \ln(\xi) \, d\xi.$$

The first term is split into

$$\int_{2\pi 2^j/3}^{8\pi 2^j/3} \left(H(t',\xi) - H(t,\xi) \right) \xi^{-(\frac{3}{2}+H(t',\xi))} \, d\xi$$

which is of negligible compared to J_j, because of (13) for $k = 1$ and into

$$\int_{2\pi 2^j/3}^{8\pi 2^j/3} H(t,\xi) \left(\xi^{-(\frac{3}{2}+H(t',\xi))} - \xi^{-(\frac{3}{2}+H(t,\xi))} \right) d\xi$$

which is also negligible compared to $J_{j,0}$, because of (14) for $k = 1$. With similar techniques it is shown that $J_{j,2}$ is of the same order as $jJ_{j,0}$.

The contribution of the term $p = 2$ in the sum (33) is estimated with similar arguments. Let us stress only one technical point. The partial derivative $\frac{\partial^2}{\partial \xi^2} \xi^{-(\frac{1}{2}+H(t,\xi))}$ leads to consider a term:

$$\int_{2\pi 2^j/3}^{8\pi 2^j/3} \left(\left| \frac{\partial H}{\partial \xi}(t',\xi) \right|^2 \left(\xi^{-(\frac{1}{2}+H(t,\xi))} - \xi^{-(\frac{1}{2}+H(t',\xi))} \right) \right) \ln(\xi) \, d\xi.$$

Because of (14) for $k = 1$ we have

$$\left| \frac{\partial H}{\partial \xi}(t',\xi) \right|^2 \leqslant C\xi^{2g-2}$$

and we have seen that the preceding integral is conveniently bounded if $2g < a$.

$\square$

4.2 Study of the Hölder regularity of the GMGP

The Hölder regularity of the GMGP X, can be studied quite similarly to that of the GMBM [2], [4] and [5]. The proof of Theorem 1 follows exactly the same lines as that of Theorem 1.1 in [2], so we will only sketch it. Let t_0 be a fixed real, and let Y_{t_0} be the Gaussian process defined for every real t by

$$Y_{t_0}(t) = \int_0^{+\infty} \frac{e^{-it.\xi} - 1}{\xi^{1/2 + H(t_0,\xi)}} W(d\xi)$$

(this is nothing but $Y(s,t)$ where the second variable is fixed).

First step: Using (28), we show that almost surely,

$$\alpha_X(t_0) \wedge \beta = \alpha_{Y_{t_0}}(t_0) \wedge \beta.$$

Second step: Using (6), we show that for arbitrarily small $\varepsilon > 0$, some constant $c > 0$ and every t, t'

$$\mathbb{E}(|Y_{t_0}(t) - Y_{t_0}(t')|^2) \leqslant c|t - t'|^{2(h(t_0) - \varepsilon)}.$$

Then a strong version of Kolmogorov criterion (see for instance Lemma 3.2 of [2]) implies that, almost surely, $\alpha_{Y_{t_0}}(t_0) \geqslant h(t_0)$.

Third step: Using (6) again, we show that for every t_0 and arbitrarily small $\varepsilon > 0$,

$$\lim_{n \to \infty} \frac{\mathbb{E}(|Y_{t_0}(t_0 + 2^{-n}) - Y_{t_0}(t_0)|^2)}{2^{-2n(h(t_0)) + \varepsilon}} = +\infty,$$

and then Lemma 2.2 of [2] implies that almost surely, $\alpha_{Y_{t_0}}(t_0) \leqslant h(t_0)$.

4.3 Comparison of W_N and V_N

In this section it is explained how estimates concerning W_N can be transferred to V_N. We postpone the proof of

$$\frac{1}{C} N^{1-2b} \leqslant \frac{1}{C} N^{1-2h_*} \leqslant W_N \leqslant C N^{1-2h_*} \tag{37}$$

to section 4.6. Let us now prove that if (37) is true we have

$$\lim_{N \to +\infty} 1/2 \left(1 - \frac{\ln(V_N)}{\ln(N)} \right) = h_*$$

which is the result we aim at. Let us rewrite (26) as

$$W_N = \sum_{p=0}^{N-2} \left[\sum_{k=0}^{2} d_k Y\left(\frac{p+k}{N}, \frac{p}{N}\right) \right]^2 \tag{38}$$

where $d_0 = 1$, $d_1 = -2$, and $d_2 = 1$; then V_N can also be written:

$$V_N = \sum_{p=0}^{N-2} \left[\sum_{k=0}^{2} d_k Y\left(\frac{p+k}{N}, \frac{p+k}{N}\right) \right]^2.$$

Hence we have

$$|W_N^{1/2} - V_N^{1/2}| \leqslant \left(\sum_{p=0}^{N-2} \left[\sum_{k=0}^{2} d_k \left(Y\left(\frac{p+k}{N}, \frac{p}{N}\right) - Y\left(\frac{p+k}{N}, \frac{p+k}{N}\right) \right) \right]^2 \right)^{1/2}$$

$$\leqslant C \left(\sum_{p=0}^{N-2} \left[\sum_{k=0}^{2} \sup_{s \in [0,1]} \left| Y\left(s, \frac{p}{N}\right) - Y\left(s, \frac{p+k}{N}\right) \right| \right]^2 \right)^{1/2}$$

and the following estimates are consequences of (28) and of (37):

$$|W_N^{1/2} - V_N^{1/2}| \leqslant CN^{1/2-\beta}$$
$$|W_N^{1/2} - V_N^{1/2}| W_N^{-1/2} \leqslant CN^{b-\beta}.$$

Since $\beta > b$ it is clear by writing

$$\ln\left(V_N^{1/2}\right) = \ln\left(1 + \left(V_N^{1/2} - W_N^{1/2}\right)W_N^{-1/2}\right) + \ln\left(W_N^{1/2}\right)$$

that

$$\lim_{N \to +\infty} 1/2 \left(1 - \frac{\ln(V_N)}{\ln(N)}\right) = \lim_{N \to +\infty} 1/2 \left(1 - \frac{\ln(W_N)}{\ln(N)}\right) = h_*.$$

4.4 Asymptotics of $\mathbb{E}\, W_N$

In order to obtain the asymptotics of W_N, our next step is to study the asymptotics of $\mathbb{E}\, W_N$. Theorem 2 will then be a consequence of the asymptotics for the variance of W_N and the Borel–Cantelli lemma. Let us rewrite W_N as in (38) with definition (25):

$$W_N = \sum_{p=0}^{N-2} \left(\int_0^{+\infty} \left(\sum_{k=0}^{2} d_k e^{-i\frac{p+k}{N}\xi} \right) \frac{W(d\xi)}{\xi^{H(p/N,\xi)+1/2}} \right)^2.$$

Because of the isometry given by the Wiener measure we have

$$\mathbb{E}\, W_N = \sum_{p=0}^{N-2} \int_0^{+\infty} 16 \sin^4\left(\frac{\xi}{2N}\right) \frac{d\xi}{\xi^{2H(p/N,\xi)+1}}. \tag{39}$$

Let us also remark that

$$16 \int_0^{+\infty} \sin^4\left(\frac{\xi}{2N}\right) \frac{d\xi}{\xi^{2h(p/N)+1}} = \frac{F\left(h(p/N)\right)}{N^{2h(p/N)}} \tag{40}$$

where

$$F(s) = \int_0^{+\infty} 16 \frac{\sin^4(\xi/2)}{\xi^{2s+1}} \, d\xi$$

is a continuous function on the interval $(0,1)$ and hence is bounded on $[a,b]$. Let us consider

$$\widetilde{W}_N = \sum_{p=0}^{N-2} \frac{F\big(h(p/N)\big)}{N^{2h(p/N)}},$$

assumption **A 1** yields

$$\frac{1}{C} N^{1-2h_*} \leqslant \widetilde{W}_N \leqslant C N^{1-2h_*}.$$

Let us show that under assumption **A 2** we have

$$\big| \mathbb{E}\, W_N - \widetilde{W}_N \big| = o(N^{1-2h_*}). \tag{41}$$

Because of (39) and (40)

$$\big| \mathbb{E}\, W_N - \widetilde{W}_N \big| \leqslant \sum_{p=0}^{N-2} \int_0^{+\infty} 16 \sin^4\!\left(\frac{\xi}{2N}\right) \big| \xi^{-2H(p/N,\xi)-1} - \xi^{-2h(p/N)-1} \big| \, d\xi. \tag{42}$$

Then the integrals in (42) are split into three parts:

$$\big| \mathbb{E}\, W_N - \widetilde{W}_N \big| \leqslant J_N^1 + J_N^2 + J_N^3.$$

The first term is

$$J_N^1 = \sum_{p=0}^{N-2} \int_0^1 16 \sin^4\!\left(\frac{\xi}{2N}\right) \big| \xi^{-2H(p/N,\xi)-1} - \xi^{-2h(p/N)-1} \big| \, d\xi$$

$$\leqslant C \sum_{p=0}^{N-2} \int_0^1 \sin^4\!\left(\frac{\xi}{2N}\right) \frac{|H(p/N,\xi) - h(p/N)|}{\xi^{2\theta(p/N,\xi)+1}} \ln(\xi) \, d\xi$$

where $\theta(p/N,\xi) \in [a,b]$. Hence

$$J_N^1 \leqslant C \frac{N-1}{N^4} \int_0^1 \xi \ln(\xi) \, d\xi \leqslant C/N^3$$

and $J_N^1 = o(N^{1-2h_*})$. Let us now consider

$$J_N^2 = \sum_{p=0}^{N-2} \int_1^{N^\alpha} 16 \sin^4\!\left(\frac{\xi}{2N}\right) \big| \xi^{-2H(p/N,\xi)-1} - \xi^{-2h(p/N)-1} \big| \, d\xi$$

$$\leqslant C \sum_{p=0}^{N-2} \int_1^{N^\alpha} \sin^4\!\left(\frac{\xi}{2N}\right) \frac{|H(p/N,\xi) - h(p/N)|}{\xi^{2\theta'(p/N,\xi)+1}} \ln(\xi) \, d\xi$$

where $\theta'(p/N, \xi) \geqslant h(p/N)$. Hence

$$
\begin{aligned}
J_N^2 &\leqslant \sum_{p=0}^{N-2} \frac{C}{N^4} \int_1^{N^\alpha} \frac{\xi^3 |H(p/N,\xi) - h(p/N)|}{\xi^{2h(p/N)}} \ln(\xi)\, d\xi \\
&\leqslant \sum_{p=0}^{N-2} \frac{C}{N^4} N^{\alpha(4-2h(p/N))} \ln(N) \\
&\leqslant C N^{4\alpha - 3 - 2h_* \alpha} \ln(N)
\end{aligned}
$$

and since $\alpha < 1$, $J_N^2 = o(N^{1-2h_*})$. For the third part we have

$$
\begin{aligned}
J_N^3 &= \sum_{p=0}^{N-2} \int_{N^\alpha}^{+\infty} 16 \sin^4\left(\frac{\xi}{2N}\right) \left| \xi^{-2H(p/N,\xi)+1} - \xi^{-2h(p/N)+1} \right| d\xi \\
&\leqslant C \sum_{p=0}^{N-2} \int_{N^\alpha}^{+\infty} \frac{|H(p/N,\xi) - h(p/N)|}{\xi^{2\theta''(p/N,\xi)+1}} \ln(\xi)\, d\xi
\end{aligned}
$$

where $\theta''(p/N, \xi) \geqslant h(p/N)$. Thus,

$$
J_N^3 \leqslant C \sum_{p=0}^{N-2} \int_{N^\alpha}^{+\infty} \frac{|H(p/N,\xi) - h(p/N)|}{\xi^{2h(p/N)+1}} \ln(\xi)\, d\xi
$$

and $J_N^3 = o(N^{1-2h_*})$ because of **A 2**. Hence

$$
\left| \mathbb{E}\, W_N - \widetilde{W}_N \right| = o\left(N^{1-2h_*}\right)
$$

and

$$
\frac{1}{C} N^{1-2h_*} \leqslant \mathbb{E}\, W_N \leqslant C N^{1-2h_*} \tag{43}
$$

is obtained.

4.5 Asymptotics of var(W_N)

Let us now show that almost surely

$$
\lim_{N \to +\infty} \frac{W_N}{\mathbb{E}\, W_N} = 1.
$$

Since the process X is Gaussian, classical formulas for the variance of Gaussian variables lead to:

$$
\begin{aligned}
\operatorname{var}(W_N) = 2 \sum_{p,p'=0}^{N-2} \mathbb{E}\Bigg(&\int_0^{+\infty} \left(\sum_{k=0}^2 d_k e^{-i\frac{p+k}{N}\xi}\right) \frac{W(d\xi)}{\xi^{H(p/N,\xi)+1/2}} \\
&\times \int_0^{+\infty} \left(\sum_{k=0}^2 d_k e^{i\frac{p'+k}{N}\xi}\right) \frac{W(d\xi)}{\xi^{H(p'/N,\xi)+1/2}} \Bigg)^2
\end{aligned} \tag{44}
$$

$$\mathrm{var}(W_N) = 2 \sum_{p,p'=0}^{N-2} \left(\int_0^{+\infty} e^{i(\frac{p'-p}{N})\xi} 16 \sin^4\left(\frac{\xi}{2N}\right) \right.$$

$$\left. \times \frac{d\xi}{\xi^{H(p/N,\xi)+H(p'/N,\xi)+1}} \right)^2. \quad (45)$$

At this point one needs similar estimates for the integrals in the previous formula to the one we had when we considered $\mathbb{E}\,W_N$. However to apply the Borel–Cantelli Lemma we aim at

$$\mathrm{var}(W_N) = O\big(N^{1+2g-4h_*}\ln^2(N)\big)$$

and we prove

$$\left| \int_0^{+\infty} e^{i(\frac{p'-p}{N})\xi} 16 \sin^4\left(\frac{\xi}{2N}\right) \frac{d\xi}{\xi^{H(p/N,\xi)+H(p'/N,\xi)+1}} \right|$$

$$\leqslant C \frac{N^g \ln(N)}{N^{h(p/N)+h(p'/N)}} \frac{1}{|p-p'|+1}$$

which yields the needed estimates. Once more the factor $\frac{1}{|p-p'|+1}$ that allows us to have a better estimate for $g < 1/2$ than $\mathrm{var}(W_N) = O(N^{2-4h_*}\ln^2(N))$ is obtained by integrations by parts in the integrals of (45). Let us compute

$$\frac{\partial}{\partial\xi}\left(\frac{\sin^4\left(\frac{\xi}{2N}\right)}{\xi^{H(p/N,\xi)+H(p'/N,\xi)+1}} \right)$$

$$= \frac{2\sin^3\left(\frac{\xi}{2N}\right)\cos\left(\frac{\xi}{2N}\right)}{N\xi^{H(p/N,\xi)+H(p'/N,\xi)+1}} - \frac{\sin^4\left(\frac{\xi}{2N}\right)\big(H(p/N,\xi)+H(p'/N,\xi)+1\big)}{\xi^{H(p/N,\xi)+H(p'/N,\xi)+2}}$$

$$- \frac{\sin^4\left(\frac{\xi}{2N}\right)\big(\frac{\partial H}{\partial\xi}(p/N,\xi)+\frac{\partial H}{\partial\xi}(p'/N,\xi)\big)\ln(\xi)}{\xi^{H(p/N,\xi)+H(p'/N,\xi)+1}}. \quad (46)$$

Then the contribution of the first term of (46) to (45) is given by

$$4 \int_0^{+\infty} \frac{e^{i(\frac{(p'-p)\xi}{N})}\sin^3\left(\frac{\xi}{2N}\right)\cos\left(\frac{\xi}{2N}\right) d\xi}{i(p-p')\xi^{H(p/N,\xi)+H(p'/N,\xi)+1}} \quad (47)$$

when $p-p' \neq 0$. This integral is then easily bounded by:

$$I_1 = \frac{4}{|p-p'|} \int_0^{+\infty} \frac{\left|\sin^3\left(\frac{\xi}{2N}\right)\right| d\xi}{\xi^{H(p/N,\xi)+H(p'/N,\xi)+1}}$$

which we split into three terms to get the required estimates. The first one is:

$$I_{1,1} = \frac{4}{|p-p'|} \int_0^1 \frac{\left|\sin^3\left(\frac{\xi}{2N}\right)\right| d\xi}{\xi^{H(p/N,\xi)+H(p'/N,\xi)+1}}$$

$$\leqslant \frac{C}{N^3|p-p'|};$$

the second one is

$$I_{1,2} = \frac{4}{|p-p'|} \int_1^N \frac{\left|\sin^3\left(\frac{\xi}{2N}\right)\right| d\xi}{\xi^{H(p/N,\xi)+H(p'/N,\xi)+1}}$$

$$\leqslant \frac{C}{N^3|p-p'|} \int_1^N \xi^{2-h(p/N)-h(p'/N)} \, d\xi$$

$$\leqslant \frac{C}{N^{h(p/N)+h(p'/N)}|p-p'|} \; ;$$

and the third one

$$I_{1,3} = \frac{4}{|p-p'|} \int_N^{+\infty} \frac{\left|\sin^3\left(\frac{\xi}{2N}\right)\right| d\xi}{\xi^{H(p/N,\xi)+H(p'/N,\xi)+1}}$$

$$\leqslant \frac{C}{\left(N^{h(p/N)+h(p'/N)}\right)(|p-p'|)} \, .$$

Hence

$$|I_1| \leqslant \left(\frac{C_1}{N^3} + \frac{C_2}{N^{h(p/N)+h(p'/N)}}\right)\left(\frac{1}{|p-p'|+1}\right) \tag{48}$$

where the easier case $p = p'$ is also considered. With similar arguments we get for the second term of (46)

$$|I_2| \leqslant \left(\frac{C_1}{N^3} + \frac{C_2}{N^{h(p/N)+h(p'/N)}}\right)\left(\frac{N^g}{|p-p'|+1}\right) \tag{49}$$

and the last term of (46) yields

$$|I_3| \leqslant \left(\frac{C_3}{N^3} + \frac{C_4}{N^{h(p/N)+h(p'/N)}}\right)\left(\frac{N^g \ln(N)}{|p-p'|+1}\right) \tag{50}$$

where (14) has been used for $k = 1$. Hence we deduce that

$$\mathrm{var}(W_N) = O(N^{1+2g-4h_*} \ln^2(N)).$$

4.6 Almost sure convergence

In order to show that

$$\lim_{N\to+\infty} \frac{W_N}{\mathbb{E}\,W_N} = 1 \qquad \text{almost surely}$$

we apply Markov inequality for $A > 0$:

$$\mathbb{P}\left(\left|\frac{W_N}{\mathbb{E}\,W_N} - 1\right|^4 \geqslant A\right) \leqslant \mathbb{E}\left|\frac{W_N}{\mathbb{E}\,W_N} - 1\right|^4 \Big/ A.$$

Moreover the same arguments as in the proof of Proposition 2 p. 43 of [9] yield that

$$\mathbb{E}\,|W_N - \mathbb{E}\,W_N|^4 \leqslant C\,\mathrm{var}(W_N)^2$$

then

$$\mathbb{P}\left(\left|\frac{W_N}{\mathbb{E}\,W_N} - 1\right|^4 \geqslant A\right) \leqslant \frac{C\,\mathrm{var}(W_N)^2}{(\mathbb{E}\,W_N)^4}.$$

For $g < 1/4$ the sequence

$$\frac{\mathrm{var}(W_N)^2}{(\mathbb{E}\,W_N)^4} \leqslant CN^{4g-2}\ln^2(N)$$

is summable and the Borel–Cantelli lemma yields the required almost sure convergence. With (43) and the results of section 4.3, this fact concludes the proof of Theorem 2.

4.7 Proof of Corollary 1

To prove the corollary, let us fix $t \in (0,1)$ and set

$$h_*(\varepsilon) = \inf_{\{|s-t|<\varepsilon\}} h(s)$$

Let also

$$\mathbf{W}_{\varepsilon,N} = \sum_{p\in V_{\varepsilon,N}(t)} \left(Y\left(\frac{p+2}{N},\frac{p}{N}\right) - 2Y\left(\frac{p+1}{N},\frac{p}{N}\right) + Y\left(\frac{p}{N},\frac{p}{N}\right)\right)^2.$$

With the same arguments as in the proof of Theorem 2

$$\frac{\varepsilon}{C}N^{1-2h_*(\varepsilon)} \leqslant \mathbb{E}\,\mathbf{W}_{\varepsilon,N} \leqslant C\varepsilon N^{1-2h_*(\varepsilon)}$$

and, if we take $\varepsilon = N^{-\gamma}$,

$$\lim_{N\to+\infty} 1/2\left(1 - \frac{\ln(\mathbb{E}\,\mathbf{W}_{N^{-\gamma},N})}{\ln N}\right) = \liminf_{s\to t} h(s).$$

Moreover

$$\mathrm{var}(\mathbf{W}_{\varepsilon,N}) = O\!\left(\varepsilon N^{1+2g-4h_*(\varepsilon)}\ln^2(N)\right)$$

and if we take $0 < \gamma < 1/2$,

$$\lim_{N\to+\infty} \frac{\mathbf{W}_{N^{-\gamma},N}}{\mathbb{E}\,\mathbf{W}_{N^{-\gamma},N}} = 1 \qquad (51)$$

almost surely. Next

$$\left|\mathbf{W}_{\varepsilon,N}^{1/2} - \mathbf{W}_{\varepsilon,N}^{1/2}\right|\mathbf{W}_{\varepsilon,N}^{-1/2} \leqslant CN^{b-\beta}$$

is still true and the corollary is proved.

References

1. Ayache, A.: 2001, Du mouvement Brownien fractionnaire au mouvement Brownien multifractionnaire, *Technique et science informatiques*, **20**, 1133–1152.
2. Ayache, A.: 2002, The generalized multifractional field: a nice tool for the study of the generalized multifractional Brownian motion. *The Journal of Fourier Analysis and Applications*, **8**, 581–601.
3. Ayache, A. and Lévy Véhel, J.: 2001, Processus à régularité locale prescrite, *Comptes Rendus Académie des Sciences de Paris*, **333**, I, 233–238.
4. Ayache, A. and Lévy Véhel, J.: 2000, The Generalized Multifractional Brownian Motion. *Statistical Inference for Stochastic Processes*, **3**, 7–18.
5. Ayache, A. and Lévy Véhel, J.: 1999, Generalized multifractional Brownian motion: Definition and preliminary results, *in* M. Dekking, J. Lévy Véhel, E. Lutton and C. Tricot (eds), *Fractals: Theory and Applications in Engineering*, Springer Verlag, 17–32.
6. Beran, J.: Statistics for Long-Memory Processes, Chapman and Hall, London, U.K., 1994.
7. Benassi, A., Bertrand, P., Cohen, S. and Istas, J.: 2000, Identification of the Hurst exponent of a step multifractional Brownian motion, *Statistical Inference for Stochastic Processes*, **3**, 101–110.
8. Benassi, A., Cohen, S. and Istas, J.: 1998a, Identifying the multifractional function of a Gaussian process, *Statistic and Probability Letters*, **39**, 337–345.
9. Benassi, A., Cohen, S., Istas, J. and Jaffard, S.: 1998b, Identification of Filtered White Noises, *Stoch. Proc. Appl.*, **75**, 31–49.
10. Benassi, A., Jaffard, S. and Roux, D.: 1997, Gaussian processes and Pseudodifferential Elliptic operators, *Revista Mathematica Iberoamericana*, **13**(1), 19–89.
11. Canus, C. and Lévy Véhel, J.: 1996, Change detection in sequences of images by multifractal analysis, *Proc. ICASSP-96, May 7-10, Atlanta, GA*.
12. Cohen, S.: 1999, From self-similarity to local self-similarity: the estimation problem, *in* M. Dekking, J. Lévy Véhel, E. Lutton and C. Tricot (eds), *Fractals: Theory and Applications in Engineering*, Springer Verlag, 3–16.
13. Cohen, S.: 2001, Champs localement auto-similaires, *in* P. Abry, P. Gonçalvès and J. Lévy Véhel (eds), *Lois d'échelles, fractales, et ondelettes*, Volume 1 Hermes.
14. Daoudi, K., Lévy Véhel, J. and Meyer, Y.:1996, Construction of continuous functions with prescribed local regularity, *Const. Approx.*, **014**, (03) 349–385.
15. Frisch, U. and Parisi, G.: 1985, Turbulence and predictability in geophysical fluid dynamics and climate dynamics, *Ghil,M. Benzi,R. and Parisi,G. eds, Amsterdam (Holland)*.
16. Kolmogorov, A.: 1940, Wienersche Spiralen und einige andere interessante Kurven in Hilbertsche Raum., *C. R. (Dokl.) Acad. Sci. URSS*, **26**, 115–118.
17. Lévy Véhel, J.: 1996, Introduction to the multifractal analysis of images, in *Fractal Geometry and Analysis,The Mandelbrot Festschrift, Curacao 1995, C.J.G. Evertsz, H.-O. Peitgen and R.F. Voss Editors, World Scientific,1996*.
18. Lévy Véhel, J. and Mignot, P.: 1994, Multifractal Segmentation of Images, *Fractals*, **2**(3), 371–378.
19. Mandelbrot, B. and Van Ness, J.: 1968, Fractional Brownian motion, fractional noises and applications, *Siam Review*, **10**, 422–437.
20. Meyer, Y.: 1990, *Ondelettes et Opérateurs*, Vol. 1, Hermann, Paris.

21. Ndoye, M.: 2001, Estimation de la fonction d'identification d'un processus gaussien multifractionnaire: le cas Hölderien, private communication.
22. Peltier, R. and Lévy Véhel, J.: 1996, Multifractional Brownian motion: definition and preliminary results, available at http://www-rocq.inria.fr/fractales/.

Failure of the Raikov Theorem
for Free Random Variables

Florent Benaych-Georges

DMA, École Normale Supérieure, 45 rue d'Ulm, 75230 Paris Cedex 05
e-mail: benaych@dma.ens.fr
http://www.dma.ens.fr/~benaych

Summary. We show that the sum of two free random variables can have a free Poisson law without any of them having a free Poisson law.

Key words: free probability theory, Raikov theorem, free Poisson distribution, free convolution, R-transform.

MSC 2000: primary 46L54, secondary 60E07.

The classical convolution $*$ of probability measures on $\mathbb{R}$ has an analogue in the theory of free probability, the free convolution $\boxplus$ (see [3]), and the structures of semi-groups defined by classical and free convolutions on the set of probability measures on $\mathbb{R}$ present many analogies. For example, we can define infinitely divisible laws in the same way for both convolutions, and there exists a natural isomorphism between the semi-group of classical infinitely divisible laws and the semi-group of free infinitely divisible laws (see [2]). This isomorphism transforms a Gaussian law into the semi-circle law with the same mean and the same variance, a Poisson law into the Marchenko–Pastur law with the same mean (see [5]). Moreover, there is a deep correspondence between limit theorems for weak convergence of sums of independent random variables and limit theorems for weak convergence of sums of free random variables (see [2]). It then seems natural to see how far the analogy goes.

Two already established properties of free convolution show that the isomorphism between the semi-group of classical infinitely divisible laws and the semi-group of free infinitely divisible laws cannot be extended to the set of probability measures. The first one is the following (see [7]): for every probability measure μ on the real line, there exists a family $(\mu_t)_{t \geqslant 1}$ of probability measures such that $\mu_1 = \mu$ and for all s, $t \geqslant 1$, one has $\mu_{s+t} = \mu_s \boxplus \mu_t$. There is no similar result when we replace $\boxplus$ by $*$ for $\mu = \frac{1}{2}(\delta_0 + \delta_1)$. The second one is the failure of the Cramér theorem for free random variables: the sum of two free random variables can be distributed according to the semi-circle law without any of them having a semi-circle law—the classical Cramér theorem states that if the sum of two independent random variables is Gaussian, then each of them is Gaussian.

We will show in this article that the analogue of Raikov's theorem for free probability is false: the sum of two free random variables can have a free Poisson law (i.e. a Marchenko–Pastur law) without any of them (even after translation) having a free Poisson law—the classical Raikov theorem states that if the sum of two independent random variables has, after translation, a Poisson law, then each of them has, after translation, a Poisson law. The proof is very close to the proof given in [4] of the failure of the Cramér theorem for free random variables.

In order to prove our result, we need to discuss the analytic method for calculating free convolutions. We will restrict ourselves to compactly supported probability measures on $\mathbb{R}$. Let $\mathbb{C}^+$ (resp. $\mathbb{C}^-$) denote the upper (resp. lower) half plane. Given μ a compactly supported measure on $\mathbb{R}$, consider the function

$$G_\mu : \mathbb{C}^+ \longrightarrow \mathbb{C}^-$$

$$z \longmapsto \int \frac{\mathrm{d}\mu(t)}{z - t}.$$

This function, called Cauchy transform of μ, is analytic, and $\lim_{|z| \to \infty} zG_\mu(z) = 1$.

Hence there exists a function K_μ that is meromorphic in a neighborhood in $\mathbb{C}^-$ of zero, with a single pole at zero such that $G_\mu(K_\mu(z)) = z$ for z close to zero. One can write $K_\mu(z) = \frac{1}{z} + R_\mu(z)$, where R_μ is an analytic function in a neighborhood of zero in $\mathbb{C}^-$ (i.e. $V \cap \mathbb{C}^-$ with V a neighborhood of zero in $\mathbb{C}$), called the R-transform of μ. It was shown in [8] (see also in [9]) that R_μ determines μ and that $R_{\mu \boxplus \nu} = R_\mu + R_\nu$.

The R-transform of the free Poisson law with mean λ, that we will denote by μ_λ, is given by the formula

$$R_{\mu_\lambda}(z) = \frac{\lambda}{1 - z}.$$

Theorem 1. *Consider $0 < \lambda < 1$, $r \in \left]0, \frac{\sqrt{\lambda}}{1-\lambda}\right[$. Let $(f_n)_{n \in \mathbb{N}}$ be a sequence of analytic functions on $\left\{z; \left|z - \frac{1}{1-\lambda}\right| > r\right\}$, such that*

(i) *for all n, for all z, $f_n(\bar{z}) = \overline{f_n(z)}$,*

(ii) *for all n, $\lim_\infty zf_n'(z) = \lim_\infty f_n(z) = 0$,*

(iii) *the sequence (f_n) converges uniformly to 0 when n tends to infinity.*

Then when n is large enough, the function $z \mapsto \frac{\lambda}{1-z} + f_n(z)$ is the R-transform of a compactly supported probability measure on $\mathbb{R}$.

Before giving the proof of the theorem, here is the corollary that has led us to establish it.

Corollary 1. *Consider* $\lambda > 0$. *There exist two compactly supported probability measures on* $\mathbb{R}$ *which, even after translation, are not free Poisson laws, and whose free convolution is the free Poisson law with mean* λ.

Proof. Consider $0 < \lambda' < \min\{1, \lambda/2\}$. It suffices to define, with ε small enough, μ_+ to be the law whose R-transform is $z \mapsto \frac{\lambda'}{1-z} + \frac{\varepsilon}{(z-1/(1-\lambda))^2}$ (such a law exists by the theorem), and μ_- to be the law whose R-transform is $z \mapsto \frac{\lambda'}{1-z} - \frac{\varepsilon}{(z-1/(1-\lambda))^2} + \frac{\lambda-2\lambda'}{1-z}$ (such a law exists by the theorem and because the sum of two R-transforms is a R-transform). Then the pair μ_+, μ_- is suitable. $\qquad\square$

As a preliminary to the proof of the theorem, here is a lemma that states the existence of a measure (see [1]).

Lemma 1. *Let G be an analytic function from the upper half plane $\mathbb{C}^+$ to the lower half plane $\mathbb{C}^-$ that satisfies:*

(i) *$zG(z)$ tends to 1 when $|z|$ tends to infinity,*

(ii) *denoting by K the inverse function of G from a neighborhood of 0 in $\mathbb{C}^-$ (i.e. from $V \cap \mathbb{C}^-$ with V a neighborhood of zero in $\mathbb{C}$) to a neighborhood of infinity in $\mathbb{C}^+$, the function $z \mapsto K(z) - \frac{1}{z}$ can be extended to a neighborhood of zero in $\mathbb{C}$.*

Then there exists a compactly supported probability measure on $\mathbb{R}$ whose Cauchy transform is G.

Here is the proof of the theorem. The symbols $\Re$ and $\Im$ will denote respectively the real and imaginary parts.

Proof. By the lemma, it is enough to show that when n is large enough, there exists a domain Ω_n of $\mathbb{C}^-$, neighborhood of 0 in $\mathbb{C}^-$, such that K_n induces a bijection from Ω_n onto $\mathbb{C}^+$, where $K_n(z) = \frac{1}{z} + \frac{\lambda}{1-z} + f_n(z)$; and that the inverse function of K_n satisfies (i) and (ii) of the lemma. For convenience, we will work with the functions $\psi_n(z) = K_n\big(\frac{\sqrt{\lambda}}{1-\lambda}z + \frac{1}{1-\lambda}\big)$, and $\psi(z) = K_{\mu_\lambda}\big(\frac{\sqrt{\lambda}}{1-\lambda}z + \frac{1}{1-\lambda}\big)$. Recall $K_{\mu_\lambda}(z) = \frac{1}{z} + \frac{\lambda}{1-z}$.

Fix $\alpha \in \,]0, 1/10[$ such that $1 - 2\alpha > \max\{r, \sqrt{\lambda}\}$ and $1 + 2\alpha < 1/\sqrt{\lambda}$, and fix $\eta \in \,]0, \alpha/2[$.

First, let us note few properties of the function ψ, that we will then extend to the functions ψ_n with n large enough, using Rouché's theorem.

We have

$$\psi(z) = \frac{z(1-\lambda)^2}{(\sqrt{\lambda}z + 1)(z + \sqrt{\lambda})},$$

so zero is the unique preimage of zero by ψ, and for every non zero Z, the set of the preimages of Z by ψ is the set of the roots of

$$X^2 + \left(\frac{1}{\sqrt{\lambda}} + \sqrt{\lambda} - \frac{(1-\lambda)^2}{Z\sqrt{\lambda}}\right)X + 1.$$

Their product is one, so ψ is one-to-one in $\mathbb{C}^+$ and in $\mathbb{C}^-$. We deduce:

(a) when n is large enough, ψ_n is one-to-one in a neighborhood of $A = \{z; 1 - \eta \leqslant |z| \leqslant 1 + \eta, |z \pm 1| \geqslant \eta, \Im z \leqslant 0\}$.

For the same reason, ψ is one-to-one in $\{z \in \mathbb{C}; 0 < |z| < 1, z \neq -\sqrt{\lambda}\}$ and in $\{z \in \mathbb{C}; 1 < |z|, z \neq -\frac{1}{\sqrt{\lambda}}\}$. Hence:

(b) for n large enough, ψ_n is one-to-one in a neighborhood of the corona $B = \{z; 1 - 2\eta \leqslant |z| \leqslant 1 - \eta\}$ and in a neighborhood of the corona $C = \{z; 1 + \eta \leqslant |z| \leqslant 1 + 2\eta\}$.

We have

$$\Im\big(\psi(z)\big) = \frac{(1-\lambda)^2\sqrt{\lambda}}{|\sqrt{\lambda}z + 1||z + \sqrt{\lambda}|}(1 - |z|^2)\,\Im\, z,$$

so $\psi(z)$ is real if and only if z is real or has modulus one, hence:

(c) for n large enough, $\Im\psi_n(z) < 0$ for $z = re^{i\theta} \in B$ with $\theta \in [-\pi + \eta, -\eta]$,

(d) for n large enough, $\Im\psi_n(z) > 0$ for $z = re^{i\theta} \in C$ with $\theta \in [-\pi + \eta, -\eta]$.

On the other hand, we have

$$\psi'(z) = (1-\lambda)^2\sqrt{\lambda}\,\frac{z^2 - 1}{\left(\sqrt{\lambda}z + 1\right)^2\left(z + \sqrt{\lambda}\right)^2}$$

so ψ' vanishes only on 1 and -1, and for all z in the unit circle such that $z \neq \pm 1$, the gradient $\nabla(\Im\psi)(z)$ is non null and orthogonal to the tangent of the circle, hence

(e) for n large enough, $\partial\Im\psi_n/\partial r > 0$ on $z = re^{i\theta} \in A$ with $\theta \in [-\pi + \eta, -\eta]$.

At last, $\psi''(1)$ and $\psi''(-1)$ are non null, so

(f) for n large enough, ψ_n' is one-to-one ψ_n is at most two-to-one in a neighborhood of each of the disks $D = \{z; |z + 1| \leqslant \alpha\}$ and $E = \{z; |z - 1| \leqslant \alpha\}$,

(g) for n large enough, ψ_n' has no zero in both disks $\{z; |z \pm 1| < \eta\}$.

For such a value of n, we actually have

(c') $\Im\psi_n(z) < 0$ for $z = re^{i\theta} \in B$ with $\theta \in \,]-\pi, 0[$,

(d') $\Im\psi_n(z) > 0$ for $z = re^{i\theta} \in C$ with $\theta \in \,]-\pi, 0[$.

Indeed, if for instance $\Im\psi_n(z)$ were positive for some $z \in B$ in the lower half plane, then we would deduce the existence of $z' \in B \cap \mathbb{C}^-$ for which $\Im\psi_n(z') = 0$. But then $\psi_n(\overline{z'}) = \overline{\psi_n(z')} = \psi_n(z')$, and this contradicts condition (b). Let u_n and v_n be the unique zeros of ψ_n' in D and E, respectively. Observe that u_n and v_n must be real. Indeed, as $\psi_n'(\overline{u_n}) = \overline{\psi_n'(u_n)} = 0$ and $\overline{u_n} \in D$, uniqueness implies $u_n = \overline{u_n}$.

Consider now the set

$$F_{n,0} = \{z \in (A \cup D \cup E); z \notin \mathbb{R}, \psi_n(z) \in \mathbb{R}\} \cup \{u_n, v_n\}.$$

We can show that $F_{n,0}$ is an analytic curve that contains a curve γ_n joining u_n to v_n in the half upper plane (the proof is the same as the one page 219 in the article by D. Voiculescu and H. Bercovici [4]).

Denote by Θ_n the part of the lower half plane below $]-\infty, u_n] \cup \gamma_n \cup [v_n, +\infty[$. Let us show that ψ_n is a one-to-one map from Θ_n onto the half upper plane.

For $R > 100$, let $\gamma_{n,R}$ be the contour that runs (anticlockwise) through

$$\gamma_n^* \cup [v_n, R] \cup \{z \in \mathbb{C}^-; |z| = R\} \cup \left[-R, -\frac{1}{\sqrt{\lambda}} - \frac{1}{R}\right]$$
$$\cup \left\{z \in \mathbb{C}^-; \left|z + \frac{1}{\sqrt{\lambda}}\right| = \frac{1}{R}\right\} \cup \left[-\frac{1}{\sqrt{\lambda}} + \frac{1}{R}, u_n\right].$$

By the residues formula, it suffices to show that for every ω in the upper half plane, the integral of $\frac{\psi_n'}{\psi_n - \omega}$ over the contour $\gamma_{n,R}$ tends to $2i\pi$ when R tends to infinity (this integral is well defined for R large enough, when n and ω are fixed, because $\lim_{\infty} \psi_n = 0$).

(1) Computation of the limit of the integral on the complementary of the semi-circles in the curve $\gamma_{n,R}$:
By our hypothesis on f_n,

$$\lim_{x \to -\infty} \psi_n(x) = 0 \quad \text{and} \quad \lim_{x \xrightarrow{<} -\frac{1}{\sqrt{\lambda}}} \psi_n(x) = +\infty,$$

$$\lim_{x \to +\infty} \psi_n(x) = 0 \quad \text{and} \quad \lim_{x \xrightarrow{>} -\frac{1}{\sqrt{\lambda}}} \psi_n(x) = +\infty.$$

The function ψ_n is real on the complement of the semi-circles, so, when R tends to $+\infty$, the integral of $\frac{\psi_n'}{\psi_n - \omega}$ on this complement tends to

$$\int_{-\infty}^{+\infty} \frac{dy}{y - \omega} = i\pi.$$

(2) Computation of the limit of the integral on the large semi-circle:
By hypothesis,

$$\lim_{R \to \infty} \left(R \sup_{|z|=R} \left|\frac{\psi_n'(z)}{\psi_n(z) - \omega}\right|\right) = 0,$$

so the integral on the large semi-circle tends to 0 when R tends to infinity.
(3) Computation of the limit of the integral on the small semi-circle:
By the same arguments as in (1) and (2), the integral of $\frac{\psi'}{\psi - \omega}$ on the complementary of the small semi-circle in the contour

$$\{z \in \mathbb{C}^-; |z| = 1\} \cup [1, R] \cup \{z \in \mathbb{C}^-; |z| = R\} \cup \left[-R, -\frac{1}{\sqrt{\lambda}} - \frac{1}{R}\right]$$
$$\cup \left\{z \in \mathbb{C}^-; |z + \frac{1}{\sqrt{\lambda}}| = \frac{1}{R}\right\} \cup \left[-\frac{1}{\sqrt{\lambda}} + \frac{1}{R}, -1\right]$$

(running anticlockwise) tends to $i\pi$ when R tends to $+\infty$. Otherwise, ψ being a one-to-one map from $\{z \in \mathbb{C}^-; |z| > 1\}$ to $\mathbb{C}^+$, the integral of $\frac{\psi'}{\psi - \omega}$ on the contour

$$\{z \in \mathbb{C}^-; |z| = 1\} \cup [1, R] \cup \{z \in \mathbb{C}^-; |z| = R\} \cup \left[-R, -\frac{1}{\sqrt{\lambda}} - \frac{1}{R}\right]$$

$$\cup \left\{z \in \mathbb{C}^-; |z + \frac{1}{\sqrt{\lambda}}| = \frac{1}{R}\right\} \cup \left[-\frac{1}{\sqrt{\lambda}} + \frac{1}{R}, -1\right]$$

(going anticlockwise) tends to $2i\pi$ when R tends to $+\infty$, so the integral of $\frac{\psi'}{\psi - \omega}$ on $\{z \in \mathbb{C}^-; |z + 1/\sqrt{\lambda}| = \frac{1}{R}\}$ tends to $i\pi$ when R tends to $+\infty$. But clearly, because of the boundedness of f_n and f_n' in a neighborhood of zero, we have

$$\lim_{R \to +\infty} \int_{|z + \frac{1}{\sqrt{\lambda}}|, \, \Im z \leqslant 0} \left(\frac{\psi_n'(z)}{\psi_n(z) - \omega} - \frac{\psi'(z)}{\psi(z) - \omega}\right) dz = 0.$$

So the integral of $\frac{\psi_n'(z)}{\psi_n(z) - \omega}$ on the small semi-circle of $\gamma_{n,R}$ tends to $i\pi$ when R tends to $+\infty$.

Thus, ψ_n is a one-to-one map from Θ_n onto the upper half plane. Hence K_n is a one-to-one map from $\left\{\frac{\sqrt{\lambda}}{1-\lambda} z + \frac{1}{1-\lambda}; \ z \in \Theta_n\right\}$, that we denote by Ω_n, onto $\mathbb{C}^+$.

Let us show that the inverse function G_n of K_n satisfies the hypothesis of the lemma.

(i) Let us show that $z G_n(z)$ tends to 1 when $|z|$ tends to infinity. The function K_n is bounded in the complementary, in Ω_n, of any open disk centered in zero with radius larger than $1 + 1/\sqrt{\lambda}$, so $G_n(z)$ tends to zero when $|z|$ tends to infinity. But $z G_n(z) = K_n(G_n(z)) G_n(z)$, so the limit of $z G_n(z)$ when $|z|$ tends to infinity is the limit of $y K_n(y)$ when y tends to zero, that is 1.

(ii) The function that associates $K_n(z) - \frac{1}{z}$ to z can be analytically extended to a neighborhood of zero in $\mathbb{C}$ because f_n is analytic in a neighborhood of zero.

So the theorem is proved.   □

The proof is very close to the proof given in [4] of the failure of the Cramér theorem for free random variables. In that article, the authors show also the superconvergence to the central limit theorem for free random variables: in addition to the weak convergence, we know that for n large enough, the measures have analytic densities, and that their densities converge uniformly to the density of the semi-circle law. The superconvergence in the free Poisson limit theorem has already been shown (using explicit computations) in [5].

Aknowledgments. The author would like to thank his advisor Philippe Biane for his comments on a draft of this paper.

References

1. Akhiezer, N.I. *The classical moment problem*, Moscou, 1961.
2. Bercovici, H., Pata, V., with an appendix by Biane, P. *Stable laws and domains of attraction in free probability theory*, Annals of Mathematics, **149** (1999), 1023–1060.
3. Bercovici, H., Voiculescu, D. *Free convolution of measures with unbounded supports*, Indiana Univ. Math. J. **42** (1993), 733–773.
4. Bercovici, H., Voiculescu, D. *Superconvergence to the central limit and failure of the Cramér theorem for free random variables*, Probability Theory and Related Fields **102** (1995), 215–222.
5. Hiai, F., Petz, D. *The semicircle law, free random variables, and entropy*, Amer. Math. Soc., Mathematical Surveys and Monographs Volume **77**, 2000.
6. Rudin, W. *Real and complex analysis*, McGraw-Hill Book Co., New York, 1987.
7. Speicher, R. *Notes of my lectures on Combinatorics of Free Probability*, IHP, Paris, (1999). Available on http://www.mast.queensu.ca/~speicher
8. Voiculescu, D.V. *Addition of certain non-commuting random variables*, J. Funct. Anal. **66** (1986), 323–346.
9. Voiculescu, D.V., Dykema, K., Nica, A. *Free random variables*, CRM Monographs Series No.1, Amer. Math. Soc., Providence, RI, 1992.

A Sharp Small Deviation Inequality
for the Largest Eigenvalue of a Random Matrix

Guillaume Aubrun

Université de Paris 6, Institut de Mathématiques, Équipe d'Analyse Fonctionnelle,
Boîte 186, 4 place Jussieu, 75005 PARIS
e-mail: aubrun@clipper.ens.fr

Summary. We prove that the convergence of the largest eigenvalue λ_1 of a $n \times n$ random matrix from the Gaussian Unitary Ensemble to its Tracy–Widom limit holds in a strong sense, specifically with respect to an appropriate Wasserstein-like distance. This unifying approach allows us both to recover the limiting behaviour and to derive the inequality $\mathbb{P}(\lambda_1 \geqslant 2 + t) \leqslant C \exp(-cnt^{3/2})$, valid uniformly for all n and t. This inequality is sharp for "small deviations" and complements the usual "large deviation" inequality obtained from the Gaussian concentration principle. Following the approach by Tracy and Widom, the proof analyses several integral operators, which converge in the appropriate sense to an operator whose determinant can be estimated.

Key words: Random matrices, largest eigenvalue, GUE, small deviations.

Introduction

Let $\mathcal{H}_n$ be the set of n-dimensional (complex) Hermitian matrices. The general element of $\mathcal{H}_n$ is denoted by $A^{(n)}$, and its entries are denoted by $(a_{ij}^{(n)})$.

We exclusively focus on the Gaussian Unitary Ensemble GUE, which can be defined by the data of a probability measure $\mathbb{P}_n$ on $\mathcal{H}_n$ which fulfills the following conditions:

1. The n^2 random variables $(a_{ii}^{(n)}), (\Re a_{ij}^{(n)})_{i<j}, (\Im a_{ij}^{(n)})_{i<j}$ are independent,
2. $\forall i$, $a_{ii}^{(n)}$ follows the Gaussian law $N(0, 1/n)$,
3. $\forall i < j$, $\Re a_{ij}^{(n)}$ and $\Im a_{ij}^{(n)}$ follow the Gaussian law $N(0, 1/2n)$.

The measure $\mathbb{P}_n$ is uniquely determined by these three conditions because of the extra symmetry constraint $a_{ij} = \overline{a_{ji}}$; it can also be made explicit. $\mathcal{H}_n$ is·a vector space on which the scalar product $\langle u, v \rangle := \operatorname{tr}(uv)$ induces a Euclidean structure, hence a Lebesgue measure. The probability measure $\mathbb{P}_n$ has a density with respect to this Lebesgue measure, which can be shown to be equal to

$$\mathrm{d}\mathbb{P}_n := \frac{1}{c_n} \exp\left(-\frac{n}{2}\operatorname{tr}M^2\right)\mathrm{d}M$$

where c_n is a normalization constant.

The GUE has the wonderful property of invariance under rotation: indeed the measure $\mathbb{P}_n$ is invariant under the conjugation action of the unitary group. This makes calculations easier and is very useful for the study of eigenvalues, which are also invariant under the same action.

From now on, "random matrix" means "element of $(\mathcal{H}_n, \mathbb{P}_n)$" seen as a probability space. Let $(\Omega, \mathbb{P})$ denote the product of all these probability spaces $\prod_{i=n}^{\infty}(\mathcal{H}_n, \mathbb{P}_n)$; an element of Ω is a sequence of random matrices, the n-th matrix being of size n. However, for all the questions we shall consider, the relationships between the $\mathcal{H}_n$'s for different n's are immaterial.

For more background, we refer the reader to the monograph [10]. Throughout the argument, $C, c, c', \dots$ will stand for positive universal constants, independent of the dimension and of any other parameters that may be involved. The values of these constants may change from place to place.

Let $\lambda_1(A^{(n)}) \geqslant \lambda_2(A^{(n)}) \geqslant \dots \geqslant \lambda_n(A^{(n)})$ be the ordered eigenvalues of a random matrix $A^{(n)}$. The global asymptotic behavior of these eigenvalues is well-known. The most famous result in this topic is the semi-circle law, which can be stated as follows: let $N(A^{(n)})$ be the probability measure on $\mathbb{R}$ derived from the random matrix $A^{(n)}$ in the following way (δ_x denotes the Dirac mass at point x)

$$N\big(A^{(n)}\big) := \sum_{k=1}^{n} \delta_{\lambda_k(A^{(n)})}.$$

Then, $\mathbb{P}$-almost surely, the sequence of probabilities $(N(A^{(n)})$ converges weakly to a deterministic measure μ_c, with a density with respect to Lebesgue measure given by

$$\mathrm{d}\mu_c := \frac{1}{2\pi}\,1_{[-2,2]}\sqrt{4 - x^2}\,\mathrm{d}x.$$

We are interested here in the asymptotic behavior of the largest eigenvalue $\lambda_1(A^{(n)})$, which is a so-called local problem. Classical results (see e.g. [1], also for precise references to the original articles) claim that

$$\lim_{n\to\infty} \lambda_1\big(A^{(n)}\big) = 2 \qquad \mathbb{P}\text{-almost surely.}$$

The asymptotic behaviour of $\lambda_1(A^{(n)})$ was further clarified by Tracy and Widom, who proved the following result: there exists a continuous decreasing function ψ_{TW} from $\mathbb{R}$ onto $(0,1)$ such that

$$\lim_{n\to\infty} \mathbb{P}_n\big(\lambda_1\big(A^{(n)}\big) \geqslant 2 + xn^{-2/3}\big) = \psi_{\mathrm{TW}}(x). \tag{1}$$

This function ψ_{TW} naturally arises as a determinant linked to the so-called "Airy kernel", which will be defined later. The most difficult point of Tracy and Widom's work was to show that this function ψ_{TW} can be written in terms of a Painlevé function (see [17]). From this point one can deduce asymptotic behavior of ψ_{TW} around $+\infty$ and find universal positive constants C, c, C', c' such that for x large enough

$$c' \exp\left(-C' x^{3/2}\right) \leqslant \psi_{\mathrm{TW}}(x) \leqslant c \exp\left(-C x^{3/2}\right). \tag{2}$$

The remainder of this article is organized as follows: in section 1, we define an appropriate Wasserstein distance and state our main theorem which asserts that Tracy–Widom convergence holds in this strong distance. In section 2, we derive from this theorem the small deviation inequality and compare it with the classical one. Section 3 introduces the needed framework of determinantal kernels, which are classical in this field, and section 4 contains the proof of the main theorem. Finally, section 5 contains an alternative simple derivation of upper bounds (2) for the Tracy–Widom distribution.

1 Convergence in Terms of a Wasserstein Distance

We call tail function of a measure μ on $\mathbb{R}$ the function $\psi_\mu : \mathbb{R} \to [0,1]$ defined by $\psi_\mu(x) := \mu((x,+\infty))$. Such a function is decreasing, left-continuous, tends to 1 at $-\infty$ and to 0 at $+\infty$. The tail function just equals 1 minus the cumulative distribution function. The function appearing in the r.h.s. of (1) is the tail function of the Tracy–Widom distribution on $\mathbb{R}$ (we denote this distribution by TW).

We want to prove that the law of the rescaled largest eigenvalue tends to the Tracy–Widom law in a strong sense. As we only focus on the upper tail, we can consider truncated laws, supported on an interval $[a,+\infty)$ for some real a. Let Λ_n^a be the probability measure with tail function defined by

$$\psi_{\Lambda_n^a}(x) = \begin{cases} \mathbb{P}_n(\lambda_1(A^{(n)}) \geqslant 2 + xn^{-2/3}) & \text{if } x \geqslant a, \\ 1 & \text{if } x < a. \end{cases}$$

Similarly, let TW^a be the truncated Tracy–Widom law defined by

$$\psi_{TW^a}(x) = \begin{cases} \psi_{\mathrm{TW}}(x) & \text{if } x \geqslant a, \\ 1 & \text{if } x <_, a. \end{cases}$$

We are going to show that for any a, Λ_n^a tends to TW^a with respect to the distance defined through a mass transportation problem in its Monge–Kantorovich formulation (see [12]).

A mass transportation problem is the question of optimizing the transshipment from a measure to another with respect to a given cost. More precisely, let μ and ν be two probability measures on the same space X, and $c : X \times X \to \mathbb{R}_+$ a symmetric function vanishing on the diagonal ($c(x,y)$ represent the price to pay to transfer a unit of mass from x to y). Ways to carry μ onto ν are represented through probability measures π on the square space $X \times X$ having μ and ν as marginals (this means that for any measurable subset A of X, $\pi(A \times X) = \mu(A)$ and $\pi(X \times A) = \nu(A)$). We denote by $\Pi(\mu, \nu)$ the space of such π.

The Wasserstein distance associated with the problem is the "minimum cost to pay", defined by

$$d(\mu, \nu) = \inf_{\pi \in \Pi(\mu,\nu)} \int_{X^2} c(x,y) \, \mathrm{d}\pi(x,y).$$

We are going to consider a very special case of this problem. Let us suppose that $X = \mathbb{R}$ and that the cost c is defined as follows

$$c(x,y) := \left| \int_x^y w(t) \, \mathrm{d}t \right| \tag{3}$$

where w is a positive function.

We can now state the main result of this note

Theorem 1. *Let $w(x) := \exp(\gamma x^{3/2})$ and let d be the Wasserstein distance associated with the cost induced by w via the formula (3). Then, for any fixed $a \in \mathbb{R}$, if $\gamma > 0$ is small enough, Λ_n^a tends to TW^a for the distance d:*

$$\lim_{n \to \infty} d(\Lambda_n^a, TW^a) = 0.$$

2 The Small Deviation Inequality

The simplest idea to get concentration inequalities for the largest eigenvalue of a GUE random matrix is to use Gaussian concentration; it is a straightforward consequence of the measure concentration phenomenon in the Gaussian space (see [1]) that

$$\forall t > 0, \, \forall n, \qquad \mathbb{P}_n\big(\lambda_1\big(A^{(n)}\big) \geqslant M_n + t\big) < \exp\big(-nt^2/2\big) \tag{4}$$

where M_n is the median of $\lambda_1(A^{(n)})$ with respect to the probability measure $\mathbb{P}_n$. One has the same upper estimate if the median M_n is replaced by the expected value $\mathbb{E}_n\lambda_1(A^{(n)})$.

The value of M_n can be controlled: for example we have $M_n \leqslant 2 + c/\sqrt{n}$. This will be a consequence of our Proposition. Plugging this into the equation (4), we get the following result, where C is a universal constant

$$\forall t > 0, \, \forall n, \qquad \mathbb{P}_n\big(\lambda_1\big(A^{(n)}\big) \geqslant 2 + t\big) < C \exp\big(-nt^2/2\big). \tag{5}$$

We ask the question whether in fact both $\mathbb{E}_n\lambda_1(A^{(n)})$ and M_n are smaller than 2. Note that since the function λ_1 is convex, its median with respect to $\mathbb{P}_n$ does not exceed its expected value ([7]). A positive answer to this question would imply that one could choose $C = 1$ in the inequality (5). The answer to the analogous question is known to be positive for the GOE (Gaussian Orthogonal Ensemble), an ensemble of real symmetric matrices defined in a similar way as GUE (see [10] for a precise definition). The argument, due to Gordon, uses a result about comparison of the supremum of Gaussian processes known as Slepian's lemma (see [1]) and doesn't carry over to the complex setting.

There are similar though not as simple results for $\mathbb{P}_n(\lambda_1(A^{(n)}) \leqslant 2 - t)$, but in this paper we will concentrate on the "upper tail" estimates.

The result of Tracy and Widom (1) shows that the majoration (5) is not optimal for very small values of t. If for example t is equal to $xn^{-2/3}$ for a fixed x, then the right-hand side in concentration inequality (5) tends to 1 when n grows to ∞, whereas the left-hand side tends to $\psi_{\mathrm{TW}}(x)$, which can be very small.

We would like to derive from our Theorem a deviation inequality which would improve the inequality (5) for small values of t. For this purpose, the uniform convergence in (1) (which, by Dini's theorem, follows formally from the pointwise convergence) is not enough. But we will prove in this section that our Theorem implies the following Proposition:

Proposition 1. *There exist positive universal constants C and c such that for every positive t and any integer n*

$$\mathbb{P}_n(\lambda_1(A^{(n)}) \geqslant 2 + t) \leqslant C \exp(-cnt^{3/2}). \qquad (6)$$

Of course, by symmetry of the law $\mathbb{P}_n$, similar results are true for the smallest eigenvalue $\lambda_n(A^{(n)})$

$$\mathbb{P}_n(\lambda_n(A^{(n)}) \leqslant -2 - t) \leqslant C \exp(-cnt^{3/2}). \qquad (7)$$

Using the fact that for a Hermitian matrix A, the norm equals the maximum absolute value of an eigenvalue, we get a similar estimate for $\|A^{(n)}\|$

$$\mathbb{P}_n(\|A^{(n)}\| \geqslant 2 + t) \leqslant C \exp(-cnt^{3/2}). \qquad (8)$$

We need the following lemma to prove the proposition, which will help us to explicitly compute Wasserstein distance

Lemma 1. *Suppose that the measures μ and ν are defined on $\mathbb{R}$, and that the cost c is defined by an integral, as in (3). If μ and ν are regular enough, for example if ψ_μ and ψ_ν are piecewise C^1, then the Wasserstein distance for the cost c equals*

$$d(\mu, \nu) = \int_{-\infty}^{\infty} w(t)|\psi_\mu(t) - \psi_\nu(t)|\, dx. \qquad (9)$$

Proof. In fact, this transportation problem is explicitly solvable. For a one-dimensional problem with a cost satisfying the Monge condition (which is always the case when the cost is defined using an integral as in (3)), the optimal transshipment is achieved through the map T defined as follows (see [12], chapter 3.1)

$$\int_{-\infty}^{x} \mathrm{d}\mu = \int_{-\infty}^{T(x)} \mathrm{d}\nu.$$

Thus, we can compute the value of $d(\mu, \nu)$

$$d(\mu, \nu) = \int_{0}^{1} c\big(\psi_{\mu}^{-1}(u), \psi_{\nu}^{-1}(u)\big)\, \mathrm{d}u.$$

Let us consider first the particular case when $\psi_{\mu} \leqslant \psi_{\nu}$. This allows us to drop the absolute values in the definition of c (see (3)) and unfold the calculations. Using the appropriate changes of variables, we come to the equality (9).

For general μ and ν, define $\mu \wedge \nu$ and $\mu \vee \nu$ using their tail functions

$$\psi_{\mu \wedge \nu}(x) = \min\big(\psi_{\mu}(x), \psi_{\nu}(x)\big) \qquad \text{and} \qquad \psi_{\mu \vee \nu}(x) = \max\big(\psi_{\mu}(x), \psi_{\nu}(x)\big).$$

We easily check that $\psi_{\mu \wedge \nu} \leqslant \psi_{\mu \vee \nu}$, $d(\psi_{\mu}, \psi_{\nu}) = d(\psi_{\mu \wedge \nu}, \psi_{\mu \vee \nu})$ and that the value of the r.h.s. of (9) does not change if we replace ψ_{μ} and ψ_{ν} by $\psi_{\mu \wedge \nu}$ and $\psi_{\mu \vee \nu}$. This yields the conclusion for the general case. $\square$

Using this lemma, we get from our theorem (with $a = 0$), using the upper bound (2) for ψ_{TW}, the uniform estimate

$$\int_{0}^{\infty} w(x)\mathbb{P}_n\big(\lambda_1\big(A^{(n)}\big) \geqslant 2 + xn^{-2/3}\big)\, \mathrm{d}x \leqslant C$$

which implies immediately for $x \geqslant 1$ (keep in mind that ψ_n is decreasing)

$$\psi_n(x) \leqslant C \exp\big(-\gamma(x - 1)^{3/2}\big) \leqslant C' \exp\big(-\gamma' x^{3/2}\big). \tag{10}$$

This is, up to the rescaling $t = xn^{-2/3}$, the content of the proposition.

Now we can also easily show that our theorem implies the Tracy–Widom limit (1): using the uniform bound (10) and Lebesgue's convergence theorem, we get from the Theorem that ψ_{TW} is the pointwise limit of the ψ_n's on $[a, +\infty)$, and thus on the whole real line if we let a go to $-\infty$.

It should be emphasized that recently (independently from and slightly preceding this work), this small deviation result has been proved by Ledoux in [8] using an argument based on the Harer–Zagier recurrence formula (see [5] for a simple proof of this formula). The same paper by Ledoux contains another proof based on hypercontractivity which gives the result up to a polynomial factor; this method works also for the Laguerre Unitary Ensemble (see [8] for the definition). However, the existence of a Tracy–Widom limit does not follow from this approach. More generally, many contributions to this and related topics either address the limit behaviour or provide dimension-free bounds, rarely combining the two. Our technique captures both phenomena in a single "stroke".

3 Relation to Determinants

The remainder of this note is devoted to the proof of the main theorem. For simplicity, we will prove only the case $a = 0$, and drop all the superscripts. The proof for a general a requires only routine modifications.

We are first going to express all involved quantities in terms of determinants of certain operators. This is quite classical work due to Gaudin and Mehta (see [10]). Part of the calculations done here are present, at least implicitly, in the paper by Tracy and Widom ([17]).

We need new notation. Let (H_n) be the Hermite polynomials, which are defined by

$$H_n(t) := (-1)^n \exp(t^2) \left(\frac{\mathrm{d}}{\mathrm{d}t} \right)^n \exp(-t^2).$$

They are orthogonal for the measure on $\mathbb{R}$ of density $\exp(-x^2)$ with respect to Lebesgue measure. Then we note

$$\phi_n(t) := \frac{1}{\sqrt{d_n}} H_n(t) \exp(-t^2/2) \tag{11}$$

where $d_n := \int_{\mathbb{R}} H_n(x)^2 \, \mathrm{d}x = 2^n n! \sqrt{\pi}$. The family (ϕ_n) is therefore orthonormal in $L^2(\mathbb{R})$. We introduce

$$k_n(x, y) := \sum_{j=0}^{n-1} \phi_j(x) \phi_j(y).$$

We can associate to k_n an integral operator K_n acting on the Hilbert space $L^2(\mathbb{R})$ in the following way

$$(K_n f)(x) := \int_{\mathbb{R}} k_n(x, y) f(y) \, \mathrm{d}y. \tag{12}$$

This operator K_n is nothing but the orthogonal projection in $L^2(\mathbb{R})$ onto the subspace spanned by $(\phi_j)_{1 \leqslant j \leqslant n}$.

This is a very general setting: if we have a measure space (X, μ) and a "kernel" $k \in L^2(X \times X)$, we can define an operator K on $L^2(X)$ using a formula similar to (12). From now on, all kernels are assumed to belong to $L^2(X \times X)$ and are denoted by small letters; associated integral operators are denoted by the corresponding capital letter.

It is straightforward to prove that Hilbert–Schmidt operators on $L^2(X)$ are exactly integral operators with a L^2 kernel. Moreover, the Hilbert–Schmidt norm of the operator and the L^2 norm of the kernel coincide. This fact is proved in [3], which is a good reference for a reader who wants more detail on integral operators. Let us just quote the formula for compositions of operators: if k and l are two kernels on the same space (X, μ), then the operator KL is an integral operator with kernel (kl):

$$(kl)(x, y) = \int_X k(x, z) l(z, y) \, d\mu(z). \tag{13}$$

The tail function of $\lambda_1(A^{(n)})$ can now be expressed using the kernel k_n. The key formula is the following (see [10])

$$\forall t \in \mathbb{R}, \qquad \mathbb{P}_n\left(\frac{\sqrt{n}}{\sqrt{2}} \lambda_1(A^{(n)}) \leqslant t\right) = \det_{[t, \infty)}(\mathrm{Id} - K_n). \tag{14}$$

In the formula (14), the right-hand side must be understood as the determinant of the operator K_n acting on the space $L^2([t, \infty))$ (or equivalently of the operator with kernel equal to is the restriction of k_n to $[t, \infty)^2$). This restricted operator is denoted $K_n^{[t]}$.

It may not be immediately obvious how to define such a determinant, as the operator involved acts on an infinite-dimensional space. However, the operator K_n that we consider here has a finite rank, hence we can define its determinant as if it were acting on a finite-dimensional space.

A problem will arise when we want to consider limits of such operators, which might fail to have a finite rank. Fortunately, a whole theory of determinants (and traces) of integral operators exists (so-called "Fredholm" determinants). In fact, there are several possible ways to extend these concepts to the infinite-dimensional case. We will focus on a more algebraic approach, due to Grothendieck (see [4] or [13] for a complete exposition), which defines determinants of a nuclear (= trace class) perturbation of identity in terms of traces of its exterior powers (here N is a nuclear operator, for which trace is well-defined):

$$\det(\mathrm{Id} + N) := 1 + \sum_{k=1}^{\infty} \mathrm{tr}\left(\Lambda^k(N)\right).$$

Of course, this definition coincides with the usual one in the finite-dimensional case.

The presence of the factor $\sqrt{n}/\sqrt{2}$ in equation (14) requires an explanation. It arose because there are several possible normalizations. We chose to define the GUE so that the first eigenvalue is about 2, while other authors, as Tracy and Widom in [17], prefer to locate it around $\sqrt{2n}$ (there are still other normalizations but an exhaustive list would be too long). As we kept their notation for the kernels k_n, a scaling factor will appear when we pass from a normalization to the other one.

To get a nontrivial limit, we must replace the t in formula (14) by the following rescaling, as for the Tracy–Widom limit (1)

$$t = \tau_n(x) := \frac{\sqrt{n}}{\sqrt{2}}\left(2 + \frac{x}{n^{2/3}}\right).$$

Let also $\tilde{k}_n$ be the rescaled kernel

$$\tilde{k}_n(x,y) := \frac{1}{\sqrt{2}n^{1/6}}\, k_n\big(\tau_n(x), \tau_n(y)\big).$$

We can see using a change of variable that $\tilde{K}_n^{[x]}$ and $K_n^{[\tau_n(x)]}$ have the same eigenvalues. More precisely, if f is an eigenfunction of $\tilde{K}_n^{[x]}$, then $f \circ \tau_n^{-1}$ is an eigenfunction of $K_n^{[\tau_n(x)]}$, with the same eigenvalue.

Plugging these renormalizations into the formula (14), we obtain

$$\mathbb{P}_n\big(\lambda_1\big(A^{(n)}\big) \leqslant 2 + x n^{-2/3}\big) = \det_{[x,+\infty)}\big(\mathrm{Id} - \tilde{K}_n\big). \tag{15}$$

Using the previous definition for the tail function ψ_n, we can write for a positive s

$$\psi_n(s) = 1 - \det\big(\mathrm{Id} - \tilde{K}_n^{[s]}\big).$$

The following result was known before Tracy and Widom's work (see for example [2])

$$\lim_{n\to\infty} \tilde{k}_n(x,y) = k(x,y) \tag{16}$$

uniformly on compact subsets in x and y.

Here k is the kernel, often called Airy kernel, defined by

$$k(x,y) := \frac{\mathrm{Ai}(x)\,\mathrm{Ai}'(y) - \mathrm{Ai}'(x)\,\mathrm{Ai}(y)}{x - y}. \tag{17}$$

The kernel k is extended by continuity to the diagonal. The function Ai is called the Airy function. It is very useful in physics and can be defined by several means. One of them is the following integral representation

$$\mathrm{Ai}(z) := \frac{1}{2\pi}\int_{-\infty}^{\infty} \exp\big(i(zt + t^3/3)\big)\, dt.$$

It can also be written as a combination of Bessel functions. It satisfies the Airy ODE

$$\frac{\partial^2}{\partial x^2} y(x) = x y(x). \tag{18}$$

The asymptotic behavior of Ai is well-known, for example [16] contains the following formula, valid when x tends to $+\infty$

$$\mathrm{Ai}(x) \sim \frac{1}{2}\, 3^{-1/4}\sqrt{\pi}\, x^{-1/4} \exp\Big(-\frac{2}{3^{3/2}} x^{3/2}\Big). \tag{19}$$

The function ψ_{TW} can be defined using this Airy kernel

$$\psi_{\mathrm{TW}}(x) := 1 - \det\big(\mathrm{Id} - K^{[x]}\big). \tag{20}$$

In [17] Tracy and Widom found another expression for ψ_{TW}. Let q be the solution of the Painlevé II ODE

$$\frac{\partial^2}{\partial x^2} q(x) = xq(x) + 2q(x)^3$$

which is determined by the asymptotics $q(x) \sim \mathrm{Ai}(x)$ for x close to $+\infty$. Then we have the representation

$$\psi_{\mathrm{TW}}(x) = 1 - \exp\left(-\int_x^\infty (t - x)q(t)^2 \, dt\right). \qquad (21)$$

It is easy to get from (19) and (21) the bounds (2) for the asymptotic behavior of ψ_{TW}. However, as we do not really need all the depth of Tracy and Widom's results and connections to Painlevé functions, we will reprove this fact in a more elementary way at the end of this note.

4 Convergence of the Operators

The convergence in (16) as determined in the existing literature is rather weak; in particular, it does not imply convergence of the associated integral operators in the Hilbert–Schmidt norm or even in the operator norm on L^2. In particular, we are not a priori allowed to exchange limit and determinant in (15) when n tends to infinity.

Our main step will be to show that $\tilde{K}_n$ tends to K with respect to the nuclear (trace class) norm. To that end we need several lemmas.

Lemma 2. *The following equality holds*

$$\left(\frac{\partial}{\partial x} + \frac{\partial}{\partial y}\right) k_n(x,y) = -\sqrt{\frac{n}{2}} \left(\phi_n(x)\phi_{n-1}(y) + \phi_{n-1}(x)\phi_n(y)\right).$$

Proof. We start with the Christoffel–Darboux formula (see [16])

$$k_n(x,y) = \sqrt{\frac{n}{2}} \frac{\phi_n(x)\phi_{n-1}(y) - \phi_{n-1}(x)\phi_n(y)}{x - y}.$$

Then we apply the operator $\partial/\partial x + \partial/\partial y$ to each term. We use the formula (11) and the following identities (those which are not obvious are shown in [16])

$$\phi_n'(x) = -\frac{\exp(-x^2/2)}{\sqrt{d_n}} \left(H_n'(x) - xH_n(x)\right),$$

$$H_{n-1}'(x) = 2xH_{n-1}(x) - H_n(x),$$

$$H_n'(x) = 2nH_{n-1}(x).$$

We obtain exactly the expected result. $\square$

Lemma 3. *The following integral representation holds*

$$\tilde{k}_n(x,y) = \frac{n^{1/6}}{2\sqrt{2}} \int_0^\infty \phi_n\big(\tau_n(x+z)\big)\phi_{n-1}\big(\tau_n(y+z)\big) \tag{22}$$
$$+ \phi_{n-1}\big(\tau_n(x+z)\big)\phi_n\big(\tau_n(y+z)\big)\, \mathrm{d}z.$$

Proof. If we apply the operator $\partial/\partial x + \partial/\partial y$ to the right-hand side of (22) (there is no trouble with interchanging the operations "$\partial/\partial x + \partial/\partial y$" and "$\int_0^\infty$" since all the functions involved are Schwartz functions), we get after standard calculations

$$\frac{n^{1/6}}{2\sqrt{2}} \left(\phi_n\big(\tau_n(x)\big)\phi_{n-1}\big(\tau_n(y)\big) + \phi_{n-1}\big(\tau_n(x)\big)\phi_n\big(\tau_n(y)\big) \right).$$

Lemma 2 asserts that we obtain exactly the same expression when we apply the operator $\partial/\partial x + \partial/\partial y$ to the left member of (22). Thus, the two members of the equation are equal modulo a function (say, α) which only depends on $x - y$. But both members tend to zero when x et y tend to infinity in an independent way. Therefore the function α has to vanish identically and the lemma is proved. □

Let us introduce some extra notation. The following kernels are defined on $[s, +\infty)^2$, where s is any positive number

$$a_n^{[s]}(x,y) := \frac{n^{1/12}}{2^{1/4}} \phi_n\big(\tau_n(x+y-s)\big),$$

$$b_n^{[s]}(x,y) := \frac{n^{1/12}}{2^{1/4}} \phi_{n-1}\big(\tau_n(x+y-s)\big),$$

$$a^{[s]}(x,y) := \mathrm{Ai}(x+y-s).$$

The equality (22) can be translated in terms of operators (this is just a consequence of the formula (13) for the composition of kernels)

$$\tilde{K}_n^{[s]} = \frac{1}{2} \big(A_n^{[s]} B_n^{[s]} + B_n^{[s]} A_n^{[s]}\big). \tag{23}$$

A similar equality for the operator K is proved (exactly in the same way) in [17]

$$K^{[s]} = \big(A^{[s]}\big)^2. \tag{24}$$

We shall subsequently show that (for a fixed s) the operators $A_n^{[s]}$ and $B_n^{[s]}$ tend to $A^{[s]}$ with respect to the Hilbert–Schmidt norm. To that end, we need estimates for ϕ_n contained in two lemmas that follow.

Lemma 4. *The functions ϕ_n, after rescaling, converge to* Ai, *uniformly on compact subsets in* y:

$$\phi_n\big(\tau_n(y)\big) 2^{-1/4} n^{1/12} \to \mathrm{Ai}(y) \quad \text{and} \quad \phi_{n-1}\big(\tau_n(y)\big) 2^{-1/4} n^{1/12} \to \mathrm{Ai}(y). \tag{25}$$

Proof. This is an immediate consequence of the following asymptotic formulae for Hermite polynomials due to Plancherel and Rotach. They can be found, in a slightly different presentation, in the book by Szegö ([16])

$$\text{If } x = \sqrt{2n+1} + \frac{y}{\sqrt{2}n^{1/6}}, \text{ then } \phi_n(x) = 2^{1/4}n^{-1/12}\big(\text{Ai}(y) + O(n^{-3/4})\big).$$

The O holds when n tends to $+\infty$, uniformly in y on compact subsets.   □

Lemma 5. *We have a bound for ϕ_n which is uniform in n: there exists a positive constant c such that for any $y > 0$ and any integer n*

$$\begin{cases} n^{1/12}\phi_n\big(\tau_n(y)\big) \leqslant C\exp\big(-cy^{3/2}\big), \\ n^{1/12}\phi_n\big(\tau_{n-1}(y)\big) \leqslant C\exp\big(-cy^{3/2}\big). \end{cases} \tag{26}$$

Proof. Let us sketch a proof of the first inequality in (26). We will use the following result, which is an exercise on page 403 of [11]. It is valid for $x \geqslant 1$

$$H_n(\nu x) \leqslant 1.13\sqrt{2\pi}\exp\big(-\nu^2/4\big)\nu^{(3\nu^2-1)/6}\exp\big(\nu^2 x^2/2\big)\Big(\frac{\zeta}{x^2-1}\Big)^{1/4}\text{Ai}\big(\nu^{4/3}\zeta\big)$$

where $\nu := \sqrt{2n+1}$ and

$$\zeta := \Big(\frac{3}{4}x\sqrt{x^2-1} - \frac{3}{4}\text{Argch}x\Big)^{2/3}.$$

Using the definition of ϕ_n given in formula (11) and Stirling's formula to estimate d_n, we obtain

$$n^{1/12}\phi_n(\nu x) \leqslant C\Big(\frac{\zeta}{x^2-1}\Big)^{1/4}\text{Ai}\big(\nu^{4/3}\zeta\big).$$

We deduce from (19) a bound for Ai, and we also use the inequality $\zeta \geqslant c(x-1)$ to get

$$n^{1/12}\phi_n\big(\sqrt{2n+1}x\big) \leqslant Cn^{-1/6}\frac{1}{(x-1)^{1/4}}\exp\big(-c(2n+1)(x-1)^{3/2}\big). \tag{27}$$

We now return to our notation through the change of variable $\sqrt{2n+1}\,x = \tau_n(y)$. We can estimate x in the following way

$$x \geqslant 1 - \frac{c}{n} + \frac{y}{\sqrt{2}n^{1/6}\sqrt{2n+1}}.$$

For y large enough, we even have

$$x \geqslant 1 + c\frac{y}{n^{2/3}}. \tag{28}$$

Combining (27) and (28) yields

$$n^{1/12}\phi_n\big(\tau_n(y)\big) \leqslant C\,\frac{1}{y^{1/4}}\exp\big(-cy^{3/2}\big).$$

The factor $y^{-1/4}$ can be deleted if c is made small enough. This inequality is only true for y large enough, but we keep in mind that convergence in (25) was uniform on compact subsets, so we can extend it to all positive y, and the inequality is proved. The same scheme of demonstration works for the second inequality, with τ_{n-1} instead of τ_n. $\square$

We are now ready to prove our main theorem.

Proof of Theorem 1. We denote by $\|\,.\,\|_{HS}$ the Hilbert–Schmidt norm and by ν the nuclear norm.

We are first going to estimate the quantity $|\psi_n(s) - \psi_{\mathrm{TW}}(s)| = |\det(\mathrm{Id} - \tilde{K}_n^{[s]}) - \det(\mathrm{Id} - K^{[s]})|$. To reach this goal, we will use the following estimate (see [4]), valid for any two nuclear operators A and B

$$|\det(\mathrm{Id}+A) - \det(\mathrm{Id}+B)| \leqslant \nu(A-B)\,\mathrm{e}^{1+\nu(A)+\nu(B)}. \tag{29}$$

It will be useful to notice that lemma 5 implies in particular the following remark: there is a positive C such that for any $s \geqslant 0$ and any integer n, all the quantities $\|A_n^{[s]}\|_{HS}$, $\|B_n^{[s]}\|_{HS}$ and $\|A^{[s]}\|_{HS}$ are bounded by C (remember that the Hilbert–Schmidt norm is just the L^2-norm of the kernel). Using inequalities (23), (24) and the noncommutative Hölder inequality, we get that $\nu(K^{[s]})$ and $\nu(\tilde{K}_n^{[s]})$ are also bounded by the constant. Hence we can drop the exponential factor in formula (29)

$$|\psi_n(s) - \psi_{\mathrm{TW}}(s)| \leqslant C\nu\big(\tilde{K}_n^{[s]} - K^{[s]}\big).$$

We need to estimate the quantity $\nu(\tilde{K}_n^{[s]} - K^{[s]})$. The key to do this is to use the equalities (23) et (24) to get

$$\tilde{K}_n^{[s]} - K^{[s]} = \frac{1}{4}\Big(\big(A_n^{[s]} - A^{[s]}\big)\big(B_n^{[s]} + A^{[s]}\big) + \big(A_n^{[s]} + A^{[s]}\big)\big(B_n^{[s]} - A^{[s]}\big)$$
$$+ \big(B_n^{[s]} + A^{[s]}\big)\big(A_n^{[s]} - A^{[s]}\big) + \big(B_n^{[s]} - A^{[s]}\big)\big(A_n^{[s]} + A^{[s]}\big)\Big).$$

The non-commutative Hölder inequality yields

$$\nu\big(\tilde{K}_n^{[s]} - K^{[s]}\big) \leqslant \frac{1}{2}\big\|A_n^{[s]} - A^{[s]}\big\|_{HS}\big\|B_n^{[s]} + A^{[s]}\big\|_{HS} \tag{30}$$
$$+ \frac{1}{2}\big\|A_n^{[s]} + A^{[s]}\big\|_{HS}\big\|B_n^{[s]} - A^{[s]}\big\|_{HS}.$$

The factors with a "+" are easy to get rid of: we can use the triangle inequality to write $\|A_n^{[s]} + A^{[s]}\|_{HS} \leqslant \|A_n^{[s]}\|_{HS} + \|A^{[s]}\|_{HS}$, which is uniformly bounded according to the remark following formula (29). We obtain

$$\nu\big(\tilde{K}_n^{[s]} - K^{[s]}\big) \leqslant C\big(\big\|A_n^{[s]} - A^{[s]}\big\|_{HS} + \big\|B_n^{[s]} - A^{[s]}\big\|_{HS}\big).$$

We can now calculate the Wasserstein distance from Λ_n to TW, using the expression given by lemma 1

$$d(\Lambda_n, TW) = \int_0^\infty \exp\big(\gamma s^{3/2}\big)|\psi_n(s) - \psi_{\mathrm{TW}}(s)|\,\mathrm{d}s$$
$$\leqslant C\int_0^\infty \exp\big(\gamma s^{3/2}\big)\big(\big\|A_n^{[s]} - A^{[s]}\big\|_{HS} + \big\|B_n^{[s]} - A^{[s]}\big\|_{HS}\big)\,\mathrm{d}s.$$

First deal with the term $\big\|A_n^{[s]} - A^{[s]}\big\|_{HS}$. Using the definition of $A_n^{[s]}$ and $A^{[s]}$ we get

$$\int_0^\infty \exp\big(\gamma s^{3/2}\big)\big\|A_n^{[s]} - A^{[s]}\big\|_{HS}\,\mathrm{d}s \tag{31}$$
$$= \sqrt{2}\int_0^\infty \exp\big(\gamma s^{3/2}\big)\bigg(\int_0^\infty z\bigg(\Big(\frac{n^{1/12}}{2^{1/4}}\phi_n \circ \tau_n - \mathrm{Ai}\Big)(z+s)\bigg)^2\mathrm{d}z\bigg)^{1/2}\,\mathrm{d}s.$$

Fix an $\varepsilon > 0$ and use the uniform bound of lemma 5: we get that for γ small enough, S large enough and any n

$$\sqrt{2}\int_S^\infty \exp\big(\gamma s^{3/2}\big)\bigg(\int_0^\infty z\bigg(\Big(\frac{n^{1/12}}{2^{1/4}}\phi_n \circ \tau_n - \mathrm{Ai}\Big)(z+s)\bigg)^2\mathrm{d}z\bigg)^{1/2}\,\mathrm{d}s \leqslant \varepsilon.$$

Similarly, for Z large enough, any s smaller than S and any n

$$\bigg(\int_Z^\infty z\bigg(\Big(\frac{n^{1/12}}{2^{1/4}}\phi_n \circ \tau_n - \mathrm{Ai}\Big)(z+s)\bigg)^2\mathrm{d}z\bigg)^{1/2} \leqslant \frac{\varepsilon}{\sqrt{2}\,S\exp(S^{3/2})}.$$

Now we can split the integral in (31) into three terms to get (remember that the convergence in lemma 4 is uniform on compact subsets):

$$\int_0^\infty \exp\big(\gamma s^{3/2}\big)\big\|A_n^{[s]} - A^{[s]}\big\|_{HS}\,\mathrm{d}s \leqslant 3\varepsilon \qquad \text{for } n \text{ large enough.}$$

We can write a similar estimate with B_n instead of A_n. We finally deduce that, for n large enough, $d(\Lambda_n, TW) \leqslant 6C\varepsilon$. Hence Λ_n tends to TW in the Wasserstein sense. This is the announced result.   □

5 An Elementary Proof of Asymptotics for ψ_{TW}

To prove our theorem, we needed the upper asymptotics (2) for ψ_{TW}. It is possible to derive them from the representation (21): keeping in mind that $q \sim \mathrm{Ai}$, we get from (19)

$$\int_x^\infty (t-x)q^2(t)\,\mathrm{d}t \leqslant C\exp\big(-cx^{3/2}\big).$$

Hence, (21) yields

$$\psi_{\mathrm{TW}}(x) \leqslant 1 - \exp\big(-C\exp(-cx^{3/2})\big) \leqslant C'\exp\big(-cx^{3/2}\big).$$

However, for sake of completeness, we are going to derive in this section this last result in a more elementary way, i.e. without using the Painlevé representation. To do this, we need some facts about integral operators. Of course, a general integral operator can fail to be nuclear (for example, any Hilbert–Schmidt operator from $L^2(X)$ into itself can be written as an integral operator). Nevertheless, there exist several "nuclearity tests", criteria ensuring that under some conditions, kernels generate nuclear operators ([3],[4]). The main result in this topic is Mercer's theorem, which enables us to expand a continuous self-adjoint kernel (i.e. the associated operator is self-adjoint) as a series of eigenfunctions of the operator. Unfortunately, these results are usually stated when dealing with a compact space of finite measure, and we have to consider half-infinite intervals $[s, +\infty)$. However, the standard proofs work also in this setting with only slight modifications.

A result which fits the present context is the following

Lemma 6. *Let $X = [s, +\infty)$, equipped with the Lebesgue measure, and k be a kernel on $X \times X$ which satisfies the following conditions:*

1. *$k \in L^2(X \times X)$*
2. *k is jointly continuous*
3. *K is positive self-adjoint as an operator on $L^2(X)$*
4. *There exists a continous positive function ϱ in $L^2(X)$ such that $|k(x,y)| \leqslant \varrho(x)\varrho(y)$ for every x, y in X.*

Then the operator K is nuclear and the trace formula holds

$$\operatorname{tr}(K) = \int_s^\infty k(x,x)\,\mathrm{d}x. \tag{32}$$

Proof. We are going to derive our result from the classical finite-measure case using a change of density trick. Let μ be the measure on X with density ϱ^2 with respect to Lebesgue measure; we have $\mu(X) < \infty$. If we (isometrically) identify $L^2(X, \mathrm{d}x)$ with $L^2(X, \mu)$ sending f to f/ϱ, the integral operator K viewed from $L^2(X, \mu)$ into itself has kernel $k(x,y)/\varrho(x)\varrho(y)$. To get the result we simply apply to the new kernel the following version of Mercer's theorem (it can be proved adapting straightforward the classical proof from [14]): if μ is a finite Borel measure on X and k a continuous bounded positive self-adjoint kernel, then the associated operator K is nuclear and its trace is equal to the integral of the kernel along the diagonal. □

Lemma 7. *The following estimation holds*

$$\exists C, c > 0, \quad \forall s > 0, \qquad \psi_{\text{TW}}(s) \leqslant C \exp\left(-cs^{3/2}\right).$$

Proof. By definition (see [13]), we have

$$\psi_{\text{TW}}(s) = \sum_{k=1}^{\infty} (-1)^{k-1} \operatorname{tr}\left(\Lambda^k\left(K^{[s]}\right)\right).$$

Using the fact that $\operatorname{tr}(\Lambda^k(K^{[s]})) \leqslant \operatorname{tr}(K^{[s]})^k/k!$, we get

$$|\psi_{\text{TW}}(s)| \leqslant \exp\left(\operatorname{tr} K^{[s]}\right) - 1.$$

Actually, formula (20) shows that ψ_{TW} is positive since we have $0 \leqslant K^{[s]} \leqslant 1$. In the end, the convexity of the exponential function on $[0, \operatorname{tr} K^{[0]}]$ yields for $s \geqslant 0$

$$\psi_{TW}(s) \leqslant C \operatorname{tr} K^{[s]}.$$

It is not hard to check that the kernel $k^{[s]}$ satisfies the hypotheses of lemma 6; to check condition 4 we can cook up a function ϱ using Ai and its derivative.

Thus we can rewrite the trace of $K^{[s]}$ as an integral

$$\psi_{TW}(s) \leqslant C \int_s^{\infty} \left((\operatorname{Ai}'(x))^2 - x \operatorname{Ai}(x)\right) dx. \tag{33}$$

The value of K on the diagonal comes from (17) and the Airy ODE (18). Using (19), we can write

$$\exists C, c > 0, \quad \forall s \geqslant 0 \qquad \operatorname{Ai}(s) \leqslant C \exp\left(-cs^{3/2}\right). \tag{34}$$

A similar majoration holds for Ai': we only need to write $\operatorname{Ai}'(s) = \int_s^{\infty} \operatorname{Ai}''(x)\, dx$ and to use formulae (18) and (34)

$$\exists C, c > 0, \quad \forall s \geqslant 0 \qquad \operatorname{Ai}'(s) \leqslant C \exp\left(-cs^{3/2}\right). \tag{35}$$

The conclusion comes when combining formulae (33), (34) and (35).   □

Possible Generalizations

Of course, we expect the inequalities analogous to (6) to be true in a much more general setting. Basically, each time a Tracy–Widom-like behavior has been proved or is suspected, we can ask whether such a uniform estimate holds.

The most natural extension would be the setting of general Wigner matrices, for which universality of Tracy–Widom limit has been proved by Soshnikov ([15]). However, the bounds on moments he obtained do not suffice to derive the small deviation inequality.

Tracy and Widom proved results similar to (1), involving a different limit law, for the matrix ensembles GOE and GSE (the real orthogonal and the symplectic cases) in [18].

Several authors investigated the behavior of the largest s-number (also called singular value) of a rectangular $m \times n$ matrix with independent entries, when the ratio m/n tends to a limit in $(0, 1)$. The paper [6] contains a result analogous to (1) for the Gaussian case (the so-called Wishart ensemble). There is strong numerical evidence indicating that a convergence on the scale $n^{-2/3}$ as in Tracy–Widom behavior occurs also universally in this case, for the largest s-number, but also for the smallest one.

Another quantity of interest is the norm of a $n \times m$ random matrix as a operator from ℓ_p^n to ℓ_q^m. Concentration results have been recently obtain in this case by Meckes (cf [9]).

In all these cases, we know concentration inequalities similar to (5), it would be interesting to prove the corresponding small deviation result.

Acknowledgements

I would like to express my sincere thanks and appreciation to my supervisor, Prof. Stanislaw J. Szarek, for our inspiring discussions and for his ongoing support. This paper will be part of my Ph.D. thesis prepared at the Équipe d'Analyse Fonctionnelle of the University Paris VI. I would also like to thank Charles-Antoine Louët for his assiduous proofreading.

References

1. DAVIDSON, K.R. and SZAREK, S.J. (2001). Local operator theory, random matrices and Banach spaces. In *Handbook of the Geometry of Banach Spaces, vol. 1*, pages 317–366.

2. FORRESTER, P.J. (1993). The spectrum edge of random matrix ensembles, *Nucl. Phys.* **B402**, 709–728.

3. GOHBERG, I. and GOLDBERG, S. (1981). *Basic operator theory.* Birkhaüser.

4. GOHBERG, I., GOLDBERG, S. and KRUPNIK, N. (1998). *Traces and Determinants of Linear Operators.* Birkhaüser.

5. HAAGERUP, U, and THORBJØRNSEN, S (2003). Random matrices with complex Gaussian entries. *Expo. Math.* **21**, no. 4, 293–337.

6. JOHNSTONE, I.M. (2001). On the distribution of the largest eigenvalue in principal component analysis. *Ann. Statist.* **29**, 295–327.

7. KWAPIEŃ, S. (1993). A remark on the median and the expectation of convex functions of Gaussian vectors. *Probabilities in Banach spaces 9 (Sandjberg, 1993), Progr. Probab.* **35**, 271–272.

8. LEDOUX, M. (2003). A remark on hypercontractivity and tail inequalities for the largest eigenvalues of random matrices. *Séminaire de probabilités XXXVII.* Lecture Notes in Mathematics, Springer.

9. MECKES, M.W. (2003). Concentration of norms and eigenvalues of random matrices. To appear in *J. Funct. Anal.*

10. MEHTA, M.L. (1991). *Random Matrices.* Academic Press, Boston, MA, second edition.

11. OLVER, F.W.J. (1974). *Asymptotics and Special Functions.* Cambridge University Press.

12. RACHEV, S.T. and RÜSCHENDORF, L. (1998). *Mass Transportation Problems.* Vol. 1, Springer-Verlag.

13. SIMON, B. (1977). Infinite determinants. *Advances in Math.* **24**, 244–273.

14. SMITHIES, F. (1959). *Integral equations.* Cambridge University Press.

15. SOSHNIKOV, A. (1999). Universality at the edge of spectrum in Wigner random matrices. *Comm. Math. Phys.* **207**, 697–733.

16. SZEGÖ, G. (1967). *Orthogonal Polynomials.* American Mathematical Society, third edition.

17. TRACY, C.A. and WIDOM, H. (1994). Level-spacing distributions and the Airy kernel. *Comm. Math. Phys.* **159**, 151–174.

18. TRACY, C.A. and WIDOM, H. (1996). On orthogonal and symplectic matrix ensembles. *Comm. Math. Phys.* **177**, 727–754.

The Tangent Space to a Hypoelliptic Diffusion and Applications

Fabrice Baudoin

Laboratoire de Probabilités et Statistiques
Université Paul Sabatier
118 route de Narbonne
31062 TOULOUSE Cedex 4 France
e-mail:fbaudoin@cict.fr, symplectik@aol.com

Summary. The goal of this article is to understand geometrically the asymptotic expansion of stochastic flows. Precisely, we show that a hypoelliptic diffusion can be pathwise approximated at each (normal) point by the lift of a Brownian motion in a graded nilpotent group with dilations. This group, called a Carnot group, appears as a tangent space in Gromov–Hausdorff's sense. We then apply this geometrical point of view in different domains:

- the study of the spectrum of regular sub-Laplacians;
- the study of Riemannian Brownian motions.

Keywords: Carnot groups, Chen expansion, Gromov–Hausdorff tangent space, Horizontal Laplacian, Hypoelliptic diffusions, Sub-Laplacian.

Mathematics Subject Classification 2000: 35H10, 58J60, 60J60, 60H10.

1 Introduction

In a famous paper, K.T. Chen [8] has shown how to deeply generalize the Baker–Campbell–Hausdorff formula in a context which can be used in the theory of flows of deterministic differential equations. The use of this Chen formal expansion of a flow is illustrated in the paper of R.S. Strichartz [28]. It has then been discovered, in particular by Ben Arous [2] and [3], that this expansion could also be applied in the theory of stochastic differential equations. Using this expansion, Ben Arous was for instance able to derive, after Rotschild and Stein [27], the asymptotic expansion of a hypoelliptic kernel on the diagonal.

In this paper, we would like to present how to use some ideas from sub-Riemannian geometry and some ideas from the Gromov metric geometry to describe the geometrical content of this Chen expansion of a stochastic flow. More precisely, consider stochastic differential equations on $\mathbb{R}^n$ of the type

$$X_t^{x_0} = x_0 + \sum_{i=1}^{d} \int_0^t V_i(X_s^{x_0}) \circ \mathrm{d}B_s^i, \qquad t \geqslant 0, \tag{1}$$

where:

1. $x_0 \in \mathbb{R}^n$;
2. $V_1, \ldots, V_d$ are C^∞ bounded vector fields on $\mathbb{R}^n$;
3. $\circ$ denotes Stratonovich integration;
4. $(B_t^1, \ldots, B_t^d)_{0 \leqslant t \leqslant T}$ is a d-dimensional standard Brownian motion.

Since $(X_t^{x_0})_{0 \leqslant t \leqslant T}$ is a strong solution of (1), we know from the general theory of stochastic differential equations that X^{x_0} is a predictable functional of B. If we want to better understand this pathwise representation, the best tool is certainly the Chen expansion of the stochastic flow associated with the SDE. Indeed, through the Chen expansion, the Itô map

$$\mathcal{I} : \mathcal{C}\big([0,T],\mathbb{R}^d\big) \longrightarrow \mathcal{C}\big([0,T],\mathbb{R}^n\big), \qquad B \longmapsto X,$$

established by the SDE can be *formally* factorized in the following manner

$$\mathcal{I} = F \circ H.$$

The map
$$H : \mathcal{C}\big([0,T],\mathbb{R}^d\big) \longrightarrow \mathcal{C}\big([0,T],\exp(\mathfrak{g}_{d,\infty})\big),$$

is a horizontal lift in $\exp(\mathfrak{g}_{d,\infty})$ where $\mathfrak{g}_{d,\infty}$ is the free Lie algebra with d generators. And F is simply a map $\exp(\mathfrak{g}_{d,\infty}) \to \mathbb{R}^n$.

This factorization can be made totally rigorous in the case where the vector fields V_i generate a nilpotent Lie algebra. Therefore, in this nilpotent case, the geometry of this Itô map stems directly from the geometry of a finite dimensional quotient of $\exp(\mathfrak{g}_{d,\infty})$. In the case where the V_i do not generate a nilpotent Lie algebra anymore but satisfy the strong Hörmander condition, then Gromov's notion of tangent space shows that it is still possible to approximate locally the geometry of the Itô map by the geometry of a finite dimensional quotient of $\exp(\mathfrak{g}_{d,\infty})$. From this point of view, the geometry of finite dimensional quotients of $\exp(\mathfrak{g}_{d,\infty})$ is thus of particular interest. The study of these quotients can be reduced to the study of the so-called free Carnot groups. They are graded nilpotent groups with dilations. Their geometry is not Riemannian but sub-Riemannian.

In the first part of the paper, we develop precisely this point of view. The second part of paper focusses on two applications:

- the study of the spectrum of regular sub-Laplacians;
- the study of Riemannian Brownian motions.

2 Motivation: Taylor and Chen Expansion of Stochastic Flows

Let $f : \mathbb{R}^n \to \mathbb{R}$ be a smooth function and denote by $(X_t^{x_0})_{t \geqslant 0}$ the solution of (1) with initial condition $x_0 \in \mathbb{R}^n$. First, by Itô's formula, we have

$$f(X_t^{x_0}) = f(x_0) + \sum_{i=1}^{d} \int_0^t (V_i f)(X_s^{x_0}) \circ dB_s^i, \qquad t \geqslant 0.$$

Now, if apply Itô's formula to $V_i f(X_s^x)$, we obtain

$$f(X_t^{x_0}) = f(x_0) + \sum_{i=1}^{d} (V_i f)(x_0) B_t^i + \sum_{i,j=1}^{d} \int_0^t \int_0^s (V_j V_i f)(X_u^{x_0}) \circ dB_u^j \circ dB_s^i.$$

We can continue this procedure to get after N steps

$$f(X_t^{x_0}) = f(x_0) + \sum_{k=1}^{N} \sum_{I=(i_1,\dots,i_k)} (V_{i_1} \dots V_{i_k} f)(x_0) \int_{\Delta^k[0,t]} \circ dB^I + \mathbf{R}_N(t),$$

for some remainder term $\mathbf{R}_N$, where we used the notations:

1.
$$\Delta^k[0,t] = \{(t_1,\dots,t_k) \in [0,t]^k, \ t_1 \leqslant \cdots \leqslant t_k\};$$

2. If $I = (i_1,\dots,i_k) \in \{1,\dots,d\}^k$ is a word with length k,

$$\int_{\Delta^k[0,t]} \circ dB^I = \int_{0 \leqslant t_1 \leqslant \cdots \leqslant t_k \leqslant t} \circ dB_{t_1}^{i_1} \circ \cdots \circ dB_{t_k}^{i_k}.$$

If we dangerously do not care about convergence questions (these questions are widely discussed in Ben Arous [3], see also the end of this section), it is tempting to let $N \to +\infty$ and to assume that $\mathbf{R}_N \to 0$. We are thus led to the nice (but formal!) formula

$$f(X_t^{x_0}) = f(x_0) + \sum_{k=1}^{+\infty} \sum_{I=(i_1,\dots,i_k)} (V_{i_1} \dots V_{i_k} f)(x_0) \int_{\Delta^k[0,t]} \circ dB^I. \qquad (2)$$

We can rewrite this formula in a more convenient way. Let Φ_t be the stochastic flow associated with the SDE (1). There is a natural action of Φ_t on smooth functions: the pull-back action given by

$$(\Phi_t^* f)(x_0) = (f \circ \Phi_t)(x_0) = f(X_t^{x_0}).$$

Formula (2) shows then that we have the following formal expansion for this action

$$\Phi_t^* = \mathrm{Id}_{\mathbb{R}^d} + \sum_{k=1}^{+\infty} \sum_{I=(i_1,\ldots,i_k)} V_{i_1} \ldots V_{i_k} \int_{\Delta^k[0,t]} \circ \, dB^I.$$

Though this formula does not make sense from an analytical point of view, at least, it shows that the *probabilistic information* contained in the stochastic flow is given by the set of Stratonovich chaos $\int_{\Delta^k[0,t]} \circ \, dB^I$. What is a priori much less clear is that the *algebraic information* which is relevant for the study of Φ_t^* is given by the structure of the Lie algebra spanned by the V_i. This is the content of Chen's expansion theorem which we now present. Denote by $\mathbb{R}[[X_1,\ldots,X_d]]$ the **non-commutative** algebra of formal series with d indeterminates.

Definition 1. *The signature of the Brownian motion* $(B_t)_{t\geqslant 0}$ *is the element of* $\mathbb{R}[[X_1,\ldots,X_d]]$ *defined by*

$$S(B)_t = 1 + \sum_{k=1}^{+\infty} \sum_{I=(i_1,\ldots,i_k)} X_{i_1} \ldots X_{i_k} \int_{\Delta^k[0,t]} \circ \, dB^I, \qquad t \geqslant 0.$$

Remark 1. Observe that the signature hence defined is the solution of the *formal* stochastic differential equation

$$S(B)_t = 1 + \sum_{i=1}^{d} \int_0^t S(B)_s X_i \circ dB_s^i, \qquad t \geqslant 0.$$

Remark 2. The element of $\mathbb{R}[[X_1,\ldots,X_d]]$ defined by

$$P_t = 1 + \sum_{k=1}^{+\infty} \sum_{I=(i_1,\ldots,i_k)} X_{i_1} \ldots X_{i_k} \, \mathbb{E}\left(\int_{\Delta^k[0,t]} \circ \, dB^I \right), \qquad t \geqslant 0.$$

is called the expectation of the signature of the Brownian motion $(B_t)_{t\geqslant 0}$. It is a pleasant exercise to show that

$$P_t = \exp\left(\frac{1}{2} t \sum_{i=1}^{d} X_i^2 \right).$$

Observe that the semigroup property of P_t, that is

$$P_{t+s} = P_t P_s,$$

stems directly from the fact that the increments of $(B_t)_{t\geqslant 0}$ are independent and stationary.

The bracket between two elements U and V of $\mathbb{R}[[X_1,\ldots,X_d]]$ is simply given by

$$[U,V] = UV - VU,$$

and it is easily checked that this bracket endows $\mathbb{R}[[X_1, \ldots, X_d]]$ with a Lie algebra structure. We denote by $\mathfrak{L}(X_1, \ldots, X_d)$ the Lie sub-algebra spanned by $X_1, \ldots, X_d$. The following theorem, which is a restatement of a result of K.T. Chen [8] and R.S. Strichartz [28], can be seen as a deep generalization of the Baker–Campbell–Hausdorff formula. Before we give this theorem, here is some notation. If $I = (i_1, \ldots, i_k) \in \{1, \ldots, d\}^k$ is a word, we denote by $|I| = k$ its length and by X_I the commutator defined by

$$X_I = [X_{i_1}, [X_{i_2}, \ldots, [X_{i_{k-1}}, X_{i_k}] \ldots]].$$

The group of permutations of the index set $\{1, \ldots, k\}$ is denoted by $\mathfrak{S}_k$. If $\sigma \in \mathfrak{S}_k$, we denote by $e(\sigma)$ the cardinality of the set

$$\{j \in \{1, \ldots, k-1\}, \, \sigma(j) > \sigma(j+1)\}.$$

Moreover for $\sigma \in \mathfrak{S}_k$, we denote by $\sigma \cdot I$ the word $(i_{\sigma(1)}, \ldots, i_{\sigma(k)})$.

Theorem 1 (Chen–Strichartz formula). *We have*

$$S(B)_t = \exp\left(\sum_{k \geqslant 1} \sum_{I=(i_1, \ldots, i_k)} \Lambda_I(B)_t X_I \right),$$

where:

$$\Lambda_I(B)_t = \sum_{\sigma \in \mathfrak{S}_k} \frac{(-1)^{e(\sigma)}}{k^2 \binom{k-1}{e(\sigma)}} \int_{\Delta^k[0,t]} \circ \, \mathrm{d}B^{\sigma^{-1} \cdot I}.$$

Remark 3. The first terms of the Chen–Strichartz expansion are:

1.

$$\sum_{I=(i_1)} \Lambda_I(B)_t X_I = \sum_{k=1}^{d} B_t^i X_i;$$

2.

$$\sum_{I=(i_1, i_2)} \Lambda_I(B)_t X_I = \frac{1}{2} \sum_{1 \leqslant i < j \leqslant d} [X_i, X_j] \int_0^t B_s^i \circ \mathrm{d}B_s^j - B_s^j \circ \mathrm{d}B_s^i.$$

Thus, thanks to this theorem, the formal expansion of the stochastic flow Φ_t^* acting on functions reads now

$$\Phi_t^* = \exp\left(\sum_{k \geqslant 1} \sum_{I=(i_1, \ldots, i_k)} \Lambda_I(B)_t V_I \right).$$

This *heuristic* discussion can actually be made rigorous and leads to the following result of Castell [7], which is an improvement of Ben Arous [3]:

Proposition 1. *Let $(X_t^{x_0})_{t \geq 0}$ be the solution of (1) associated with the initial condition $X_0^{x_0} = x_0$. Then for all integer $N \geq 2$,*

$$X_t^{x_0} = \left[\exp\left(\sum_{k=1}^{N-1} \sum_{I=(i_1,\ldots,i_k)} \Lambda_I(B)_t V_I \right) \right](x_0) + t^{N/2} \mathbf{R}_N(t), \qquad t \geq 0,$$

where the remainder term $\mathbf{R}_N$ is bounded in probability when $t \to 0$. More precisely, $\exists\ \alpha,\ c > 0$ such that $\forall A > c$,

$$\lim_{t \to 0} \mathbb{P}\left(\sup_{0 \leq s \leq t} s^{N/2} |\mathbf{R}_N(s)| \geq A t^{N/2} \right) \leq \exp\left(-\frac{A^\alpha}{c} \right).$$

All this clearly shows how the Lie algebra $\mathfrak{L} = \mathrm{Lie}(V_1, \ldots, V_d)$ which is generated by the vector fields V_i comes naturally into the study of X^{x_0}. If we want now to understand more deeply how the properties of this Lie algebra determine the geometry of X^{x_0}, it is wiser to begin with the simplest cases. In a way, the simplest Lie algebras are the nilpotent ones. In that case, i.e. if $\mathfrak{L}$ is nilpotent, then the sum

$$\sum_{k \geq 1} \sum_{I=(i_1,\ldots,i_k)} \Lambda_I(B)_t V_I$$

is actually finite and we are going to show that the solutions of equation (1) can be represented from the lift of the Brownian motion $(B_t)_{t \geq 0}$ in a graded free nilpotent Lie group with dilations. These groups called the free Carnot groups are now introduced and their geometries are discussed.

3 Carnot Groups and Nilpotent SDE's

We introduce now the notion of Carnot group. Carnot groups are to sub-Riemannian geometry what Euclidean spaces are to Riemannian geometry. Numerous papers and some books are devoted to the analysis of these groups (see for example [11] and [15]).

Definition 2. *A Carnot group of depth N is a simply connected Lie group $\mathbb{G}$ whose Lie algebra can be written*

$$\mathcal{V}_1 \oplus \cdots \oplus \mathcal{V}_N,$$

where

$$[\mathcal{V}_i, \mathcal{V}_j] = \mathcal{V}_{i+j}$$

and

$$\mathcal{V}_s = 0, \qquad s > N;$$

We consider in all this section a Lie group $\mathbb{G}$ which satisfies the hypothesis of the above definition. Notice that the vector space $\mathcal{V}_1$, which is called the basis of the Carnot group $\mathbb{G}$, Lie-generates $\mathfrak{g}$, where $\mathfrak{g}$ denotes the Lie algebra of $\mathbb{G}$. Since $\mathbb{G}$ is nilpotent and simply connected, the exponential map is a diffeomorphism and the Baker–Campbell–Hausdorff formula therefore completely characterizes the group law of $\mathbb{G}$ because for $U, V \in \mathfrak{g}$,

$$\exp U \exp V = \exp\big(P(U,V)\big)$$

for some universal Lie polynomial P. On $\mathfrak{g}$ we can consider the family of linear operators $\delta_t : \mathfrak{g} \to \mathfrak{g}$, $t \geqslant 0$ which act by scalar multiplication t^i on V_i. These operators are Lie algebra automorphisms due to the grading. The maps δ_t induce Lie group automorphisms $\Delta_t : \mathbb{G} \to \mathbb{G}$ which are called the canonical dilations of $\mathbb{G}$.

Example 1 (Heisenberg Group). The Heisenberg group $\mathbb{H}$ can be represented as the set of 3×3 matrices:

$$\begin{pmatrix} 1 & x & z \\ 0 & 1 & y \\ 0 & 0 & 1 \end{pmatrix}, \qquad x, y, z \in \mathbb{R}.$$

The Lie algebra of $\mathbb{H}$ is generated by the matrices

$$D_1 = \begin{pmatrix} 0 & 1 & 0 \\ 0 & 0 & 0 \\ 0 & 0 & 0 \end{pmatrix}, \qquad D_2 = \begin{pmatrix} 0 & 0 & 0 \\ 0 & 0 & 1 \\ 0 & 0 & 0 \end{pmatrix}, \qquad D_3 = \begin{pmatrix} 0 & 0 & 1 \\ 0 & 0 & 0 \\ 0 & 0 & 0 \end{pmatrix},$$

for which the following equalities hold

$$[D_1, D_2] = D_3, \qquad [D_1, D_3] = [D_2, D_3] = 0.$$

Thus

$$\mathfrak{h} \sim \mathbb{R}^2 \oplus [\mathbb{R}, \mathbb{R}],$$

and, therefore, $\mathbb{H}$ is a (free) two-step Carnot group.

Now take a basis $U_1, \ldots, U_d$ of the vector space $\mathcal{V}_1$. The vectors U_i can be seen as left invariant vector fields on $\mathbb{G}$ so that we can consider the following stochastic differential equation on $\mathbb{G}$:

$$\mathrm{d}\tilde{B}_t = \sum_{i=1}^{d} \int_0^t U_i\big(\tilde{B}_s\big) \circ \mathrm{d}B_s^i, \qquad t \geqslant 0, \tag{3}$$

which is easily seen to have a unique (strong) solution $(\tilde{B}_t)_{t \geqslant 0}$ associated with the initial condition $\tilde{B}_0 = 0_{\mathbb{G}}$.

Definition 3. *The process $(\tilde{B}_t)_{t \geqslant 0}$ is called the lift of the standard Brownian motion $(B_t)_{t \geqslant 0}$ in the group $\mathbb{G}$ with respect to the basis $(U_1, \ldots, U_d)$.*

Notice that $(\tilde{B}_t)_{t \geqslant 0}$ is a Markov process with generator $\frac{1}{2} \sum_{i=1}^{d} U_i^2$. This second-order differential operator is, by construction, left-invariant and hypoelliptic. As a direct consequence of the Chen–Strichartz formula expansion,

Proposition 2. *We have*

$$\tilde{B}_t = \exp\left(\sum_{k=1}^{N} \sum_{I=(i_1,\ldots,i_k)} \Lambda_I(B)_t U_I\right), \qquad t \geqslant 0.$$

Remark 4. We have the following scaling property, for every $c > 0$,

$$\left(\tilde{B}_{ct}\right)_{t \geqslant 0} \overset{\text{law}}{=} \left(\Delta_{\sqrt{c}} \tilde{B}_t\right)_{t \geqslant 0},$$

which stems from the elementary fact

$$\left(\Lambda_I(B)_{ct}\right)_{t \geqslant 0} \overset{\text{law}}{=} \left(c^{|I|/2} \Lambda_I(B)_t\right)_{t \geqslant 0}.$$

This scaling property leads directly to the following property of the density $\tilde{p}_t$ of $\tilde{B}_t$ with respect to any Haar measure of $\mathbb{G}$:

$$p_t(0_\mathbb{G}) = \frac{C}{t^{D/2}}, \qquad t > 0,$$

where $C > 0$ and $D = \sum_{i=1}^{N} i \dim \mathcal{V}_i$.

We now turn to the geometry of $\mathbb{G}$. The Lie algebra $\mathfrak{g}$ can be identified with the set of left-invariant vector fields on $\mathbb{G}$. From this identification and from the decomposition

$$\mathfrak{g} = \mathcal{V}_1 \oplus \cdots \oplus \mathcal{V}_N,$$

we deduce a decomposition of the tangent space $\mathbf{T}_x \mathbb{G}$ to $\mathbb{G}$ at x:

$$\mathbf{T}_x \mathbb{G} = \mathcal{V}_1(x) \oplus \cdots \oplus \mathcal{V}_N(x),$$

where $\mathcal{V}_i(x)$ is the fibre at x of the left-invariant distribution spanned by $\mathcal{V}_i$. This decomposition endows naturally $\mathbb{G}$ with a left-invariant $(0,2)$-tensor g. Precisely, for $x \in \mathbb{G}$, we define g_x as being the scalar product on $\mathbf{T}_x \mathbb{G}$ such that:

1. The vectors $U_1(x), \ldots, U_d(x)$ form an orthonormal basis;
2. $g_x|_{\mathcal{V}_i(x) \times \mathcal{V}_j(x)} = 0$, if i or j is different from 1.

An absolutely continuous curve $c : [0,1] \to \mathbb{G}$ is called horizontal if for almost every $s \in [0,1]$ we have $c'(s) \in \mathcal{V}_1(c(s))$. The length of a horizontal curve c with respect to g is defined by

$$l(c) = \int_0^1 \sqrt{g_{c(s)}\big(c'(s), c'(s)\big)} \, \mathrm{d}s.$$

We can now state the basic result on the geometry of Carnot groups: Chow's theorem.

Theorem 2. *Given two points x and $y \in \mathbb{G}$, there is at least one horizontal absolutely continuous curve $c : [0,1] \to \mathbb{G}$ such that $c(0) = x$ and $c(1) = y$.*

Proof. Denote by G the subgroup of diffeomorphisms $\mathbb{G} \to \mathbb{G}$ generated by the one-parameter subgroups corresponding to $U_1, \ldots, U_d$. The Lie algebra of G can be identified with the Lie algebra generated by $U_1, \ldots, U_d$, i.e. $\mathfrak{g}$. We deduce that G can be identified with $\mathbb{G}$ itself, so that it acts transitively on $\mathbb{G}$. It means that for every $x \in \mathbb{G}$, the map $G \to \mathbb{G}$, $g \to g(x)$ is surjective. Thus, every two points in $\mathbb{G}$ can be joined by a piecewise smooth horizontal curve where each piece is a segment of an integral curve of one of the vector fields U_i. $\square$

Remark 5. In the above proof, the horizontal curve constructed to join two points is not smooth. Nevertheless, it can be shown that it is always possible to connect two points with a smooth horizontal curve (see Gromov [16] p. 120).

The Carnot–Carathéodory distance between x and y and denoted by $d_g(x, y)$ is defined as being the infimum of the lengths of all the horizontal curves joining x and y. It is easily checked that this distance satisfies $d_g(\Delta_c x, \Delta_c y) = cd_g(x, y)$, for every $c > 0$, x, $y \in \mathbb{G}$.

Remark 6. The distance d_g depends on the choice of a basis for $\mathcal{V}_1$. Nevertheless, it can be shown that all the Carnot–Carathéodory distances that can be constructed are bi-Lipschitz equivalent.

A horizontal curve with length $d_g(x, y)$ is called a sub-Riemannian geodesic joining x and y. The topology of the metric space $(\mathbb{G}, d_g)$ is really of interest. Indeed, though the topology given by the distance d_g is compatible with the natural topology of the Lie group $\mathbb{G}$,

Proposition 3. *The Hausdorff dimension of the metric space $(\mathbb{G}, d_g)$ is equal to*

$$D = \sum_{j=1}^{N} j \dim \mathcal{V}_j.$$

We conclude now our short presentation of the Carnot groups with the free Carnot groups. The Carnot group $\mathbb{G}$ is said to be free if $\mathfrak{g}$ is free. In that case, $\dim \mathcal{V}_j$ is the number of Hall words of length j in the free algebra with d generators. We thus have, according to Bourbaki [6] (see also Reutenauer [26] p. 96):

$$\dim \mathcal{V}_j = \frac{1}{j} \sum_{i \mid j} \mu(i) d^{j/i}, \qquad j \leqslant N,$$

where μ is the Möbius function. We easily deduce from this that when $N \to +\infty$,

$$\dim \mathfrak{g} \sim \frac{d^N}{N}.$$

An important algebraic point is that there are many algebraically non isomorphic Carnot groups having the same dimension (even uncountably many for $n \geqslant 6$), but up to isomorphism there is one and only one free Carnot with a given depth and a given dimension for the basis. Actually, as in the theory of vector spaces, we can reduce the study of the free Carnot groups to standard numerical models. Set $m = \dim \mathbb{G}$. Choose now a Hall family and consider the $\mathbb{R}^m$-valued process $(B_t^*)_{t \geqslant 0}$ obtained by writing the components of $(\ln(\tilde{B}_t))_{t \geqslant 0}$ in the corresponding Hall basis of $\mathfrak{g}$. It is easily seen that $(B_t^*)_{t \geqslant 0}$ solves a stochastic differential equation that can be written

$$B_t^* = \sum_{i=1}^{d} \int_0^t D_i(B_s^*) \circ \mathrm{d}B_s^i,$$

where the D_i are polynomial vector fields on $\mathbb{R}^m$ (for an explicit form of the D_i, which depend of the choice of the Hall basis, we refer to Vershik–Gershkovich [13], p. 27). With these notations, we have the following proposition which stems from our very construction.

Proposition 4. On $\mathbb{R}^m$, there exists a unique group law $\circledast$ which makes the vector fields $D_1, \ldots, D_d$ left invariant. This group law is unimodular[1], polynomial of degree N and we have moreover

$$(\mathbb{R}^m, \circledast) \sim \mathbb{G}.$$

The group $(\mathbb{R}^m, \circledast)$ is called the free Carnot group of step N over $\mathbb{R}^d$. It will be denoted by $\mathbb{G}_N(\mathbb{R}^d)$. The process B^* will be called the lift of B in $\mathbb{G}_N(\mathbb{R}^d)$.

Remark 7. Notice that, by construction, $\mathbb{G}_N(\mathbb{R}^d)$ is endowed with the basis of vector fields $(D_1, \ldots, D_d)$. These vector fields agree at the origin with $(\partial/\partial x_1, \ldots, \partial/\partial x_d)$.

The universality of $\mathbb{G}_N(\mathbb{R}^d)$ is the following. If $\mathbb{G}_1$ and $\mathbb{G}_2$ are two Carnot groups, a Lie group morphism $\phi : \mathbb{G}_1 \to \mathbb{G}_2$ is said to be a Carnot group morphism if for any $t \geqslant 0$, $g \in \mathbb{G}_1$,

$$\phi(\Delta_t^{\mathbb{G}_1} g) = \Delta_t^{\mathbb{G}_2} \phi(g),$$

where $\Delta^{\mathbb{G}_1}$ (resp. $\Delta^{\mathbb{G}_2}$) denote the canonical dilations on $\mathbb{G}_1$ (resp. $\mathbb{G}_2$).

Proposition 5. Let $\mathbb{G}$ be a Carnot group. There exists a surjective morphism of Carnot groups $\pi : \mathbb{G}_N(\mathbb{R}^d) \to \mathbb{G}$, where d is the dimension of the basis of $\mathbb{G}$ and N its depth.

---

[1] A group law on $\mathbb{R}^m$ is said to be unimodular if the translations leave the Lebesgue measure invariant.

Proof. Let $U_1, \ldots, U_d$ be a basis of the basis of $\mathbb{G}$. Since $\mathbb{G}_N(\mathbb{R}^d)$ is free, there exists a unique surjective morphism of Lie algebras $\mathrm{d}\pi : \mathfrak{g}_N(\mathbb{R}^d) \to \mathfrak{g}$ such that $\mathrm{d}\pi(D_i) = U_i$, $i = 1, \ldots, d$. We can now define a surjective morphism of Carnot groups $\pi : \mathbb{G}_N(\mathbb{R}^d) \to \mathbb{G}$ by $\pi(e^g) = e^{\mathrm{d}\pi(g)}$, $g \in \mathfrak{g}_N(\mathbb{R}^d)$. Observe that it defines π in a unique way because in Carnot groups the exponential map is a diffeomorphism. □

After this quite long digression on Carnot groups, we now come back to the study of the SDE (1) and assume that the Lie algebra $\mathfrak{L} = \mathrm{Lie}(V_1, \ldots, V_d)$ is nilpotent of depth N, i.e., that every commutator constructed from the V_i with length greater than N is 0.

Theorem 3. *There exists a smooth map*

$$F : \mathbb{R}^n \times \mathbb{G}_N(\mathbb{R}^d) \longrightarrow \mathbb{R}^n$$

such that, for $x_0 \in \mathbb{R}^n$, the solution $(X_t^{x_0})_{0 \leqslant t \leqslant T}$ of the SDE (1) can be written

$$X_t^{x_0} = F(x_0, B_t^*),$$

where B^* is the lift of B in the group $\mathbb{G}_N(\mathbb{R}^d)$.

Proof. An iteration of Itô's formula shows that the process

$$\left[\exp\left(\sum_{k=1}^{N} \sum_{I=(i_1,\ldots,i_k)} \Lambda_I(B)_t V_I \right) \right](x_0),$$

solves the equation (1) (for further details, we refer to Castell [7] or Strichartz [28]). We deduce hence by pathwise uniqueness property that

$$X_t^{x_0} = \left[\exp\left(\sum_{k=1}^{N} \sum_{I=(i_1,\ldots,i_k)} \Lambda_I(B)_t V_I \right) \right](x_0).$$

The definition of $\mathbb{G}_N(\mathbb{R}^d)$ shows that we can therefore write

$$X_t^{x_0} = F(x_0, B_t^*). \qquad \square$$

The above theorem shows the universal property of $\mathbb{G}_N(\mathbb{R}^d)$ in theory of nilpotent stochastic flows. Here is its counterpart in the theory of second order hypoelliptic operators (this property is implicitly pointed out in the seminal work of Rotschild and Stein [27]).

Proposition 6. *Let*

$$\mathcal{L} = \sum_{i=1}^{d} V_i^2$$

be a second order differential operator on $\mathbb{R}^n$*. Assume that the Lie algebra* $\mathfrak{L} =$ Lie$(V_1, \ldots, V_d)$ *which is generated by the vector fields* V_i *admits a stratification*

$$\mathcal{E}_1 \oplus \cdots \oplus \mathcal{E}_N,$$

with $\mathcal{E}_1 = \operatorname{span}(V_1, \ldots, V_d)$*,* $[\mathcal{E}_1, \mathcal{E}_i] = \mathcal{E}_{i+1}$ *and* $[\mathcal{E}_1, \mathcal{E}_N] = 0$*. Then, there exists a submersion map* $\pi : \mathbb{R}^m \to \mathbb{R}^n$*, with* $m = \dim \mathbb{G}_N(\mathbb{R}^d)$ *such that for every smooth* $f : \mathbb{R}^n \to \mathbb{R}$*,*

$$\Delta_{\mathbb{G}_N(\mathbb{R}^d)}(f \circ \pi) = (\mathcal{L}f) \circ \pi,$$

where $\Delta_{\mathbb{G}_N(\mathbb{R}^d)} = \sum_{i=1} D_i^2$*, is the canonical sub-Laplacian on* $\mathbb{G}_N(\mathbb{R}^d)$*.*

4 Nilpotentization of a differential System and canonical Approximation of a hypoelliptic Diffusion

The previous section has shown the fundamental role played by the Lie algebra $\mathfrak{L} = \operatorname{Lie}(V_1, \ldots, V_d)$ in the study of the solution X^{x_0}. In particular, if $\mathfrak{L}$ is nilpotent, we have represented X^{x_0} from the lift of B in a free Carnot whose depth is the degree of nilpotence of $\mathfrak{L}$. In this section we show how to extend these results when $\mathfrak{L}$ is not nilpotent anymore. To make this extension possible we shall assume in the sequel that the following assumption is satisfied:

Strong Hörmander Condition. *For every* $x \in \mathbb{R}^n$*, we have:*

$$\operatorname{span}\left\{ V_I(x), \ I \in \bigcup_{k \geqslant 1} \{1, \ldots, d\}^k \right\} = \mathbb{R}^n.$$

We recall that if $I = (i_1, \ldots, i_k) \in \{1, \ldots, d\}^k$ is a word, we denote by V_I the commutator defined by

$$V_I = [V_{i_1}, [V_{i_2}, \ldots, [V_{i_{k-1}}, V_{i_k}] \ldots]].$$

Let us mention that in the sub-Riemannian litterature, the strong Hörmander condition is more often refered to as Chow's condition or bracket generating condition. In that case, in some sense made precise later it is possible to *approximate* $\mathfrak{L}$ at each (regular) point by a nilpotent Lie algebra. Therefore, by using the results of the previous chapter, this approximation leads to a canonical pathwise approximation of X^{x_0} in small times.

First, we have to introduce some concepts of differential geometry. The set of linear combinations with smooth coefficients of the vector fields $V_1, \ldots, V_d$ is called the differential system (or sheaf) generated by these vector fields. It will be denoted by $\mathcal{D}$ in the sequel. Notice that $\mathcal{D}$ is naturally endowed with a structure of $\mathcal{C}_\infty(\mathbb{R}^n, \mathbb{R})$-module. For $x \in \mathbb{R}^n$, we put

$$\mathcal{D}(x) = \{X(x), X \in \mathcal{D}\}.$$

If the integer $\dim \mathcal{D}(x)$ does not depend on x, then $\mathcal{D}$ is said to be a distribution. Observe that the Lie bracket of two distributions is not necessarily a distribution, so that we really have to work with differential systems. The Lie brackets of vector fields in $\mathcal{D}$ generates a flag of differential systems,

$$\mathcal{D} \subset \mathcal{D}^2 \subset \cdots \subset \mathcal{D}^k \subset \cdots,$$

where $\mathcal{D}^k$ is recursively defined by the formula

$$\mathcal{D}^k = \mathcal{D}^{k-1} + [\mathcal{D}, \mathcal{D}^{k-1}].$$

As a module, $\mathcal{D}^k$ is generated by the set of vector fields V_I, where I describes the set of words with length k. Moreover, due to the Jacobi identity, we have $[\mathcal{D}^i, \mathcal{D}^j] \subset \mathcal{D}^{i+j}$. This flag is called the *canonical flag* associated with the differential system $\mathcal{D}$. Hörmander's strong condition, which we supposed to hold, states that for each $x \in \mathbb{R}^n$, there is a smallest integer $r(x)$ such that $\mathcal{D}^{r(x)} = \mathbb{R}^n$. This integer is called the degree of non holonomy at x. Notice that r is an upper continuous function, that is, $r(y) \leqslant r(x)$ for y near x. For each $x \in \mathbb{R}^n$, the canonical flag induces a flag of vector subspaces,

$$\mathcal{D}(x) \subset \mathcal{D}^2(x) \subset \cdots \subset \mathcal{D}^{r(x)}(x) = \mathbb{R}^n.$$

The integer list $\left(\dim \mathcal{D}^k(x)\right)_{1 \leqslant k \leqslant r(x)}$ is called the growth vector of $\mathcal{D}$ at x. The point x is said to be a regular point of $\mathcal{D}$ if the growth vector is constant in a neighborhood of x. Otherwise, we say that x is a singular point. On a Carnot group, due to the homogeneity, all points are regular.

We are now able to define in a purely algebraic manner what will be relevant for us: the nilpotentization and the tangent space of $\mathcal{D}$ at a regular point. Later, we shall see that this tangent space also can be constructed in a purely metric manner. Let $\mathcal{V}_i = \mathcal{D}^i / \mathcal{D}^{i-1}$ denote the quotient differential systems, and define

$$\mathcal{N}(\mathcal{D}) = \mathcal{V}_1 \oplus \cdots \oplus \mathcal{V}_k \oplus \cdots$$

The Lie bracket of vector fields induces a bilinear map on $\mathcal{N}(\mathcal{D})$ which respects the grading: $[\mathcal{V}_i, \mathcal{V}_j] \subset \mathcal{V}_{i+j}$. Actually, $\mathcal{N}(\mathcal{D})$ inherits the structure of a sheaf of Lie algebras. Moreover, if x is a regular point of $\mathcal{D}$, then this bracket induces a $r(x)$-step nilpotent graded Lie algebra structure on $\mathcal{N}(\mathcal{D})(x)$. Observe that the dimension of $\mathcal{N}(\mathcal{D})(x)$ is equal to n and that from the definition, $(V_1(x), \ldots, V_d(x))$ Lie-generates $\mathcal{N}(\mathcal{D})(x)$.

Definition 4. *If x is a regular point of $\mathcal{D}$, the $r(x)$-step nilpotent graded Lie algebra $\mathcal{N}(\mathcal{D})(x)$ is called the nilpotentization of $\mathcal{D}$ at x. This Lie algebra is the Lie algebra of a unique Carnot group which will be written $\mathbf{Gr}(\mathcal{D})(x)$ and called the tangent space to $\mathcal{D}$ at x.*

Remark 8. The notation $\mathbf{Gr}$ is for Gromov.

Definition 5. *If x is a regular point of $\mathcal{D}$, we say that x is a normal point $\mathcal{D}$ if there exists a neighborhood U of x such that:*

1. for every $y \in U$, y is a regular point of $\mathcal{D}$;
2. for every $y \in U$, there exists a Carnot algebra isomorphism $\psi : \mathcal{N}(\mathcal{D})(x) \to \mathcal{N}(\mathcal{D})(y)$, such that $\psi(V_i(x)) = V_i(y)$, $i = 1, \ldots, d$.

Let us mention that it may happen that $\mathcal{N}(\mathcal{D})(x)$ is not constant in a neighborhood of x even if x is regular (see [32]). At this point, it may be useful to give several examples.

Example 2. Let $\mathbb{G}$ be a Carnot group with Lie algebra $\mathfrak{g}$ and consider for $\mathcal{D}$ the left invariant differential system which is generated by the basis of $\mathbb{G}$. Then, $\mathcal{D}$ satisfies the strong Hörmander condition and it is immediate that for every $x \in \mathbb{G}$,

$$\mathcal{N}(\mathcal{D})(x) = \mathfrak{g},$$

$$\mathbf{Gr}(\mathcal{D})(x) = \mathbb{G}.$$

Example 3. Let $\mathbb{M}$ be a manifold of dimension d. Assume that there exists on $\mathbb{M}$ a family of vector fields $(V_1, \ldots, V_d)$ such that for every $x \in \mathbb{M}$, $(V_1(x), \ldots, V_d(x))$ is a basis of the tangent space at x . Denote by $\mathcal{D}$ the differential system generated by $(V_1, \ldots, V_d)$ (it is actually a distribution). Then, $\mathcal{D}$ satisfies the strong Hörmander condition and for every $x \in \mathbb{M}$,

$$\mathcal{N}(\mathcal{D})(x) = \mathbb{R}^d,$$

$$\mathbf{Gr}(\mathcal{D})(x) = \mathbb{R}^d.$$

Example 4. Consider the Lie group $\mathbf{SO}(3)$, i.e., the group of 3×3, real, orthogonal matrices of determinant 1. Its Lie algebra $\mathfrak{so}(3)$ consists of 3×3, real, skew-adjoint matrices of trace 0. A basis of $\mathfrak{so}(3)$ is formed by

$$V_1 = \begin{pmatrix} 0 & 1 & 0 \\ -1 & 0 & 0 \\ 0 & 0 & 0 \end{pmatrix}, \qquad V_2 = \begin{pmatrix} 0 & 0 & 0 \\ 0 & 0 & 1 \\ 0 & -1 & 0 \end{pmatrix}, \qquad V_3 = \begin{pmatrix} 0 & 0 & 1 \\ 0 & 0 & 0 \\ -1 & 0 & 0 \end{pmatrix}$$

Observe that the following commutation relations hold

$$[V_1, V_2] = V_3, \qquad [V_2, V_3] = V_1, \qquad [V_3, V_1] = V_2,$$

so that the differential system $\mathcal{D}$ which is generated by V_1 and V_2 satisfies the strong Hörmander condition. The group $\mathbf{SO}(3)$ can be seen as the orthonormal frame bundle of the unit sphere $\mathbb{S}^2$ and, via this identification, $\mathcal{D}$ is generated by the horizontal lifts of vector fields on $\mathbb{S}^2$. Therefore, in a way, the sub-Riemannian geometry associated with $\mathcal{D}$ is the geometry of the holonomy on $\mathbb{S}^2$. This example will be generalized in this paper; actually many interesting examples of sub-Riemannian geometries arise from principal bundles. In that case, it is easily checked that for every $x \in \mathbf{SO}(3)$,

$$\mathcal{N}(\mathcal{D})(x) = \mathfrak{g}_2(\mathbb{R}^2),$$

$$\mathbf{Gr}(\mathcal{D})(x) = \mathbb{G}_2(\mathbb{R}^2).$$

Example 5. Consider the Lie group $\mathbf{SU}(2)$, i.e., the group of 2×2, complex, unitary matrices of determinant 1. Its Lie algebra $\mathfrak{su}(2)$ consists of 2×2, complex, skew-adjoint matrices of trace 0. A basis of $\mathfrak{su}(2)$ is formed by

$$V_1 = \frac{1}{2} \begin{pmatrix} i & 0 \\ 0 & -i \end{pmatrix}, \qquad V_2 = \frac{1}{2} \begin{pmatrix} 0 & 1 \\ -1 & 0 \end{pmatrix}, \qquad V_3 = \frac{1}{2} \begin{pmatrix} 0 & i \\ i & 0 \end{pmatrix}.$$

Note the commutation relations

$$[V_1, V_2] = V_3, \qquad [V_2, V_3] = V_1, \qquad [V_3, V_1] = V_2,$$

so that the differential system $\mathcal{D}$ which is generated by V_1 and V_2 satisfies the strong Hörmander condition. Let us mention that there exists an explicit homomorphism $\mathbf{SU}(2) \to \mathbf{SO}(3)$ which exhibits $\mathbf{SU}(2)$ as a double cover of $\mathbf{SO}(3)$, so that this example is actually a consequence of the previous one. Therefore, for every $x \in \mathbf{SU}(2)$,

$$\mathcal{N}(\mathcal{D})(x) = \mathfrak{g}_2(\mathbb{R}^2),$$

$$\mathbf{Gr}(\mathcal{D})(x) = \mathbb{G}_2(\mathbb{R}^2).$$

A really striking fact is that the tangent space $\mathbf{Gr}(\mathcal{D})(x)$ at a regular point is not only a differential invariant but also a purely *metric invariant*. Actually Gromov discovered that it is possible, in a very general way, to define a notion of tangent space to an abstract metric space. This point of view is widely developed in [16] and [17], and is the starting point of the so-called metric geometry.

The Gromov–Hausdorff distance between two metric spaces $\mathbb{M}_1$ and $\mathbb{M}_2$ is defined as follows: $\delta_{\mathbf{GH}}(\mathbb{M}_1, \mathbb{M}_2)$ is the infimum of real numbers ρ for which there exists isometric embeddings of $\mathbb{M}_1$ and $\mathbb{M}_2$ in a same metric space $\mathbb{M}_3$, say $i_1 : \mathbb{M}_1 \to \mathbb{M}_3$ and $i_2 : \mathbb{M}_2 \to \mathbb{M}_3$, such that the Hausdorff distance of $i_1(\mathbb{M}_1)$ and $i_2(\mathbb{M}_2)$ as subsets of $\mathbb{M}_3$ is lower than ρ. Thanks to this distance, we have now a convenient notion of limit of a sequence metric spaces.

Definition 6. *A sequence of pointed metric spaces $(\mathbb{M}_n, x_n)$ is said to Gromov–Hausdorff converge to the pointed metric space $(\mathbb{M}, x)$ if for any positive R*

$$\lim_{n \to +\infty} \delta_{\mathbf{GH}} \big(\mathbf{B}_{\mathbb{M}_n}(x_n, R), \mathbf{B}_{\mathbb{M}}(x, R) \big) = 0,$$

where $\mathbf{B}_{\mathbb{M}_n}(x_n, R)$ is the open ball centered at x_n with radius R in $\mathbb{M}_n$. In that case we shall write

$$\lim_{n \to +\infty} (\mathbb{M}_n, x_n) = (\mathbb{M}, x).$$

If $\mathbb{M}$ is a metric space and $\lambda > 0$, we denote by $\lambda \cdot \mathbb{M}$ the new metric space obtained by multiplying all distances by λ.

Definition 7. *Let $\mathbb{M}$ be a metric space and $x_0 \in \mathbb{M}$. If the Gromov–Hausdorff limit*

$$\lim_{n \to +\infty} (n \cdot \mathbb{M}, x_0)$$

exists, then this limit is called the tangent space to $\mathbb{M}$ at x_0.

To apply now this concept of tangent space in our context, we first construct a Carnot–Carathéodory distance associated with our differential system $\mathcal{D}$. This can be done using Chow's theorem in its full generality (see chapter 2 of Montgomery [23]).

Theorem 4. *Let $(x, y) \in \mathbb{R}^n \times \mathbb{R}^n$. There exists at least one absolutely continuous curve $c : [0,1] \to \mathbb{R}^n$ such that:*

1. For almost all $s \in [0,1]$, $c'(s) \in \mathrm{span}(V_1(c(s)), \ldots, V_d(c(s)))$;
2. $c(0) = x$ and $c(1) = y$.

Thanks to this theorem, it is possible, as we did it in the case of Carnot groups, to define a distance on $\mathbb{R}^n$. The starting point of Gromov's metric geometry is then the following question: What can be said about the geometry of the differential system $\mathcal{D}$ by using only this Carnot–Carathéodory distance?

For instance, we have the following theorem due to Mitchell [22] (see also Pansu [24]).

Theorem 5. *Let $x_0 \in \mathbb{R}^n$ be a regular point of $\mathcal{D}$, then the tangent space at x_0 in the Gromov–Hausdorff sense exists and is equal to $\mathbf{Gr}(\mathcal{D})(x_0)$.*

Actually, even if x_0 is not a regular point of $\mathcal{D}$, the tangent space in the Gromov–Hausdorff sense exists. Therefore, from this theorem, it is possible to define $\mathbf{Gr}(\mathcal{D})(x_0)$ at any point of $\mathcal{D}$. Nevertheless, if x_0 is not a regular point, then $\mathbf{Gr}(\mathcal{D})(x_0)$ is not a Lie group (see [1]).

Example 6. Consider in $\mathbb{R}^2$, the two vector fields

$$V_1 = \frac{\partial}{\partial x}, \qquad \text{and} \qquad V_2 = x \frac{\partial}{\partial y}.$$

These vector fields span $\mathbb{R}^2$ everywhere, except along the line $x = 0$, where adding

$$[V_1, V_2] = \frac{\partial}{\partial y}$$

is needed. So, the distribution $\mathcal{D}$ generated by V_1 and V_2 satisfies the strong Hörmander condition. The sub-Riemannian geometry associated with $\mathcal{D}$ is called the geometry of the Grusin plane. In that case, for every $(x, y) \in \mathbb{R}^2$,

$$\mathbf{Gr}(\mathcal{D})(x, y) = \mathbb{R}^2,$$

if $x \neq 0$, whereas

$$\mathbf{Gr}(\mathcal{D})(x, y) = \mathbb{G}_2(\mathbb{R}^2)/\exp(\mathbb{R}V_2),$$

if $x = 0$.

We now apply the concepts introduced above to study the SDE (1).

Theorem 6. *Let* $x \in \mathbb{R}^n$ *be a normal point of* $\mathcal{D}$. *Let* $(X_t^x)_{t \geqslant 0}$ *denote the solution of (1) with initial condition* x. *There exist a surjective Carnot group morphism*

$$\pi_x : \mathbb{G}_{r(x)}(\mathbb{R}^d) \longrightarrow \mathbf{Gr}(\mathcal{D})(x)$$

and a local diffeomorphism

$$\psi_x : U \subset \mathbf{Gr}(\mathcal{D})(x) \longrightarrow \mathbb{R}^n$$

such that

$$X_t^x = \psi_x \left(\pi_x B_t^* \right) + t^{\frac{r(x)+1}{2}} \mathbf{R}(t), \qquad 0 < t < T$$

where:

1. U *is an open neighborhood of the identity element of* $\mathbf{Gr}(\mathcal{D})(x)$;
2. B^* *is the lift of* B *in the free Carnot group* $\mathbb{G}_{r(x)}(\mathbb{R}^d)$;
3. T *is an almost surely non negative stopping time*;
4. $\mathbf{R}$ *is bounded in probability when* $t \to 0$.

Proof. From Castell's result [7],

$$X_t^x = \left[\exp \left(\sum_{k=1}^{r(x)} \sum_{I=(i_1,\ldots,i_k)} \Lambda_I(B)_t V_I \right) \right](x) + t^{\frac{r(x)+1}{2}} \mathbf{R}(t), \qquad t \geqslant 0,$$

where the remainder term $\mathbf{R}$ is bounded in probability when $t \to 0$. Now, since x is a normal point of $\mathcal{D}$, we can write

$$\left[\exp \left(\sum_{k=1}^{r(x)} \sum_{I=(i_1,\ldots,i_k)} \Lambda_I(B)_t V_I \right) \right](x) = \psi_x(\hat{B}_t), \qquad t < T,$$

where

1. $\psi_x : U \subset \mathbf{Gr}(\mathcal{D})(x) \to \mathbb{R}^n$ is a local diffeomorphism;
2. $\hat{B}$ is the lift of B in the Carnot group $\mathbf{Gr}(\mathcal{D})(x)$, with respect to the family $(V_1(x),\ldots,V_d(x))$ (recall that by construction, this family Lie-generates $\mathcal{N}(\mathcal{D})(x)$) ;
3. T is an almost surely non negative stopping time.

Now, since $\mathfrak{g}_{r(x)}(\mathbb{R}^d)$ is free, there exists a unique Lie algebra surjective homomorphism $\alpha_x : \mathfrak{g}_{r(x)}(\mathbb{R}^d) \rightarrow \mathcal{N}(\mathcal{D})(x)$ such that $\alpha_x(D_i) = V_i(x)$. Since Carnot groups are simply connected nilpotent groups for which the exponential map is a diffeomorphism, there exists a unique Carnot group morphism $\pi_x : \mathbb{G}_{r(x)} \rightarrow \mathbf{Gr}(\mathcal{D})(x)$ such that $d\pi_x = \alpha_x$ (see also proposition 5). We have $\pi_x(B_t^*) = \hat{B}_t$ which concludes the proof. $\square$

Remark 9. Observe that $\pi_x B_t^*$ is a lift of B in the Carnot group $\mathbf{Gr}(\mathcal{D})(x)$.

Remark 10. We stress that theorem (6) is not true in general if x is not a normal point of $\mathcal{D}$. Indeed, let us assume that the nilpotentization $\mathcal{N}(\mathcal{D})(x)$ is not constant in a neighborhood of x and that there exists a bi-Lipschitz map $\psi_x : U \subset \mathbf{Gr}(\mathcal{D})(x) \rightarrow \mathbb{R}^n$. In that case a sub-Riemannian extension extension of Rademacher's theorem due to [21] would imply that ψ_x is almost everywhere Pansu differentiable and the derivatives would provide Carnot group morphisms between groups which are not isomorphic.

An immediate corollary of this theorem is the behaviour in small times of a hypoelliptic heat kernel on the diagonal. This behaviour, first discovered by Rotschild and Stein [27] using the parametrix method (see also [31]), has then been understood in a probabilistic way by Ben Arous [2] and Léandre [18], [19].

Corollary 1. *Let x_0 be a normal point of $\mathcal{D}$. Let p_t, $t > 0$, denote the density of X^{x_0} with respect to the Lebesgue measure. We have,*

$$p_t(x_0) \underset{t \to 0}{\sim} \frac{C(x_0)}{t^{D(x_0)/2}},$$

where $C(x_0)$ is a non negative constant depending smoothly of x_0 and $D(x_0)$ the Hausdorff dimension of the tangent space $\mathbf{Gr}(\mathcal{D})(x_0)$.

Remark 11. At a regular point, even if $\mathcal{N}(\mathcal{D})(x_0)$ is not constant in a neighborhood of x_0, the previous asymptotic development holds.

5 Regular self-adjoint sub-Laplacians on compact manifolds

Let $\mathbb{M}$ be a connected compact smooth manifold. We consider on $\mathbb{M}$ a second order differential operator

$$\mathcal{L} = \sum_{i=1}^{d} V_i^2,$$

which satisfies the strong Hörmander condition. Let $\mathcal{D}$ denote the differential system generated by $V_1, \ldots, V_d$. We assume that there exists a Carnot group $\mathbb{G}$ such that for every $x \in \mathbb{M}$,

$$\mathbf{Gr}(\mathcal{D})(x) = \mathbb{G}.$$

If the above assumptions are satisfied, then $\mathcal{L}$ will be said to be regular. Since the manifold $\mathbb{M}$ is assumed to be compact, it is possible to develop a spectral theory for $\mathcal{L}$ which is similar to the spectral theory of elliptic operators. More precisely, if we denote by X the diffusion associated with $\mathcal{L}$, by using Bony's strong maximal principle for hypoelliptic operators (see [5]), it is possible to show that X is Harris recurrent. That is, there is a Borel measure m on $\mathbb{M}$ which is invariant for X such that for every Borel set $A \subset \mathbb{M}$:

1. $m(A) > 0$ implies that for any $x \in \mathbb{M}$, and $t > 0$,

$$\mathbb{P}(\exists s > t, \, X_s \in A \mid X_0 = x) = 1;$$

2.

$$\int_{\mathbb{M}} \mathbb{P}(X_t \in A \mid X_0 = x) \, m(\mathrm{d}x) = m(A), \qquad t > 0.$$

Observe that m is a solution of the equation

$$\mathcal{L}^* m = 0,$$

so that, by Hörmander's theorem, it admits a smooth density. We shall now assume furthermore that $\mathcal{L}$ is self-adjoint with respect to m, i.e., for any smooth functions $f, g : \mathbb{M} \to \mathbb{R}$

$$\int_{\mathbb{M}} g(\mathcal{L}f) \, \mathrm{d}m = \int_{\mathbb{M}} (\mathcal{L}g)f \, \mathrm{d}m.$$

In that case, $e^{t\mathcal{L}}$ is a compact selfadjoint operator in $\mathbf{L}^2(\mathbb{M}, m)$. We deduce that $\mathcal{L}$ has a discrete spectrum tending to $-\infty$. Denote by $\mathbf{Sp}(\mathcal{L})$ the set of eigenvalues of $\mathcal{L}$ repeated according to multiplicity.

Theorem 7. *For $\lambda > 0$, let*

$$\mathbf{N}(\lambda) = \mathrm{Card}\big(\mathbf{Sp}(\mathcal{L}) \cap [-\lambda, 0]\big).$$

We have

$$\mathbf{N}(\lambda) \underset{\lambda \to +\infty}{\sim} C(\mathcal{L}, \mathbb{M})\lambda^{D/2},$$

where $C(\mathcal{L}, \mathbb{M})$ is a non negative constant and D the Hausdorff dimension of $\mathbb{G}$.

Proof. The asymptotic expansion of the heat semigroup $e^{t\mathcal{L}}$ on the diagonal leads to

$$\mathrm{Tr}\big(e^{t\mathcal{L}}\big) = \int_{\mathbb{M}} p_t(x, x) \, m(\mathrm{d}x) \underset{t \to 0}{\sim} \frac{K}{t^{D/2}},$$

where K is a non negative constant. On the other hand,

$$\mathrm{Tr}\big(e^{t\mathcal{L}}\big) = \sum_{k=0}^{+\infty} N_k \, e^{-\mu_k t},$$

where

1. $\{-\mu_k\}$ is the set of eigenvalues of $\mathcal{L}$;
2. $N_k = \dim\{f \mid \mathcal{L}f = -\mu_k f\}$.

Therefore,

$$\sum_{k=0}^{+\infty} N_k\, e^{-\mu_k t} \underset{t \to 0}{\sim} \frac{K}{t^{D/2}}.$$

The result follows then from the following theorem of Karamata: if μ is a Borel measure on $[0, \infty)$, $\alpha \in (0, +\infty)$, then

$$\int_0^{+\infty} e^{-t\lambda}\, \mu(d\lambda) \underset{t \to 0}{\sim} \frac{1}{t^\alpha},$$

implies

$$\int_0^x \mu(d\lambda) \underset{x \to +\infty}{\sim} \frac{x^\alpha}{\Gamma(1+\alpha)}.$$

$\square$

Remark 12. We believe that the constant $C(\mathcal{L}, \mathbb{M})$ is an interesting invariant of the sub-Riemannian geometry that $\mathcal{L}$ induces on $\mathbb{M}$ (recall that in the Riemannian case, it is simply, up to scale, the Riemannian volume of the manifold). For instance, it would be interesting to know if $C(\mathcal{L}, \mathbb{M})$ is the Hausdorff measure of $\mathbb{M}$.

To conclude this section, it may be interesting to study carefully an example. Consider the Lie group $\mathbf{SU}(2)$. As already seen, a basis of $\mathfrak{su}(2)$ is formed by

$$V_1 = \frac{1}{2}\begin{pmatrix} i & 0 \\ 0 & -i \end{pmatrix}, \qquad V_2 = \frac{1}{2}\begin{pmatrix} 0 & 1 \\ -1 & 0 \end{pmatrix}, \qquad V_3 = \frac{1}{2}\begin{pmatrix} 0 & i \\ i & 0 \end{pmatrix},$$

and the commutation relations hold

$$[V_1, V_2] = V_3, \qquad [V_2, V_3] = V_1, \qquad [V_3, V_1] = V_2. \tag{4}$$

We want to study the regular sub-Laplacian

$$\mathcal{L} = V_2^2 + V_3^2.$$

Actually, we shall study the following family of operators defined for $\varepsilon \in [0, 1]$,

$$\mathcal{L}^\varepsilon = \varepsilon V_1^2 + V_2^2 + V_3^2.$$

Observe that each $\mathcal{L}^\varepsilon$ is self-adjoint with respect to the normalized Haar measure of $\mathbf{SU}(2)$. For $\varepsilon > 0$, $\mathcal{L}^\varepsilon$ is elliptic so that

$$\mathrm{Card}\big(\mathbf{Sp}(\mathcal{L}^\varepsilon) \cap [-\lambda, 0]\big) \underset{\lambda \to +\infty}{\sim} C_\varepsilon \lambda^{3/2},$$

whereas

$$\mathrm{Card}\big(\mathbf{Sp}(\mathcal{L}) \cap [-\lambda, 0]\big) \underset{\lambda \to +\infty}{\sim} C\lambda^2.$$

Therefore, this is interesting to understand the spectrum when $\varepsilon \to 0$.

Proposition 7. *Let $\varepsilon \in [0,1)$. The set of eigenvalues of $\mathcal{L}^\varepsilon$ is the set*

$$\{-\lambda_{n,m}^\varepsilon, \, n \in \mathbb{N}, \, 0 \leqslant m \leqslant n\},$$

where

$$\lambda_{n,m}^\varepsilon = \varepsilon \frac{n^2}{4} + \frac{n}{2} - (1 - \varepsilon)m^2 + (1 - \varepsilon)mn.$$

Moreover, the multiplicity of $\lambda_{n,m}^\varepsilon$ is equal to $n + 1$.

Proof. We use the theory of representations of $\mathbf{SU}(2)$; for a detailed account on it, we refer to Taylor [29], chapter 2. Let $\varepsilon \in [0,1)$. Thanks to the relations (4), note that $\mathcal{L}^\varepsilon$ commutes with $\mathcal{L}^1 = V_1^2 + V_2^2 + V_3^2$. Therefore $\mathcal{L}^\varepsilon$ acts on each eigenspace of $\mathcal{L}^1$. We can examine the spectrum of $\mathcal{L}^\varepsilon$ by decomposing $\mathbf{L}^2(\mathbf{SU}(2))$ into eigenspaces of $\mathcal{L}^1$, which is equivalent to decomposing it into subspaces irreducible for the regular action of $\mathbf{SU}(2) \times \mathbf{SU}(2)$ given by

$$((g,h) \cdot f)(x) = f(g^{-1}xh), \qquad f \in \mathbf{L}^2(\mathbf{SU}(2)), \qquad g, h, x \in \mathbf{SU}(2).$$

Now, it is known (see for instance Taylor [29]) that, up to equivalence, for every $k \in \mathbb{N}$, there exists one and only one irreducible representation $\pi_k : \mathbf{SU}(2) \to \mathbb{C}^{k+1}$. Thus, by the Peter–Weyl theorem, the irreducible spaces of $\mathbf{L}^2(\mathbf{SU}(2))$ for the regular action are precisely the spaces of the form

$$\mathcal{V}_k = \text{span}\{\pi_k^{i,j}, \, 1 \leqslant i, j \leqslant k+1\},$$

where $\pi_k^{i,j}$ denotes the components of the representation π^k in a chosen orthonormal basis of $\mathbb{C}^{k+1}$. Observe now that each space $\mathcal{V}_k$ is an eigenspace of $\mathcal{L}^1$. The associated eigenvalue is $-k(k+2)/4$. If we consider now the left regular representation

$$(g \cdot f)(x) = f(g^{-1}x), \qquad f \in \mathbf{L}^2(\mathbf{SU}(2)), \qquad g, x \in \mathbf{SU}(2),$$

then $\mathcal{V}_k$ is a direct sum of $k + 1$ irreducible representations of $\mathbf{SU}(2)$, each equivalent to π_k:

$$\mathcal{V}_k = \bigoplus_{l=1}^{k+1} \mathcal{V}_{k,l},$$

where $\mathcal{V}_{k,l} = \text{span}\{\pi_k^{i,l}, 1 \leqslant i \leqslant k+1\}$. Each $\mathcal{V}_{k,l}$ splits into one-dimensional eigenspaces for V_1:

$$\mathcal{V}_{k,l} = \bigoplus_\mu \mathcal{V}_{k,l,\mu},$$

where,

$$\mu \in \left\{ -\frac{k}{2}, -\frac{k}{2} + 1, \ldots, \frac{k}{2} \right\},$$

and on $\mathcal{V}_{k,l,\mu}$,

$$V_1 = i\mu.$$

Since $\mathcal{L}^\varepsilon = \mathcal{L}^1 - (1-\varepsilon)V_1^2$, we have

$$\mathcal{L}^\varepsilon = -\left(\frac{k(k+2)}{4} - (1-\varepsilon)\mu^2\right),$$

on $\mathcal{V}_{k,l,\mu}$. $\square$

From this, we deduce immediately:

$$\mathrm{Card}\big(\mathbf{Sp}(\mathcal{L}^\varepsilon)\cap[-\lambda,0]\big) \underset{\lambda\to+\infty}{\sim} \frac{8}{3}\sqrt{\varepsilon}\,\lambda^{3/2},$$

whereas

$$\mathrm{Card}\big(\mathbf{Sp}(\mathcal{L})\cap[-\lambda,0]\big) \underset{\lambda\to+\infty}{\sim} 2\lambda^2.$$

6 Application to the study of Brownian motions on Riemannian manifolds

To conclude the paper, we provide a very general and natural geometric framework in which regular hypoelliptic operators appear.

Let $(\mathbb{M}, g)$ be a d dimensional connected compact Riemannian manifold. We assume that the Riemannian curvature tensor on $(\mathbb{M}, g)$ is nowhere degenerate. We denote by Δ the Laplace–Beltrami operator on $\mathbb{M}$ (for us Δ is negative). The tangent bundle to $\mathbb{M}$ is denoted by $T\mathbb{M}$ and $T_m\mathbb{M}$ is the tangent space at m: We have hence $T\mathbb{M} = \bigcup_n T_n\mathbb{M}$. The orthonormal frame bundle of $\mathbb{M}$ is denoted by $\mathcal{O}(\mathbb{M})$. Therefore, $(\mathcal{O}(\mathbb{M}), \mathbb{M}, \mathcal{O}_d(\mathbb{R}))$ is a principal bundle on $\mathbb{M}$ with structure group the group $\mathcal{O}_d(\mathbb{R})$ of $d \times d$ orthogonal matrices. We denote by π the canonical surjection $\mathcal{O}(\mathbb{M}) \to \mathbb{M}$. The horizontal fundamental vector fields of $\mathcal{O}(\mathbb{M})$ are denoted by $(H_i)_{i=1,\ldots,d}$. The Bochner horizontal Laplacian, i.e. the lift of Δ, is then given by

$$\Delta_{\mathcal{O}(\mathbb{M})} = \sum_{i=1}^d H_i^2.$$

We denote by $\mathcal{D}$ the differential system on $\mathcal{O}(\mathbb{M})$ generated by the horizontal vector fields $H_1, \ldots, H_d$, that is, the horizontal distribution for the Levi-Civita connection on $\mathbb{M}$. We have the following proposition (compare to Chernyakov's theorem, see [13] p. 22):

Proposition 8. *The distribution $\mathcal{D}$ satisfies the strong Hörmander condition. Moreover, any point $x_0 \in \mathcal{O}(\mathbb{M})$ is a regular point of $\mathcal{D}$ and $\mathbf{Gr}(\mathcal{D})(x_0) = \mathbb{G}_2(\mathbb{R}^d)$.*

Proof. From Cartan's formula, we get

$$\Theta([H_i, H_j]) = H_i\Theta(H_j) - H_j\Theta(H_i) - d\Theta(H_i, H_j),$$

where Θ is the tautological one-form on $\mathcal{O}(M)$. Now, from the first structural equation, we have

$$d\Theta = -\Theta \wedge \omega$$

where ω is the Levi-Civita connection form. This implies,

$$\Theta([H_i, H_j]) = 0.$$

Thus, the commutator of two fundamental vector fields is vertical. Using the second structure equation

$$d\omega = -\omega \wedge \omega + \Omega,$$

where Ω is the curvature form, we obtain in a similar way with Cartan's formula

$$\omega([H_i, H_j]) = -\Omega(H_i, H_j).$$

This ensures that for any $x_0 \in \mathcal{O}(M)$ the family $(H_i(x_0), [H_j, H_k](x_0))$ is always a basis of the tangent space because of the assumed non-degenerence of Ω. □

By applying theorem 6 and using the identification $\mathbb{G}_2(\mathbb{R}^d) \simeq \mathbb{R}^d \times \mathbb{R}^{d(d-1)/2}$, we get hence

Corollary 2. *Let* $(B_t)_{t \geq 0}$ *be a Brownian motion on* M. *There exist a local diffeomorphism*

$$\psi : U \subset \mathbb{R}^d \times \mathbb{R}^{d(d-1)/2} \longrightarrow \mathcal{O}(M)$$

and a standard linear Brownian motion $(\beta_t)_{t \geq 0}$ *on* $\mathbb{R}^d$, *such that for any smooth function* $f : M \to \mathbb{R}$,

$$f(B_t) = f\left((\pi \circ \psi)\left(\beta_t, \left(\int_0^t \beta_s^i \, d\beta_s^j - \beta_s^j \, d\beta_s^i \right)_{1 \leq i < j \leq d} \right) \right) + t^{3/2} \mathbf{R}(t, f),$$

for $0 < t < T$, *where:*

1. U *is an open neighborhood of* 0 *in* $\mathbb{R}^d \times \mathbb{R}^{d(d-1)/2}$;
2. $\pi : \mathcal{O}(M) \to M$ *is the bundle projection;*
3. T *is an almost surely non negative stopping time;*
4. $\mathbf{R}$ *is bounded in probability when* $t \to 0$.

Remark 13. We believe that the same analysis can be performed in other bundles than orthonormal bundles. More precisely, we have in mind spinor bundles. In that case, an analogue of the above approximation theorem seems closely related to Bismut's proof of the Atiyah–Singer theorems (see [4]).

7 Conclusion and Comments

- Let us mention that the geometry of the Chen expansion is also of particular importance in the "rough paths" theory of T. Lyons [20]. Indeed, Lyons' fundamental theorem shows that for any Itô map $B \to X$, coming from a hypoelliptic system or not, there exists a continuous (in a convenient topology) map $F : \mathcal{C}([0,T], \mathbb{G}_2(\mathbb{R}^d)) \to \mathcal{C}([0,T], \mathbb{R}^n)$ such that $X = F(B^*)$, where B^* is the lift of B in the free two-step Carnot group $\mathbb{G}_2(\mathbb{R}^d)$.
- As a conclusion, we would like to say that we think that many ideas of Gromov's metric geometry could be applied in probability theory.

References

1. A. Bellaïche: The tangent space in sub-Riemannian geometry, In Sub-Riemannian Geometry, edited by A. Bellaïche and J.J. Risler, Birkhäuser (1996).
2. G. Ben Arous: Développement asymptotique du noyau de la chaleur hypoelliptique sur la diagonale, Annales de l'Institut Fourier **39**, 73–99 (1989).
3. G. Ben Arous: Flots et séries de Taylor stochastiques, Probability Theory and Related Fields **81**, 29–77 (1989).
4. J.M. Bismut: The Atiyah-Singer Theorems: A Probabilistic Approach, J. Func. Anal., Part I: **57** (1984), Part II: **57**, 329–348 (1984).
5. J.M. Bony: Principe du maximum, inégalité de Harnack et unicité du problème de Cauchy pour les opérateurs elliptiques dégénérés, Annales de l'Institut Fourier **19**, 277–304 (1969).
6. N. Bourbaki: Groupes et Algèbres de Lie, Chap. 1–3, Hermann (1972).
7. F. Castell: Asymptotic expansion of stochastic flows, Probability Theory and Related Fields **96**, 225-239 (1993).
8. K.T. Chen: Integration of paths, Geometric invariants and a Generalized Baker–Hausdorff formula, Annals of Mathematics **65** (1957).
9. W.L. Chow: Uber system von lineare partiellen Differentialgleichungen erster Ordnung, Math. Ann. **117** (1939).
10. M. Fliess, D. Normand-Cyrot: Algèbres de Lie nilpotentes, formule de Baker–Campbell–Hausdorff et intégrales itérées de K.T. Chen, in Séminaire de Probabilités XVI, LNM **920**, Springer-Verlag (1982).
11. G. B. Folland, E. M. Stein: Hardy spaces on homogeneous groups. Princeton University Press (1982).
12. B. Gaveau: Principe de moindre action, propagation de la chaleur et estimées sous-elliptiques sur certains groupes nilpotents. Acta Math. **139**, 95-153 (1977).
13. V. Ya. Gershkovich, A. M. Vershik: Nonholonomic Dynamical Systems, Geometry of Distributions and Variational Problems, In *Dynamical Systems VII, Encyclopaedia of Mathematical Sciences*, Vol. **16**, Eds. V.I. Arnold, S.P. Novikov (1994).
14. C. Golé, R. Karidi: A note on Carnot geodesics in nilpotent Lie groups. Jour. Control and Dynam. Systems **1**, 535–549 (1995).
15. N. Goodman: Nilpotent Lie groups, Springer Lecture Notes in Mathematics, Vol. **562**, (1976).

16. M. Gromov: Carnot–Carathéodory spaces seen from within, In Sub-Riemannian Geometry, edited by A. Bellaïche and J.J. Risler, Birkhäuser (1996).

17. M. Gromov: Metric structures for Riemannian and non-Riemannian spaces, with appendices by M. Katz, P. Pansu and S. Semmes. Birkhäuser (1999).

18. R. Léandre: Développement asymptotique de la densité d'une diffusion dégénérée, Forum Math. **4**, 45–75 (1992).

19. R. Léandre: Intégration dans la fibre associée à une diffusion dégénérée, Probability Theory and Related Fields **76**, 341–358 (1987).

20. T. Lyons: Differential Equations Driven by Rough Signals.Revista Mathemàtica Iberio Americana **14**, 215–310 (1998).

21. G. Margulis, G.D. Mostow: The differential of a quasi-conformal mapping of a Carnot–Carathéodory space, Geom. Funct. Anal. **5**, 402–433 (1995).

22. J. Mitchell: On Carnot–Carathéodory metrics. J. Differential Geom. **21**, 35–45 (1985).

23. R. Montgomery: A Tour of Subriemannian Geometries, Their Geodesics and Applications, Mathematical Surveys and Monographs **91**, AMS (2002).

24. P. Pansu: Métriques de Carnot–Carathéodory et quasi-isométries des espaces symétriques de rang un, Ann. Math. **129**, 1–60 (1989).

25. C.B. Rayner: The exponential map for the Lagrange problem on differential manifolds, Phil. Trans. of the Roy. Soc. of London, Ser. A, Math. and Phys., **1127**, v. **262**, 299–344 (1967).

26. C. Reutenauer: Free Lie algebras, London Mathematical Society Monographs, New series **7** (1993).

27. L.P. Rotschild, E.M. Stein: Hypoelliptic differential operators and Nilpotent Groups, Acta Mathematica **137**, 247–320, (1976).

28. R.S. Strichartz: The Campbell-Baker–Hausdorff-Dynkin formula and solutions of differential equations, Jour. Func. Anal. **72**, 320–345 (1987).

29. M. E. Taylor: Noncommutative Harmonic Analysis, Mathematical Surveys and Monographs **22**, AMS (1986).

30. M. E. Taylor: Partial Differential Equations, Qualitative Studies of Linear Equations, Springer, Applied Mathematical Sciences **116** (1996).

31. S. Takanobu: Diagonal short time asymptotics of heat kernels for certain degenerate second order differential operators of Hörmander type. Publ. Res. Inst. Math. Sci. **24**, 169–203 (1988).

32. A.M. Vershik and V. Ya. Gershkovich: Non holonomic problems and the theory of distributions, Acta Appl. Math., **12**, 181–209, (1988).

33. Y. Yamato: Stochastic differential equations and nilpotent Lie algebras. Z. Wahrscheinlichkeitstheorie verw. Geb., **47**, 213–229 (1979).

34. J.C. Watkins: Donsker's invariance principle for Lie 'groups, Annals of Probability **17**, 1220–1242 (1989).

Homogenization of a Diffusion with Locally Periodic Coefficients

Abdellatif Benchérif-Madani and Étienne Pardoux

Université Ferhat Abbas, Fac. Sciences, Dépt. Maths., Sétif 19000, Algeria.
Laboratoire d'Analyse, Topologie, Probabilités UMR 6632, CMI,
39 rue Joliot-Curie, Marseille cedex 13, France.
e-mail: lotfi_madani@yahoo.fr, pardoux@cmi.univ-mrs.fr

Summary. We present a result of homogenization for a class of second order parabolic partial differential equations with locally periodic coefficients, and highly oscillating potential. Our method of proof is mainly probabilistic. We deduce the homogenization result from weak convergence for a class of diffusion processes.

1 Introduction

In this paper we deal with the problem of the homogenization property of the following singular parabolic PDE in $\mathbf{R}^d$ with locally periodic coefficients, understood in the strong sense of Gilbarg and Trudinger [6] chapter 9, with a Cauchy type boundary condition $u^\varepsilon(0,x) = g(x)$, for all $\varepsilon > 0$,

$$\partial_t u^\varepsilon(t,x) = L^\varepsilon u^\varepsilon(t,x) + \lambda^\varepsilon\left(x, \frac{x}{\varepsilon}\right) u^\varepsilon(t,x), \tag{1}$$

where

$$L^\varepsilon(\,.\,) = \frac{1}{2} \sum_{i,j=1}^{d} a_{ij}\left(x, \frac{x}{\varepsilon}\right) \partial^2_{x_i x_j}(\,.\,) + \sum_{i=1}^{d}\left(\varepsilon^{-1} b_i\left(x, \frac{x}{\varepsilon}\right) + c_i\left(x, \frac{x}{\varepsilon}\right)\right) \partial_{x_i}(\,.\,), \tag{2}$$

and $\lambda^\varepsilon(x) = \varepsilon^{-1} e(x, x/\varepsilon) + f(x, x/\varepsilon)$. The matrix $a(x,y)$ is assumed to be symmetric, for all $x, y \in \mathbf{R}^d$, and all the coefficients are periodic with respect to the second variable with period one in each direction in $\mathbf{R}^d$. The latter is called the fast variable as opposed to the first slow component. The operator L^ε is supposed to be uniformly elliptic. That is $\exists \beta$ strictly positive and finite s.t. for all x, y and ξ in $\mathbf{R}^d$,

$$\beta\|\xi\|^2 \leqslant (a(x,y)\xi, \xi). \tag{3}$$

In studying homogenization, it is required to find the form of the limit operator L, if any, to the sequence $L^\varepsilon + \lambda^\varepsilon I$ whereby the real sequence $u^\varepsilon(t,x)$ also converges to the solution $u(t,x)$ of

$$\partial u(t,x)/\partial t = Lu(t,x),$$

subject to the same Cauchy initial boundary condition $u(0,x) = g(x)$. That L is also a differential operator is well accounted for in, for example, Allaire [1]. Homogenization thus appears as a kind of a reverse process to standard approximation technics in functional analysis. We intend here to substantially weaken the known regularity constraints on the coefficients $a_{ij}(x,y)$, $b_i(x,y)$, $c_i(x,y)$, $e(x,y)$, $f(x,y)$, for x and y in $\mathbf{R}^d$ and $i,j = 1,\ldots,d$, that were observed to ensure the validity of the homogenization property. In Bensoussan and al. [2], for example, it was assumed that all the coefficients were twice continuously differentiable with bounded partial derivatives, including those of order zero.

1.1 Probabilistic approach

After the pioneer work of Freidlin [4], the probabilistic way of handling the problem has been exposed in Chapter 3 of Bensoussan, Lions and Papanicolaou [2]. Pardoux [8] used that approach in order to solve the fully periodic case with a highly oscillating potential. For the same approach applied to random homogenization, we refer the reader to the survey of Olla [7]. The idea of the probabilistic approach is to build on some probability space $(\Omega, \mathcal{F}, \mathcal{F}_t, B_t, X_t^\varepsilon, \mathbf{P})$, where B_t is under $\mathbf{P}$ an $\mathcal{F}_t$ Brownian motion, the diffusions X_t^ε in $\mathbf{R}^d$ solutions to the stochastic differential equations for $0 \leqslant t \leqslant T$,

$$X_t^\varepsilon = x + \int_0^t \left(\varepsilon^{-1}b(X_s^\varepsilon, X_s^\varepsilon/\varepsilon) + c(X_s^\varepsilon, X_s^\varepsilon/\varepsilon)\right) \mathrm{d}s + \int_0^t \sigma(X_s^\varepsilon, X_s^\varepsilon/\varepsilon) \, \mathrm{d}B_s, \quad (4)$$

where the matrix σ satisfies $\sigma\sigma^*(x,y) = a(x,y)$, whenever sufficient regularity conditions are exhibited. The operator L^ε turns out to be the generator of the Markov process X_t^ε. Conditions for relative compactness of the laws P_{X^ε} in $\mathcal{C}[0,T]$ are looked for. Uniqueness of the limit point P^0 follows from weak uniqueness of the associated SDE, see Stroock and Varadhan [11]. The generator L^0 that corresponds to the law P^0 is considered to be the limit of the sequence L^ε. To finish off, a Girsanov argument coupled to the Feynman–Kac formula, which yields an effective (probabilistic) formula for $u^\varepsilon(t,x)$, deals with the perturbation due to $\lambda^\varepsilon I$. In the totally periodic case, a crucial idea due to Freidlin [4], see Pardoux [8] page 499, is to transform, by means of the scaling property of the Brownian motion, the short space scale into a long time behaviour of a diffusion on the compact torus $\mathbb{T}^d$. By the works of Doob in the forties concerning ergodic theory, a limit at $t = \infty$ does exist, irrespective of the starting point x/ε. Unfortunately, it is not obvious how to extend this argument to the locally periodic case and a different approach is called for. At this stage, we give a few comments on the method of [2] for the convenience of the reader and for later reference. The ideas pertaining to ergodic theory are obviously still in force. The idea is to freeze the slow component in (2) and consider the family of operators on the compact torus $\mathbb{T}^d$, indexed by x and acting on y,

$$L_{x,y}(\,.\,) = \frac{1}{2} \sum_{i,j=1}^{d} a_{ij}(x,y)\, \partial^2_{y_i y_j}(\,.\,) + \sum_{i=1}^{d} b_i(x,y)\, \partial_{y_i}(\,.\,). \tag{5}$$

The coefficient c, though present in the limit, plays no asymptotic role as far as ergodicity is concerned and the leading term is b, that is why the coefficient c does not appear in (5). To these operators correspond the following diffusions parametrized by x, with transition densities $p_t(x,y,y')$,

$$Y_t^x = y + \int_0^t b(x, Y_s^x)\, ds + \int_0^t \sigma(x, Y_s^x)\, dB_s.$$

In fact they may be thought of as diffusions on the torus $\mathbb{T}^d$, i.e.,

$$\dot{Y}_t^x = y + \int_0^t b(x, \dot{Y}_s^x)\, ds + \int_0^t \sigma(x, \dot{Y}_s^x)\, dB_s,$$

with transition densities

$$\dot{p}_t(x,y,y') = \sum_{k_1 \ldots k_d} p_t\left(x, y, y' + \sum_{i=1}^{d} k_i e_i\right),$$

e_i being the canonical basis of $\mathbf{R}^d$ and k_i integers. In what follows we shall drop the dots when no ambiguity arises. As already pointed out, these diffusions possess invariant probability measures $\mu(x, dy)$ with densities $p_\infty(x,y)$. By standard elementary mechanisms of ergodic theory, to hope for a convergence as $\varepsilon \to 0$, we need the following centering condition on the singular coefficient b, which is rather reminiscent of "passing through the eye of a hurricane", for all x,

$$\int_{\mathbb{T}^d} b(x,y)\, \mu(x, dy) = 0. \tag{6}$$

We can then solve the Poisson equation

$$L_{x,y}\widehat{b}(x,y) = -b(x,y). \tag{7}$$

Given enough regularity on the coefficients, for example so as to make classical calculations, see [2], a solution $\widehat{b}(x,y)$ exists and is also periodic in y. Moreover the $\widehat{b}$ Itô formula can be simultanuously applied to both the slow and the fast variables. As a result of this, the singularity ε^{-1} is not only lifted (by (7)) but the process X^ε itself is recovered "unthorned", at the small cost of introducing asymptotically small terms. This is in essence the method of Freidlin [4]. A martingale problem method now leads to the identification of the unique limit point law P^0 that was seen to exist by tightness. The crucial point here is the need for a locally periodic version of the totally periodic ergodic theorem of Pardoux [8], Proposition 2.4. This is carried out again by means of the Itô formula applied in the obvious way, though with less intuition. Our point is that even if we have less regularity, the freezing procedure above can be pushed further to effectively stop the slow component on sufficiently small time intervals Δt. We can achieve this thanks to tightness. In other words, we fall back, loosely speaking, to the totally periodic case which needs a single variable and hence less regularity. Our fundamental task therefore is to assume enough regularity to ensure tightness. Our technical manipulations on $\widehat{b}$ and related functions are based on those of Pardoux and Veretennikov [10] who considered ergodic properties on the whole of $\mathbf{R}^d$ for a similar two-component homogenization problem. However, a lot of additional work is needed, since our regularity assumptions are much weaker than those in [10].

In Section 2 we state our conditions on the coefficients and prove tightness by means of an auxiliary technical Lemma which we prove in Section 4. In Section 3 the limit generator L^0 is identified within a technical Lemma left to Section 5. The convergence of $u^\varepsilon(t,x)$ to $u(t,x)$ is treated in Section 6.

1.2 Notation

Let ξ denote any component of a coefficient a, b, c, e, f. Expressions like $\xi(X^\varepsilon_s, X^\varepsilon_t/\varepsilon)$ will appear many times below. For the sake of clarity, we shall systematically write instead $\xi(s,t)$. There should be no confusion between $\xi(s,t)$ meaning the above quantity, and $\xi(x,y)$, since we use different letters for space and time variables. Moreover we shall systematically write

$$\Delta_{s,t} f(\,\cdot\,)$$

to denote the difference

$$f(t) - f(s).$$

The space of k times continuously differentiable functions ($k \leqslant \infty$) on an open domain D of $\mathbf{R}^d$ is designated by $\mathcal{C}^k(D)$ and those with compact support in D by $\mathcal{C}^k_K(D)$. The linear space of $\mathbf{R}^d$ valued continuous functions on $[0,T]$ is denoted by $\mathcal{C}[0,T]$. If $u(x)$ is a function of x in $\mathbf{R}^d$, we shall write $\partial_x u(x)$ to denote the d-dimensional vector whose i-th coordinate is $\partial_{x_i} u(x)$; similarly $\partial_x^2 u(x)$ will denote a $d \times d$ matrix and so on. The integral part of real number x is denoted by $[x]$. The letters R and r, with possibly subscripts and superscripts, are reserved to indicate unimportant remainder quantities against leading terms. Unimportant constants will invariably be designated by c the value of which may vary from line to line while proofs are in process but when there are many constants within a string of relations, we will use c, c', $\ldots$

2 Tightness

We first have to get rid of the singularity ε^{-1} in (4). The function $\widehat{b}(x,y)$ need not behave regularly in the couple (x,y) so that the joint $\widehat{b}$ Itô formula cannot be applied a priori and the method outlined in the Introduction will not carry over. However, thanks to the regularity theory exhibited by elliptic equations, the Poisson equation (7) is apt to yield enough regularity for the use of the Itô formula on the fast variable, given sufficient regularity of the coefficients a and b.

2.1 Assumptions on the coefficients

Our standing assumptions are, next to (3), in which β is fixed, and (6),

Condition 1. *Global Lipschitz condition : there exists a constant c s.t. for any $\xi = a,\ b,\ c$ and e,*

$$\|\xi(x,y) - \xi(x',y')\| \leqslant c(\|x - x'\| + \|y - y'\|), \qquad \forall x, x' \in \mathbf{R}^d,\ y, y' \in \mathbb{T}^d.$$

The function f is continuous and for some $p > d$ and all $x \in \mathbf{R}^d$, $f(x, \,.\,) \in W^{1,p}(\mathbb{T}^d)$, and moreover there exists $c(p) < \infty$ such that

$$\|f(x, \,.\,)\|_{W^{1,p}(\mathbb{T}^d)} \leqslant c(p).$$

Condition 2. *The partial derivatives $\partial_x \xi(x,y)$ as well as the mixed derivatives $\partial^2_{xy} \xi(x,y)$ exist and are continuous, $\xi = a,\ b$ and c, $x \in \mathbf{R}^d$, $y \in \mathbb{T}^d$.*

Condition 3. *The coefficients are bounded, i.e. there exists a constant c s.t. for any $\xi = a,\ b,\ c,\ e$ and f,*

$$\|\xi(x,y)\| \leqslant c, \qquad x \in \mathbf{R}^d,\ y \in \mathbb{T}^d.$$

2.2 Removing the singularity

We have to control the highly oscillating terms. We have

Lemma 1. *Let $h(x,y)$ be a continuous bounded function on $\mathbf{R}^d \times \mathbb{T}^d$ such that for all $x \in \mathbf{R}^d$,*

$$\int_{\mathbb{T}^d} h(x,y)\,\mu(x,\mathrm{d}y) = 0.$$

Then we have

$$\varepsilon^{-1} \int_s^t h(r,r)\,\mathrm{d}r = \varepsilon^{-1} \int_s^t \left(\Delta_{s,r} h(\,.\,,r) + \Delta_{s,r} L_{\,.\,,r}\widehat{h}(s,r)\right)\mathrm{d}r$$

$$+ \int_s^t \partial_y \widehat{h}(s,r)c(r,r)\,\mathrm{d}r + \int_s^t \partial_y \widehat{h}(s,r)\sigma(r,r)\,\mathrm{d}B_r + \varepsilon\Delta_{t,s}\widehat{h}(s,\,.\,).$$

(8)

Proof. Let $\widehat{h}$ denote the solution of the Poisson equation (see Lemma 7 below). From the Itô–Krylov formula (see the Appendix in [9]) we have

$$
\begin{aligned}
\varepsilon \Delta_{s,t} \widehat{h}(s, \, . \,) &= \varepsilon \widehat{h}(s,t) - \varepsilon \widehat{h}(s,s) \\
&= \varepsilon^{-1} \int_s^t L_{r,r} \widehat{h}(s,r) \, dr \\
&\quad + \int_s^t \partial_y \widehat{h}(s,r) \big(c(r,r) \, dr + \sigma(r,r) \, dB_r \big).
\end{aligned}
$$

It follows from (7) that

$$
\begin{aligned}
\varepsilon^{-1} \int_s^t h(r,r) \, dr = \varepsilon^{-1} \int_s^t h(r,r) \, dr &- \varepsilon^{-1} \int_s^t h(s,r) \, dr \\
&- \varepsilon^{-1} \int_s^t L_{s,r} \widehat{h}(s,r) \, dr.
\end{aligned}
$$

The Lemma follows, by adding these two identities.

Let us take a fine enough equidistant subdivision, ultimately depending on ε, of the interval $[0,T]$ by means of the points $t_i, i = 0, \ldots, [T/\Delta t] = N$, where $t_0 = 0$, $\Delta t_i = t_{i+1} - t_i$. We denote by t_* the largest t_i below t, by t^* the least t_i above t and by N_t the integer $[t/\Delta t]$, for $t \leqslant T$. Applying the preceding lemma to $b(x,y)$ on each Δt_i we can derive a representation of X_t^ε in which the singularity is removed by introducing a multiplicative small corrector term. That is, we use the idea of freezing the slow component that was alluded to in the Introduction. Indeed, removing the singularity ε^{-1} amounts to introducing a partition t_i of $[0,T]$ and a sort of localization procedure. This reduction phenomenon lies at the root of any asymptotic manipulation on X_t^ε which has to be carried out by controlling the out-flow Δt_i. Let us first define, for $0 \leqslant s \leqslant T$,

$$
\begin{aligned}
R^{0,\varepsilon}(s_*,s) &= \Delta_{s_*,s} b(\, . \, , s) + \Delta_{s_*,s} L_{. \, ,s} \widehat{b}(s_*,s) \\
F^{0,\varepsilon}(s_*,s) &= \big(I + \partial_y \widehat{b}(s_*,s) \big) c(s,s) \\
G^{0,\varepsilon}(s_*,s) &= \big(I + \partial_y \widehat{b}(s_*,s) \big) \sigma(s,s)
\end{aligned}
$$

and state

Corollary 1. *With the notations above, we have for $0 \leqslant t \leqslant T$*

$$
\begin{aligned}
X_{t_*}^\varepsilon &= x + \int_0^{t_*} F^{0,\varepsilon}(s_*,s) \, ds + \int_0^{t_*} G^{0,\varepsilon}(s_*,s) \, dB_s + \varepsilon^{-1} \int_0^{t_*} R^{0,\varepsilon}(s_*,s) \, ds \\
&\quad + \varepsilon \sum_{i=0}^{N_t - 1} \Delta_{t_{i+1},t_i} \widehat{b}(t_i, \, . \,).
\end{aligned}
$$

Before establishing tightness, we need estimates on the growth of the functions $\widehat{b}(x,y)$, $\partial_x\widehat{b}(x,y)$, $\partial_y\widehat{b}(x,y)$, $\partial_y^2\widehat{b}(x,y)$ and $\partial_{xy}^2\widehat{b}(x,y)$ for x in $\mathbf{R}^d$ and y in $\mathbf{T}^d$. We will see in Section 3 that these functions arise from the last term in the decompostion of $X_{t_*}^\varepsilon$ above, quite like in the case where $\widehat{b}(x,y)$ exhibits more joint regularity, see the Introduction and [2].

Lemma 2. *Under the conditions above, there exists a constant $c > 0$ s.t. for all x in $\mathbf{R}^d$ and y in $\mathbf{T}^d$*

$$\left\|\widehat{b}(x,y)\right\| + \left\|\partial_x\widehat{b}(x,y)\right\| + \left\|\partial_y\widehat{b}(x,y)\right\| + \left\|\partial_y^2\widehat{b}(x,y)\right\| + \left\|\partial_{xy}^2\widehat{b}(x,y)\right\| \leqslant c,$$

and these derivatives are continuous.

The proof of this Lemma will be postponed till section 4. For now suppose it is true.

2.3 Establishing tightness

Let us first prove the

Lemma 3. *There exists a constant c s.t. for all $\varepsilon > 0$ and $0 \leqslant s < t \leqslant T$,*

$$\mathbf{E}\left(\sup_{s \leqslant v \leqslant t} \|X_v^\varepsilon - X_s^\varepsilon\|^4\right) \leqslant c\big[(t-s)^2 + \varepsilon^4\big].$$

Proof. Let t_i be as in corollary 1 and $0 \leqslant s \leqslant v \leqslant t \leqslant T$, we can write

$$\|X_v^\varepsilon - X_s^\varepsilon\| \leqslant \|X_{v_*}^\varepsilon - X_{s_*}^\varepsilon\| + \|X_v^\varepsilon - X_{v_*}^\varepsilon\| + \|X_s^\varepsilon - X_{s_*}^\varepsilon\|.$$

We need to provide a bound on $\mathbf{E}(\sup_{r_* \leqslant v \leqslant r} \|X_v^\varepsilon - X_{r_*}^\varepsilon\|)$ for $0 \leqslant r \leqslant T$. By Lemma 1 and Lemma 2,

$$\|X_v^\varepsilon - X_{r_*}^\varepsilon\| \leqslant c\bigg(\varepsilon^{-1}\int_{r_*}^v \|X_u^\varepsilon - X_{r_*}^\varepsilon\|\,\mathrm{d}u + (v - r_*)$$
$$+ \left\|\int_{r_*}^v G^{0,\varepsilon}(r_*, u)\,\mathrm{d}B_u\right\| + \varepsilon\bigg).$$

Therefore by Hölder and convexity,

$$\|X_v^\varepsilon - X_{r_*}^\varepsilon\|^4 \leqslant c\bigg(\varepsilon^{-4}(v - r_*)^3\int_{r_*}^v \|X_u^\varepsilon - X_{r_*}^\varepsilon\|^4\,\mathrm{d}u + (v - r_*)^4$$
$$+ \left\|\int_{r_*}^v G^{0,\varepsilon}(r_*, u)\,\mathrm{d}B_u\right\|^4 + \varepsilon^4\bigg).$$

Hence

$$\mathbf{E}\left(\sup_{r_* \leqslant v \leqslant r} \|X_v^\varepsilon - X_{r_*}^\varepsilon\|^4 \right) \leqslant c\left[\varepsilon^{-4}(r - r_*)^3 \int_{r_*}^r \mathbf{E}\left(\sup_{r_* \leqslant v \leqslant u} \|X_v^\varepsilon - X_{r_*}^\varepsilon\|^4 \right) du \right.$$

$$\left. + (r - r_*)^4 + (r - r_*)^2 + \varepsilon^4 \right].$$

By the Gronwall–Bellman lemma,

$$\mathbf{E}\left(\sup_{r_* \leqslant v \leqslant r} \|X_v^\varepsilon - X_{r_*}^\varepsilon\|^4 \right) \leqslant c\left((r - r_*)^2 + (r - r_*)^4 + \varepsilon^4 \right) e^{c'\varepsilon^{-4}(r - r_*)^4}. \quad (9)$$

We now choose $\Delta t_i = \varepsilon^2$. With this choice

$$\mathbf{E}\left(\sup_{r_* \leqslant v \leqslant r} \|X_v^\varepsilon - X_{r_*}^\varepsilon\|^4 \right) \leqslant c\varepsilon^4.$$

Since from Lemma 2 the function $\widehat{b}(\,.\,,y)$ is Lipschitz on $\mathbf{R}^d$ uniformly in $y \in \mathbf{T}^d$, we have by convexity for $s \leqslant v \leqslant t$

$$\mathbf{E}\left(\sup_{s \leqslant v \leqslant t} \left\| \sum_{i=N_s+1}^{N_v-1} \Delta_{t_{i-1},t_i} \widehat{b}(\,.\,,t_i) \right\|^4 \right) \leqslant c\,\mathbf{E}\left(\sum_{i=N_s+1}^{N_t-1} \|X_{t_i}^\varepsilon - X_{t_{i-1}}^\varepsilon\| \right)^4$$

$$\leqslant c\left(\frac{t-s}{\Delta t_i} \right)^4 \varepsilon^4.$$

Hence

$$\mathbf{E} \sup_{s \leqslant v \leqslant t} \left\| \varepsilon \sum_{N_s+1}^{N_v-1} \Delta_{t_{i-1},t_i} \widehat{b}(\,.\,,t_i) \right\|^4 \leqslant c(t-s)^4.$$

These estimates clearly yield by corollary 1

$$\mathbf{E} \sup_{s \leqslant v \leqslant t} \|X_{v_*}^\varepsilon - X_{s_*}^\varepsilon\|^4 \leqslant c\left((t_* - s_*)^4 + (t_* - s_*)^2 + \varepsilon^4 \right),$$

which implies the result, provided $t - s \leqslant T$.

We can now state the

Theorem 1. *Under our assumptions on the coefficients, the family of processes $\{X^\varepsilon, 0 < \varepsilon \leqslant 1\}$ is tight in $\mathcal{C}[0, T]$.*

Proof. By theorem 8.3 in Billingsley [3], it suffices to check that for any α and $\delta > 0$, there exist $0 < \varepsilon_0 \leqslant 1$ and $0 \leqslant \theta \leqslant T$ such that

$$\theta^{-1}\mathbf{P}\left(\sup_{s \leqslant v \leqslant s+\theta} \|X_v^\varepsilon - X_s^\varepsilon\| > \delta \right) < \alpha,$$

for all $s \leqslant T - \theta$ and $\varepsilon \leqslant \varepsilon_0$. This follows easily from Chebychev's inequality and the last Lemma.

3 Identification of the limit

As pointed out in the Introduction, we can recover our process as a main term, which converges in law, plus asymptotically small terms, thanks to tightness. We will proceed to the decompostion of our process X_t^ε by degrees starting from Corollary 1. Let us take again a subdivision t_i depending on ε. We know by Corollary 1 that

$$X_{t_*}^\varepsilon = x + \int_0^{t_*} \left(I + \partial_y \widehat{b}(s_*, s)\right) c(s, s)\, ds + \int_0^{t_*} \left(I + \partial_y \widehat{b}(s_*, s)\right) \sigma(s, s)\, dB_s$$

$$+ \varepsilon \sum_{i=0}^{N_t - 1} \Delta_{t_{i+1}, t_i} \widehat{b}(t_i, \cdot\,) + \sum_{i=0}^{N_t - 1} \varepsilon^{-1} \int_{\Delta t_i} \left(\Delta_{t_i, s} b(\cdot\,, s) + \Delta_{t_i, s} L_{\cdot\,, s} \widehat{b}(t_i, s)\right) ds.$$

Denote by $F_2^{1,\varepsilon}(s)$ the process $(I + \partial_y \widehat{b}(s_*, s)) c(s, s)$, by $G^{1,\varepsilon}(s)$ the process $(I + \partial_y \widehat{b}(s_*, s)) \sigma(s, s)$ and define

$$R_{N_t}^{1,\varepsilon} = \sum_{i=0}^{N_t - 1} \varepsilon^{-1} \int_{\Delta t_i} \left(\Delta_{t_i, s} b(\cdot\,, s) + \Delta_{t_i, s} L_{\cdot\,, s} \widehat{b}(t_i, s)\right) ds. \qquad (10)$$

Let us first deal with the last but one sum in $X_{t_*}^\varepsilon$ above. We have

$$S_{N_t}^{1,\varepsilon} = \varepsilon \sum_{i=0}^{N_t - 1} \Delta_{t_{i+1}, t_i} \widehat{b}(t_i, \cdot\,) = \varepsilon \sum_{i=1}^{N_t - 1} \Delta_{t_{i-1}, t_i} \widehat{b}(\cdot\,, t_i)$$

$$+ \varepsilon \left(\widehat{b}(0, 0) - \widehat{b}(t_{N_t - 1}, t_{N_t})\right).$$

Define

$$S_{N_t}^{2,\varepsilon} = \varepsilon \sum_{i=1}^{N_t - 1} \Delta_{t_{i-1}, t_i} \widehat{b}(\cdot\,, t_i)$$

$$R_{N_t}^{2,\varepsilon} = \varepsilon \left(\widehat{b}(0, 0) - \widehat{b}(t_{N_t - 1}, t_{N_t})\right). \qquad (11)$$

We clearly have by Lemma 2

$$S_{N_t}^{2,\varepsilon} = \varepsilon \sum_{i=1}^{N_t - 1} \int_0^1 \left\langle \partial_x \widehat{b}(X_{t_{i-1}}^\varepsilon + \ell \Delta_{t_{i-1}, t_i} X_\cdot^\varepsilon, X_{t_i}^\varepsilon / \varepsilon), \Delta_{t_{i-1}, t_i} X_\cdot^\varepsilon \right\rangle d\ell$$

$$= \sum_{i=1}^{N_t - 1} \int_{t_{i-1}}^{t_i} \partial_x \widehat{b}(t_{i-1}, t_i) b(s, s)\, ds + \varepsilon \sum_{i=1}^{N_t - 1} \int_{t_{i-1}}^{t_i} \partial_x \widehat{b}(t_{i-1}, t_i) c(s, s)\, ds$$

$$+ \varepsilon \sum_{i=1}^{N_t - 1} \partial_x \widehat{b}(t_{i-1}, t_i) \int_{t_{i-1}}^{t_i} \sigma(s, s)\, dB_s$$

$$+ \varepsilon \sum_{i=1}^{N_t - 1} \int_0^1 \left\langle \partial_x \widehat{b}(X_{t_{i-1}}^\varepsilon + \ell \Delta_{t_{i-1}, t_i} X_\cdot^\varepsilon, X_{t_i}^\varepsilon / \varepsilon) - \partial_x \widehat{b}(t_{i-1}, t_i), \Delta_{t_{i-1}, t_i} X_\cdot^\varepsilon \right\rangle d\ell.$$

Let us denote by $F_1^{2,\varepsilon}(s)$ the process defined on $]t_{i-1}, t_i]$ by $\partial_x \widehat{b}(t_{i-1}, t_i)$
$b(s, s)$ and define

$$
\begin{aligned}
R_{N_t}^{3,\varepsilon} &= \varepsilon \sum_{i=1}^{N_t-1} \int_0^1 \left\langle \partial_x \widehat{b}\left(X_{t_{i-1}}^\varepsilon + \ell \Delta_{t_{i-1}, t_i} X_.^\varepsilon, X_{t_i}^\varepsilon / \varepsilon\right) - \partial_x \widehat{b}(t_{i-1}, t_i), \Delta_{t_{i-1}, t_i} X_.^\varepsilon \right\rangle \mathrm{d}\ell \\
&\quad + \varepsilon \sum_{i=1}^{N_t-1} \int_{t_{i-1}}^{t_i} \partial_x \widehat{b}(t_{i-1}, t_i) c(s, s) \, \mathrm{d}s \\
&= R_{N_t}^{3,\varepsilon,1} + R_{N_t}^{3,\varepsilon,2}.
\end{aligned}
\tag{12}
$$

Put

$$
S_{N_t}^{3,\varepsilon} = \varepsilon \sum_{i=1}^{N_t-1} \partial_x \widehat{b}(t_{i-1}, t_i) \int_{t_{i-1}}^{t_i} \sigma(s, s) \, \mathrm{d}B_s,
$$

then look at the stochastic term

$$
\begin{aligned}
S_{N_t}^{3,\varepsilon,i} &= \varepsilon \, \partial_x \widehat{b}(t_{i-1}, t_i) \int_{t_{i-1}}^{t_i} \sigma(s, s) \, \mathrm{d}B_s \\
&= \varepsilon \Delta_{t_{i-1}, t_i} \partial_x \widehat{b}(t_{i-1}, \cdot) \int_{t_{i-1}}^{t_i} \sigma(s, s) \, \mathrm{d}B_s + \varepsilon \int_{t_{i-1}}^{t_i} \partial_x \widehat{b}(t_{i-1}, t_{i-1}) \sigma(s, s) \, \mathrm{d}B_s.
\end{aligned}
$$

We have again by Lemma 2

$$
\begin{aligned}
\varepsilon \left(\partial_x \widehat{b}(t_{i-1}, t_i) - \partial_x \widehat{b}(t_{i-1}, t_{i-1}) \right) &= \partial_{xy}^2 \widehat{b}(t_{i-1}, t_{i-1}) \Delta_{t_{i-1}, t_i} X_.^\varepsilon \\
&\quad + \int_0^1 \left(\partial_{xy}^2 \widehat{b}\left(X_{t_{i-1}}^\varepsilon, (X_{t_{i-1}}^\varepsilon + \ell \Delta_{t_{i-1}, t_i} X_.^\varepsilon)/\varepsilon\right) - \partial_{xy}^2 \widehat{b}(t_{i-1}, t_{i-1}) \right) \Delta_{t_{i-1}, t_i} X_.^\varepsilon \, \mathrm{d}\ell.
\end{aligned}
$$

Let us set $U^i = \partial_{xy}^2 \widehat{b}(t_{i-1}, t_{i-1})$,

$$
\begin{aligned}
S_{N_t}^{4,\varepsilon} &= \sum_{i=1}^{N_t-1} \int_{t_{i-1}}^{t_i} U^i \sigma(s, s) \, \mathrm{d}B_s \int_{t_{i-1}}^{t_i} \sigma(s, s) \, \mathrm{d}B_s \\
&\quad + \varepsilon^{-1} \sum_{i=1}^{N_t-1} \int_{t_{i-1}}^{t_i} U^i b(s, s) \, \mathrm{d}s \int_{t_{i-1}}^{t_i} \sigma(s, s) \, \mathrm{d}B_s \\
&\quad + \sum_{i=1}^{N_t-1} \int_{t_{i-1}}^{t_i} U^i c(s, s) \, \mathrm{d}s \int_{t_{i-1}}^{t_i} \sigma(s, s) \, \mathrm{d}B_s
\end{aligned}
$$

and

$$
\begin{aligned}
R_{N_t}^{4,\varepsilon} &= \sum_{i=1}^{N_t-1} \int_0^1 \left(\Delta_i^\ell \partial_{xy}^2 \widehat{b} \right) \Delta_{t_{i-1}, t_i} X_.^\varepsilon \, \mathrm{d}\ell \int_{t_{i-1}}^{t_i} \sigma(s, s) \, \mathrm{d}B_s \\
&\quad + \varepsilon \sum_{i=1}^{N_t-1} \int_{t_{i-1}}^{t_i} \partial_x \widehat{b}(t_{i-1}, t_{i-1}) \sigma(s, s) \, \mathrm{d}B_s \\
&= R_{N_t}^{4,\varepsilon,1} + R_{N_t}^{4,\varepsilon,2},
\end{aligned}
\tag{13}
$$

where

$$\Delta_i^\ell \partial_{xy}^2 \widehat{b} = \partial_{xy}^2 \widehat{b}\big(X_{t_{i-1}}^\varepsilon, (X_{t_{i-1}}^\varepsilon + \ell \Delta_{t_{i-1}, t_i} X_{\cdot}^\varepsilon)/\varepsilon\big) - \partial_{xy}^2 \widehat{b}(t_{i-1}, t_{i-1}).$$

The third part of the drift of the Itô main process in $X_{t_*}^\varepsilon$ is now lurking in the background. Indeed the first term in $S_{N_t}^{4,\varepsilon}$ induces us to write $F_3^{3,\varepsilon}(s) = \mathrm{Tr}\,\partial_{xy}^2 \widehat{b}a(s,s)$. We now define $R_{N_t}^{5,\varepsilon} = R_{N_t}^{5,\varepsilon,1} + R_{N_t}^{5,\varepsilon,2}$, where

$$R_{N_t}^{5,\varepsilon,1} = \varepsilon^{-1} \sum_{i=1}^{N_t-1} \int_{t_{i-1}}^{t_i} U^i b(s,s)\,\mathrm{d}s \int_{t_{i-1}}^{t_i} \sigma(s,s)\,\mathrm{d}B_s$$

$$R_{N_t}^{5,\varepsilon,2} = \sum_{i=1}^{N_t-1} \int_{t_{i-1}}^{t_i} U^i c(s,s)\,\mathrm{d}s \int_{t_{i-1}}^{t_i} \sigma(s,s)\,\mathrm{d}B_s. \tag{14}$$

Before gathering all the main terms and remainders so far obtained over three steps above, let us smooth out the irregularities in the construction of the main parts $F_1^{2,\varepsilon}(s)$ and $F_2^{1,\varepsilon}(s)$. Note that for the sake of clarity the level at which a quantity appears is recorded in the superscript and that step zero served only to derive tightness. Indeed, let us write the following definition

$$F(x,y) = \big(\partial_x \widehat{b}b + (I + \partial_y \widehat{b})c + \mathrm{Tr}\,\partial_{xy}^2 \widehat{b}a\big)(x,y) \tag{15}$$
$$= (F_1 + F_2 + F_3)(x,y)$$

and

$$G(x,y) = \big((I + \partial_y \widehat{b})\sigma\big)(x,y). \tag{16}$$

Note that F and G are continuous.

These operations involve the extra rests

$$R_{N_t}^{6,\varepsilon} = \sum_{i=1}^{N_t-1} \int_{t_{i-1}}^{t_i} \big(\partial_x \widehat{b}(t_{i-1},t_i) - \partial_x \widehat{b}(s,s)\big)b(s,s)\,\mathrm{d}s, \tag{17}$$

$$R_{N_t}^{7,\varepsilon} = \int_0^{t_*} (I + \partial_y \widehat{b}(s_*,s))c(s,s)\,\mathrm{d}s - \int_0^{t_*} (I + \partial_y \widehat{b}(s,s))c(s,s)\,\mathrm{d}s, \tag{18}$$

$$R_{N_t}^{8,\varepsilon} = \sum_{i=1}^{N_t-1} \int_{t_{i-1}}^{t_i} U^i \sigma(s,s)\,\mathrm{d}B_s \int_{t_{i-1}}^{t_i} \sigma(s,s)\,\mathrm{d}B_s - \int_0^{t_*} \big(\mathrm{Tr}\,\partial_{xy}^2 \widehat{b}a\big)(s,s)\,\mathrm{d}s \tag{19}$$

and

$$R_{N_t}^{9,\varepsilon} = \int_0^{t_*} (I + \partial_y \widehat{b}(s_*,s))\sigma(s,s)\,\mathrm{d}B_s - \int_0^{t_*} (I + \partial_y \widehat{b}(s,s))\sigma(s,s)\,\mathrm{d}B_s. \tag{20}$$

We have thus proved the

Lemma 4. *For any subdivision t_i with constant step Δt_i we have*

$$X^{\varepsilon}_{t_*} = \overline{X}^{\varepsilon}_{t_*} + R^{\varepsilon}_{N_t},$$

where

$$\overline{X}^{\varepsilon}_{t_*} = x + \int_0^{t_*} F(s,s)\,\mathrm{d}s + \int_0^{t_*} G(s,s)\,\mathrm{d}B_s,$$

and the remainder $R^{\varepsilon}_{N_t}$ is the sum of the residual quantities, i.e.

$$R^{\varepsilon}_{N_t} = \sum_{i=1}^{9} R^{i,\varepsilon}_{N_t}.$$

We can now state the following theorem, which deals with the order of magnitude of the remainder $R^{\varepsilon}_{N_t}$. The proof is defered to section 5.

Theorem 2. *With the notations above, we have the decomposition*

$$X^{\varepsilon}_t = \overline{X}^{\varepsilon}_t + R^{\varepsilon}_t, \qquad t \geqslant 0,$$

where $\overline{X}^{\varepsilon}_t$ is the Itô process

$$\overline{X}^{\varepsilon}_t = x + \int_0^t F(s,s)\,\mathrm{d}s + \int_0^t G(s,s)\,\mathrm{d}B_s,$$

and the remainder term R^{ε}_t satisfies

$$\mathbf{P}\left(\sup_{t \leqslant T} \|R^{\varepsilon}_t\| > \delta\right) \longrightarrow 0, \qquad \forall \delta > 0.$$

Therefore, in order to identify the limit points of $\mathbf{P}_{X^{\varepsilon}}$, it suffices to do so for those of $\mathbf{P}_{\overline{X}^{\varepsilon}}$. We use a martingale problem approach. Let $\varphi(x)$ be a function in $C_K^{\infty}(\mathbf{R}^d)$ and apply the Itô formula, we have for $t \geqslant t_0$

$$\varphi(\overline{X}^{\varepsilon}_t) - \varphi(\overline{X}^{\varepsilon}_s) = \int_s^t \partial_x \varphi(\overline{X}^{\varepsilon}_r)\big(F(X^{\varepsilon}_r, X^{\varepsilon}_r/\varepsilon)\,\mathrm{d}r + G(X^{\varepsilon}_r, X^{\varepsilon}_r/\varepsilon)\,\mathrm{d}B_r\big)$$
$$+ \frac{1}{2}\int_s^t \mathrm{Tr}\,\partial_x^2 \varphi(\overline{X}^{\varepsilon}_r)(GG^*)(X^{\varepsilon}_r, X^{\varepsilon}_r/\varepsilon)\,\mathrm{d}r.$$

Now let $\Phi_{t_0}(\,.\,)$, $t_0 \leqslant T$, be a bounded continuous functional on the Wiener space $\mathcal{C}[0,T]$ which depends only on the past up to t_0. Define $GG^* = \Lambda$. We have

$$\mathbf{E}\left[\left(\Delta_{t_0,t}\varphi(\overline{X}^{\varepsilon}_\cdot) - \int_{t_0}^t \left(\partial_x\varphi(\overline{X}^{\varepsilon}_r)F(r,r) + \frac{1}{2}\,\mathrm{Tr}\,\partial_x^2\varphi(\overline{X}^{\varepsilon}_r)\Lambda(r,r)\right)\mathrm{d}r\right)\Phi_{t_0}(\,.\,)\right] = 0. \tag{21}$$

Let us now homogenize F and Λ by setting

$$\overline{F}(x) = \int_{\mathbb{T}^d} F(x, y)\, \mu(x, dy)$$

$$\overline{\Lambda}(x) = \int_{\mathbb{T}^d} \Lambda(x, y)\, \mu(x, dy) \tag{22}$$

The relation (21) becomes

$$\mathbf{E}\left[\left(\Delta_{t_0,t}\varphi(\overline{X}^\varepsilon_\cdot) - \int_{t_0}^t \left(\partial_x\varphi(\overline{X}^\varepsilon_r)\overline{F}(X^\varepsilon_r) + \frac{1}{2}\operatorname{Tr}\partial_x^2\varphi(\overline{X}^\varepsilon_r)\overline{\Lambda}(X^\varepsilon_r)\right)dr\right)\Phi_{t_0}(\,.\,)\right]$$

$$= \mathbf{E}\left[\left(\int_{t_0}^t \left(\partial_x\varphi(\overline{X}^\varepsilon_r)h_F(r,r) + \frac{1}{2}\operatorname{Tr}\partial_x^2\varphi(\overline{X}^\varepsilon_r)h_\Lambda(r,r)\right)dr\right)\Phi_{t_0}(\,.\,)\right], \quad (23)$$

where $h_F(x, y) = F(x, y) - \overline{F}(x)$ and $h_\Lambda(x, y = \Lambda(x, y) - \overline{\Lambda}(x)$. Now observe that

Lemma 5. *If* $x_n \to x$, *then* $\mu_{x_n} \Rightarrow \mu_x$.

Proof. The collection of probability measures $\{\mu(x_n,\,.\,),\ n = 1, 2, \ldots\}$ on the compact set $\mathbb{T}^d$ is tight. Then there exists a subsequence, still denoted $\{x_n\}$, and a probability measure μ_0 on $\mathbb{T}^d$, s. t. $\mu(x_n,\,.\,) \Rightarrow \mu_0$.
For any $\varphi \in C(\mathbb{T}^d)$, $t \geq 0$,

$$\int_{\mathbb{T}^d} \varphi(y)\, \mu(x_n, dy) = \int_{\mathbb{T}^d} \mathbf{E}^y[\varphi(Y^{x_n}_t)]\, \mu(x_n, dy),$$

hence

$$\int_{\mathbb{T}^d} \varphi(y)\, \mu(x_n, dy) = \int_{\mathbb{T}^d} \mathbf{E}^y[\varphi(Y^x_t)]\, \mu(x_n, dy)$$
$$+ \int_{\mathbb{T}^d} \mathbf{E}^y[\varphi(Y^{x_n}_t) - \varphi(Y^x_t)]\, \mu(x_n, dy). \tag{24}$$

But since $\mu(x_n,\,.\,)$ has a density which is bounded uniformly with respect to n, see e.g. [8],

$$\left|\int_{\mathbb{T}^d} \mathbf{E}^y[\varphi(Y^{x_n}_t) - \varphi(Y^x_t)]\, \mu(x_n, dy)\right| \leq c \int_{\mathbb{T}^d} \left|\mathbf{E}^y[\varphi(Y^{x_n}_t) - \varphi(Y^x_t)]\right| dy,$$

which tends to zero, as $n \to \infty$. Consequently, passing to the limit in (24), we obtain the identity

$$\int_{\mathbb{T}^d} \varphi(y)\mu_0(dy) = \int_{\mathbb{T}^d} \mathbf{E}^y[\varphi(Y^x_t)]\mu_0(dy),$$

for all φ and all $t \geq 0$. Hence $\mu_0 = \mu(x,\,.\,)$, and the whole sequence $\{\mu(x_n,\,.\,)\}$ converges to $\mu(x,\,.\,)$.

Lemma 6. *The functions* $\overline{F}(\,.\,)$ *and* $\overline{G}(\,.\,)$ *are continuous.*

Proof. Let $x_n \to x_0$. We have

$$\int_{\mathbb{T}^d} F(x_n, y)\,\mu(x_n, \mathrm{d}y) = \int_{\mathbb{T}^d} \big(F(x_n, y) - F(x_0, y)\big)\,\mu(x_n, \mathrm{d}y)$$
$$+ \int_{\mathbb{T}^d} F(x_0, y)\,\mu(x_n, \mathrm{d}y).$$

Now

$$\left\| \int_{\mathbb{T}^d} \big(F(x_n, y) - F(x_0, y)\big)\,\mu(x_n, \mathrm{d}y) \right\| \leqslant \sup_{y \in \mathbb{T}^d} \|F(x_n, y) - F(x_0, y)\|,$$

which converges to zero when $x_n \to x_0$. On the other hand it follows from the previous Lemma that

$$\int_{\mathbb{T}^d} F(x_0, y)\,\mu(x_n, \mathrm{d}y) \longrightarrow \int_{\mathbb{T}^d} F(x_0, y)\,\mu(x_0, \mathrm{d}y),$$

and the continuity of $\overline{F}$ follows. The same argument yields the continuity of $\overline{G}$.

We need now a locally periodic ergodic theorem to deal with the expected value on the right in (23).

Theorem 3. *Let $h(x, y)$ be a continuous bounded function on $\mathbb{R}^d \times \mathbb{T}^d$ such that for all $x \in \mathbb{R}^d$,*

$$\int_{\mathbb{T}^d} h(x, y)\,\mu(x, \mathrm{d}y) = 0,$$

$h(x, \cdot) \in W^{1,p}(\mathbb{T}^d)$ *for some $p > d$, and moreover there exists $c(p) < \infty$ such that*

$$\|h(x, \cdot)\|_{W^{1,p}(\mathbb{T}^d)} \leqslant c(p).$$

Suppose moreover that $\varphi \in \mathcal{C}_K^\infty(\mathbb{R}^d)$. Then

$$H^\varepsilon(t) = \int_0^t \varphi(\overline{X}_s^\varepsilon) h(X_s^\varepsilon, X_s^\varepsilon/\varepsilon)\,\mathrm{d}s$$

converges to zero in $L^1(\Omega)$ for any $0 < t \leqslant T$.

Proof. Let us take yet again a subdivision $\{t_i\}$ with $\Delta t_i = \varepsilon$ and write, by virtue of Lemma 1,

$$\int_{\Delta t_i} \varphi(\overline{X}_{t_i}^\varepsilon) h(t_i, s)\,\mathrm{d}s = \int_{\Delta t_i} \varphi(\overline{X}_{t_i}^\varepsilon) \Delta_{t_i, s} L_{\cdot, s} \widehat{h}(t_i, s)\,\mathrm{d}s$$
$$+ \varepsilon \int_{\Delta t_i} \varphi(\overline{X}_{t_i}^\varepsilon) \partial_y \widehat{h}(t_i, s) c(s, s)\,\mathrm{d}s + \varepsilon \int_{\Delta t_i} \varphi(\overline{X}_{t_i}^\varepsilon) \partial_y \widehat{h}(t_i, s) \sigma(s, s)\,\mathrm{d}B_s$$
$$+ \varepsilon^2 \Delta_{t_{i+1}, t_i} \varphi(\overline{X}_{t_i}^\varepsilon) \widehat{h}(t_i, \cdot).$$

On the other hand for $t \leqslant T$ we can write

$$H^\varepsilon(t) = \int_0^t \varphi(\overline{X}_{s_*}^\varepsilon) h(s_*, s) \, \mathrm{d}s + R_H^\varepsilon,$$

where

$$R_H^\varepsilon = H^\varepsilon(t) - \sum_{i=0}^{N_t-1} \int_{\Delta t_i} \varphi(\overline{X}_{t_i}^\varepsilon) h(t_i, s) \, \mathrm{d}s - \int_{t_*}^t \varphi(\overline{X}_{t_*}^\varepsilon) h(t_*, s) \, \mathrm{d}s.$$

We first have

$$\mathbf{E}\left| \int_0^{t_*} \varphi(\overline{X}_{s_*}^\varepsilon) h(s_*, s) \, \mathrm{d}s \right|$$

$$\leqslant c \sum_{i=0}^{N_t-1} \left(\mathbf{E} \int_{\Delta t_i} \|\Delta_{t_i,s} X_\cdot^\varepsilon\| \left(\sup_{j,k} |\partial_{y_j, y_k} \widehat{h}(t_i, s)| + \sup_j |\partial_{y_j} \widehat{h}(t_i, s)| \right) \mathrm{d}s \right.$$

$$+ \varepsilon \, \mathbf{E} \left\| \int_{\Delta t_i} \partial_y \widehat{h}(t_i, s) c(s, s) \, \mathrm{d}s \right\| + \varepsilon \, \mathbf{E} \left\| \int_{\Delta t_i} \partial_y \widehat{h}(t_i, s) \sigma(s, s) \, \mathrm{d}B_s \right\|$$

$$\left. + \varepsilon^2 \, \mathbf{E} |\Delta_{t_{i+1}, t_i} \widehat{h}(t_i, \cdot)| \right).$$

By Lemma 15 below (whose proof is easily adapted to our h here) and Lemma 3, the first term in the right hand side of the inequality above is bounded by

$$c \sum_{i=0}^{N_t-1} \int_{\Delta t_i} \mathbf{E} \|\Delta_{t_i,s} X_\cdot^\varepsilon\| \, \mathrm{d}s \leqslant c' \left((\Delta t_i)^{1/2} + \varepsilon \right) = c' \left(\varepsilon^{1/2} + \varepsilon \right).$$

The second term is easier to render small. The third term which involves a stochastic integral is treated by the Burkholder–Davis–Gundy inequality and is bounded by $c\varepsilon^{1/2}$. The last term is easy given ε^2 in front and the choice of $\Delta t_i = \varepsilon$. It remains to deal with R_H^ε. We have for s in $]t_i, t_{i+1}]$

$$\varphi(\overline{X}_s^\varepsilon) h(s, s) - \varphi(\overline{X}_{t_i}^\varepsilon) h(t_i, s) = \varphi(\overline{X}_{t_i}^\varepsilon) \Delta_{t_i, s} h(\cdot, s) + \Delta_{t_i, s} \varphi(\overline{X}_\cdot^\varepsilon) h(s, s).$$

It suffices to prove that R_H^ε converges to zero in probability, since it is bounded. Now let $\delta > 0$ and $\alpha > 0$. If we denote by τ^ε the first exit time of the process X_t^ε from the ball $B(0, M)$, where M is large enough so that $M \geqslant \|x\|$, then by virtue of tightness, M can be chosen s.t. for all sufficiently small ε,

$$\mathbf{P}(\tau^\varepsilon < T) \leqslant \alpha.$$

Now

$$\mathbf{P}(|R_H^\varepsilon| > \delta) \leqslant \mathbf{P}\left(\sum_{i=0}^{N_t} |\varphi(\overline{X}_{t_i}^\varepsilon)| \int_{\Delta t_i} |\Delta_{t_i, s} h(\cdot, s)| \, \mathrm{d}s > \delta/2 \right)$$

$$+ \mathbf{P}\left(\sum_{i=0}^{N_t} \int_{\Delta t_i} |\Delta_{t_i, s} \varphi(\overline{X}_\cdot^\varepsilon) h(s, s)| \, \mathrm{d}s > \delta/2 \right).$$

As $h(x, y)$ is continuous, it possesses a modulus of continuity denoted by γ_M on $\overline{B}(0, M) \times \mathbb{T}^d$. We have

$$\mathbf{P}\left(\sum_{i=0}^{N_t} |\varphi(\overline{X}_{t_i}^\varepsilon)| \int_{\Delta t_i} |\Delta_{t_i,s} h(\,.\,, s)| \, ds > \delta/2\right)$$

$$\leqslant \mathbf{P}\left(\sum_{i=0}^{N_t} c \sup_{s \in \Delta t_i} \gamma_M(\|\Delta_{t_i,s} X_.^\varepsilon\|)\Delta t_i > \delta/2 \text{ and } T \leqslant \tau^\varepsilon\right) + \alpha$$

$$\leqslant \mathbf{P}\left(ct^* \sup_i \sup_{s \in \Delta t_i} \gamma_M(\|\Delta_{t_i,s} X_.^\varepsilon\|) > \delta/2\right) + \alpha$$

$$\leqslant \mathbf{P}\left(ct^* \gamma_M\left(\sup_i \sup_{s \in \Delta t_i} \|\Delta_{t_i,s} X_.^\varepsilon\|\right) > \delta/2\right) + \alpha.$$

As the function γ_M is continuous the probability above is $\leqslant 2\alpha$. On the other hand,

$$\mathbf{P}\left(\sum_{i=0}^{N_t} \int_{\Delta t_i} |\Delta_{t_i,s} \varphi(\overline{X}_.^\varepsilon) h(s, s)| \, ds > \delta/2\right)$$

$$\leqslant \mathbf{P}\left(\sum_{i=0}^{N_t} c \sup_{s \in \Delta t_i} \|\Delta_{t_i,s} \overline{X}_.^\varepsilon\| \int_{\Delta t_i} |h(s, s)| \, ds > \delta/2\right)$$

$$\leqslant \mathbf{P}\left(c \sup_i \sup_{s \in \Delta t_i} \|\Delta_{t_i,s} \overline{X}_.^\varepsilon\| \int_0^{t^*} |h(s, s)| \, ds > \delta/2\right) \leqslant \alpha,$$

for all sufficiently small ε.

We finally have the

Corollary 2. *There is only one limit point of the family X^ε as $\varepsilon \to 0$, namely X^0, the Itô process solution of the SDE*

$$X_t^0 = x + \int_0^t \overline{F}(X_s^0) \, ds + \int_0^t \overline{\Lambda}^{1/2}(X_s^0) \, dB_s,$$

where $\overline{F}(x)$ and $\overline{\Lambda}(x)$ are defined in (22) (see also (15'), (16)).

Proof. Since $\overline{F}(x)$ and $\overline{\Lambda}(x)$ are continuous, it suffices, in order to be able to apply Stroock and Varadhan's uniqueness theorem (see Stroock and Varadhan [11]), to show that for each $x \in \mathbf{R}^d$, the matrix $\overline{\Lambda}(x)$ is positive definite. Suppose this is not the case. Then there exists $x \in \mathbf{R}^d$ and $\xi \in \mathbf{R}^d \backslash \{0\}$ s.t.

$$\langle \overline{\Lambda}(x)\xi, \xi \rangle = 0.$$

But

$$\langle \overline{A}(x)\xi, \xi \rangle = \int_{\mathbb{T}^d} \langle a(I + \partial_y \widehat{b})^* \xi, (I + \partial_y \widehat{b})^* \xi \rangle(x, y) \, \mu(x, dy)$$

$$\geqslant \beta \int_{\mathbb{T}^d} \| (I + \partial_y \widehat{b})^* \xi \|^2 (x, y) \, \mu(x, dy).$$

Our assumption implies that $(I + \partial_y \widehat{b})^* \xi(x, y) = 0$ y a.e., or in other words for all $1 \leqslant i \leqslant d$, and almost all y,

$$\partial_{y_i} \langle \widehat{b}, \xi \rangle(x, y) = -\xi_i.$$

But there exists $1 \leqslant i \leqslant d$ such that $\xi_i \neq 0$, and this contradicts the periodicity of the mapping $y \to \langle \widehat{b}, \xi \rangle(x, y)$.

4 Proof of lemma 2

The following result, adapted to our needs here, has already been proved in Pardoux and Veretennikov [9].

Lemma 7. *Consider the Poisson equation for the operator $L_{x,y}$ on the torus $\mathbb{T}^d$ with parameter x*

$$L_{x,y} \widehat{h}(x, y) = -h(x, y),$$

where $h(x, \, .)$ is continuous and satisfies

$$\int_{\mathbb{T}^d} h(x, y) \, \mu(x, dy) = 0.$$

Then for all x the Poisson equation has a solution $\hat{h}(x, \, .)$ in $\cap_{p \geqslant 1} W^{2,p}(\mathbb{T}^d)$. The solution is unique in each $W^{2,p}(\mathbb{T}^d)$, if we impose that

$$\int_{\mathbb{T}^d} \hat{h}(x, y) \, \mu(x, dy) = 0.$$

Moreover

$$\widehat{h}(x, y) = \int_0^\infty dt \int_{\mathbb{T}^d} h(x, y') p_t(x, y, y') \, dy'.$$

We need to trace up the dependence of $\widehat{b}$, and its derivatives, on x. For some fixed $t_0 > 0$, we should expect

$$\partial \widehat{b}(x, y) = \int_0^{t_0} dt \int_{\mathbb{T}^d} \partial b(x, y') p_t(x, y, y') \, dy' + \int_0^{t_0} dt \int_{\mathbb{T}^d} b(x, y') \partial p_t(x, y, y') \, dy'$$

$$+ \int_{t_0}^\infty dt \int_{\mathbb{T}^d} \partial b(x, y') (p_t(x, y, y') - p_\infty(x, y')) \, dy'$$

$$+ \int_{t_0}^\infty dt \int_{\mathbb{T}^d} b(x, y') \partial (p_t(x, y, y') - p_\infty(x, y')) \, dy',$$

where ∂ stands for derivatives with respect to x, y. We are led to investigate the asymptotic behaviour of the derivatives of orders zero and one of the difference inside the third and fourth integral through a few lemmas. There is no point rewriting the proof of corollary 2.3 on p. 502 of Pardoux [8] which gives an exponential bound on the semi-group of the diffusion Y_t^x on $\mathbb{T}^d$. Note that this estimate has already been derived by Doob in the forties by different methods. We have the so called spectral gap Lemma.

Lemma 8. *There are positive constants c and ϱ s.t. for all $t \geqslant 0$,*

$$\left\| \mathbf{E}^y f(Y_t^x) - \int_{\mathbb{T}^d} f(y') p_\infty(x, y') \, \mathrm{d}y' \right\|_{L^\infty(\mathbb{T}^d)} \leqslant c\|f\|_{L^\infty(\mathbb{T}^d)} \mathrm{e}^{-\varrho t},$$

where f is any bounded measurable function on the torus.

Due to the boundedness condition on the coefficients, it is easily seen that the constants c and ϱ can be chosen independent of x. Hence the above can be rewritten as

Lemma 9. *There are positive constants c and ϱ s.t. for all $t \geqslant 0$ and $x \in \mathbf{R}^d$, $y \in \mathbb{T}^d$,*

$$\int_{\mathbb{T}^d} |p_t(x, y, y') - p_\infty(x, y')| \, \mathrm{d}y' \leqslant c\mathrm{e}^{-\varrho t}.$$

On the other hand the method of Friedman [5] chap.1 and chap.9.4. theorem 2 gives the following estimates on the derivatives of the transition densities of Y_t^x considered on the whole of $\mathbf{R}^d$ (the estimate on the derivatives with respect to the starting point requires the coefficients a and b to satisfy at least a Lipschitz condition on y)

Lemma 10. *For $n = 0, 1$ and $t \leqslant T$, there are constants c and c' s.t. for all x*

$$\left| \partial_{y'}^n p_t(x, y, y') \right| \leqslant ct^{-(n+d)/2} \mathrm{e}^{-c'\|y'-y\|^2/t}$$

$$\left| \partial_y^n p_t(x, y, y') \right| \leqslant ct^{-(n+d)/2} \mathrm{e}^{-c'\|y'-y\|^2/t}$$

and these derivatives are continuous.

We deduce from this the fundamental estimate

Lemma 11. *Let $n = 0, 1$, then there exists a constant c s.t. for all x*

$$\sup_y \left| \partial_{y'}^n p_t(x, y, y') - \partial_{y'}^n p_\infty(x, y') \right| \leqslant c\mathrm{e}^{-\varrho t}$$

Proof. By the Chapman–Kolmogorov identity

$$\partial_{y'}^n p_t(x, y, y') = \int_{\mathbb{T}^d} p_{t-1}(x, y, y'') \partial_{y'}^n p_1(x, y'', y') \, \mathrm{d}y''$$

$$\partial_{y'}^n p_\infty(x, y') = \int_{\mathbb{T}^d} p_\infty(x, y'') \partial_{y'}^n p_1(x, y'', y') \, \mathrm{d}y'',$$

therefore

$$\left|\Delta_{t,\infty}\partial_{y'}^n p.(x,y,y')\right| \leqslant \int_{\mathbb{T}^d} \left|\Delta_{t-1,\infty}p.(x,y,y'')\right| \left|\partial_{y'}^n p_1(x,y'',y')\right| dy'',$$

and the Lemma follows from Lemma 9 and Lemma 10.

The following exponential estimate on the derivative with respect to the starting point of the transition densities makes matters easier to deal with than in the case of the $\mathbf{R}^d$ ergodic theory considered in [10].

Lemma 12. *We have for some c and all x, y, $t \geqslant 1$,*

$$\left|\partial_y p_t(x,y,y')\right| \leqslant ce^{-\varrho t}.$$

Proof. Clearly

$$\left|\partial_y p_t(x,y,y')\right| = \left|\int_{\mathbb{T}^d} \partial_y p_1(x,y,y'')(p_{t-1}(x,y'',y') - p_\infty(x,y')) \, dy''\right|$$
$$\leqslant ce^{-\varrho t}.$$

Now, we are in the position to prove the

Lemma 13. *There exists a constant c s.t. for all x, y*

$$\left\|\widehat{b}(x,y)\right\| \leqslant c.$$

Proof. We have for any t_0 fixed

$$\widehat{b}(x,y) = \int_0^\infty dt \int_{\mathbb{T}^d} b(x,y')(p_t(x,y,y') - p_\infty(x,y')) \, dy'.$$

By Lemma 9 and the boundedness of $b(x,y)$ we arrive at the desired result.

Returning to the study of $\partial\widehat{b}(x,y)$ we can state

Lemma 14. *For some constant c, we have for all x, y*

$$\left\|\partial_y\widehat{b}(x,y)\right\| \leqslant c.$$

Proof. The result follows from Lemma 12 and

$$\partial_y\widehat{b}(x,y) = \int_0^\infty dt \int_{\mathbb{T}^d} b(x,y')\partial_y p_t(x,y,y') \, dy'.$$

Consider now the second partial derivatives $\partial_y^2\widehat{b}(x,y)$.

Lemma 15. *For some constant c we have for all x, y,*

$$\left\|\partial_y^2\widehat{b}(x,y)\right\| \leqslant c.$$

Proof. As the coefficients a and b are Lipschitz with repect to y, we have from (7)

$$L_{x,y}\partial_y\widehat{b}(x,y) = -\partial_y b(x,y) - \partial_y L_{x,y}\widehat{b}(x,y).$$

It then follows from a variant of Lemma 9.17 in [6] that there exists c s.t. for all $x \in \mathbf{R}^d$ and $p \geqslant 1$,

$$\sup_{i,j,k}\left\|\partial^3_{y_iy_jy_k}\widehat{b}(x,\cdot)\right\|_{L^p(\mathbf{T}^d)} \leqslant c.$$

Since we can choose $p > d$, we have from the Sobolev embedding theorem that for some c,

$$\sup_{i,j}\left\|\partial^2_{y_iy_j}\widehat{b}(x,y)\right\|_{L^\infty(\mathbf{R}^{2d})} \leqslant c.$$

Let us now deal with the partial derivatives of $\widehat{b}(x,y)$ with respect to x.

Lemma 16. *We have for some constant c and all x, y,*

$$\left\|\partial_x\widehat{b}(x,y)\right\| \leqslant c.$$

Proof. We should have

$$
\begin{aligned}
\partial_x\widehat{b}(x,y) &= \int_0^{t_0} dt \int_{\mathbf{T}^d} \partial_x b(x,y')p_t(x,y,y')\,dy' \\
&\quad + \int_0^{t_0} dt \int_{\mathbf{T}^d} b(x,y')\partial_x p_t(x,y,y')\,dy' \\
&\quad + \int_{t_0}^\infty dt \int_{\mathbf{T}^d} \partial_x b(x,y')\bigl(p_t(x,y,y') - p_\infty(x,y')\bigr)\,dy' \\
&\quad + \int_{t_0}^\infty dt \int_{\mathbf{T}^d} b(x,y')\bigl(\partial_x p_t(x,y,y') - \partial_x p_\infty(x,y')\bigr)\,dy' \\
&= I^1_x(x,y) + I^2_x(x,y) + I^{\infty,1}_x(x,y) + I^{\infty,2}_x(x,y).
\end{aligned}
$$

We will use now an explicit formula for the derivative of the transition densities with respect to x, which is denoted $q_t(x,y,y')$, that was established in Pardoux and Veretennikov [10]. Indeed, by writing $\Delta p_t(\cdot,y,y')$ as the solution of a parabolic PDE, $q_t(x,y,y')$ can be directly represented by the well known formula for the solutions for a Cauchy initial boundary condition ; namely we have

$$q_t(x,y,y') = \int_0^t ds \int_{\mathbf{T}^d} \partial_x L^*_{x,y''}p_{t-s}(x,y,y'')p_s(x,y'',y')\,dy''. \qquad (25)$$

Now we pass on to

Lemma 17. *Consider the full transition densities $p_t(x,y,y')$ on $\mathbf{R}^d$. Then for any $T > 0$, there exist constants c and c' such that for all $0 \leqslant t \leqslant T$, $x,y,y' \in \mathbf{R}^d$,*

$$\left|q_t(x,y,y')\right| \leqslant ct^{-d/2}e^{-c'\|y'-y\|^2/t},$$

and these derivatives are continuous.

Proof. We have obviously

$$L^*_{x,y}(\,.\,) = 1/2 \sum_{i,j} \partial_{y_i}\big(a_{ij}(x,y)\partial_{y_j}(\,.\,)\big) - \sum_i \partial_{y_i}\big(\widetilde{b}_i(x,y)(\,.\,)\big),$$

where $\widetilde{b}_i(x,y) = b_i(x,y) - \frac{1}{2}\sum_j \partial_{y_j}a_{ij}(x,y)$. Therefore

$$q_t(x,y,y')$$

$$= -\sum_i \int_0^t ds \int_{\mathbb{T}^d} 1/2 \sum_j \partial_x a_{ij}(x,y'')\partial_{y''_j}p_{t-s}(x,y,y'')\partial_{y''_i}p_s(x,y'',y')\,dy''$$

$$+ \sum_i \int_0^t ds \int_{\mathbb{T}^d} \partial_x \widetilde{b}_i(x,y'')p_{t-s}(x,y,y'')\partial_{y''_i}p_s(x,y'',y')\,dy''.$$

By standard properties of independent Gaussian laws, the Lemma follows easily.

The relevant bounds on $I^1_x(x,y)$, $I^2_x(x,y)$ and $I^{\infty,1}_x(x,y)$ are now straightforward.

We carry on with

Lemma 18. *Given the same hypotheses as in the previous Lemma, we have*

1) $q_\infty(x,y') = \lim_{t\to\infty} q_t(x,y,y') = \partial_x p_\infty(x,y')$;

2) *for some c we have* $|q_t(x,y,y') - q_\infty(x,y')| \leqslant (c+c't)e^{-\varrho t}$.

Proof. The first point is in [10] theorem 6. On the other hand, we want an estimate on $q_t(x,y,y') - q_\infty(x,y')$ as a function of t, which is small enough to allow for a further integral convergence when we come to derivatives of $\widehat{b}(x,y)$ expressed by the Pardoux–Veretennikov formula in Lemma 7 above. We therefore have to consider the following two parts

$$\Delta_{\infty,t}q.(x,y,y') = \int_0^t ds \int_{\mathbb{T}^d} \partial_x L^*_{x,y''}\Delta_{\infty,t-s}p.(x,y,y'')p_s(x,y'',y')\,dy''$$

$$- \int_t^\infty ds \int_{\mathbb{T}^d} \partial_x L^*_{x,y''}p_\infty(x,y'')p_s(x,y'',y')\,dy''.$$

Let us consider the first term on the right hand side above. We have

$$\int_0^t ds \int_{\mathbb{T}^d} \partial_x L^*_{x,y''}\Delta_{\infty,t-s}p.(x,y,y'')p_s(x,y'',y')\,dy''$$

$$= -\sum_i \int_0^t ds \int_{\mathbb{T}^d} 1/2 \sum_j \partial_x a_{ij}(x,y'')\partial_{y''_j}\Delta_{\infty,t-s}p.(x,y,y'')$$

$$\times \partial_{y''_i}p_s(x,y'',y')\,dy''$$

$$+ \sum_i \int_0^t ds \int_{\mathbb{T}^d} \partial_x \widetilde{b}_i(x,y'')\Delta_{\infty,t-s}p.(x,y,y'')\partial_{y''_i}p_s(x,y'',y')\,dy'',$$

hence, this term is bounded in absolute value by

$$c \int_0^t e^{-\varrho(t-s)} e^{-\varrho s} \, ds = cte^{-\varrho t}.$$

To consider the second term we need to write

$$\partial_{y''} p_\infty(x, y'') = \int_{\mathbb{T}^d} p_\infty(x, y') \partial_{y''} p_1(x, y', y'') \, dy'.$$

We note that

$$|\partial_{y''} p_\infty(x, y'')| \leqslant \int_{\mathbb{T}^d} p_\infty(x, y') |\partial_{y''} p_1(x, y', y'')| \, dy' \leqslant c.$$

Therefore

$$\left| \sum_i \int_t^\infty ds \int_{\mathbb{T}^d} 1/2 \sum_j \partial_x a_{ij}(x, y'') \partial_{y_j''} p_\infty(x, y'') \partial_{y_i''} p_s(x, y'', y') \, dy'' \right|$$

$$\leqslant c \int_t^\infty e^{-\varrho s} \, ds = c' e^{-\varrho t}.$$

The other integration by parts gives the same result. Consequently, the bound on $I_x^{\infty,2}(x, y)$ follows.

In the end, we look at the mixed derivatives.

Lemma 19. *We have for some constant c and all x, y,*

$$\left\| \partial_{xy}^2 \widehat{b}(x, y) \right\| \leqslant c.$$

Proof. Let us write

$$\partial_{xy} \widehat{b}(x, y) = \int_0^\infty dt \int_{\mathbb{T}^d} \partial_x b(x, y') \partial_y p_t(x, y, y') \, dy'$$

$$+ \int_0^\infty dt \int_{\mathbb{T}^d} b(x, y') \partial_y q_t(x, y, y') \, dy'$$

$$= I_{xy}^{\infty,1}(x, y) + I_{xy}^{\infty,2}(x, y).$$

As in the case of $\partial_y \widehat{b}(x, y)$, the integral $I_{xy}^{\infty,1}(x, y)$ is easily seen to be bounded by c. In order to estimate the other term, we write

$$\partial_y q_t(x, y, y')$$

$$= \int_0^t ds \int_{\mathbb{T}^d} \partial_x L_{x,y''}^* \partial_y p_{t-s}(x, y, y'') p_s(x, y'', y') \, dy''$$

$$= - \sum_i \int_0^t ds \int_{\mathbb{T}^d} 1/2 \sum_j \partial_x a_{ij}(x, y'') \partial_{yy_j}^2 p_{t-s}(x, y, y'') \partial_{y_i''} p_s(x, y'', y') \, dy''$$

$$+ \sum_i \int_0^t ds \int_{\mathbb{T}^d} \partial_x \tilde{b}_i(x, y'') \partial_y p_{t-s}(x, y, y'') \partial_{y_i''} p_s(x, y'', y') \, dy''$$

$$\tag{26}$$

and remark that

$$\partial^2_{yy''} p_{t-s}(x, y, y'') = \int_{\mathbb{T}^d} \partial_y p_{t-s-1}(x, y, y') \partial_{y''} p_1(x, y', y'') \, dy'.$$

It now follows from Lemma 12 that

$$\left| \partial_y p_{t-s-1}(x, y, y') \right| \leqslant c e^{-\varrho(t-s-1)}$$
$$\left| \partial_{y''} p_s(x, y'', y') \right| \leqslant c e^{-\varrho s}$$
$$\left| \partial_y p_{t-s}(x, y, y'') \right| \leqslant c e^{-\varrho(t-s)}.$$

Replacing this into (26), we see that

$$\left| \partial_y q_t(x, y, y') \right| \leqslant c t e^{-\varrho t},$$

so that the bound on $I^{\infty,2}_{xy}(x, y)$ follows immediately.

Note that the above considerations show that under our conditions, all our partial derivatives are continuous.

5 Proof of theorem 2

Fix $\delta > 0$ and $t \leqslant T$. Let us choose the optimal subdivision $\Delta t_i = \varepsilon^2$. Now let $\alpha > 0$. Let again τ^ε be the exit time from the ball $B(0, M)$ by the process X^ε_t. There exists an M so large that for all $0 < \varepsilon \leqslant 1$,

$$\mathbf{P}(\tau^\varepsilon < T) \leqslant \alpha.$$

The fundamental subdivision above will serve as a lever that controls the new subdivisions $\Delta_{i_k} t = \varepsilon^2 / k$, k being a natural number which depends on α and is to be chosen below. In this way we can effectively control R^ε_t in which there is a wild discrepancy between the behaviour of $R^{1;\varepsilon}_{N_t}$ in (10) and $R^{3;\varepsilon}_{N_t}$ in (12). In what follows, manipulations with respect to the finer subdivision t_{i_k}, $i_k = 0, 1, 2, \ldots$, will be specified by a subscript k. In particular, $N_{k,t} = [kt/\varepsilon^2]$. We have from Lemma 4

$$R^\varepsilon_{N_t} = X^\varepsilon_{t_*} - \overline{X}^\varepsilon_{t_*} = \left(X^\varepsilon_{t_*} - X^\varepsilon_{t_{k,*}} \right) + \left(X^\varepsilon_{t_{k,*}} - \overline{X}^\varepsilon_{t_*} \right).$$

On the other hand

$$X^\varepsilon_{t_{k,*}} = R^\varepsilon_{N_{k,t}} + \overline{X}^\varepsilon_{t_{k,*}},$$

hence

$$\mathbf{P}\left(\left\| R^\varepsilon_{N_t} \right\| > \delta \right) \leqslant \mathbf{P}\left(\left\| X^\varepsilon_{t_*} - X^\varepsilon_{t_{k,*}} \right\| > \delta/3 \right)$$
$$+ \mathbf{P}\left(\left\| \overline{X}^\varepsilon_{t_{k,*}} - \overline{X}^\varepsilon_{t_*} \right\| > \delta/3 \right) + \mathbf{P}\left(\left\| R^\varepsilon_{N_{k,t}} \right\| > \delta/3 \right).$$

By tightness, the first term of the right hand side of the last inequality is dominated by $c\delta^{-1}\varepsilon$. Since F and G are bounded, the second term is also dominated by $c\delta^{-1}\varepsilon$. Let us turn to

$$\mathbf{P}\big(\|R^{\varepsilon}_{N_{k,t}}\| > \delta/3\big) \leqslant \sum_{i=1}^{9} \mathbf{P}\big(\|R^{i,\varepsilon}_{N_{k,t}}\| > c\delta\big).$$

Let us begin with the critical probabilities. The estimate (9), which is fine enough for tightness, is however insufficient here and we need the

Lemma 20. *For some $c > 0$, we have for any $0 \leqslant r \leqslant T$*

$$\mathbf{E}\|X^{\varepsilon}_r - X^{\varepsilon}_{r_*}\| \leqslant c\big[\varepsilon^{-1}(r - r_*) + r - r_* + (r - r_*)^{1/2}\big].$$

Proof. Due to the boundedness of the integrands, it is an immediate consequence of (4).

We have

$$\mathbf{P}\big(\|R^{1,\varepsilon}_{N_{k,t}}\| > c\delta\big) \leqslant c'\delta^{-1}\varepsilon^{-1} \sum_{i_k=0}^{N_{k,t}-1} \int_{\Delta t_{i_k}} \mathbf{E}\|\Delta_{t_{i_k},s}X^{\varepsilon}_{\cdot}\| \, ds$$

$$\leqslant c\delta^{-1}\big(\varepsilon^{-2}\Delta t_{i_k} + \varepsilon^{-1}\Delta t_{i_k} + \varepsilon^{-1}(\Delta t_{i_k})^{1/2}\big)$$

$$= c\delta^{-1}\big(k^{-1} + \varepsilon k^{-1} + k^{-1/2}\big)$$

$$\leqslant \alpha,$$

provided $\varepsilon \leqslant 1$ and k is large enough but fixed, which we assume from now on. On the other hand

$$\mathbf{P}\big(\|R^{3,\varepsilon,1}_{N_{k,t}}\| > c\delta\big) \leqslant \mathbf{P}\big(\|R^{3,\varepsilon,1}_{N_{k,t}}\| > c\delta \text{ and } T \leqslant \tau^{\varepsilon}\big) + \alpha.$$

We clearly have

$$\|R^{3,\varepsilon,1}_{N_{k,t}}\| \leqslant \varepsilon \sup_{i_k} \gamma_{\partial_x \widehat{b}}\big(\|X^{\varepsilon}_{t_{i_k}} - X^{\varepsilon}_{t_{i_k-1}}\|\big) \bigg(\sum_{i_k=1}^{N_{k,t}-1} \|\Delta_{t_{i_k-1},t_{i_k}}X^{\varepsilon}_{\cdot}\|\bigg),$$

where $\gamma_{\partial_x \widehat{b}}$ is a modulus of continuity of $\partial_x \widehat{b}$ on $\overline{B}(0,M) \times \mathbb{T}^d$. Now put

$$B_{\varepsilon,k,1} = k^{1/2} \sup_{i_k} \gamma_{\partial_x \widehat{b}}\big(\|X^{\varepsilon}_{t_{i_k}} - X^{\varepsilon}_{t_{i_k-1}}\|\big)$$

and

$$B_{\varepsilon,k,2} = k^{-1/2}\varepsilon \sum_{i_k=1}^{N_{k,t}-1} \|\Delta_{t_{i_k-1},t_{i_k}}X^{\varepsilon}_{\cdot}\|.$$

We obviously have

$$\|R_{N_{k,t}}^{3,\varepsilon,1}\| \leqslant B_{\varepsilon,k,1} \times B_{\varepsilon,k,2}.$$

Now by Cauchy–Schwarz,

$$\mathbf{E}\big(B_{\varepsilon,k,2}^2\big) \leqslant c,$$

and $B_{\varepsilon,k,1}$ tends to zero in probability, as $\varepsilon \to 0$, hence we can choose $0 < \varepsilon_0 \leqslant 1$ such that for all $0 < \varepsilon \leqslant \varepsilon_0$,

$$\mathbf{P}\big(\|R_{N_{k,t}}^{3,\varepsilon,1}\| > c\delta\big) \leqslant c'\alpha.$$

The other remainders are treated similarly because no new fundamental asymptotic phenomenon arises. Indeed, the rests $R_{N_{k,t}}^{2,\varepsilon}$, $R_{N_{k,t}}^{3,\varepsilon,2}$ and $R_{N_{k,t}}^{4,\varepsilon,2}$ are easily treated. The quantity $R_{N_{k,t}}^{4,\varepsilon,1}$ and $R_{N_{k,t}}^{5,\varepsilon,1}$ behave exactly as $R_{N_{k,t}}^{1,\varepsilon}$ does. Indeed we use the following equation obtained clearly from (7)

$$L_{x,y}\partial_x\widehat{b}(x,y) = -\partial_x b(x,y) - \partial_x L_{x,y}\widehat{b}(x,y),$$

which garanties a Lipschitz condition on the function $\partial_{xy}^2\widehat{b}(x, \cdot)$.

The term $R_{N_{k,t}}^{5,\varepsilon,2}$ is simple. The remaining rests on the list are $R_{N_{k,t}}^{6,\varepsilon}$ to $R_{N_{k,t}}^{9,\varepsilon}$, which clearly tend to zero in probability, as ε tends to zero.

Obviously, all the above estimates are uniform in $t \leqslant T$.

6 Convergence of $u^\varepsilon(t,x)$ to $u(t,x)$

For us, the solution $u^\varepsilon(t,x)$ to equation (1) subject to the Cauchy initial value problem $u^\varepsilon(0,x) = g(x)$, for all $\varepsilon > 0$, where $g(x)$ is supposed to be continuous with at most polynomial growth at infinity is given by the Feynman–Kac formula

$$u^\varepsilon(t,x) = \mathbf{E}\, g(X_t^\varepsilon) \exp\left(\int_0^t \lambda^\varepsilon(s,s)\, \mathrm{d}s\right). \tag{27}$$

Clearly, $e(x,y)$ must be centered, i.e. we assume that

$$\int_{\mathbb{T}^d} e(x,y)\, \mu(x,\mathrm{d}y) = 0, \qquad \forall x \in \mathbf{R}^d. \tag{28}$$

Let us first define a few items. Set $Y_t^\varepsilon = \int_0^t \lambda^\varepsilon(s,s)\, \mathrm{d}s$ and recall that the functions $F(x,y)$ and $G(x,y)$ were defined in (15) and (16). Furthermore put

$$F_e(x,y) = \big(\partial_x\widehat{e}b + \partial_y\widehat{e}c + \mathrm{Tr}\,\partial_{xy}^2\widehat{e}a\big)(x,y)$$

and

$$G_e(x,y) = (\partial_y\widehat{e}\sigma)(x,y).$$

Fix $t \leqslant T$ and x in $\mathbf{R}^d$. It follows from Corollary 1 and Theorem 2 that

$$X_t^\varepsilon = \overline{X}_t^\varepsilon + R_t^\varepsilon,$$
$$Y_t^\varepsilon = \overline{Y}_t^\varepsilon + r_t^\varepsilon,$$

where

$$\overline{X}_t^\varepsilon = x + \int_0^t F(s,s)\,\mathrm{d}s + \int_0^t G(s,s)\,\mathrm{d}B_s,$$
$$\overline{Y}_t^\varepsilon = \int_0^t (f + F_e)(s,s)\,\mathrm{d}s + \int_0^t G_e(s,s)\,\mathrm{d}B_s,$$

and both R_t^ε and r_t^ε converge to zero in probability (uniformly in $t \leqslant T$), as $\varepsilon \to 0$.

We now use a Girsanov transform to get rid of the stochastic integral in the argument of the exponential in (27) above. Indeed, let $\widetilde{\mathbf{P}}$ denote a new probability measure s.t. (with the notation $\Lambda_e = G_e G_e^*$)

$$\frac{\mathrm{d}\widetilde{\mathbf{P}}}{\mathrm{d}\mathbf{P}}\bigg|_{\mathcal{F}_t} = \exp\left(\int_0^t G_e(s,s)\,\mathrm{d}B_s - 1/2\int_0^t \Lambda_e(s,s)\,\mathrm{d}s\right).$$

Then (27) becomes

$$u^\varepsilon(t,x) = \widetilde{\mathbf{E}}\left[g\big(\overline{X}_t^\varepsilon + R_t^\varepsilon\big)\exp\left(\int_0^t (f + F_e + 1/2\Lambda_e)(s,s)\,\mathrm{d}s + r_t^\varepsilon\right)\right], \quad (29)$$

where

$$\overline{X}_t^\varepsilon = x + \int_0^t \left(F + \big(I + \partial_y\widehat{b}\big)a\partial_y\widehat{e}^*\right)(s,s)\,\mathrm{d}s + \int_0^t G(s,s)\,\mathrm{d}\widetilde{B}_s,$$

in which $(\widetilde{B}_t,\ t \geqslant 0)$ is a $\widetilde{\mathbf{P}}$ Brownian motion. The stochastic integrals and finite variation processes inside R_t^ε and r_t^ε undergo similar changes. It is clear that under $\widetilde{\mathbf{P}}$, these remainders converge to zero in probability. Moreover, define

$$\widetilde{f}(x,y) = (f + F_e + 1/2\Lambda_e)(x,y),$$
$$C(x) = \int_{\mathbb{T}^d} \widetilde{f}(x,y)\,\mu(x,\mathrm{d}y),$$

and let us write

$$\int_0^t \widetilde{f}(s,s)\,\mathrm{d}s = \int_0^t C(X_s^0)\,\mathrm{d}s + \widetilde{r}_t^\varepsilon,$$

where

$$\widetilde{r}_t^\varepsilon = \int_0^t \big(\widetilde{f}(s,s) - C(X_s^\varepsilon)\big)\,\mathrm{d}s + \int_0^t \big(C(X_s^\varepsilon) - C(X_s^0)\big)\,\mathrm{d}s.$$

Clearly, $\widetilde{r}_t^\varepsilon$ converges to zero in probability (under $\widetilde{\mathbf{P}}$). Now we are able to state the

Theorem 4. *Let $t \leqslant T$ and x in $\mathbf{R}^d$, then*

$$u^\varepsilon(t, x) \to u(t, x),$$

as $\varepsilon \to 0$, and the limiting PDE is the following equation

$$\partial_t u(t, x) = 1/2 \sum_{i,j=1}^{d} A_{ij}(x)\, \partial^2_{x_i x_j} u(t, x) + \sum_{i=1}^{d} B_i(x)\, \partial_{x_i} u(t, x) + C(x) u(t, x),$$

where

$$A(x) = \int_{\mathbf{T}^d} \left(I + \partial_y \widehat{b}\right) a \left(I + \partial_y \widehat{b}\right)^* (x, y)\, \mu(x, dy)$$

and

$$B(x) = \int_{\mathbf{T}^d} \left(F + \left(I + \partial_y \widehat{b}\right) a \partial_y \widehat{e}^*\right)(x, y)\, \mu(x, dy).$$

Proof. It suffices to check that the family of random variables given by

$$\varsigma^\varepsilon = g\left(\overline{X}_t^\varepsilon + R_t^\varepsilon\right) \exp\left(\int_0^t \widetilde{f}(s, s)\, ds + r_t^\varepsilon\right)$$

is $\widetilde{\mathbf{P}}$ uniformly integrable for $0 < \varepsilon \leqslant 1$. Since $\widetilde{f}$ is bounded and g grows at most polynomially at infinity, the required uniform integrability follows from the following three Lemmas:

Lemma 21. *For any $t \leqslant T$ and $p > 0$, there exists a constant c s.t. for all $\varepsilon > 0$,*

$$\widetilde{\mathbf{E}}\|\overline{X}_t^\varepsilon\|^p \leqslant c.$$

Lemma 22. *For any $t \leqslant T$ and $p > 0$, there exists a constant c s.t. for all $\varepsilon > 0$,*

$$\widetilde{\mathbf{E}}\|R_t^\varepsilon\|^p \leqslant c.$$

Lemma 23. *For any $t \leqslant T$ and $\varrho > 0$, there exists a constant c s.t. for all $\varepsilon > 0$,*

$$\widetilde{\mathbf{E}}\exp(\varrho r_t^\varepsilon) \leqslant c.$$

Since R^ε and r^ε have similar forms, Lemma 23 is strictly more difficult to establish than Lemma 22. Hence we will prove only Lemma 21 and Lemma 23.

Proof of Lemma 21. Since F and G are bounded, it suffices to apply the Cauchy–Schwarz (resp. the Burkholder–Davis–Gundy) inequality to the Lebesgue (resp. the stochastic) integral term in the expression of $\overline{X}_t^\varepsilon$ to yield the desired result.

Proof of Lemma 23. The only estimates which are not straightforward are those that exhibit the behaviour of

$$r_{N_t}^{1,\varepsilon} = \sum_{i=0}^{N_t-1} \varepsilon^{-1} \int_{\Delta t_i} \left(\Delta_{t_i,s} e(\,.\,,s) + \Delta_{t_i,s} L_{.,s} \widehat{e}(t_i, s) \right) \mathrm{d}s,$$

with $\Delta t_i = \varepsilon^2$. Hence, in the spirit of section 5, if we can show that for any $\varrho > 0, t > 0$,

$$\sup_{\varepsilon > 0} \sup_{0 \leqslant s \leqslant t} \widetilde{\mathbf{E}} \exp\left(\frac{\varrho}{\varepsilon} \|X_s^\varepsilon - X_{s_*}^\varepsilon\| \right) < \infty. \tag{30}$$

Then we would write

$$r_t^{1,\varepsilon} \leqslant c t_* \varepsilon^{-1} \int_0^{t_*} \|X_s^\varepsilon - X_{s_*}^\varepsilon\| \frac{\mathrm{d}s}{t_*},$$

hence we deduce from Jensen's inequality that

$$\widetilde{\mathbf{E}} \exp(\varrho r_t^{1,\varepsilon}) \leqslant \int_0^{t_*} \widetilde{\mathbf{E}} \exp\left(\frac{c \varrho t_*}{\varepsilon} \|X_s^\varepsilon - X_{s_*}^\varepsilon\| \right) \frac{\mathrm{d}s}{t_*}$$

$$\leqslant \sup_{0 \leqslant s \leqslant t} \widetilde{\mathbf{E}} \exp\left(\frac{c \varrho t}{\varepsilon} \|X_s^\varepsilon - X_{s_*}^\varepsilon\| \right)$$

$$< \infty,$$

where we have used (30) for the last line.

It thus remains to prove (30). Since $s - s_* \leqslant \varepsilon^2$, we have for some $c > 0$ and all $s \leqslant t$,

$$(1 - c\varepsilon) \sup_{s_* \leqslant v \leqslant s} \|X_v^\varepsilon - X_{s_*}^\varepsilon\| \leqslant c \left(\varepsilon + \varepsilon^2 + \sup_{s_* \leqslant v \leqslant s} \left\| \int_{s_*}^v G^{0,\varepsilon}(s_*, u) \, \mathrm{d}\widetilde{B}_u \right\| \right).$$

Choosing ε small enough, we deduce

$$\frac{c}{\varepsilon} \|X_s^\varepsilon - X_{s_*}^\varepsilon\| \leqslant c' + \sup_{s_* \leqslant v \leqslant s} \frac{c''}{\varepsilon} \left\| \int_{s_*}^v G^{0,\varepsilon}(s_*, u) \, \mathrm{d}\widetilde{B}_u \right\|.$$

Hence our result will follow from the

Lemma 24. *Let* $\{M_t, t \geqslant 0\}$ *be a d-dimensional continuous martingale of the form*

$$M_t = \int_0^t \varphi_s \, \mathrm{d}B_s,$$

where $\{B_t, t \geqslant 0\}$ *is a k-dimensional standard Brownian motion, and* $\{\varphi_t, t \geqslant 0\}$ *is a d × k-dimensional adapted stochastic process, such that* $\|\varphi_t\| \leqslant K$, *a.s., for all* $t \geqslant 0$. *Then there exists a constant c depending only on the dimension d such that*

$$\mathbf{E} \exp\left(\sup_{0 \leqslant s \leqslant t} \|M_s\| \right) \leqslant c \exp(cK^2 t).$$

Proof. We have

$$\exp\left(\sup_{0\leqslant s\leqslant t}\|M_s\|\right) \leqslant \prod_{i=1}^{d}\exp\left(\sup_{0\leqslant s\leqslant t}|M_s^i|\right)$$

$$\leqslant \prod_{i=1}^{d}\left[\exp\left(\sup_{0\leqslant s\leqslant t}M_s^i\right) + \exp\left(\sup_{0\leqslant s\leqslant t}(-M_s^i)\right)\right].$$

Consequently

$$\mathbf{E}\left(e^{\sup_{0\leqslant s\leqslant t}\|M_s\|}\right) \leqslant \left(\prod_{i=1}^{d}\mathbf{E}\left(e^{\sup_{0\leqslant s\leqslant t}M_s^i} + e^{\sup_{0\leqslant s\leqslant t}(-M_s^i)}\right)^d\right)^{1/d}$$

$$\leqslant 2^{d-1}\left(\prod_{i=1}^{d}\left(\mathbf{E}\,e^{d\sup_{0\leqslant s\leqslant t}M_s^i} + \mathbf{E}\,e^{d\sup_{0\leqslant s\leqslant t}(-M_s^i)}\right)\right)^{1/d}.$$

$$(31)$$

It thus remains to estimate $\mathbf{E}\exp(d\sup_{0\leqslant s\leqslant t}M_s)$ when M is one-dimensional. In that case,

$$M_t = \int_0^t \varphi_s\,\mathrm{d}B_s = W_{\int_0^t |\varphi_s|^2\,\mathrm{d}s},$$

where $\{W_t,\ t\geqslant 0\}$ is a standard one-dimensional Brownian motion. Let $T = K^2 t$. We have

$$\sup_{0\leqslant s\leqslant t}M_s \leqslant \sup_{0\leqslant r\leqslant T}W_r.$$

But $\sup_{0\leqslant r\leqslant T}W_r$ has the same law as $|W_T|$. Hence,

$$\mathbf{E}\exp\left(d\sup_{0\leqslant r\leqslant T}W_t\right) = \mathbf{E}\exp(d|W_T|)$$

$$\leqslant 2\,\mathbf{E}\exp(dW_T) = 2\exp\left(\frac{d^2 T}{2}\right) = 2\exp\left(\frac{d^2 K^2 t}{2}\right). \quad (32)$$

The result follows from (31) and (32).

Acknowledgement. The first author is thankful for the hospitality and support of the CMI of the Université de Provence, Marseille, France.

References

1. ALLAIRE, G. (1992): Homogenization and two-scale convergence. SIAM J. Math. Anal., **23**, 1482–1518.
2. BENSOUSSAN, A., LIONS, J.-L. and PAPANICOLAOU, G. (1978): Asymptotic analysis of periodic structures. North-Holland Publ. Comp., Amsterdam.

3. BILLINGSLEY, P. (1968): Convergence of probability measures. Wiley, New York.

4. FREIDLIN, M. (1964): The Dirichlet problem for an equation with periodic coefficients depending on a small parameter. Teor. Primenen, **9**, 133–139.

5. FRIEDMAN, A. (1964): Partial differential equations of parabolic type. Prentice–Hall, Englewood Cliffs, N. J.

6. GILBARG, D. and TRUDINGER, N. S. (1998): Elliptic partial differential equations of second order. Revised third printing. Springer-Verlag, New York.

7. OLLA, S. (2002): Notes on the central limit theorems for tagged particles and diffusions in random fields, in "Milieux aléatoires", F. Comets and E. Pardoux eds., Panoramas et Synthèses **12**, SMF.

8. PARDOUX, E. (1999): Homogenization of linear and semilinear second order parabolic PDEs with periodic coefficients : a probabilistic approach. J. Funct. Anal., **167**, 498-520.

9. PARDOUX, E. and VERETENNIKOV, A. Yu. (2001): On the Poisson equation and diffusion approximation, I. The Annals of Proba., **29**, 3, 1061-1085.

10. PARDOUX, E. and VERETENNIKOV, A. Yu. (2003): On the Poisson equation and diffusion approximation, II. The Annals of Proba., **31**, 3.

11. STROOCK, D. W. and VARADHAN, S. R. S. (1979): Multidimensional diffusion processes. Springer-Verlag, New York.

NOTE TO CONTRIBUTORS

Contributors to the Séminaire are reminded that their articles should be formatted for the Springer Lecture Notes series.

Manuscripts should preferably be prepared with LaTeX version 2e, using the macro packages provided by Springer for multi-authored books. These files can be downloaded using your web browser from:

`ftp://ftp.springer.de/pub/tex/latex/mathegl/mult`

VISIT THE WEB SITE OF THE SÉMINAIRE!

The *Séminaire de Probabilités* maintains a web site: you can visit us at

`http://www-irma.u-strasbg.fr/irma/semproba/e_index.shtml`

or, if you prefer French, at

`http://www-irma.u-strasbg.fr/irma/semproba/index.shtml`

You will not find there the articles published in the 35 volumes, but you will find a historical account of the Séminaire, and an on-line index by authors, or volumes. You have also access to notices on the content of the articles, with hyperlinks between the notices. A search engine allows you to explore these notices by keywords or classification.

The database consisting of the notices for all articles is still under construction; about one half of the entire collection has been analyzed so far.

All your comments and suggestions are most welcome. As explained on the site, you can help us, in particular with the content of the notices.

Printing and Binding: Strauss GmbH, Mörlenbach